Laser Parameter Measurements Handbook

WILEY SERIES IN PURE AND APPLIED OPTICS

Advisory Editor
STANLEY S. BALLARD, University of Florida

Lasers, BELA A. LENGYEL

Ultraviolet Radiation, second edition, LEWIS R. KOLLER

Introduction to Laser Physics, BELA A. LENGYEL

Laser Receivers, MONTE ROSS

The Middle Ultraviolet: Its Science and Technology, A.E.S. GREEN, *Editor*

Optical Lasers in Electronics, EARL L. STEELE

Applied Optics, A Guide to Optical System Design/Volume 1, LEO LEVI

Laser Parameter Measurements Handbook, HARRY G. HEARD

Laser Parameter Measurements Handbook

H. G. HEARD
RESALAB, INC.

John Wiley & Sons, Inc. New York/London/Sydney

Library of Congress Catalog Card Number: 67–31181
GB 471 36665X
Printed in the United States of America

Contributing Authors

Chapter 3

G. D. BALDWIN, Westinghouse Corporation
I. T. BASIL, Westinghouse Corporation
M. BIRNBAUM, Aerospace Corporation*
J. H. CULLOM, Westinghouse Corporation
E. S. DAYHOFF, Naval Ordnance Laboratories
K. E. ERICKSON, Keuffel and Esser Company
G. E. FRANCOISE, Stanford University
J. R. KERR, Sylvania EDL
C. M. STICKLEY, Air Force Cambridge Research Laboratories
R. H. WAYANT, Westinghouse Corporation

Chapter 4

J. A. ACKERMAN, Aircraft Armaments, Inc.
J. H. BOYDEN, Korad
R. C. HONEY, Stanford Research Institute*
B. H. SOFFER, Korad
M. SUBRAMANIAN, Purdue University

Chapter 5

M. L. BAUMIK, Electro-Optical Systems
G. COURVILLE, Fairleigh Dickinson University
T. M. HOLZBERLEIN, University of Oklahoma
T. C. MARSHALL, Columbia University
P. B. MAUER, Eastman Kodak Company
A. A. VUYLSTEKE, Martin Company

P. J. WALSH, Fairleigh Dickinson University
J. H. WASKO, Fairleigh Dickinson University
C. B. ZAROWIN, Lasers, Inc.

Chapter 6

H. KLEIMAN, University of California*

Chapter 7

R. L. AAGARD, Minneapolis-Honeywell
M. HERCHER, University of Rochester*

Chapter 8

K. E. GILLILAND, National Bureau of Standards
K. D. MIELENZ, National Bureau of Standards*
K. F. NEFFLEN, National Bureau of Standards
R. B. STEPHENS, National Bureau of Standards

Chapter 9

E. O. AMMAN, Sylvania EDL
C. F. BUHRER, General Telephone & Electronics Laboratories
D. E. CADDES, Sylvania EDL
H. A. HAUS, Massachusetts Institute of Technology
J. R. KERR, Sylvania EDL
C. M. MCINTYRE, Sylvania EDL
G. A. MASSEY, Sylvania EDL

*Consulting authors.

Preface

The *Laser Parameter Measurements Handbook* is a compendium of measurement that encompasses the laser technology. It includes a wealth of information gleaned from more than 650 articles surveyed in an exhaustive search of the literature that reviewed American as well as foreign scientific journals and government reports. This book contains the contributions of 37 authors whose writings were edited to conform with the text and abridged to eliminate redundancy. It is believed that it treats all of the significant laser measurement techniques published to date in the areas of beam sampling, beam parameters, power, energy, gain, wavelength, bandwidth, coherence, and frequency stability. The techniques of modulation and the methods of measurement are discussed, as are the communication aspects of noise in the laser signal source.

Quantum electronics is a multidisciplinary endeavor that borrows from the fields of classical and modern physical optics, spectroscopy, radiometry, thermodynamics, and theoretical physics. Many of the foundations for communications applications of lasers derive from microwave and radio engineering practice. Associated with each of these disciplines is a system of nomenclature and symbology. When it became obvious that it would be a disservice to the reader to create a new set of symbols that would preclude completely the ambiguous description of all of the variables of interest, it was decided that each chapter would contain its own list of principal symbols. Thus, for example, the g-factor of spectroscopy and the g-factor of mirror geometries are well distinguished. An attempt has been made, of course, to add continuity by using the same symbol throughout the book. To assist the reader further, symbols are defined in each chapter in the text that accompanies initial usage.

This effort was jointly supported by the U.S. Air Force through the Rome Air Development Center (RADC) and Ohio Steel Foundry's subsidiary,

RESALAB, INC. (*h nu systems*). It is a pleasure to acknowledge the assistance of Richard A. Robinson of RADC, who reviewed and proofread the entire book.

Illustrations for the *Handbook* were prepared by H. Mizote. The monumental job of typing and the many tasks that required careful attention were handled with efficiency and dispatch by Miss J. Bond, Mrs. D. Farrand, Mrs. B. Strassmeyer, and Mrs. K. Turley. The names of the contributing authors are listed separately in the pages that follow.

A work of this magnitude represents not only the effort of those who prepared the articles but an equally great number of individuals who by their support contributed to this compilation. I should like to express my appreciation here for the encouragement of management, as well as the patience and understanding of my associates' wives and husbands. Their interest made this *Handbook* possible.

H. G. Heard

Menlo Park, California
April 1968

Contents

Laser Parameter Measurements Handbook

1

Laser Parameters and Measurements

1.0 INTRODUCTION

This book is concerned with the measurement of laser parameters. The acronym laser, referring to *l*ight *a*mplification by *s*timulated *e*mission of *r*adiation, is used in the broadest sense to include the entire electromagnetic spectrum. Lasers today span the wavelength range from approximately 200 nm to 400μm. Judging from the rate of discovery of new wavelengths, the extremes of the spectrum will be expanded to include the millimeter to x-ray range and beyond as the technology develops. The spectral density of laser lines will be found to increase. In the interim the application of this breakthrough to many fields of endeavor will occupy the genius of a large number of people in the scientific and technical community. For these developments to have any continuity and meaning, however, measurements are necessary. The words of Lord Kelvin are as timely today as they were a century ago when he is reputed to have said, "When you can measure what you are speaking about and express it in numbers, you know something about it . . ."

In this chapter the basic laser parameters that are discussed in subsequent sections of the book are identified. Because this book is a treatise on parameter measurements, this chapter also summarizes some of the principles of measurement and some statistics of data analysis. Those highly skilled in scientific laboratory technique are advised to proceed directly to Chapters 3 through 9. Reference to Chapter 2 may, however, be advisable in view of the novel characteristics of lasers and the need to modify laboratory techniques to insure meaningful measurements of this new source of electromagnetic radiation.

1.1 PRINCIPAL LASER PARAMETERS

If a single property can distinguish a laser from other conventional thermal light sources, it is radiance. The measurement of radiance, that is, the power density per unit solid angle, per unit frequency interval, includes the parametric measurement of time-dependent power, power density, beam divergence, polarization, and spectral content. To the list of external-beam parameters that must be measured to determine radiance must be added coherence. Table 1.1 lists the external-beam parameters of interest.

TABLE 1.1 EXTERNAL BEAM PARAMETERS

1. Power
   (a) Power density in the near and far field
   (b) Power distribution within the beam
2. Energy
   (a) Energy density in the near and far field
   (b) Energy distribution within the beam
3. Angular Divergence and beam spot size
   (a) Variation of angular divergence in the near and far field
4. Output wavelength
   (a) Discrete spectrum and time-dependent character thereof
5. Coherence
   (a) Spatial coherence
   (b) Temporal coherence
   (c) Mutual coherence
6. Polarization

In many cases parameter determinations must be made on a time scale of nanoseconds to milliseconds. The extension of conventional optical laboratory techniques to these time domains and to extremely high power densities requires both the refinement and extension of old techniques as well as the development of new techniques.

Significant internal laser parameters include mode spectrum, gain, noise, and modulation capability. These laser parameters are detailed in Table 1.2. Measurement of the internal- and external-beam parameters involve the use of signal, power, and energy sensors, beam couplers, and attenuators. These measurements must be based on standards of lenght and energy, for calibration against standards is of the utmost importance.

Above all, laser parameter measurements depend strongly on good laboratory techniques. Therefore, wherever it is feasible in the chapters that follow the text contains a brief introduction to the laser parameter to be measured, including establishment of the parameter on a theoretical basis, followed by a survey of state-of-the-art techniques and instrumentation. Specific methods are detailed for making measurements in which the techniques have been developed and reported. Included in these treatments are the arrangement of the experimental apparatus, a description of typical

TABLE 1.2 INTERNAL LASER PARAMETERS

1. Mode spectrum
   (a) Discrete frequencies within the spectral line
   (b) Band-width of an isolated mode
   (c) Absolute and relative values of frequency
   (d) Stability, short and long term, and limits
   (e) Frequency resettability and limits
2. Gain
   (a) Parameters that determine the existence thereof
   (b) Losses within the median
   (c) Losses within the optical cavity
   (d) Methods of output coupling
   (e) Saturation effects and line narrowing
   (f) Efficient methods of utilizing stored energy
3. Noise
   (a) Intrinsic and extrinsic sources
   (b) Spectral distribution and properties thereof
4. Modulation
   (a) Types and techniques
   (b) Detection

instrumentation, and, in many cases, a detailed description of the measurements procedure, followed by a discussion of both the sources of error and special precautions that will enable more meaningful measurements.

The remainder of this chapter is concerned with the principles of measurements and the analysis of data.

1.2 TREATMENT OF EXPERIMENTAL DATA

1.2.1 Precision and Accuracy

The precision of measurements concerns itself with the care taken with and the refinement of the technique of making measurements. Precision is frequently judged on the basis of the small spread in data obtained from repeated measurements of the same quantity. Precision is usually associated with the magnitude of random errors that are produced by a large number of unpredictable or unknown variations in a given measurement. A lack of precision in measurements may be caused by fluctuations in temperature, illumination, line voltage, mechanical vibration, fluctuation in pressure, and so on.

Accuracy of measurements is associated with the deviation of the final data from a standard and reveals the closeness with which the measurements represent the true value of the parameter being measured. Accuracy invariably depends on the skill and experience of the investigator. It requires that he avoid systematic errors and that he calibrate his instruments against standards. High accuracy is associated with a measurement that has small systematic errors, whereas high precision is associated with a measurement that has small random errors.

1.2.2 Types of Error

Just as precision and accuracy have distinctly different meanings, so do errors and mistakes or blunders. Blunders include taking incorrect data, such as false meter readings, or recording data incorrectly and making mistakes in addition. These mistakes are not properly classified as errors. Four classes of error are identified: accidental, systematic, short-term, and constant errors[1]. Accidental errors arise as a result of estimates in meter readings, inaccuracies in setting values in instruments, and parallax in reading, among others. The errors are as likely to be positive as negative and tend to follow a random pattern. Accidental errors can best be avoided by taking multiple readings. Doubling the number of readings will reduce the effect of accidental errors by the square root of 2. Multiple readings can be reduced by the method of least squares to reduce their effect on the measure of the variables being studied.

Systematic errors include such items as change in calibration with environment, incorrect setting of the zero of an instrument, and incorrect determination of a reference position. Systematic errors may be included in a list of mechanical or electrical deficiencies that affect a reading. The minimization of systematic errors requires the accumulation of large quantities of data, their interpretation and interrelation. Considerable ingenuity on the part of the investigator is often required to discover and eliminate those sources of error that affect readings.

Short-term errors arise as the result of line voltage fluctuations, changes in temperature or humidity, and changes in pressure, among others. The accumulation of a large number of measurements tend to make short-term errors stand out. The systematic investigator can frequently identify sources of short-term error if the data that are obtained are plotted immediately on acquisition.

Constant errors can arise as a result of improper calibration of instrumentation, drift in calibration, and equipment malfunction.

1.2.3 Significant Figures

It is good laboratory practice, when recording the results of measurements, to retain all of the figures in which the investigator has confidence. This may include many ciphers, for the latter can be significant if the measurement is made with high precision. It is conventional to indicate the precision of data points by recording the number of significant figures, that is, the number of meaningful figures irrespective of the decimal point. In indirect measurements the result is usually calculated by some established relationship between the variables.

To ensure uniformity, quantities that are to be added or subtracted should be measured accurately to the same decimal place irrespective of whether they have the same number of significant figures or not, whereas quantities that are to be multiplied or divided should be measured to the same number of significant figures.

1.2.4 Propagation of Errors

If a series of observations is necessary to determine the value of a physical variable and if the observations each have errors associated with them, the results of the measurements will also be in error by an amount that depends on the errors of the individual observations. Suppose, for example, it is necessary to measure a quantity Λ, where Λ is a function of a number of variables, a, b, c, d, ... The error in measuring Λ resulting from the independent errors can be represented as

$$\Delta\Lambda = \frac{\partial\Lambda}{\partial a}\Delta a + \frac{\partial\Lambda}{\partial b}\Delta b + \frac{\partial\Lambda}{\partial c}\Delta c + \cdots \tag{1.1}$$

and the fractional error may be represented as

$$\frac{\Delta\Lambda}{\Lambda} = \frac{1}{\Lambda}\frac{\partial\Delta}{\partial a}\Delta a + \frac{1}{\Lambda}\frac{\partial\Lambda}{\partial b}\Delta b + \frac{1}{\Lambda}\frac{\partial\Lambda}{\partial c}\Delta c + \cdots. \tag{1.2}$$

It is frequently of value to express the fraction error or logarithmic derivative of a functional relationship to identify the way in which the various sources of error interact.

1.2.5 Mean Values and Deviations

Two mean values are of interest in measurements. One is the mean or average value of a number of measurements expressed mathematically as

$$\bar{x} = \frac{1}{N}\sum_{i=1}^{N} x_i. \tag{1.3}$$

If the results of a group of measurements have a weighting factor applied that relates for example, to the degree of certainty attached to the measurement, the weighted mean has value in the following definitions:

$$\bar{x} = \frac{\sum_{i=1}^{N} g_i x_i}{\sum_{i=1}^{N} g_i}, \tag{1.4}$$

where g_i represents the weighting factor.

It is frequently of value to determine the average scatter or dispersion of data. This can be obtained from the deviation of individual measurements from the mean value and can be expressed mathematically as

$$\varphi = \frac{1}{N} \sum_{i=1}^{N} (x_i - \bar{x}), \tag{1.5}$$

where $\bar{x}$ is defined in 1.3. A more meaningful measure of the precision of measurements, denoted the standard deviation, is obtained by the root-mean-square value of the deviations. The standard deviation is defined as

$$\sigma = \left[\frac{1}{N} \sum_{i=1}^{N} (x_i - \bar{x})^2 \right]^{1/2}. \tag{1.6}$$

The square of the standard deviation σ^2 is called the variance. It is unlikely that the mean value of the quantity observed will be in error by as much as the standard deviation if the number of observations is large. In many cases the error in the mean value is unlikely to be greater than the standard deviation divided by the square root of a number of measurements. Obviously, more data points will give a more reliable mean value. In comparing results with a standard error curve, one deviation will include 68 percent of the readings taken, whereas two deviations include 95 percent[2], [3].

1.2.6 Gauss Distribution

The Gauss distribution or normal error function describes the distribution of random errors in many kinds of physical measurements. It can be shown in fact that even if the individual errors do not follow this distribution, the averages of the groups of such errors are distributed in a manner that approaches the Gauss distribution for very large groups[4]. If it can be assumed that the Gauss distribution describes the results, it is possible to use the mean deviation to determine the standard deviation, for it can be shown that

$$\sigma = 1.25\varphi. \tag{1.7}$$

This enables an economy in reducing the data in that the square root of the sums of the squares need not be computed.

1.2.7 Probable Error

Probable error in a series of observations may be defined as the mean error from which there are as many larger errors as there are smaller errors. The probable error is defined as 0.6745σ.

1.2.8 Method of Least Squares

If we can assume that the mean value of an infinite number of measurements is the true value of that quantity, the best estimate that we can make of the true value of the quantity in a finite set of N measurements is the mean value of that set. The principle of least squares holds that the most probable value of a quantity, obtained from a set of N measurements, results from choosing the value that minimizes the sums of the squares of the deviations of these measurements. A mathematical statement of the above is

$$\frac{d}{dx}\sum_{i=1}^{N}(x - x_i)^2 = 0. \tag{1.8}$$

Differentiating the sum term by term and nulling the result, we obtain

$$\sum_{i=1}^{N}\frac{d}{dx}(x - x_i)^2 = 2Nx - 2\sum_{i=1}^{N}x_i = 0. \tag{1.9}$$

The value of x that satisfies this relation is

$$x = \frac{1}{N}\sum_{i=1}^{N}(x - x_i)^2 \tag{1.10}$$

that is the mean value of x obtained from the N measurements.

Stated otherwise, the "best" value of the measured quantity that can be obtained in a set of N measurements is the most probable value. The latter maximizes the probability that the measurements are distributed normally; that is, if the observations have a Gaussian distribution, the probability of obtaining the best value is maximized by minimizing the sums of the squares of the deviations. This is accomplished if the least-squares sum

$$\Psi(x) = \sum_{i=1}^{N}\frac{(x_i - x)^2}{2\sigma^2} \tag{1.11}$$

is minimized.

The standard deviation of the mean of a set of N measurements is just the variance of measurements divided by the number of measurements:

$$\sigma_m{}^2 = \frac{\sigma^2}{N}, \tag{1.12}$$

where $\sigma_m{}^2$ is the variance of the mean. In a set of measurements of Λ, where Λ is defined in (1.2), it can be shown that

$$\sigma_{m_1}{}^2 = \left(\frac{\partial\Lambda}{\partial a}\right)^2 \sigma_{ma}{}^2 + \left(\frac{\partial\Lambda}{\partial b}\right)^2 \sigma_{mb}{}^2 + \cdots. \tag{1.13}$$

If, for example, $c = ab$,

$$\sigma_{m\Lambda}{}^2 = \sigma_{ma}{}^2 b^2 + \sigma_{mb}{}^2 a^2. \tag{1.14}$$

Corresponding to (1.2) the fractional standard deviations of the means for $c = a^m b^n$ become

$$\left(\frac{\sigma_{mc}}{c}\right)^2 = m^2\left(\frac{\sigma_{ma}}{a}\right)^2 + n^2\left(\frac{\sigma_{mb}}{b}\right)^2 \tag{1.15}$$

As an example of the application of the method of least squares consider the problem of interpreting the results of a calorimeter measurement

$$U = mc_p \Delta T, \tag{1.16}$$

where the relation between U and ΔT is known but the value of mc_p is to be determined.

Assuming that all the errors in ΔT are random and have the same distribution and same variance, we find that the error in determining ΔT is

$$\epsilon_i = \Delta T_i - \frac{U_i}{mc_p}. \tag{1.17}$$

The least-squares sum becomes

$$\Psi = \sum_{i=1}^{N} \frac{\epsilon_i}{2\sigma^2} \tag{1.18}$$

and

$$\frac{\partial \Psi}{\partial (mc_p)} = \frac{1}{\sigma^2 (mc_p)^2} \sum_{i=1}^{N} U_i \left[\Delta T_i - (U_i/mc_p)\right] = 0 \tag{1.19}$$

or

$$mc_p = \frac{\sum_{i=1}^{N} U_i^2}{\sum_{i=1}^{N} U_i \Delta T_i}. \tag{1.20}$$

The variance of mc_p is

$$\frac{1}{\sigma_{mc_p}^2} = \frac{1}{\sigma^2 (mc_p)^4} \sum_{i=1}^{N} U_i^2, \tag{1.21}$$

where the value of mc_n used is the most probable value. The value for mc_p should agree with independent measurements if the calorimeter absolute calibration is to have meaning.

A frequent problem occurs in comparing results of measurements of a physical variable obtained at two different laboratories. It can be shown that in computing the average of several quantities whose variances are known the most probable value is a weighted average in which each weight is given by the reciprocal variance of the set of measurements:

$$g_i = \frac{1}{\sigma_i^2};$$

that is

$$x = \frac{\sum_{i=1}^{N} x_i/\sigma_i^2}{\sum_{i=1}^{N} 1/\sigma_i^2} \tag{1.22}$$

and the mean is the weighted mean.

Suppose two laboratories measure the wavelength of the Kr^{86} average line and obtain the result

$$\lambda_1 = 6458.0720 \pm 0.0002 \text{ Å}$$

$$\lambda_2 = 6458.0728 \pm 0.0004 \text{ Å}$$

where the $\pm$ refer to standard deviations of the mean. Each observation is weighted according to $1/\sigma^2$, so that the most probable value is

$$\lambda = \frac{(4)\lambda_1 + (1)\lambda_2}{4+1} = \frac{4}{5}\lambda_1 + \frac{1}{5}\lambda_2$$

$$= 6458.0721 \text{ Å}.$$

The new standard deviation becomes

$$\frac{1}{\sigma^2} = \frac{1}{\sigma_1^2} + \frac{1}{\sigma_2^2} \quad \text{or} \quad \sigma = \pm 0.00018.$$

The new value for λ becomes

$$\lambda = 6458.0721 \pm 0.0002 \text{ Å}.$$

The method of least squares can be extended to determine the most probable value of a quantity that results from the measurement of several unknowns. It also can be applied to the cases where the relation among the observables are nonlinear. The treatment of these more complex cases is straightforward and is found in the literature[5], [6].

1.3 REFERENCES

[1] R. Kingslake, *Applied Optics and Optical Engineering*, Academic, New York, 1965, Vol. 1, p. 390.
[2] T. N. Whitehead, *The Design and Use of Instruments and Accurate Mechanism*, Dover, New York, 1954.
[3] A. C. McNish, *Precision Measurement and Calibration*, NBS Handbook 77, U. S. Department of Commerce, Washington, D.C., Vol. III, pp. 14/636–20/643.
[4] H. D. Young, *Statistical Treatment of Experimental Data*, McGraw-Hill, New York, 1962.
[5] L. G. Parratt, *Probability and Experimental Errors in Science*, Wiley, New York, 1961.
[6] H. Jeffreys, *Theory of Probability*, Oxford, New York, 1948.

1.4 PRINCIPAL SYMBOLS, NOTATIONS, AND ABBREVIATIONS

Symbol	*Meaning*
a, b, c	Arbitrary functional variables
g_i	Statistical weight for x_i
mc_p	Mass-specific-heat product
N	Number of observations or measurements
T	Absolute temperature
U, U_i	Energy, specific value thereof, see text
x, x_i	General variable, specific value thereof, see text
$\bar{x}$	Mean value of x
φ	Dispersion or scatter in the value of a variable as compared with its mean value
Δx	Change in variable x
ϵ_i	Difference between the measured value of a variable and a standard of reference
Λ	A quantity whose errors are of interest
λ, λ_i	Wavelength, specific value thereof, see text
σ	Standard deviation
σ_m	Standard deviation of the mean
$\sigma_{m\Lambda}, \sigma_{ma}$	Standard deviation of the mean of Λ, or a, and so on, see text
Ψ	Least-squares sum

2

Beam-sampling Techniques

2.0 INTRODUCTION

The meaningful application of instruments and apparatus in experiments and systems often necessitates the sampling of some characteristic of a laser beam, although it is a difficult problem to ensure that a quantitative, representative sample of the laser beam is obtained. Two sampling techniques that are frequently used involve monitoring the characteristics of the laser beam while observing the effect of radiation upon some object or sensor and reducing the magnitude of beam intensity. Some laser beams possess sufficient intensity or energy to cause destruction of the sensor if used at full strength.

The extremely high energy and power densities available in laser beams today frequently lead to nonlinear-optical effects that are not encountered in sampling of conventional optical radiation. Proper care must therefore be taken to ensure that beam interaction with the sampling devices is minimized. When this cannot be prevented, it is necessary to understand the properties of the sampler to account for the various effects of interest.

Although the laser beam may very well interact with the sampling device, the inverse case, that of the sampling device interacting with the laser, may also occur. If the optical geometry of the beam sampler and laser allows the reflection of energy back to the laser cavity (in an appropriate phase, amplitude, and polarization), the sampler may perturb the cavity modes and change the laser output.

To the conventional techniques of beam sampling, amplitude splitting and beam scattering, must be added induced fluorescence, resonant-transition monitoring, harmonic conversion, and photochemical decomposition. These newer techniques, capable of providing quantitative data, are, at their present state of development, also extremely useful in providing qualitative information about a laser.

2.1 AMPLITUDE DIVISION

Glass and fused-silica beam splitters are frequently used as Fresnel reflectors to sample a portion of the beam. In its basic form a beam splitter is a flat plate of transparent dielectric material, which, for use in the visible or near infrared, is usually made of quartz and, for use in the infrared, is frequently made of germanium. Because these beam splitters will interact with the radius of curvature of the wave front, they must be smooth, extremely flat, and have negligible wedge if precision monitoring is to be attempted. Because Fresnel reflection occurs at the dielectric interface, it is important that the orientation of the beam sampling plate be such that the reflected light does not interact with the basic laser cavity. Furthermore, as shown later, the angle of the beam sampling plate with respect to the beam direction and polarization must be known if a qualitative sample of the beam is to be obtained.

A small portion of the incident beam is reflected towards a sensor, the fraction that is sampled depends on a number of factors. Thus, for example, consider the amplititude of reflected and transmitted waves that result when a plane wave is incident upon a homogeneous, isotropic beam splitter of zero conductivity. Assuming unit relative magnetic permeability and letting A' be the amplitude of the electric vector of the incident optical field (where A' is complex), the amplitudes of the transmitted (refracted) and reflected fields can be calculated. Resolving A' into parallel and perpendicular components and using T' and R' as the complex amplitudes of the transmitted and reflected waves, we obtain [1]

$$T'_{\parallel} = \frac{2n_1 \cos\theta_i}{n_2 \cos\theta_i + n_1 \cos\theta_t} A'_{\parallel}, \tag{2.1a}$$

$$T'_{\perp} = \frac{2n_1 \cos\theta_i}{n_1 \cos\theta_i + n_2 \cos\theta_t} A'_{\perp}, \tag{2.1b}$$

$$R'_{\parallel} = \frac{n_2 \cos\theta_i - n_1 \cos\theta_t}{n_2 \cos\theta_i + n_1 \cos\theta_t} A'_{\parallel}, \tag{2.2a}$$

$$R'_{\perp} = \frac{n_1 \cos\theta_i - n_2 \cos\theta_t}{n_1 \cos\theta_i + n_2 \cos\theta_t} A'_{\perp}. \tag{2.2b}$$

Equations 2.1 and 2.2 are the familiar Fresnel formulas, which are often written in a more compact alternate form by use of the law of refraction

$$\frac{\sin\theta_i}{\sin\theta_t} = \frac{n_2}{n_1}, \tag{2.3}$$

from which we obtain

$$T'_{\parallel} = \frac{2 \sin \theta_t \cos \theta_i}{\sin(\theta_i + \theta_t) \cos(\theta_i - \theta_t)} A'_{\parallel}, \tag{2.4a}$$

$$T'_{\perp} = \frac{2 \sin \theta_t \cos \theta_i}{\sin(\theta_i + \theta_t)} A'_{\perp} \tag{2.4b}$$

$$R'_{\parallel} = \frac{\tan(\theta_i - \theta_t)}{\tan(\theta_i + \theta_t)} A'_{\parallel} \tag{2.5a}$$

$$R'_{\perp} = -\frac{\sin(\theta_i - \theta_t)}{\sin(\theta_i + \theta_t)} A'_{\perp} \tag{2.5b}$$

Because the angles of incidence and refraction respectively, θ_i and θ_t, are real, the phase of each component of the reflected and transmitted wave is either equal to the phase of the corresponding component of the incident wave or differs from it by π. It will be noted that the parallel and perpendicular components of the transmitted wave have the same phase as the parallel and perpendicular components of the incident wave. The phase of the transmitted wave will be equal to that of the incident wave. The phase of the reflected wave, however, will depend upon the relative magnitudes of the angle of the incidence and the angle of refraction. When the material of the beam splitter is denser than the surrounding medium (the usual case), the angle of refraction will always be less than the angle of incidence, the perpendicular component of the reflected wave and the perpendicular component of the incident wave will be different, and the phases will therefore differ by π. Whenever the sum of the angle of incidence and the angle of refraction is greater than $\pi/2$, the phase of the parallel component of the refracted wave and the parallel component of the incident wave will differ by π. Whenever the angle of incidence is zero, that is, normal incidence, the distinction between parallel and perpendicular components disappears, and we obtain

$$T'_{\parallel} = \frac{2}{n_2/n_1 + 1} A'_{\parallel}, \tag{2.6a}$$

$$T'_{\perp} = \frac{2}{n_2/n_1 + 1} A'_{\perp}, \tag{2.6b}$$

$$R'_{\parallel} = \frac{n_2/n_1 - 1}{n_2/n_1 + 1} A'_{\parallel}, \tag{2.7a}$$

$$R'_{\perp} = -\frac{n_2/n_1 - 1}{n_2/n_1 + 1} A'_{\perp}. \tag{2.7b}$$

The amount of energy in the incident, reflected, and transmitted waves may be expressed as follows:

$$U_i = \frac{cn_1}{4\pi} |A'|^2 \cos \theta_i, \tag{2.8a}$$

$$U_r = \frac{cn_1}{4\pi} |R'|^2 \cos \theta_i, \tag{2.8b}$$

$$U_t = \frac{cn_2}{4\pi} |T'|^2 \cos \theta_t, \tag{2.8c}$$

where the reflectivity R is given by

$$R = \frac{|R'|^2}{|A'|^2} \tag{2.9}$$

and the transmissivity T by

$$T = \frac{n_2 \cos \theta_t}{n_1 \cos \theta_i} \frac{|T'|^2}{|A'|^2}. \tag{2.10}$$

Neglecting scattering and absorption for the moment, we see that the law of conservation of energy requires

$$R + T = 1 \qquad \text{(for no absorption or scattering).} \tag{2.11}$$

If α_i is the angle between the electric vector of the incident wave and the plane of incidence, the energy and the parallel and perpendicular components of the incident and reflected waves can be expressed as

$$U_{i\parallel} = \frac{cn_1}{4\pi} |A'_{\parallel}|^2 \cos \theta_i = U_i \cos^2 \alpha_i, \tag{2.12a}$$

$$U_{i\perp} = \frac{cn_1}{4\pi} |A'_{\perp}|^2 \cos \theta_i = U_i \sin^2 \alpha_i, \tag{2.12b}$$

$$U_{r\parallel} = \frac{cn_1}{4\pi} |R'_{\parallel}|^2 \cos \theta_i, \tag{2.13a}$$

$$U_{r\perp} = \frac{cn_1}{4\pi} |R'_{\perp}|^2 \cos \theta_i. \tag{2.13b}$$

The reflectivity therefore for parallel components may be expressed as

$$R_{\parallel} = \frac{\tan^2(\theta_i - \theta_t)}{\tan^2(\theta_i + \theta_t)} \tag{2.14}$$

and

$$R_\perp = \frac{\sin^2(\theta_i - \theta_t)}{\sin^2(\theta_i + \theta_t)} \tag{2.15}$$

for perpendicular components.
The respective parallel and perpendicular components of the transmitted wave may be expressed as

$$T_\parallel = \frac{\sin 2\theta_i \sin 2\theta_t}{\sin^2(\theta_i + \theta_t)\cos^2(\theta_i - \theta_t)}, \tag{2.16a}$$

$$T_\perp = \frac{\sin 2\theta_i \sin 2\theta_t}{\sin^2(\theta_i + \theta_t)}, \tag{2.16b}$$

where again it may be shown that [1] the sum of the transmissivity and reflectivity for the parallel components is equal to the sum of the transmissivity and reflectivity of the perpendicular components, that sum being unity. Thus

$$R_\parallel + T_\parallel = R_\perp + T_\perp = 1. \tag{2.17}$$

For normal incidence the distinction between parallel and perpendicular components again disappears, and we have the reduced equations

$$R = \left(\frac{(n_2/n_1 - 1)}{n_2/n_1 + 1}\right)^2, \tag{2.18}$$

$$T = \frac{4(n_2/n_1)}{(n_2/n_1 + 1)^2}. \tag{2.19}$$

The denominators of (2.14), 2.15), and (2.16) are finite provided that the sum of the angle of incidence and the angle of refraction is $\pi/2$. For this case

$$\begin{aligned} \theta_i + \theta_t &= 2\pi, \\ R_\parallel = 0 \qquad &\text{as } \tan(\theta_i + \theta_t) \to \infty. \end{aligned} \tag{2.20}$$

In this case the reflected and transmitted rays are perpendicular to one another. It follows from the law of refraction that

$$\tan\theta_i = \frac{n_2}{n_1}, \tag{2.21}$$

which is the polarizing or Brewster angle.

From the discussions above it is clear that if the light is linearly polarized, the reflected and transmitted components will also be linearly polarized, for the phase will only change by zero or π. The directions of the electric vector in the reflected and transmitted components of the light, as compared with the incident light, are, however, reversed. If the incident light is elliptically polarized, the reflectivity, being a function of angle of incidence, will cause the degree of polarization to be changed upon reflection. A corresponding change occurs in transmission, so that the degree of polarization of the beam sample and the transmitted light will be altered by the measurements process. From these relations and the measured data the actual polarization of the beam can be inferred.

If the beam-splitter surfaces are perfectly plane and parallel, the transmitted beam, after refraction, will propagate parallel to the direction of the incident beam but will be displaced by an amount

$$x = \zeta(\sin\theta_i - \cos\theta_i \tan\theta_t). \tag{2.22}$$

For $\theta_i \sim 0$

$$x \cong \left(\frac{n_2/n_1 - 1}{n_2/n_1}\right) \zeta \sin\theta_i, \tag{2.23}$$

where x is the linear beam displacement and ζ is the plate thickness, as shown in Figure 2.1. If the beam splitter has a finite wedge, that is the entrance and exit surfaces make some small angle with respect to one another, the beam will be displaced and rotated at an angle. A slight realignment of the overall system will be required.

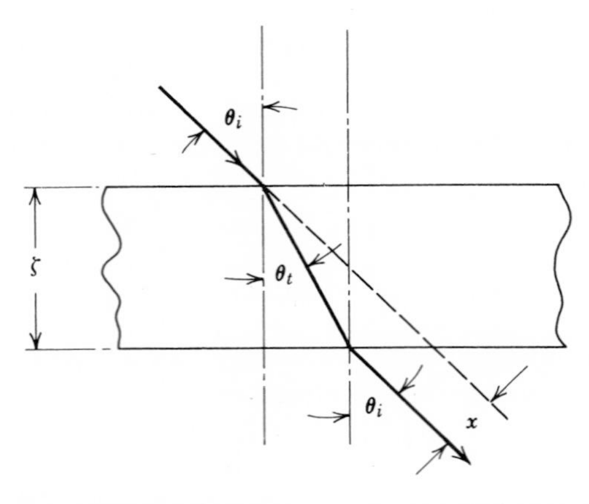

Figure 2.1. Displacement of a beam by a parallel beam splitter.

2.1.1 Technique of Measuring Optical Wedge with a Single-frequency Laser

The single-frequency, helium-neon gas laser may be conveniently used for direct display of the wedge angle between two surfaces, if it is placed in an arrangement that can be termed an inverse Twyman-Green interferometer [2], [3]. The residual wedge angle γ of an optical flat can readily be determined from the spacing of two fringes ℓ as long as the fringe spacing is smaller than the diameter d of the flat and the diameter of the collimated light beam:

$$\tan \gamma = \frac{\lambda}{2nl} \cong \gamma \qquad \gamma \ll \frac{\pi}{2}. \tag{2.24}$$

The beams from the front and rear surfaces of a wedged flat will interfere in space. If the resultant beam is displayed on a screen, the path difference between the front and rear surfaces of the flat will produce an interference pattern of uniformly spaced fringes (for a uniform wedge). If the flat is tilted by an angle β, the path difference δ between the two reflected components changes such that

$$\delta(\beta) = 2nl\left[1 - \left(\sin\frac{\beta}{n}\right)^2\right]^{1/2}. \tag{2.25}$$

To carry out the measurement, the laser beam is enlarged and collimated for which a fixture provided with a vertical axis of rotation and a suitably apertured screen are required. The latter should be provided with a scale. The flat is rotated about the laser beam axis until the projected fringes are parallel to the axis of the fixture holding the flat. Using the scale on the screen the position of the deflected beam with respect to the aperture is determined for two different angles, $2\beta_1$ and $2\beta_2$, for which the fringe line coincides with the edges of the beam. The wedge angle may be computed from these measurements and the relations described above.

2.1.2 Thin-film Beam Splitters for High Peak-power Lasers

Quantitative measurements of the peak-power output of a high-power, Q-switched laser are difficult to make because the sensor can be damaged in a single pulse. The use of conventional neutral-density filters is also prohibited by the vulnerability of the filters to laser damage. It has been found that thin celluloid beam splitters consisting of pellicles approximately 8μ thick [4] can be used to attenuate the beam without damage to the thin film.

2.1.3 Amplitude Division with a Grating

A transmission (or reflection) grating can be used as an amplitude attenuator. The degree of attenuation achieved depends upon the grating constant and the spectral order being observed. Considerable attenuation can be obtained if the grating is coarse. The efficiency of the grating depends upon whether or not the electric vector of the plane polarized light is parallel or perpendicular to the grooves. It can be shown that [2] in a slit grating the ratio of the illumination in the *m*th order to that in zero order is

$$\frac{I_m}{I_0} = \frac{1}{m^2\pi^2}. \tag{2.26}$$

Only about 10 percent of the light appears in the first order, about 2 percent in the second order, and so on. Once a coarse grating has been calibrated for different orders, it may be used to make semiprecision intensity measurements over a very large dynamic range and may be quite useful in attenuating the laser beam amplitude to a level appropriate for detector sensitivity studies.

2.1.4 Granularity of Reflected Coherent Light

When the output beam of a laser impinges upon a surface, the plane wave is decomposed into a series of wavelets that represent a phase map of the beam sampling surface. At a distant point in space the interference of the portions of this decomposed wave front yield a spatial Fourier transform of the reflecting surface. This results in a periodic reinforcement and cancellation of the wave fronts producing a speckled pattern. It is important that the sensor used to determine the characteristics of the beam be placed sufficiently close to the reflecting surface so that it can average the sampled signal over a large number of granules.

2.1.5 Interference Effect in Quarter-wave Plates

Because of the difference in optical thickness for fast and slow waves in a quarter-wave plate, it is possible that the transmittance for two rays may, due to internal interference effects, be significantly different. In the analysis of elliptically polarized light the transmittance as a function of the polarizer azimuth may vary by a factor of 2, as is obvious from the analysis presented below. The transmission ceofficient for a plate of thickness d and refracted index n with no absorption is [5]

$$T^2 = \frac{8n^2}{1 + n^4 + 6n^2 - (1 - n^2)^2 \cos 4\pi(nd/\lambda)}, \tag{2.27}$$

and the phase change in transmittance through the film is given by

$$\gamma = \tan^{-1}\left(\frac{1 + n^2}{2n} \tan \frac{2\pi nd}{\lambda}\right). \tag{2.28}$$

When the optical thickness for one ray is

$$n_1 d = \frac{2m\lambda}{4},$$

$$\gamma_1 = m\pi, (m = (1, 2, 3, \ldots), \tag{2.29}$$

that of the other will be

$$(2m \pm 1)\frac{\lambda}{4}, \quad \gamma_2 = (2m \pm 1)\frac{\pi}{2}.$$

The ratio of the two transmittances for $n = 1.6$ will be 1:0.81. For other optical thicknesses of the plate the ratio of the two transmittances will be less than this extreme value, but only for

$$n_1 d = (2m + 0.46)\frac{\lambda}{4},$$

$$\gamma = 180\, m + 45^\circ, \tag{2.30}$$

and

$$n_2 d = (2m - 0.46)\frac{\lambda}{4},$$

$$\gamma = 180\, m - 45^\circ, \tag{2.31}$$

will the two transmittances be equal ($T^2 = 0.898$).

If mica is used in the analysis of elliptically polarized light, the mica is set with its principal directions parallel to the axes of the ellipse. The azimuth of the restored, linearly polarized beam then gives the ratio of the ellipse axes. As the transmittance of the two principal directions can vary as much as 0.81:1, this measurement can be in considerable error. It is therefore essential that the transmittance ratio of any quarter wave plate be measured before its use so that appropriate correction can be made.

2.2 SAMPLING BY BEAM SCATTERING

Although it is more conventional to sample laser beams by amplitude splitting, sampling by beam scattering should by no means be ignored. Highly attenuated qualitative samples of beam intensity can be obtained from light that is scattered from dust particles in the air through which the beam passes. A more reproducible, attenuated sample of a high-power laser beam that is more readily calibrated can be obtained if the light that is scattered in a dielectric slab is directed to a sensor. The scattering centers in an optical wedge represent an ideal attenuated sample of the beam. As long as the beam is transmitted through the same spatial volume the intensity of the scattered beam is directly related to the transmitted beam.

A small fraction of the total output of a pulsed solid-state laser may be sampled by fine metallic wire placed in front of a sensor. Although this technique is capable of withstanding high peak-power pulses, it is also semi-quantitative in that the shape of the solid-state laser-beam pattern varies from pulse to pulse in a way that is not represented by the fraction of the light scattered from the edge of the beam.

2.3 INDUCED FLUORESCENCE

Only an extremely narrow portion of the spectral range for which lasers exist can be observed by visual techniques. It is of value to extend the effective visible range by employing techniques that will enable us to see the laser beam even though its wavelength range is beyond the visible. Some of these techniques are discussed briefly below.

2.3.1 Monitoring Ultraviolet Beams with Fluorescent Materials

Many materials, especially those of organic origin, will fluoresce in the visible when irradiated with ultraviolet light. One of the most convenient wavelength converters to use in the laboratory for observing ultraviolet laser-beam spots is ordinary white paper, for the whitening agents used in the manufacture of white papers have a relatively high fluorescent efficiency.

Another readily available fluorescent material, which is frequently used as a wavelength converter for sensitizing photomultipliers, is ordinary stopcock grease. Probably the most widely used wavelength converter is sodium salicylate. Its chief values are its very wide absorption range (40 to 360 nm). Its fluorescent output is in the blue-green portion of the spectrum where the quantum efficiency of most phototubes is large.

2.3.2 Monitoring Infrared Laser Beams with Cadmium-based Phosphors

Several cadmium-based phosphors (available from U.S. Radium Corp.) have been found to fluoresce in the wavelength range from 700 to approximately 1200 nm. These phosphors, when appropriately applied to a substrate, form valuable tools for detecting the output of 1.06 μm pulsed, solid-stage lasers. They are also quite useful for aligning helium-neon gas lasers at 1.15 μm.

2.3.3 Monitoring CW-argon Laser Beams by Fluorescence

When a high-power argon laser beam is transmitted through optical-quality acrylic resins, as well as some clear glasses, it causes fluorescence in the orange portion of the spectrum. Because the fluorescent decay time is long, being of the order of 1 sec, it provides an excellent source of attenuated radiation that is proportional to the average power output of the laser. The scattered blue-green light from the argon laser is easily distinguished from the fluorescent light on the basis of color.

2.4 HARMONIC CONVERSION

Optical harmonic generation techniques are of value in beam monitoring, when it is necessary to double or triple the output frequency of the laser to bring its wavelength into a range more suitable for detection and sensing [6]. The output frequency of a Q-switched CO_2 laser operating at 10.6 μm can be doubled with tellurium to bring its wavelength into the range of sensitive indium antimonide photodetectors. When this detector is terminated in a low impedance, it is capable of resolving the output pulse shape with excellent time resolution. The second harmonic conversion monitoring technique tends to give an optimistic measure of the pulse shape of the laser, for the second harmonic power output of the converter is quadratically related to the fundamental power output of the laser.

Qualitative visual verification of the output wavelength of pulsed nitrogen and pulsed helium neon lasers (in a wavelength range of 1 to 1.2 μm) has been obtained by doubling their output with KDP crystals. Although the overall efficiency of the harmonic conversion process is relatively low, particularly for peak power levels of the order of 100 W, adequate flux is produced to enable visual detection.

2.5 PHOTODECOMPOSITION

Techniques for sampling the output beam of molecular gas lasers in the wavelength range of 5 to 10 μm are currently under development. One of the techniques that has proved useful for observing the output beam of a CO_2 laser at 10.6 μm consists of passing the beam through a cell containing gaseous ammonia. Presence of the laser beam is indicated by a green fluorescence, which is attributed to photodecomposition of the gas. (The intensity of the fluorescent output light at constant laser intensity decreases slowly with time.) Research currently in process in a number of industrial laboratories will undoubtedly result in the development of many other multiple-photon type wavelength converters that will enable visual sampling of infrared laser beams.

2.6 RESONANT TRANSITION MONITORING IN INFRARED GAS LASERS

The population densities of excited levels in gas lasers are known to change markedly when laser action takes place. These effects are easily observed by placing the grating of the spectrograph near the plasma tube.

The spectra output of the laser in the visible range is determined under conditions where the Fabry-Perot cavity is resonant and nonresonant. A typical example of this method of beam monitoring occurs in helium-neon lasers. The green transition that may be observed in a spontaneous emission from the side of the plasma tube can be observed to change in intensity, depending on whether the 3.39-μm laser cavity is aligned. Once the visibility of the associated transitions has been established, a pocket spectroscope may be used to tune the laser for maximum power output.

2.7 THE SOURCES OF ERROR IN BEAM MONITORING

Probably one of the most serious blunders that can be made during measurements in beam sampling is to try to determine the beam power or energy at or near the laser housing. All lasers emit an enormous flux of spontaneous radiation as part of their total energy output. If the beam is sampled near the laser, adequate precautions must be taken to filter the beam sample to exclude the direct spontaneous emission from the laser cavity, pump light, and, if we are working in the infrared, extraneous sources of heat such as incandescent lamps. Overly optimistic measures of the output energy of ruby lasers have been obtained as a result of irradiating calorimeters and other energy-measuring devices in close proximity to the laser cavity. It is of value to measure energy and power at several places along the beam axis. Any indication of change in beam intensity or energy should be checked, for the energy in the laser beam should not vary with distance.

As we have already mentioned, care must be taken to ensure that the reflected or scattered beam from the sampling device or apparatus which receives the laser beam does not interfere with the operation of the laser; that is, good isolation is necessary. Particularly in high-gain infrared lasers, aperature scattering can be a serious source of error.

Considerable care must be taken in interpreting samples of the output of solid-state lasers (or gas lasers) equipped with flat ends. The polarization of the output beam may vary with time. The sampling process will change the percentage polarization and, of course, an account of this fact must be taken in the interpretation of data.

Great care must be used in the selection of attenuators to determine their suitability for use in the energy flux being measured. (Techniques for calibrating laser attenuators are discussed later.)

Care must be taken that interference effects, which often occur as a result of multiple reflections between nearly parallel surfaces or within optical flats, do not cause errors. The chance of error increases when measurements are being made outside the visible spectrum because the eye cannot assist in the identification of experimental anomalies. A typical example of a measuring situation that can result in errors develops when power measurements are being made in the infrared. A radiation thermopile, because of its relative insensitivity to wavelength (see Chapter 4), is used to measure average power. Most radiation thermopiles are multiple junction devices, and their calibration depends on measuring the average power in a plane wave front. If the beam being measured is uniform, the results are capable of unambiguous interpretation. Consider the output-power measurement of a CW-infrared laser, in which the power to be measured exceeds the maximum rating of the thermopile. An attenuator must be used to reduce the beam power to a satisfactory value. The attenuator can be placed directly in front of the thermopile or it can be placed at the laser. Now the thermopile is usually placed 10 to 50 feet from the laser, so that the beam spot will uniformly illuminate the aperture of the thermopile. If a high-quality attenuator is placed at the laser, there is a possibility that a Fabry-Perot cavity will be formed by the attenuator, which will lead to interference bands in the beam. Should this occur, the thermopile will be illuminated by a wave front with periodic fringes and serious errors in measurement can result (8:1). This difficulty can usually be circumvented by placing the attenuator at the junction thermopile.

2.8 REFERENCES

[1] M. Born and E. Wolf, *Principles of Optics*, Pergamon, New York, 1959.
[2] R. W. Ditchburn, *Light*, Interscience, New York, 1963, 2nd ed., pp. 129, 336–339.
[3] V. Met, *Appl. Opt*. **5**, 1242 (1966).
[4] R. W. Waynant, *et al., Beam Divergence Measurements in Q-Switched Lasers*, AF30(602) 3332 Report.
[5] O. S. Heavens, *Optical Properties of Thin Films*, Butterworth, London, 1955, pp. 58, 90.
[6] R. W. Terhune, P. D. Maker, and C. M. Savage, *Phys. Rev. Letters*, 8, 404 (1962).

2.9 PRINCIPAL SYMBOLS, NOTATIONS, AND ABBREVIATIONS

Symbol	*Meaning*
$A'_{\parallel}$, $A'_{\perp}$	Amplitude of the parallel and perpendicular incident components of an elliptically polarized wave
c	Free-space velocity of light
d	Thickness of quarter-wave plate
ζ	Thickness of a beam-sampling plate
l	Spacing of two fringes
m	Integer
n_1, n_2	Index of refraction in medium 1, 2
R	Power reflectivity of a surface

$R'_{\parallel}$, $R'_{\parallel}$	Amplitude of the parallel and perpendicular reflected components of plane wave
$R_{\parallel}$, $R_{\perp}$	Reflectivity associated with the parallel and perpendicular reflected components of a plane wave
T	Power transmissivity of a surface
$T'_{\parallel}$, $T'_{\perp}$	Amplitude of the electric field of the parallel and perpendicular transmitted (refracted components of a plane wave
$T_{\parallel}$, $T_{\perp}$	Transmissivity of the parallel and perpendicular transmitted (refracted) components of a plane wave
U_i, U_t, U_r	Incident, transmitted, and reflected energies of a wave incident upon a boundary between two media
x	Linear displacement of the laser beam
α_i, α_t	Angle of incidence and transmission of the electric vector of the incident plane wave and the plane of incidence
2β	Angle between the incident and reflected light beams in wedge measurements
λ	Free-space wavelength
θ_i, θ_t	Angles in the plane of incidence between the surface normal and the incident and transmitted (refracted) rays of light
γ	Wedge angle of an optical flat

3

Measurement of Beam Parameters

3.1 INTRODUCTION

Operating lasers now span the wavelength region of approximately 200 nm to 0.4 mm, although most are extremely narrow band oscillators. The spectral range is in large part without laser wavelengths, even though hundreds of transitions have now been discovered. Many experimental techniques are needed to determine laser beam parameters over such a large spectral range. It is evident that in some regions of the spectrum limitations on techniques are imposed by the types of detector that are available. The spectral region from 0.25 to approximately 1 μm embraces the most familiar part of the electromagnetic spectrum. It is in this range that a great variety of detectors has long been available. These detectors include photographic films, photoemissive and photoconductive detectors, and a wide variety of actinometric materials. In the wavelength region of 0.7 to 1000 μm, generally referred to as the infrared, many advances have been made and many detectors have been developed. The important properties of some of the most widely used infrared detectors are described in Chapter 4.

The operating characteristics of a particular laser sometimes play an important role in determining which measurement techniques are most suitable for parameter determination. The types of operation can be classified as follows: continuous operating or CW, modulated or pulsed-CW, random-spiking, mode-locked, and the *Q*-switched. Examples of each type of lasers are the CW, helium-neon laser; the pulsed-CW, gallium arsenide semiconductor diode laser; the pulsed ruby-crystal laser, operated in the spiking mode; the mode-locked argon ion laser; and the *Q*-switched or giant-pulse neodymium glass laser. It will become evident that the precision of measurement of the laser-beam parameters varies widely for the different types of operation; for example, in the case of multimode solid-state crystal lasers (ruby) operating in the spiking mode the laser switches modes rapidly (less than 10^{-6} sec) and continuously during a millisecond pulse. Thus the observed beam parameters will be an average over many different oscillating modes.

Because the measurement of laser-beam parameters is intimately connected with the mode structure of the laser material and its optical cavity, a section of this chapter treats some of the most important types of optical resonators and their modes (resonances). Methods for the determination of the near-field and far-field radiation patterns associated with the modes of the resonators are illustrated with examples of actual measurements. These illustrations serve to clarify the descriptions and provide a framework for comparison of the attainable precisions. Also, special attention is given to instrumentation that has been developed specifically for the measurement of laser parameters and to novel methods of application of existing instruments.

3.1.1 Mode Spectrum of Laser Resonators

The radiation pattern of a laser is closely linked to the characteristics of the laser resonator. There are seven basic open-cavity configurations used in lasers: parallel-plane, large-radius, confocal, spherical, concave-convex, hemispherical, and folded confocal. All laser resonators have a common characteristic; they are open resonators that do not require sidewalls. When dielectric interfaces become a part of the active laser medium, as in fibers, cubes, spheres, and ring structures, the coupling effects play an important role in the mode structure of the output radiation.

In general, two conditions must be satisfied to obtain a resonator system with a high Q. First there must be a closed set or group of rays that can repeatedly strike the reflectors after a number of consecutive reflections, however large. Second the Fresnel number, defined as

$$N = \frac{a_1 a_2}{d\lambda} = \frac{a^2}{d\lambda} \qquad \text{if } (a_1 = a_2), \tag{3.1}$$

where a_1, a_2 are the radii of the limiting apertures at the ends of the laser N, should be large if diffraction losses are to be kept small. The first condition is imposed by geometrical optics and the second, by physical optics. The electric-field configurations of the various modes that can be supported by the resonators have been the subject of extensive theoretical and experimental investigations [1] to [7]. The information presented in the next section draws heavily on these sources.

3.1.1.1 Plane Parallel Fabry-Perot, Square Aperture The geometry of the plane-parallel Fabry-Perot (PPFP) system consists of two very flat plane and parallel reflectors of rectangular or square shape. The formulas in this section are given for the case of gas lasers, for which the symbol λ = wavelength refers to the vacuum value.

The resonances of a rectangular conducting box of dimensions a and d in the x, y, and z directions, respectively, are given by

$$\left(\frac{2}{\lambda}\right)^2 = \left(\frac{r}{a}\right)^2 + \left(\frac{s}{c}\right)^2 + \left(\frac{q}{d}\right)^2 \tag{3.2}$$

where r, s, and q are integers and d is the cavity length along the z direction. The actual resonances of the open-resonator structure with a square aperture closely approximate those given by (3.2). The transverse-mode field distribution is given (for a particular polarization) by

$$\frac{E(x, y)}{E_0} = \sin\left(\frac{\pi r x}{2a}\right) \sin\left(\frac{\pi s y}{2c}\right). \tag{3.3}$$

These modes can be expected to be good approximations to the actual modes in the case of a large Fresnel number, the usual situation in optical resonators.

The mode separation is given by

$$\Delta\left(\frac{1}{\lambda}\right) = \frac{1}{2d}\left[(q_2 - q_1) + \frac{d\lambda}{16a^2}(r_2{}^2 - r_1{}^2 + s_2{}^2 - s_1{}^2)\right], \tag{3.4}$$

where $r, s < < q$, $q \sim 2d/\lambda$. The axial mode separation corresponding to changing the number of half-wavelengths in the cavity by one, $\Delta q = 1$ is given by $\Delta\lambda = \lambda/q$, the spectral range of the interferometer.

3.1.1.2 PPFP, Circular Aperture The resonances of a cylindrical metallic cavity are given by

$$(f)^2(2a)^2 = \left(\frac{c p_{nm}}{\pi}\right)^2 + \left(\frac{cq}{2}\right)^2 \left(\frac{a}{d}\right)^2, \tag{3.5}$$

where $2a$ = diameter, p_{nm} = mth root of the nth-order Bessel function (J_n) for the TM modes and the derivative of the Bessel function (J_n') for the TE modes, and q = number of half-period variations along d.

The transverse-field distribution for a particular direction of polarization is given by

$$\frac{E(\rho, \theta)}{E_0} = J_n\left(\frac{\rho_{nm}\rho}{a}\right) e^{-jn\theta}, \tag{3.6}$$

where ρ and θ are the usual polar coordinates, and $j = (-1)^{1/2}$.

The mode separations for both the TE and TM modes are given by

$$\Delta\left(\frac{1}{\lambda}\right) = \left(\frac{1}{2d}\right)(q_2 - q_1) + \frac{\lambda}{8\pi^2 a^2}(p_2{}^2 - p_1{}^2), \tag{3.7}$$

where $p_2 = p_{n_2 m_2}$ and $p_1 = p_{n_1 m_1}$.

In the case of fiber optic lasers the frequency difference between modes is essentially identical to those given by (3.7).

3.1.1.3 PPFP, Slight Spherical Curvature, Square-aperture Resonator Consider the PPFP with a small perturbation which consists of a slight curvature of the end plates[8]. The curvature of the concave mirrors is given by λ/Q. The mode separation is given by

$$\Delta\left(\frac{1}{\lambda}\right) = \frac{1}{2d}\left[(q_2 - q_1) + \frac{b\lambda}{16a^2}({r_2}^2 + {s_2}^2 - {r_1}^2 - {s_1}^2) + \frac{4}{\pi^2 Qd}\left(\frac{1}{{r_1}^2 + {s_1}^2} - \frac{1}{{r_2}^2 + {s_2}^2}\right)\right]. \tag{3.8}$$

For (3.8) to be valid the reflector curvature is restricted according to the inequality $Q > N$, where N is the Fresnel number given by

$$N = \frac{a^2}{d\lambda}. \tag{3.9}$$

This condition can also be stated as $b'/d > > N^2$ where b' is the radius of curvature of the reflector. The field distributions (transverse) agree closely with those of the PPFP, square aperture (3.3).

3.1.1.4 Spherical Resonators, Large Square Aperture The variation of the electric field of a given polarization over the surface of a spherical reflector is given by

$$\frac{E(x, y, \pm d/2)}{E_0} = H_m\left(\frac{x\sqrt{2}}{w_s'}\right) H_n\left(\frac{y\sqrt{2}}{w_s'}\right) \exp\left[-\frac{(x^2 + y^2)}{{w_s'}^2}\right], \tag{3.10}$$

where H_m is the Hermite polynomial of order m. The resonant modes of this structure are designated TEM_{mnq}, which stands for transverse electric and magnetic wave. The parameter w'_s denotes the spot size of the TEM_{00q} mode at the mirror and is given by

$$w_s' = \left(\frac{\lambda b'}{\pi}\right)^{1/2}\left(\frac{d}{2b' - d}\right)^{1/4}. \tag{3.11}$$

In (3.10) a Cartesian system of coordinates is used with the z axis coinciding with the resonator axis and the x, y axes on the curved resonator surface.

When the resonator employs reflectors with unequal radii of curvature

b_1 and b_2 and a reflector spacing d, the spot sizes w_1 and w_2 at the respective reflectors are given by

$$\left(\frac{w_1}{w_2}\right)^2 = \left(\frac{b_1}{b_2}\right)\left(\frac{b_2 - d}{b_1 - d}\right), \tag{3.12}$$

$$(w_1 w_2)^2 = \left(\frac{\lambda}{\pi}\right)^2 \frac{b_1 b_2 d}{b_1 + b_2 - d}. \tag{3.13}$$

The field patterns for a particular direction of polarization of the TEM_{mnq} modes are given by (3.10), by substituting w_1 or w_2 for the w' of (3.10). The functions of (3.10) occur in the quantum mechanical solution of the harmonic oscillator problem and consequently are known as harmonic oscillator functions[9]. The light-intensity patterns corresponding to the field distributions of (3.10) are given by the square of the electric field.

The mode separation of the "quasiconfocal" resonator is given by

$$\Delta\left(\frac{1}{\lambda}\right) = \frac{1}{2d}\left[(q_2 - q_1) + \frac{1}{\pi d}(\Delta m + \Delta n) \cos^{-1}\left\{\left(1 - \frac{d}{b_1}\right)\left(1 - \frac{d}{b_2}\right)\right\}\right]^{1/2} \tag{3.14}$$

The term "quasiconfocal" is used to indicate that the reflectors are nearly confocal but may have different radii of curvature and, within limits, an arbitrary spacing. For the case of equal radii of curvature, b' nearly equal to d, the mode spacing is given by

$$\Delta\left(\frac{1}{\lambda}\right) = \frac{1}{2d}\left[(q_2 - q_1) + \frac{1}{\pi}(\Delta m + \Delta n)\left(\frac{\pi}{2} - \frac{b' - d}{b'}\right)\right]. \tag{3.15}$$

The theory of Boyd and Gordon[2] for the nonconfocal resonator does not hold for the limiting case of plane parallel reflectors. This is indicated by (3.11) for the spot size of the TEM_{00q} mode, which approaches infinity as the radius of curvature b' is permitted to go to infinity. This shows that an approximate limit for the validity of (3.10) occurs when the spot size is of the order of the mirror dimension a. The range of applicability of (3.10) is obtained by requiring that

$$a > \left(\frac{\lambda b'}{\pi}\right)^{1/2} \cdot \left(\frac{d}{2b' - d}\right)^{1/4} \tag{3.16}$$

3.1.1.5 Spherical Reflectors, Large Circular Aperture The case of circular mirrors with a spherical curvature has been analyzed by Boyd and

Kogelnik[3]. The field patterns for a particular direction of polarization have been found to be the TEM_{plq} modes given by

$$\frac{E(\rho, \theta, \pm d/2)}{E_0} = \left(\frac{\rho\sqrt{2}}{w_s'}\right)^l L_p^l \left(\frac{2\rho^2}{w_s'^2}\right) \exp\left(-\frac{p^2}{w_s'^2}\right) \cos l\theta, \tag{3.17}$$

where the L_p^l are known as the associated Laguerre polynomials. The remainder of the symbols in (3.17) have been previously defined. The condition for the range of validity of (3.17) is given by

$$g > \left(\frac{\lambda b'}{\pi}\right)^{1/2} \cdot \left(\frac{d}{2b' - d}\right)^{1/4}, \tag{3.18}$$

where g is the radius of the mirror.

The TEM_{plq} modes spacing is given by

$$\Delta\left(\frac{1}{\lambda}\right) = \frac{1}{2d}\left[(q_2 - q_1) + \frac{1}{\pi}(2\Delta p + \Delta l)\cos^{-1}\left(1 - \frac{d}{b'}\right)\right] \tag{3.19}$$

in which it has been assumed that $b_1 = b_2 = b'$. If the separation of the reflectors corresponds to the confocal condition, the spectrum given by (3.19) is the same as that given by (3.15), except that $(\Delta m + \Delta n)$ is replaced by $(2\Delta p + \Delta l)$.

For reference purposes the expansions for the Hermite and associated Laguerre polynomials are given below[9]:

$$H_n(x) = (2x) - \frac{(2x)^{n-2}n(n-1)}{1!} + \frac{(2x)^{n-4}n(n-1)(n-2)(n-3)}{2!} - \cdots \tag{3.20}$$

where $H_n(x)$ is the Hermite polynomial of degree n;

$$L_{n+l}^{2l+1}(\zeta) = \sum_{k=0}^{n-l-1} (-1)^{k+1} \frac{[(n+l)!]^2 \zeta^k}{(n-l-1-k)!(2l+1+k)!k!} \tag{3.21}$$

where L_{n+l}^{2l+1} is the associated Languerre polynominal and the symbol

$$k\Delta = k(k-1)(k-2)\cdots(1).$$

3.1.1.6 Resonators, Uniformly Filled with Dielectric The description of the solid-state lasers is practically identical to that given for gas lasers (Sections 3.1.1.1 through 3.1.1.5) if the space between the reflectors is filled with a material of uniform dielectric constant, ϵ. Equations 3.4, 3.8, 3.14,

3.15, and 3.19, which give the mode separation, apply here, with the modification that the velocity of light in the dielectric, or

$$c' = \frac{c}{(\epsilon)^{1/2}} = \frac{c}{n}. \tag{3.22}$$

Note that the presence of appreciable dispersion (dependence of n on frequency) may remove the degeneracy between the frequency difference of the modes (3.14), (3.15), and (3.19). Dispersion occurs in the case of the semiconductor lasers.

The equations that describe the field distribution apply, provided that w' is suitably modified. All that is required is to replace λ, the wavelength in a vacuum, with λ', the wavelength in the dielectric, in the formulas which give w'. In general, the modifications consist of using the wavelength and velocity of light appropriate for the dielectric in the equations presented for the different resonator configurations.

3.1.1.7 Resonators, Not Uniformly Filled with a Dielectric The simplest case of a resonator not uniformly filled with a dielectric is that of a normal ruby laser in which the laser crystal is a dielectric rod with plane parallel end faces but with detached dielectric mirrors. The frequency difference between the resonant modes in this case is given by (3.4) where d is replaced by $d' = d_1 + d_2 + d_3$, where d_2 is the laser rod length, d_1 d_3 are the spaces between the mirrors, and n is the index of refraction of the rod.

3.1.1.8 Resonators, Uniformly Filled with a Dielectric Slab The geometry of the semiconductor laser is that of a resonator uniformly filled with a dielectric slab. The dielectric is assumed to have negligible loss, which is only approximately the case for semiconductors[10]. Nevertheless the form of the solutions is similar to that of waves guided along dielectric slabs. Those modes are designated TM that have a single magnetic field component H_y and the electric-field components E_x and E_z. The TE modes have the roles of E and H interchanged, and thus have components E_y, H_x, and H_z. Propagation is assumed to be in the z direction. In both cases, solutions are of the symmetrical type (even) and of the antisymmetrical type (odd), depending upon the variation of the fields about the symmetry plane $x = 0$.

For the TM modes the even solutions are of the form

$$E_x \propto H_y \propto \begin{cases} \cos(hx)\exp(-j\beta z) & |x| \leq t, \quad (3.23) \\ \exp[-(p\,|x| + t)]\exp(-j\beta z) & |x| \geq t. \quad (3.24) \end{cases}$$

For the odd modes [omitting the phase factor $\exp(-j\beta z)$];

$$E_x \propto H_y \propto \begin{cases} \exp[-(p(x-t)] & x \geq t, & (3.25) \\ \csc(ht)\sin(hx) & -t \leq x \leq t, & (3.26) \\ \exp[p(x+t)] & x \leq -t, & (3.27) \end{cases}$$

The symmetrical (even) TE modes can be obtained by the substitution of E_y for H_y in (3.25), (3.26), and (3.27) and H_x for E_x. The antisymmetric TE modes are given by (omitting the phase factor)

$$E_x \propto H_x \propto \begin{cases} \exp[-p(|x|-t)] \; x \geq t, & (3.28) \\ \sec(ht)\sin(hx) \quad x \geq t. & (3.29) \end{cases}$$

The field distribution in the x direction exhibits an exponential fall-off in the regions outside of the bounded slab. The axial-mode resonance of the semiconductor laser with nominally plane and parallel end reflectors is given by (3.2), which must be modified to take into account the dielectric constant of the semiconductor, and also the dispersion of the semiconductor because the oscillations occur close to the resonance of the lowest interband transition. The wavelength difference ($\Delta\lambda$) between the axial modes is given by

$$(q_2 - q_1)\frac{\Delta\lambda}{\lambda} = \frac{1}{2n\left[1 - \frac{\lambda}{n}\frac{dn}{d\lambda}\right]}, \tag{3.30}$$

where $dn/d\lambda$ is the rate of change of the index of refraction with wavelength, evaluated at the wavelength of the mode.

3.1.1.9 Modes in Closed Resonators (Bouncing-ball Modes) The frequency spectrum [11] of the unperturbed modes of a rectangular ruby prism, with all surfaces functioning as ideal reflectors, is given by

$$\frac{\omega^2{}_{mn\nu}\epsilon_r}{c^2} = \left(\frac{m\pi}{a}\right)^2 + \left(\frac{m\pi}{b}\right)^2 + \left(\frac{\nu\pi}{l}\right)^2 \tag{3.31}$$

where ω_{mnv} = angular frequency; ϵ_r = dielectric constant of ruby; m, n, and v = small integers; a and b are the x and y dimensions of the prism end face and ℓ is the prism length.

If one of the end reflectors $R\ell$ is assumed to have resistive losses, the modes with the highest Q are emitted in directions given by

$$\tan\theta = \left(\frac{a^2N_x{}^2 + b^2Ny^2}{2lN_z{}^2}\right)^{1/2},$$

$$\tan\phi = \frac{bN_y}{aN_x}(N_x N_y N_z), \qquad (3.32)$$

where N_x, N_y, and N_z are small integers. The preferred directions can be shown to correspond to the directions of motion of a ball bounding elastically from the reflecting walls of the cavity. In these paths the ball bounces on the X, Y, and Z surfaces N_x, N_y, and N_z times, respectively, and then passes through its original point, thus retracing its path. The field distributions have not yet been derived, although it appears evident that the distributions for a rectangular crystal laser will resemble those of a resonant rectangular cavity.

3.1.1.10 Fiber Optic Resonators The mode structure of the fiber optic wave guide is more complex than the modes of the open optical resonators discussed above[12]. The cylindrically symmetric modes are transverse electric (TE_{nm}) or transverse magnetic (TM_{nm}) modes and have been obtained for $n = 0$. The subscript n refers to the order of the Bessel function (3.33), (3.34), and (3.35), and the subscript m refers to the successive roots of the Bessel function. The higher-order modes are hybrid, that is, they have nonvanishing values for both E_z and H_z. They are designated by the symbols EH_{nm} and HE_{nm} and result from the two possible values that p can assume for fixed n and m.

The HE_{nm} modes are distinguished from the others in two important respects. They are the only modes with light intensity at the center of the guide and thus possess an on-axis lobe in the radiation pattern. The lowest-order member, the HE_{1m} mode, does not have a cutoff wavelength.

The field distributions that correspond to various dielectric waveguide modes are given using the notation of Snitzer. Using cylindrical coordinates, r, θ, z, free-space permeability $= \mu$, core index of refraction (higher than the cladding) $= n_1$, and core radius $= a$, the field components are

$$E_z = J_n \frac{ur}{a} F_c, \qquad (3.33)$$

$$E_r = j\frac{ha}{u}\left(\frac{1-P}{2} J_{n-1} - \frac{1+P}{2} J_{n+1}\right) F_c, \qquad (3.34)$$

$$E_\theta = j\frac{ha}{u}\left(-\frac{1-P}{2}J_{n-1} - \frac{1+P}{2}J_{n+1}\right)F_s, \tag{3.35}$$

$$H_z = \frac{-h}{\mu\omega}PJ_nF_s, \tag{3.36}$$

$$H_r = -j\frac{k_1{}^2}{\mu\omega}\frac{a}{u}\left(-\frac{1-Ph^2/k_1{}^2}{2}J_{n-1} - \frac{1+Ph^2/k_1{}^2}{2}J_{n+1}\right)F_s, \tag{3.37}$$

$$H_\theta = j\frac{k_1{}^2}{\mu\omega}\frac{a}{u}\left(\frac{1-Ph^2/k_1{}^2}{2}J_{n-1} - \frac{1+Ph^2/k_1{}^2}{2}J_{n+1}\right)F_c, \tag{3.38}$$

where

$$F_c = A\cos(n\theta + \psi_n)\exp[j(hz - \omega t)], \tag{3.39}$$

$$F_s = A\sin(n\theta + \psi_n)\exp[j(hz - \omega t)], \tag{3.40}$$

$$k_1 = \frac{2\pi n_1}{\lambda}. \tag{3.41}$$

In the cladding, which has an index of refraction n_2, the field is given by equations similar to (3.33) to (3.37), but in place of the propagation constant k_1, k_2, is substituted, and the radial dependence is given by modified Hankel functions in place of the Bessel functions J_n. The phase ψ_n can take the values 0 and $\pi/2$. The parameters u and h are related by

$$\left(\frac{u}{a}\right)^2 = k_1{}^2 - h^2. \tag{3.42}$$

The values of the parameters u, h, and P can be found from the boundary conditions.

3.1.1.11 Ring Patterns A number of studies of the far-field patterns of solid-state lasers, particularly ruby lasers, have shown a system of rings in addition to the central light spot. This phenomenon is often observed in studying the near-field and far-field radiation patterns. The observed rings are similar to those found when the laser light is analyzed by means of Fabry-Perot etalons. When the laser rod itself is the FP interferometer, that is, a rod with coated end faces, the size of the observed ring pattern is given by

$$R_n^2 - R_{n-1}{}^2 = \frac{f^2\lambda_0 n}{d}, \tag{3.43}$$

where R_n is the radius of the nth ring; f is the focal length of the lens, λ_0 is the free-space wavelength, n is the index of refraction of the host crystal, and d is the length of the host crystal.

The coherence of the light from different rings and from different parts of the same ring has been tested by observing the interference fringes obtained. In this way it has been ascertained that the light in the ring patterns possesses coherence properties similar to the laser light in the central spot. It has been suggested, therefore, that the rings are produced by the scattering of laser light by the optical inhomogeneities of the host crystal. Thus the ring patterns are analogous to those obtained when the laser light is analyzed with an external FP etalon.

3.2 MEASUREMENT OF THE TRANSVERSE INTENSITY DISTRIBUTION AND THE BEAM DIVERGENCE

The transverse-mode structure or intensity distribution of the output beam from a laser can be examined basically in either of two ways:

1. The beam may be photographed.
2. The beam may be scanned by an optical detector that is sensitive to a small transverse portion of the beam at any given instant.

The specific techniques to be used in any given situation will depend on the output characteristics of the particular laser under study (e.g., CW output for gas lasers or pulsed-spiking output for ruby and other solid-state lasers), the expected mode structure, the power level and wavelength, and the physical structure of the laser. Photographic methods may simply be used to indicate the gross mode structure, or they may be used in a sophisticated fashion actually to determine the quantitative transverse-intensity distribution. High-speed photography may be applied to the determination of the time-varying characteristics of complex laser outputs.

The beam divergence is determined by measuring the mode structure at more than one transverse plane. For example, a measurement may be made in the near field (i.e., very close to the partially transparent reflector on the laser), followed by a measurement in the far field, from which the gross beam divergence may be determined. Alternatively, a more nearly three-dimensional determination of the mode structure may be made.

3.2.1 Laser Beam Photography

In the simplest application of photography to determine the mode structure of a laser beam a photographic film or plate is inserted directly into a laser beam to determine the qualitative nature of the transverse intensity distribution at a given distance from the laser. In addition, a lens may be used to focus the near or far field of the radiation onto the film. The film and exposure time must be chosen with regard to the duration, power, energy, and frequency of the laser radiation. In this simple case the information obtained is simply a nonquantitative pattern that indicates in a gross manner the size and shape of the mode structure. Care must be taken to prevent burning of the film, especially when lenses are employed. The aperture of the camera must be larger than the gross size of the beam to preclude spurious diffraction effects.

3.2.1.1 Image-converter Tubes Image-converter tubes are devices that may be used to convert infrared or ultraviolet radiation into visible light. Image photographs may be made to render infrared or ultraviolet laser mode patterns directly observable. Image converters are valuable tools to have when laser adjustments are being made. The range of wavelengths now covered by photographic films includes almost the entire range covered by image-converter tubes (although Polaroid films are not yet available in the infrared), but the use of such films (rather than an image converter) necessitates the processing of the film each time an adjustment is made so that the resulting mode pattern can be determined. If the goal in making the adjustments is to secure a pure transverse mode, the photographic procedure is quite cumbersome.

The basic structure of an image-converter system is shown in Figure 3.1. The incident radiation strikes a photocathode, generating a beam of photoelectrons which are accelerated by a high voltage and refocussed by suitable electric and magnetic fields to form an electronic real image on a phosphor screen. Infrared image converters are available which provide coverage to 1.2μm. Image converters may be applied in high-speed photography.

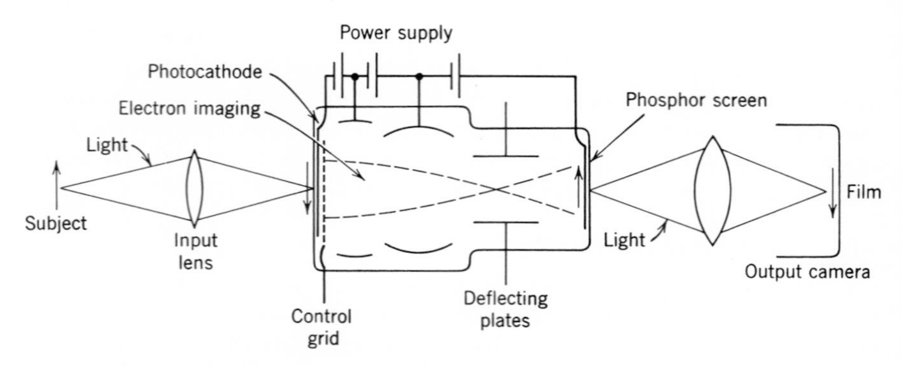

Figure 3.1. Basic structure of an image-converter camera.

TABLE 3.1 COMPARISON OF IMAGE-FORMING RECEIVERS

Type of Receiver	Physical Process	Long Wavelength Threshold (μm)	Response Time (sec)	Relative Comparative Efficiency
Cesium cathode-image tube	External photoeffect	1.2	10^{-8}	1.00
Phosphorescent image tube	Quenching of the phosphor	1.2	0.1 to 10^4	0.005
IR photography	Photographic	1.2	1/60	0.002
Evaporation from a surface at low pressure (evaporograph)	Thermal absorption process	16	1/60	0.001

3.2.1.2 Evaporographs The evaporograph permits the extension of photographic techniques out to 18 μm. Image converters, direct infrared photography, and the evaporograph are compared in Table 3.1, and a schematic of the Czerny evaporograph is shown in Figure 3.2. The side facing the infrared source consists of a nitrocellulose membrane coated with a layer of material of high infrared absorption (blackened aluminum, zinc, or bismuth). The inner surface of the membrane is coated with paraffin oil or some other easily volatilized substance evaporated onto the membrane in a layer a few tenths of a micron in thickness. The oil layer exhibits a vivid interference color that changes markedly on exposure to infrared radiation. These changes can be recorded photographically. Operation is slow, but the detectability is good.

3.2.1.3 Photometry The photographic method may be applied to the quantitative determination of the transverse-intensity distribution in a laser beam. The method is known as photographic photometry, and it is based on the fact that the density of a photographic negative is a linear function of the logarithm of the exposure, over a substantial exposure range[13]. A definition of terms and a plot of a typical "characteristic curve" are Shown in Figure 3.3. If the slope (usually designated γ) of the straight-line portion of the characteristic curve is known, the exposure, hence the intensity of the light, can be inferred from measurements of the density of the photographic negative. The γ of the film is determined by the negative

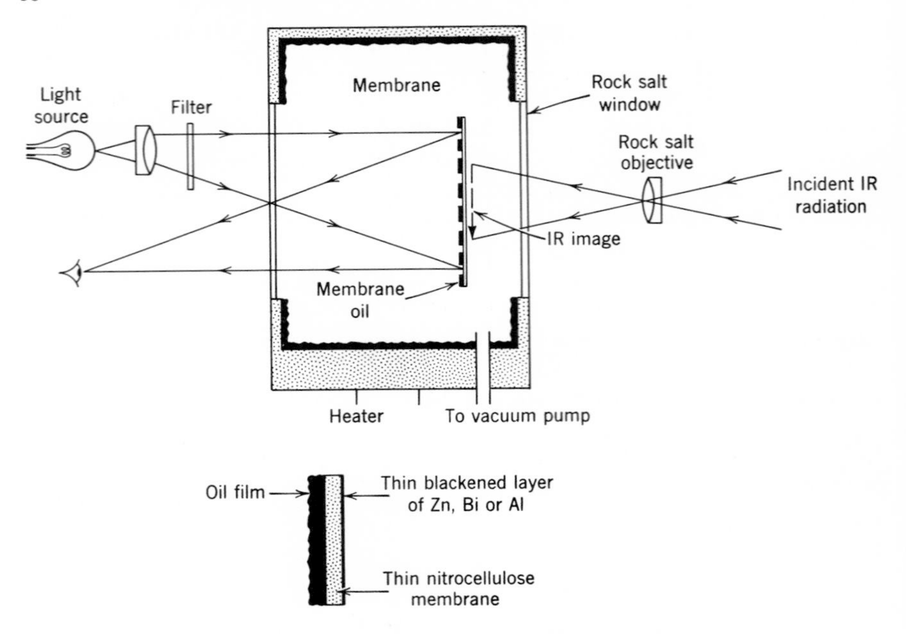

Figure 3.2. Schematic of Czerny evaporograph.

material and by the development of the negative[13]. The dependence of γ on these parameters is shown for typical cases in Figures 3.4 and 3.5. The procedures for obtaining negative materials with a predetermined γ are described in numerous photographic manuals. It is evident that if a large

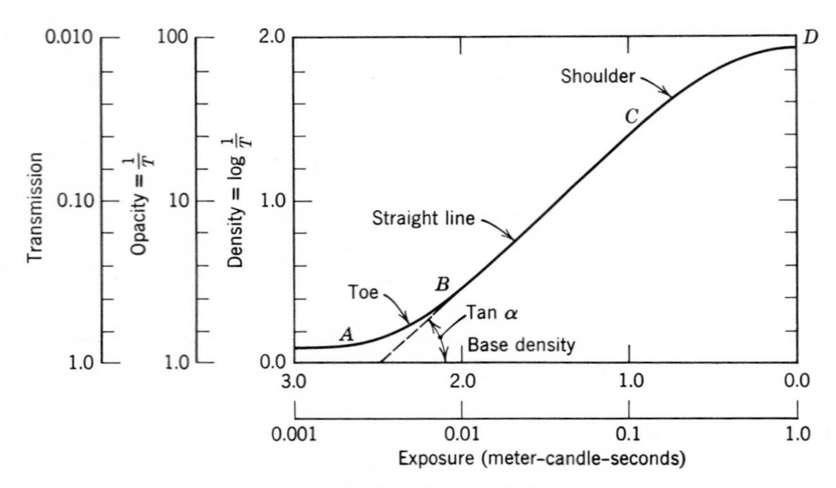

Figure 3.3. The film-characteristic curve.

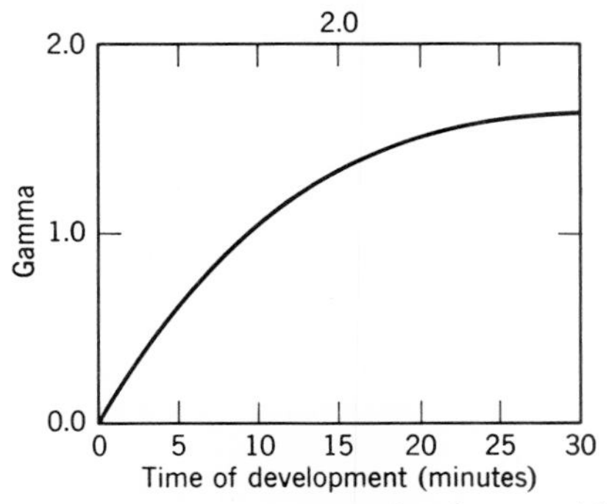

Figure 3.4. Time-gamma curve of a photographic plate.

range of intensities is to be covered, a series of different exposures will be required.

To translate the density variations of the negative material into light-intensity variations, it is convenient to use a microdensitometer[14]. There are several commercial versions of these instruments, and the most convenient and accurate are the automatic recording microdensitometers. A diagram of the optical system of a typical microdensitometer is shown in Figure 3.6. The film is moved across the slit at a constant velocity that is synchronized with the output of a high-speed pen recorder. This recorder then plots the transmittance of the photographic film versus the position on the film. The trace on the chart can be easily related to the area under scan and thus the corresponding light intensity can be determined.

In measurements of the transverse field distributions only the relative values of the light intensities are required. This simplifies the calibration problem and improves the accuracy with which the determinations can be made. Suppose that the transmission T is known at points 1 and 2 on the film and that the corresponding abscissas x_1 and x_2 have been determined

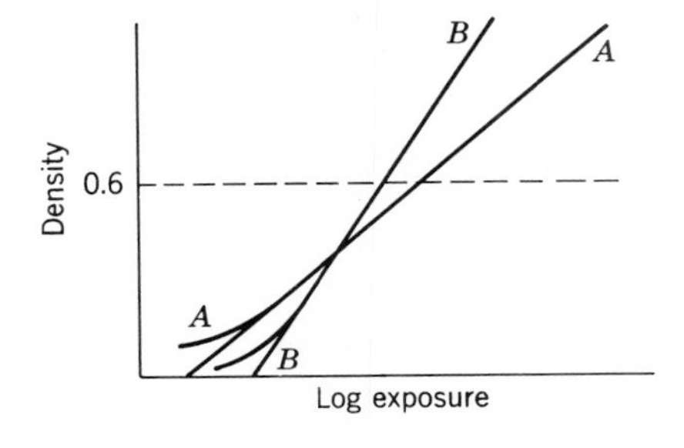

Figure 3.5 Characteristic curves illustrating the relative speeds of two materials.

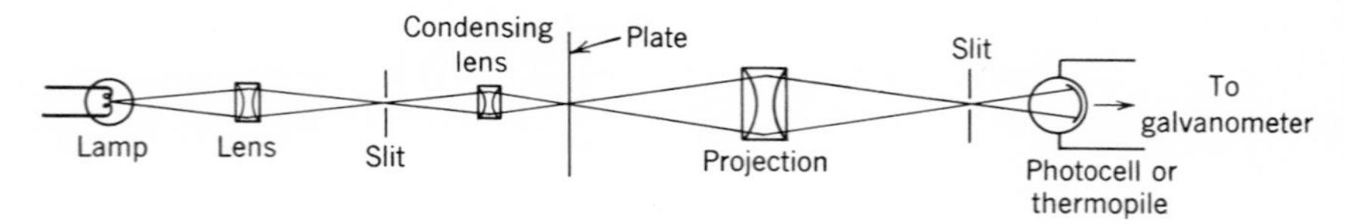

Figure 3.6. The optical system of a typical microdensitometer.

on the axes of the characteristic curve [x = density or log $(1/T)$]; then the ratio of the corresponding intensities (or exposures) is given by

$$\log \frac{I_1}{I_2} = \log(I_1) - \log(I_2) = \frac{x_1 - x_2}{\gamma}, \tag{3.44}$$

where I_1 and I_2 are the intensities: (or exposures). Thus only γ, the slope of the characteristic curve, is needed. This presumes, of course, that we confine ourselves to the linear portion of the curve.

As an additional check on the calibration of the characteristic curves, calibration marks may be made on the photometric plates by means of standard filters. The observed variations in film densities are then compared with the variations obtained with the known filters.

The accuracy of the photometric method is principally limited by variations in the value of film γ and nonlinearities in the film characteristic curve. At densities in the typical range between 0.3 and 1.6, a mean error of 0.5 to 2 percent can be made.

Films are available that cover a wavelength range from 0.25 to 1.2 μm [13]. Various types of Kodak film developed for selected portions of the spectrum are illustrated in Figure 3.7.

3.2.1.4 High-speed Photography of Solidstate Lasers Several techniques have been used for photographing laser beams, although photography of solid-state lasers represents several interesting photographic problems. The photography of low-power, gas-laser beam spots is simple. Mode structures are stable and simple, for the Fresnel number is usually much smaller. High-speed photography is required in the study of time-dependent mode structures in solid-state lasers.

3.2.1.4.1 Near- and Far-field Photography If the laser source is viewed as an ideal emitter of an apertured plane wave, as is shown for a solid-state laser in Figure 3.8, the radiating fields at distances very large compared to $D = a^2/2\ \lambda$ can be best represented by the Fraunhofer or "far-field" approximation. The value of a should be the radius enclosing, say, 95 percent of the radiant brightness of the radiating mode, and, in the case of solid-state lasers, may be much smaller than the radius of the laser rod itself. The Fresnel or "near-field" approximation is valid for distances very small compared to D. For typical solid-state lasers we find the characteristics given in the table ($\lambda = 694$ nm).

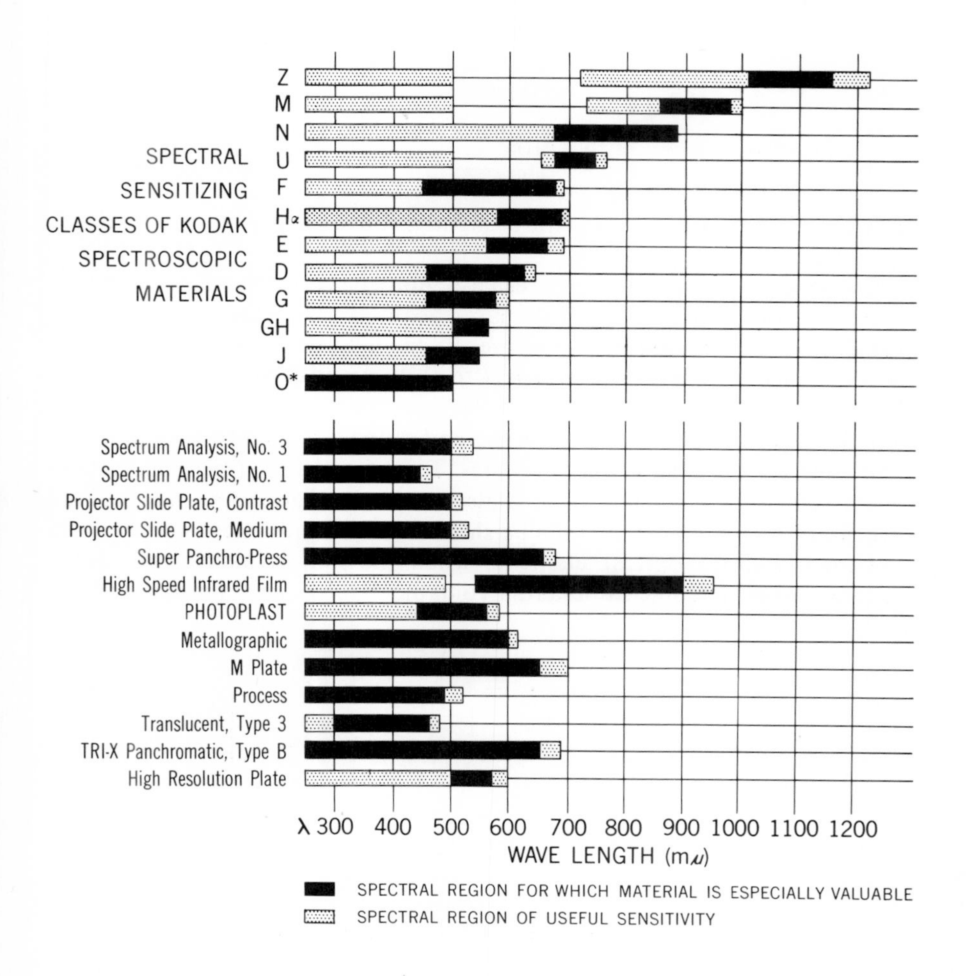

Figure 3.7. Spectroscopic-sensitivity ranges of different emulsion classes. (Reproduced with permission from Eastern Kodak Co.)

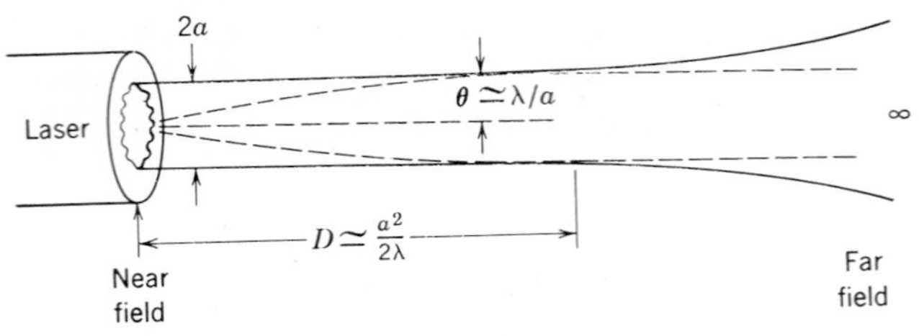

Figure 3.8. Beam divergence in a solid-state laser.

a	D	*Near-field*	*Far-field*
1mm filament mode	71cm	$D < 7$ cm	$D > 7$m
10 mm ideal case	71m	$D < 7$ m	$D > 700$m

For actual nonideal lasers with a slightly convex wave front, D must be increased.

To observe the near-field (Fresnel-region) pattern we may in principle place a photographic plate in the laser beam at a distance very small with respect to D; conversely to observe the far field we must place the plate at a distance very large with respect to D. In practice the near field can be observed by focussing a camera on the laser rod end, as is shown in Figure 3.9. The lens may be focussed on the end of the laser rod by very precisely focussing on small flaws and scratches. To observe the far field the camera lens is set to focus at infinity as is also shown in Figure 3.9. The intensity distribution on the film can then be shown[17] to be identical to the pattern obtained without optical elements when the film is placed an infinite distance away. In some cameras unavoidable shadows and vignetting effects

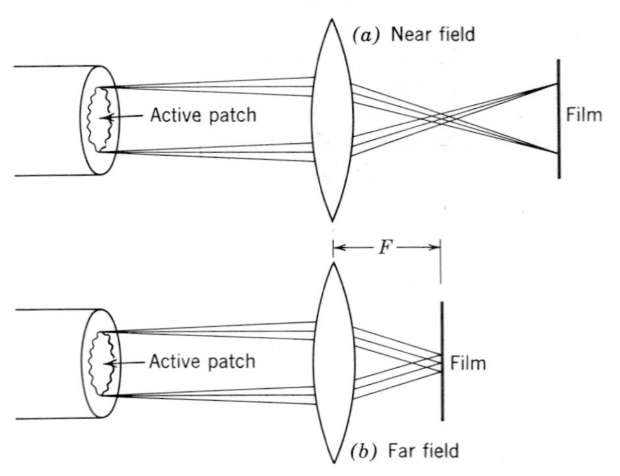

Figure 3.9. Focusing conditions for photographic recording of near- and far-field laser pattern.

are caused by internal apertures, which may obscure parts of the pattern. These effects may be interpreted by ray diagrams.

3.2.1.4.2 f-Number Considerations In photography of ordinary sources the f number of the camera lens, defined as the ratio of focal length to lens diameter, is a measure related to the solid angle of rays from each point on the source that will enter through the lens and reach a focus on the film[19]. Because most subjects are matte or diffuse reflectors or emitters, this number measures the brightness of the real image formed on the film in terms of the source brightness. When photographing laser sources, however, it is common to use cameras whose lenses are much larger in diameter than the laser beam. The f number of the lens system has a different meaning. As shown in Figure 3.10, all of the laser light in the primary beam is useful. If, however, one wished to photograph far off-axis modes ("bouncing-ball" modes) simultaneously, large-aperature optics and large f numbers would be needed, not for optical speed but merely for angular coverage.

3.2.1.4.3 Film Effects with Laser Sources When a direct recording camera is used to transfer all of the laser beam into a focused image, the density of energy on the film surface is $1/M^2$ times the energy density at the laser surface, where M is the linear magnification of the optical system; M is usually less than unity, so that the energy density at the film surface is greater than that at the laser surface, and thermal destruction of parts of the film may occur. At slightly lower power densities harmonic generation or unusual multiple excitation may be possible in the film emulsion, resulting

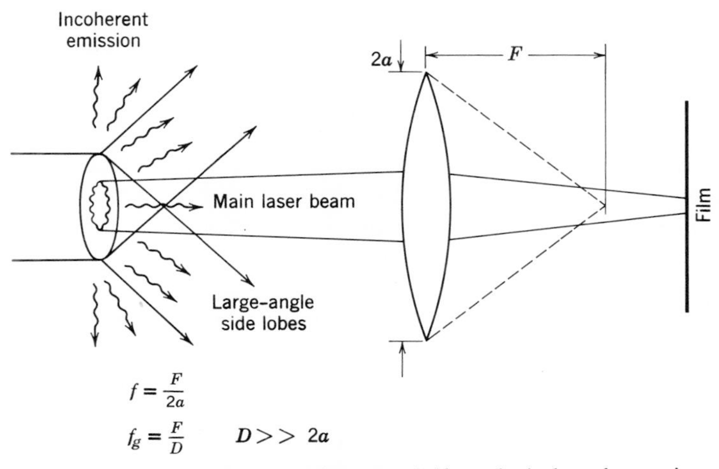

Figure 3.10. Relation between camera-lens speed (f number), laser-beam size, and lens diameter for recording the Gaussian beam and side lobes.

in the formation of images in films expected to be completely insensitive at the laser-emission wavelength. The short exposure time for laser photography puts the film into a region where the failure of the reciprocity-law is quite severe. For these reasons quantitative intensity work with films requires that calibrations be made under similar illumination conditions.

3.2.1.4.4 Observing Average Polarization By placing a prism such as the Wollaston prism in front of the laser face two images will appear in the camera, each with a different polarization direction. Because this actually creates two laser beams slightly displaced from the original in space, it is necessary that the camera aperture be large enough to accept both beams; otherwise, vignetting effects can occur. The prism should be masked, so that no higher angle, uncollimated laser light leaks around the edge to form a third undisplaced image [16], [22].

3.2.1.4.5 Divergence Effects and Sampling Mirrors Experience with ruby lasers[23] indicates that the collimation of light from individual small patches on the emitting face of the laser is better than that of the whole beam; that is, rays from one part of the crystal face may form a narrow conical bundle that does not intersect similar bundles from other radiating patches anywhere in the near-field region. Thus, if we sample the beam by reflecting a small part of it into a camera by means of a small opaque mirror, the camera will show light being radiated from only a part of the surface. The other parts are radiating ray bundles that miss the mirror. Incoherent light from the entire crystal face, however, will be equally sampled by such a small mirror. Thus it is possible to observe the outline of the entire face produced by pump light but the laser action of only a part of the laser. For this reason cameras that dissect the laser beam by multifacetted mirrors, using one facet for each sequential frame, cannot really be used for sequence photography of lasers.

3.2.1.5 Types of Cameras Suited to the High-speed Photography of Solid-state Lasers In the operation of solid-state lasers using pulsed pumping it is generally found that rapid and irregular pulsations occur in the output. To study the kinds of modes operating in individual pulses it is necessary to use high framing-rate photography with framing times comparable to the duration of the light spikes. The data to be obtained include the near-field pattern, that is, the pattern of light appearing on the laser rod end seen by forming an image from all angular rays emitted from every point, and the far-field pattern, that is, the distribution of intensity as a function of angle summed over all emitting points on the rod end. Highspeed photography is also required to observe the effects produced by very short laser light pulses. Photographs of the plume, produced when the output of a Q-switched laser interacts with a surface, require high-speed photography techniques. Laser-effects photography is little different from ordinary high-speed photography of incandescent sources[19], whereas photography of the laser itself must take into account the properties of the laser as a light source.

3.2.1.5.1 High-speed Nonelectric Framing Cameras Only stationary-film cameras are presently fast enough to be useful in laser studies. The basic principle of these cameras is illustrated Figure 3.11. A rotating optical element, generally a mirror, deflects the incoming light beam successively into a series of individual stationary cameras. A real image is formed near the surface of the rotating mirror and each output camera, though focused on it, does not receive light until reflected from the mirror. By driving the rotating mirror with a turbine to speeds limited ultimately by its bursting speed, shuttering rates of some 5,000,000 per sec for a number of frames (generally somewhat less than 100) can be obtained[20], [25]. Within their limits, stationary-film cameras are suitable for laser diagnostic photography[16], [22], [23].

A nonelectronic framing camera can generate a pulse to trigger the laser-pump lamp, but because of the relatively long and variable delay before the onset of lasing, it is usually impossible to predict which frame, if any, will show the beginning. Also, because of the interframe deadtime, many spikes will not be photographed. It is not possible to synchronize the individual frames with individual spikes.

The design of rotating-mirror cameras generally requires the presence of deadtime between frames, commonly about two-thirds of the center-to-center frame time. By using narrower stops the "open" time can be decreased and the deadtime increased without changing the framing rate.

A framing-camera mirror will rotate to its initial position in the time of one rotation (or less for multifaceted mirrors), and a new series of images will be superimposed on the original set if the object being photographed is still luminous. Although shutters of several types are available to prevent

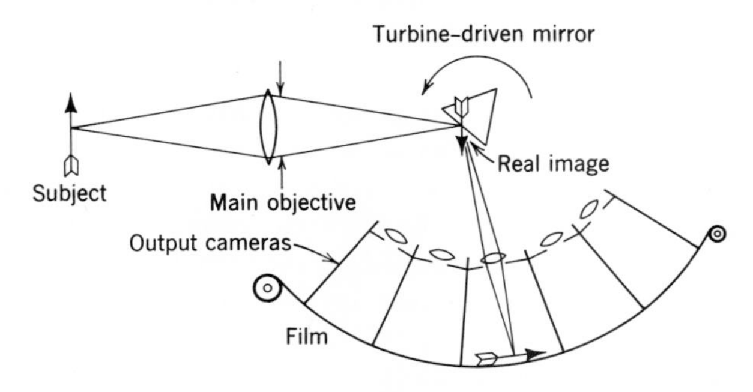

Figure 3.11. Schematic, high-speed nonelectronic framing Camera.

this, the laser can often be operated so that emission has virtually ceased before the mirror can come into position for rewrite.

The entrance-pupil framing camera is ordinarily diamond-shaped with the long axis parallel to the axis of mirror rotation. The laser beam should enter the center of the entrance pupil to preclude being eclipsed by the narrow direction of the diamond. Internal diamond-shaped stops are used in each output camera to effect shuttering by imaging the entrance pupil on each subcamera stop successively. When making far-field patterns, the internal stops may cause shadowing of part of the pattern. The effective *f*-number for these cameras is typically *f*/25.

Some rotating-mirror cameras are actually double cameras with two diamond-shaped entrance pupils appearing in the objective lens. Each entrance pupil leads to a different set of output cameras, the two sets being used successively. Each of these two sets forms its image with light leaving the laser in a different direction. Therefore the laser coherent emission cannot be photographed on both film strips. The incoherent emission from the laser, however, enters both entrance pupils from each point on the laser face and will be recorded on both film strips.

Because the image magnification and orientation are subject to small changes from frame to frame, and variations accumulate from one end of the film strip to the other, quantitative work requires provision of a reference mark in each picture.

3.2.1.5.2 Use of Streak or Smear Cameras There are three applications of streak cameras in laser diagnostic research: (a) mode diagnostic studies, (b) spectral diagnostic studies, and (c) framing.

When the active regions of a laser face have been determined by other means, a slit aperture may be placed across the region of interest. If the streak camera is then aligned to streak in the direction normal to the slit, a record is obtained of the spatial development of the mode patterns in time for that part of the crystal which is viewed. Resolved time intervals in the nanosecond range can be obtained with ruby lasers, enabling the direct observations of transverse-mode beats. For longer wavelength lasers in which the film sensitivity is less the slits may have to be widened, thus preventing us from obtaining such high time resolution[18],[20].

If the laser output is first passed through a Fabry-Perot etalon and then transmitted to the entrance slit of a streak camera, the time dependence of the spectral output may be studied. These techniques have been used to record mode jumping and changes in the refractive index of the laser crystal.

Laser spikes often have a duration (between half-power points) of a small fraction of a microsecond. If a streak camera is used without a slit aperture, we obtain a sequential record of mode patterns. The laser must be pumped to a low value of excess over threshold energy so that the spikes will not overlap or be too closely spaced[15].

3.2.1.5.3 Electronic Image Converters A real optical image of an event is formed on a semiopaque photoelectric cathode of an image converter. An efficient output camera then projects the resulting light emission into a real image on a photographic film. Several advantages can accrue from this approach. The electronic image forming process can be gated on and off at extremely high rates by purely electronic means; image recording can be synchronized with the event. Moreover, it is possible in some image-converter instruments to deflect the images to several separate frames with the same image tube. The input optics of an image-converter camera are relatively simple and can be quite flexible. In multiframe systems an arbitrary and easily varied delay may be placed between frames. The open time for a single frame may be made as short as 5 nsec. A deflection-type image-converter camera may easily be converted into a streak camera.

Although image-converter cameras are ordinarily triggered by the event to be photographed, the camera trigger circuitry may cause a delay of the order of 50 nsec. To photograph the beginning of the event, therefore, it is necessary either to obtain an earlier trigger, or to introduce an optical delay in the event. If the event is a laser spike, an optical delay can be obtained by taking the trigger pulse from a beam splitter in the laser beam near the laser itself. The image is delayed by focusing the emitted beam after it has traveled a round-trip distance of the order of 50 ft to a plane mirror and back.

There are severe limitations in the use of the electronic image-converter camera, in its application to laser photographic diagnostics. Thus the number of resolved information elements per frame is low (currently less than is available with nonelectronic cameras). The overall dynamic brightness range for high-contrast images is less than that provided directly by films. Electron-spreading effects reduce the resolution obtained in comparison to that obtained with the all-optical camera. Because the spectral response of the photocathode cannot be altered, a change of spectral response can be obtained only by changing the image converter also. The choice of spectral response is limited compared with that possible from the use of film. The image-converter tube is fragile. If the laser source power becomes too high, the photocathode will be damaged. The image tube, however, warns of high power levels by producing saturation effects in its output before the photocathode input reaches damaging level. In direct photography excess power simply burns the film, the replacement cost of which is small.

In one type of image-converter camera a battery of image-converter tubes looks at different facets of a multifaceted mirror behind a single large objective lens. This camera offers no advantage with a collimated laser beam source, for not more than one camera tube can project a laser beam and consequently only one action frame can be obtained no matter how many tubes are in the cluster.

3.2.1.5.4 Image-dissection Cameras The initial optical image of an image-dissecting camera is reduced to a number of elements (strips or small spots), each of which can be recorded in the manner of a smear camera [26]. The original image may be reconstructed by viewing the processed film through a "reconstructor," which performs the inverse operation. Successful laser photography with an image-dissecting camera requires that the elements being projected do not individually restrict the collimation of the radiated beam; otherwise the previously discussed sampling mirror geometry would apply.

3.2.1.5.5 Electronic Shutters A Kerr cell has long been used as a high-speed camera shutter[21]. The spectral sensitivity of a camera shuttered by a Kerr cell is determined by the combined transmission of the Kerr cell liquid and the film sensitivity. The image-resolving power of the electric Kerr cell shutter can be quite high compared to devices that use electron imaging. Although the angular acceptance of Kerr cells is small, this does not constitute a limitation for laser photography because the light beam is already highly collimated. To obtain more than one frame with a Kerr-cell-shuttered camera without using sampling mirrors, it is necessary to use a series of beam splitters, one for each Kerr cell. Although the length of the beam-splitter assembly reduces the f number of the objective lenses which can be used, this does not impair the photography of laser sources. One camera configuration that overcomes this objection distributes the light to a multiplicity of Kerr cell (or image converter) sections with a multifaceted prism behind an objective lens. This sampling-mirror geometry does not allow more than one frame of a collimated laser source to be obtained. It is possible to use a sampling-mirror camera with some sacrifice in image quality if the laser beam forms an image on a ground-glass screen or matte reflector. The camera then photographs the image of the scattered light. This technique sacrifices image brightness and resolving power and introduces image scintillation due to the spatial Fourier transformation at the surface and the resulting interference phenomena.

3.2.2 Scanning Techniques

If a photodetector, which samples only a small portion of the radiation pattern at any given instant, is mechanically scanned through a laser beam, an electronic signal may be derived that corresponds to the intensity distribution of the beam. This technique may be used to provide a plot of either the near-field or the far-field pattern of a continuous or repeatable laser beam pattern. The scan rate is obviously limited by mechanical considerations.

Because of their physical size and sensitivity to mechanical shock solid-state lasers are usually kept stationary. Semiconductor lasers, however, which are quite small and less sensitive to shock, may be mounted on a movable table and their output beam scanned past a detector kept in a fixed position. The electronic signal may be displayed on a chart recorder. The mechanical movements obviously must be precisely calibrated and repeatable.

Spatial resolution will depend on the use of a photodetector with a very small sensitive area or the use of appropriate masks. Because photodetectors have highly reproducible characteristics and are inherently linear over a large intensity range, they can be used to obtain more accurate measurement of laser-beam patterns than photographic films. Detectors that are available to provide spectral coverage anywhere in the range in which lasers operate are discussed in Chapter 4[31], [30].

Laser patterns can be scanned with a servopositioner that scans the photodetector through the beam horizontally at a given vertical position. The detector output is fed into the Y input of the plotter, whose X input is proportional to horizontal displacement. A plot of intensity versus horizontal position is thus obtained. The scan may be repeated for a number of vertical positions to obtain a complete two-dimensional intensity distribution for the beam[33], [34].

Another means of beam scanning involves the gradual insertion of a knife-edge into the beam. A lens is used beyond the knife-edge to focus the image of the knife-edge on to the photodetector, thus reducing the dimensions of the diffraction pattern produced by the edge. It is important that the lens introduce negligible diffraction of its own. The output of the detector may then be plotted as a function of the position of the knife-edge to obtain a curve of the integrated one-dimensional intensity distribution in the beam. Alternatively, a slit system can be scanned past the laser spot to derive the beam intensity distribution directly[34].

3.2.2.1 Measurement of Mode Patterns in Gas Lasers Measurements of the transverse-field distributions were first observed for the case of the CW helium-neon gas laser operating at 1.15 μm[33]. The qualitative near-field patterns were obtained by using an image-converter tube and a camera. The resonator, with a nearly hemispherical mirror geometry, produced photographs that corresponded closely to those predicted by electromagnetic theory outlined earlier. Pure modes were isolated by adjusting the gain of the laser, the reflector spacing, manipulating the mirrors, and Brewster-angle windows to utilize the irregularities in their optical properties to obtain any desired transverse mode.

Unless special precautions are taken several axial modes (different values of the axial mode number q) are always found to be simultaneously oscillating; these, however, produce the same transverse-field distribution and are not resolved by gross intensity measurements. Mode patterns may not exhibit azimuthal symmetry if astigmatism is introduced by the Brewster-angle windows or nonspherical mirror surfaces.

The purity of the individual transverse modes may be tested by determining if difference frequencies corresponding to the presence of more than one transverse mode are absent. The intensity distributions found in mode patterns agreed closely with the values calculated from (3.10).

Mode patterns possessing azimuthal symmetry have been observed with a helium-neon gas laser operating at 1.15 μm using a hemispherical resonator. The modes were isolated with wires crossing the optic axis of the resonator and with diaphragms having circular holes of various radii[34]. The mode purity was verified by noting the absence of beats between the different transverse modes when the pattern corresponded to that of a pure transverse mode. The mode patterns were observed and photographed by means of an infrared image-converter tube. Measurements of the intensity distributions of the pattern were in good accord with the predictions made in (3.10).

Scanning techniques were used to determine the transverse-intensity distribution of the output beam of the first optically pumped cesium gas laser[35]. Oscillation occurred at 7.18 μm in a hemispherical resonator. The spatial distribution of the output radiation, being beyond the wavelength range of films and photoemissive detectors, was measured by scanning the laser output with a gold-doped Ge detector.

The theoretical expression for the normalized intensity distribution for the lowest order mode is taken from Section 3.1.1 by

$$I(\omega^2 = x^2 + y^2) = \exp\left[-\frac{2k(\omega^2)}{b(1+\xi^2)}\right], \tag{3.45}$$

where $k = 2\pi/\lambda$; b = confocal spacing; $\xi = 2z_o/b$; z_o is the distance from the center of the confocal cavity to the surface considered; and x and y are distances measured in the plane normal to the cavity. The fields of the confocal resonator over the surfaces of constant phase at an arbitrary distance from the center of the resonator are given by

$$\frac{E(\omega^2)}{E_0} = \left[\frac{2}{(1+\xi)^2}\right]^{1/2} H_m(X\beta)\, H_n(Y\beta) \exp - \left(\frac{\pi\omega^2\beta}{b'\lambda}\right), \tag{3.46}$$

where $X = x(2\pi/b'\lambda)^{1/2}$, $Y = y(2\pi/b'\lambda)^{1/2}$, $\xi = 2z_0/b'$, and $\beta = 2/(1 + \xi)2$. Omitted from this expression is a phase factor, which describes the periodic variation of intensity in the z direction. The measured values of the intensity distribution at 7.18 μm agreed with the theoretical prediction within $\pm$ 3 percent. Agreement is considered excellent.

The characteristics of a TEM_{00} beam-intensity distribution have been very carefully investigated for a helium-neon laser operated at 633 nm. The CW beam was precisely scanned with a silicon solar cell mounted behind a brass plate containing a 0.0015-in. diameter hole. Horizontal scans were repeated at vertical intervals of 0.1 mm. Synchronous detection was used to obtain signal-to-noise ratios in excess of 10^3 with a 1 H_z band width even though the hole accepted only 10^{-6} W in the center of the beam. The scanner was mounted on a Kulicke and Soffa three-dimensional micro-positioner. A stepmotor drove the horizontal micrometer in increments of 0.001 in. Limit switches were used to command a relay that reversed the direction of the motor and provided a fast return while suppressing the detector output in order to obtain a reference for zero signal. A complete round trip over a $\frac{1}{4}$-in. scan lasted about 30 sec. The vertical micrometer was driven continuously by a small synchronous motor.

In addition to the intensity distribution, the total optical power was determined. Because accurate electronic squaring and integrating devices were not available, the output of the detector was fed into a dc-to-frequency converter. Pulses, the frequency of which represented beam intensity, were recorded on magnetic tape and then transferred to computer cards. A digital computer was used to obtain the average value, as well as a contour plot of the power density at equal increments across the beam.

The results of the precise measurements from the studies described above are in agreement with theoretical predictions of beam-intensity distributions predicted by (3.46).

For certain applications of gas lasers it is necessary that the cavity be operated in a very pure TEM_{00} mode. Measurements of the beam shape of these lasers have been found insufficient for determining mode purity to the degree required; phase-sensitive techniques, such as those used for measuring the intensity distribution in the diffraction pattern of a circular aperture, have been found to yield the needed precision. Although the phase structure of the beam is not extremely critical in the formation of diffraction-limited beam widths, it is of paramount importance when a laser is used as a light source for simulation of microwave antenna patterns. In the latter case it has been found that verification of the Gaussian shape of a laser-output beam, characteristic of the TEM_{00} mode, does not suffice to ensure uniform phase output. Small departures (i.e., cavity 1 percent short) from a hemi-spherical cavity can lead to mode impurities that are disturbing for simulation work. The technique for measuring the diffraction patterns follows[35].

A system of microscope and telescope objectives is arranged to collimate further the 5-mm diameter near-field collimated output of a gas laser. These optics increase the beam to 40 mm in diameter. A 18-mm focal length microscope objective 49 mm in diameter is coupled to a 48-mm focal-length telescope objective that is 343 mm in diameter. The diffraction patterns produced by the 40-mm diameter beam interacting with a centrally located 5-mm diameter aperture are examined, after they are passed through a similar set of lenses to bring the pattern to a focus, by scanning the resultant output light with a 931-A photomultiplier the input of which is limited by a 0.4-mm diameter aperture. Two cases were studied, one with the laser operated in a hemispherical cavity geometry with the maximum spacing that would yield stable operation, the other with the cavity shortened by 1 percent.

The diffraction pattern observed from the hemispherical cavity used in the procedure described above exhibited deep, well-defined nulls between concentric rings; it was found to be in excellent agreement with the theoretical diffraction pattern of a circular aperture illuminated by a plane wave of uniform phase and amplitude. In contrast, the foreshortened cavity produced circular patterns without nulls. Only a 5-dB peak-to-valley ratio was found in contrast to an 18-dB ratio for hemispherical operation. Because the degree of null fill-in was of the order of 15 dB below the power in the TEM_{00} mode output, it is not surprising that simple measurements of beam shape are incapable of detecting the effect.

The null fill-in results mentioned above are interpreted as follows. Higher order modes, of different beam shapes, provide different amplitude distributions across the diffracting aperture and result in diffraction patterns of wider beam widths and therefore noncoincident nulls. Because the higher order modes are at different frequencies and all differ in frequency from the axial modes, there is no coherence.

The theoretical shape of the intensity pattern may be predicted as follows. Let a point on the circular aperture of radius have plane coordinates ρ, θ and a point on the diffraction pattern have plane coordinates ω, φ. The Kirchhoff diffraction integral then expresses the amplitude of the electric field as[36].

$$E \propto \int_0^a \int_0^{2\pi} \exp[-jk\rho\omega \cos(\theta - \alpha)] \rho \, d\rho \, d\theta. \tag{3.47}$$

This may be evaluated in terms of Bessel functions as[37]

$$E \propto \left[\frac{2J_1(kx\omega)}{kx\omega} \right], \tag{3.48}$$

where J_1 is the Bessel function of first order. The intensity becomes ($I \propto E^2$)

$$I = I_0\left[\frac{2J_1(kx\omega)}{kx\omega}\right]^2. \tag{3.49}$$

Nulls are predicted at the zeros of J_1, which occur at $\omega = 0.61\lambda/x, 1.116\lambda/x, 1.619\lambda/x$, and so on.

3.2.2.2 Ruby Lasers (Crystal Lasers) Pulsed-laser action was first observed in ruby crystals that emit at 694.3 nm in the red, but, although ruby lasers have been extensively studied, the data obtained have been difficult to interpret with a precision and detail comparable to that which can be obtained from the study of gas lasers. This situation is typical of all solid-state lasers, for multimode laser action, which is observed in most solid-state lasers, is characteristic of a laser medium that is optically inhomogeneous, has an inhomogeneous distribution of excited ions (resulting from nonuniform pumping geometry), and long spatial cross-relaxation times. Furthermore, studies have shown that mode jumping occurs in solid-state lasers in times less than 10^{-6} sec. Despite these difficulties it has been possible to use many experimental techniques to study ruby lasers because a great variety of detectors is available.

Near-field patterns that are reasonably simple have been photographed for ruby lasers operated very close to threshold[38]. The simple patterns that accompany laser action have been identified in some instances with the emission of a single spike that lasts from 0.5 to 1 μsec. When the pump power is increased, many output spikes of laser light appear and the simple mode patterns are destroyed. Furthermore, patterns have been obtained that show only a 180° axis of symmetry, although most patterns show neither linear nor circular polarization.

To compare these results with theoretical predictions, measurements were made on a two-lobe pattern. In the $y = 0$ plane the maxima are separated by

$$x_m = \left[\frac{b\lambda(1+\xi^2)}{4\pi}\right]^{1/2}. \tag{3.50}$$

If b' is the actual radius of curvature of the resonator end plates, then

$$b = (2b'd - d^2)^{1/2}; \qquad \xi = \frac{2z_0}{b}. \tag{3.51}$$

The separation of the maxima is given by

$$x_m = \left[\frac{\lambda d}{4\pi}\left(1 + \frac{4z_0{}^2}{2db' - d^2}\right)\right]^{1/2}\left(\frac{2b'}{d} - 1\right)^{1/4} \tag{3.52}$$

Because the resonator is filled with dielectric, 1 (3.52) becomes

$$x_m = \left[\frac{\lambda d}{4\pi n}\left(1 + \frac{4z_0^2}{2db' - d^2}\right)\right]^{1/2}\left(\frac{2b'}{d} - 1\right)^{1/4} \tag{3.53}$$

where n is the dielectric constant.

A comparison of the observed near-field two-lobe patterns with those obtained theoretically was found to be in good agreement. Although the end faces were nominally flat and parallel (PPFP), the results indicated agreement with experiment even if the end mirrors were assumed to be slightly curved.

Far-field patterns have been photographed for a ruby laser with external spherical mirrors, operated near threshold. The resulting patterns agree with the calculation of (3.10). To obtain far-field patterns, it is necessary to locate an aperture near the axis of the resonator at one of the mirrors. Results were obtained for the $TEM_{00,\ 01,\ 02,\ 04}$ transverse modes. The separation between the maxima were carefully measured and were found to agree closely with the theoretical predictions of (3.10).

A rotating-mirror camera has photographed mode hopping in ruby lasers [39]. A cylindrical ruby rod 5 cm long and 0.6 cm in diameter, with a 90° c-axis orientation, was used. High-reflectivity multilayer dielectric mirrors were coated onto the polished end faces of the rod. Typical photographs showed mode hopping, and some indicated that adjacent parts of the rod operated simultaneously in different transverse modes.

Transient phenomena in ruby lasers have been examined extensively with streak cameras[40]. The image of the slit was focussed onto the face of an image-converter tube in a streak camera. Writing times of from 5×10^{-8} to 10^{-5} sec were encountered. The light gain was better than $50 \times$ from the photocathode to the photoanode; that is, the f number was considerably lower than that of the rotating mirror camera. Incidentally, this technique is capable of displaying photobeats and showing time-dependent laser action from various parts of a laser rod.

An image-converter camera has been used in a framing mode to examine near-field patterns from a ruby laser that was 3.81 cm long and 4.74 mm in diameter. The c-axis was orientated at 57°, and the lateral surfaces were polished. The end faces were freshly silvered to provide high-reflectivity mirrors. The exposure time of each frame was typically 10^{-7} sec, and the frames were separated by 0.5×10^{-6} sec. In the framing mode an image of the laser spot was focused on the photocathode by a suitable optical system, and the electron beam within the tube was switched on and off repetitively at the proper times to provide the necessary shutter action. The patterns obtained were analyzed and found to agree with the calculated near-field patterns.

3.2.2.3 Semiconductor Lasers Because the light-emitting junction in semiconductor lasers is only a few microns thick, application of photographic techniques is hampered by junction size, photographic film response in the infrared, and the requirement of operating at liquid nitrogen temperatures.

The mode structure of diode lasers has been studied experimentally, and it was found that when the diodes are cooled to a temperature of 78° K and excited with current pulses of a duration of about 5×10^{-7} sec, at a repetition rate of 100 to 1000 Hz, the threshold current densities range from 800 to 10,000 A/cm^2. To maintain the diodes at 78° K, the diodes were immersed in a bath consisting of a mixture of liquid nitrogen and liquid oxygen, which in turn was in thermal contact with a reservoir of liquid nitrogen. This procedure was found to be necessary to prevent the coolant from bubbling at the front surface of the diodes when they are operated at high current densities.

Photographs of the near-field patterns emitted by the GaAs diodes have been obtained using an infrared image-converter tube[42]. Preliminary observations indicated that the different bright spots were usually—but not always—emitted coherently.

The diode-radiation patterns were examined by a scanning technique. The diodes were mounted on a turntable, which provided rotation about an axis perpendicular to the junction plane or about an axis in the junction plane and parallel to the polished surfaces. The radiation was also sampled by the slit of a monochromator that provided an angular resolution of 0.1° or less, and a spectral resolution of 0.1 nm or less [44].

The complexity of the diode-radiation patterns could not be completely explained because the reasons for the appearance and number of lobes could not be interpreted simply. The widths of the lobes in the two planes can be qualitatively understood on the basis of the small thickness of the junction as contrasted with the width of the diode.

The diode-radiation patterns exhibited the essential features of a diffraction pattern from a rectangular slit. The width of the (horizontal) major lobe was about 3/4°, indicating coherence for a distance of about 50 wavelengths, a small fraction of the diode width. Other radiation patterns indicated that even for a single axial mode, the radiation was emitted in several lobes. Near-field patterns of the radiation emitted along the junction plane showed the existence of nodes and antinodes. Groups of these spots are sometimes periodically spaced, and are similarly polarized. The results may be understood if we consider a model that includes modes that are excited in the diode and interact with an intensity and phase distribution that varies periodically along the junction. The extent of the emitting region is only a few microns perpendicular to the junction but up to 50 microns in the junction plane[42].

3.2.2.4 Bouncing-ball Modes in Semiconductor Lasers Studies of the directionality of light emitted by GaAs diodes uncovered effects that are most simply interpreted as being the result of radiation due to bouncing-ball modes[43]. The scanning method was used to interpret the radiation pattern with the diode supported on an accurately calibrated turntable. The radiation was detected with an infrared photomultiplier filtered with a Corning 7-69 filter that passed only wavelengths in the simulated emission wavelength range of about 840 nm. The photomultiplier aperture, at a distance of 1 m from the diode, had an angular resolution of 0.2°. The sample was fabricated so that all four faces of the diodes perpendicular to the plane of the junction were smooth. The observations were obtained with the sample immersed in liquid nitrogen. The radiation pattern emitted in the horizontal plane showed peaks that were found to correlate well with the standing-wave patterns calculated for bouncing-ball modes[43].

3.2.2.5 Bouncing-ball Modes in Rectangular Ruby Lasers Far-field photographs of the emission pattern of a rectangular ruby laser rod have been found to exhibit bouncing-ball modes. The ruby was cut and polished on all sides in the configuration of a rectangular parallelepiped. The ends were silvered, and the ruby was pumped with a helical flash lamp. A pattern of four spots was observed when four thin strips of the silver were removed from the output end of the rod. Two of the strips were parallel to the vertical and two were parallel to the horizontal axes of the rectangular rod.

3.2.2.6 Ring Patterns (Bouncing-ball Modes) in Cylindrical Ruby Lasers A photograph of the far-field pattern of a cylindrical rod with uniformly silvered end faces has been studied. Prominent rings were observed in addition to the usual axial emission. The rings were found to be analogous to (002), (043), (032), (021), (031), and (061) "line" emission from a rectangular rod.

3.2.2.7 Fiber Optic Lasers The fiber optic dielectric wave guide has been analyzed theoretically and its properties have been demonstrated experimentally[12]. A number of fiber optic lasers have been constructed using Nd^{3+} in glass (wavelength 1.06 μm) as the active medium, but these devices have thus far not played a conspicuous role in the development of laser technology. The photographic method has been employed to study the output of fiber optic laser radiation patterns.

A number of fiber optic lasers have been constructed using Nd^{3+} in glass as the core material. Near-field laser mode patterns have been obtained for a fibre with a core diameter of 8 μ and a length of 4 m [12]. The observed mode patterns were more complex than would be expected on the basis of simple dielectric waveguide theory. It was calculated that the fiber was capable of supporting at least 20 different transverse modes. Thus the patterns did not correspond to pure dielectric wave-guide modes but rather to linear combinations thereof.

Observations of the near-field patterns excited in passive optic fibers have been obtained for visible radiation and have been found to agree accurately with the theory of dielectric waveguides[12].

3.2.3 Beam-divergence Measurement

The radiation emitted from the spherical mirror end of a gas laser operated in the TEM_{00} mode in the hemispherical cavity configuration is essentially a spherical wave with its center in the plane mirror. The normalized amplitude of this wave decreases from the center as a Gaussian function:

$$A(r) = \exp\left(-\frac{4r^2}{D^2}\right), \tag{3.54}$$

where r is the radial distance measured normal to the beam axis and $D/2$ is the value of r for which $A(r)$ decreases to $1/e$ of its central value

$$A(r) = \frac{1}{e} \quad \text{at} \quad r = \frac{D}{2}. \tag{3.55}$$

The beam spreads as though generated by a point source in the plane mirror. By using a lens placed so that its focal length F coincides with the plane mirror, the beam may be collimated to the diffraction limit of the Fraunhofer region, namely

$$\theta = \frac{2\lambda}{\pi D}, \tag{3.56}$$

where θ is the half-angle spread to the point at which the beam amplitude falls to $1/e$ of its central value.

If the beam is focused to the diffraction limit by a lens or focal length F and the lens diameter h is large compared with the beam spot size, a Gaussian image is formed (which is larger than the Airy disk) that may be expressed as

$$A = \exp\left(-\frac{\pi^2 r^2}{4\lambda^2 f_g{}^2}\right), \tag{3.57}$$

where

$$f_g = \frac{F}{D} = \frac{\text{focal length of lens}}{\text{diameter of beam spot at entrance to lens}}$$

The distinction between the Gaussian f_g number and the usual f number for producing an Airy disk is fundamental. If the beam spot fills the lens, the Airy disk (smaller) and its associated diffraction rings will be formed.

The spot diameter d defined by the $1/e$ amplitude points is given by

$$d = \frac{4\lambda f_2}{\pi} = \frac{4\lambda}{\pi}\left(\frac{F}{D}\right). \tag{3.58}$$

Consider a 633-nm gas laser in which $D = 4$ mm. A lens of $h = 20$-mm diameter and focal length $F = 40$ mm will produce a diffraction limited spot of

$$d = \frac{(4)(6.33 \times 10^{-5})}{\pi}\left(\frac{4}{0.4}\right) = 7.92\ \mu m$$

The fraction of the total normalized beam power W contained within a radius r of the beam center is

$$W(r) = \int_0^r \int_0^{2\pi} A^2(r) r\, dr\, d\phi = 2\pi \int_0^r \exp\left(\frac{-8r^2}{D^2}\right) r\, dr$$

$$= \frac{\pi D^2}{4}\left[1 - \exp\left(\frac{-8r^2}{D^2}\right)\right] \tag{3.59}$$

so that the fraction of the power contained within a radius r is

$$\frac{W(r)}{W} = \left[1 - \exp\left(\frac{-8r^2}{D^2}\right)\right]. \tag{3.60}$$

Two values of r are of interest: $r = D/2$ and $r = D/2\sqrt{2}$. For these values the amplitude of A is $1/e$ and the power is $1/e$ of the respective values at beam center.

If we assume that a laser produces a diffraction-limited beam and we use a lens whose diameter equals that of the beam to produce a spot, the image formed will have an intensity distribution described by the Airy-disk diffraction pattern:

$$I(r) = I_0\left[\frac{2J_1(\pi r/\lambda f)}{\pi r/\lambda f}\right]^2, \tag{3.61}$$

where r is the radius in the focal plane and f is the ordinary f number of the focusing lens, defined as the ratio of the focal length F to the lens diameter.

The first zero of the Bessel function J_1 is zero for $\pi r/\lambda f \simeq 3.83$. Consequently, the radius of the first dark band in the Fraunhoffer diffraction pattern occurs at

$$r = 1.22\lambda f \tag{3.62}$$

or in angular measure

$$\theta = \frac{1.22\lambda}{d}. \tag{3.63}$$

Few solid-state lasers are of sufficient optical quality to produce a diffraction-limited beam without constraining mode-limiting optics (such as pinholes).

It may be desirable to obtain a measure of beam divergence. Two photographic methods are appropriate for pulsed solid-state lasers. In the first technique a photographic plate is exposed at a distance of at least ten times the diameter of the radiating aperture. The diameter of the circle which effectively circumscribes the exposed area is divided by the distance from the laser to obtain a measure of beam divergence

$$\theta \sim \frac{D}{R}. \tag{3.64}$$

A more accurate measure of beam divergence may be estimated if a long focal length telephoto lens is used to photograph the beam. If the lens is focussed at infinity, a true far-field image of the beam is obtained. The full width of the beam divergence at $1/e$ points becomes

$$\theta = \frac{l}{F}, \tag{3.65}$$

where l is the spot diameter at $1/e$ intensity points and F is the lens focal length. It makes little sense to try to match the lens diameter to the spot in a solid-state laser because the radiation pattern exposed on the film will be a composite of a large number of random modes. At least the lens diameter should be larger than the beam-spot size at the lens entrance surface.

The advantage of a long-focal-length lens will be obvious from the following calculation. Suppose $\theta = 10$ mrad and $F = 50$ cm. Then

$$l = F\theta = (50)(10^{-2}) = 0.5\ cm.$$

Unless care is taken the film will be burned or at least badly overexposed. The total energy may be reduced in the case of a gas laser, but the exposure problem remains as $\theta \sim 10^{-4}$ rad and the spot must be examined with a traveling microscope.

Lasers employing the PPFP geometry should produce beams whose angular spread for the lowest order transverse mode will conform closely to the theoretical value given by Fraunhofer diffraction theory for an aperture-limited plane wave. For a uniformly illuminated aperture the predicted beam spread is approximately

$$\theta = \frac{n\lambda}{d}, \tag{3.66}$$

where θ is the full width of the beam, n is the index of refraction of the host crystal, and d is the diameter of the aperture (approximately the diameter of the emitting area in the near-field pattern). Comparison of the measured angle with the theoretically predicted angle indicates that lasers are approaching ideal performance.

When the laser is operating in the lowest order transverse mode, the intensity distribution over a reflector approximates the Gaussian function closely. If the limits of the Fraunhofer diffraction integral can be extended to infinity (a valid approximation in most cases), the integral can be evaluated exactly. The result obtained is that the intensity distribution in the diffracted beam also has the Gaussian functional form. The Fraunhofer diffraction integrals of the field patterns, given in (3.10), can be evaluated exactly so that the intensity distribution of the diffracted beam has a functional form identical with the near field pattern.

For confocal resonators operating in the TEM_{00} mode, the spot size at the reflector is given by

$$w_s'\left(\frac{b}{2}\right) = \left(\frac{\lambda b}{\pi}\right)^{1/2}. \tag{3.67}$$

The intensity distribution at an arbitrary distance from the center of the resonator is given by (3.60), from which the spot size as a function of distance from the resonator may be calculated. The half-angle of the far-field pattern is $\tan^{-1}(\lambda/\pi w'_s)^{1/2}$, that is $\theta_{\frac{1}{2}} = (\lambda/2d)^{1/2}$. This will be recognized as the diffraction angle of a nearly plane wave limited to an area of $\approx \pi(w'_s)^2$. These considerations are valid for small angles, or alternately $w_s' >> \lambda$. This discussion neglects the fact that, at least in gas lasers, the wave front radius of curvature is reduced to $1/n$ of its value at the reflector surface. The beam will thus diverge more than indicated if the back surface of the mirror is flat.

When the resonators are filled with material of dielectric constant n, (3.11), (3.12), (3.13) still hold for the spot size, provided that, instead of b', b'/n is substituted.

In the case of gas lasers employing the PPFP geometry, the observed beam divergence has been in close agreement with the value predicted by diffraction theory when the output mirror surface is properly contoured.

The observed divergence in solid-state lasers that employ the PPFP geometry and in which the optical quality of the laser material has been improved, has been found to approach more closely the limit set by diffraction. Although better optical quality has been developed in ruby, significant improvements must still be made if ruby laser beam quality is to approach that of gas lasers.

Although the usual method for measurement of beam divergence is to intercept the laser beam on a screen and photograph the pattern for a series of measurements at increasing distances from the laser, an alternate method has been used in the study of the two-lobe pattern of a ruby laser operated near threshold. The half-separation between a two-lobe pattern as a function of position along the beam is given by

$$x_m = \left(\frac{d\lambda}{4\pi n}\right)^{1/2}\left(1 + \frac{2z_0 n}{d}\right)^{1/2}. \tag{3.68}$$

The value of x_m was measured at several distances from the ruby laser. The results were compared with (3.67), and the observations found to agree within a few percent[38]. The theoretical curve predicted by (3.68) is not in so good agreement with the experimental results as with the curve computed from Fraunhofer diffraction theory for radiation from two lobes 180° out of phase.

3.2.3.1 Beam Divergence Measurement for Q-Switched Ruby Lasers A photographic study has been made of the intensity distribution and divergence of a Q-switched laser beam [44]. The extremely high peak intensity and short duration of the pulse created several problems; direct exposure to the full intensity of the beam may cause direct damage to the film or at least serious overexposure. The short duration of the pulse causes reciprocity failure in the photographic film. The use of beam samplers or attenuators is problematical in that the beam splitters or neutral-density filters are damaged by the laser beam in a single pulse.

Damage to beam-measuring components may be circumvented if dielectrically coated pellicles are used to attenuate the beam; they reflect about 8 percent of the incident energy. Successive reflections from several pellicles enables a reduction in laser beam intensity to levels appropriate for use with conventional attenuators, filters, and photographic film.

The far-field intensity pattern of a Q-switched laser beam was recorded by placing the film at the focal point of a long-focal-length lens (136cm) and allowing the entire duration of the attenuated pulse to fall on the film. The long focal length resulted in a sufficiently large far-field beam spot for precise measurement. The photographic-density distribution over the film surface was obtained by scanning the film in narrow strips with a microdensitometer, using the smallest slit dimensions consistent with a full-scale reading, so that as much detail as possible was preserved during the readout. These narrow two-dimensional strips were then assembled in the proper sequence to give a plot of the density distribution appearing on the film. Contours of constant intensity were obtained by passing a plane through the intensity distribution at appropriate levels.

Because photographs of high-intensity light sources with short exposure durations comparable to those obtained from a Q-switched ruby laser had never been taken before, no information was available on film response. To obtain the necessary gamma, the film was calibrated using the Q-switched laser under examination, along with pellicles and neutral-density filters. Using filters to vary the intensity, the gamma was obtained under the same conditions of pulse duration, intensity, and optical frequency as those that were maintained in the subsequent measurements.

The film used was the new extended range (XR) film marketed by Edgertion, Gremehausen and Grier, Inc. The extended range (wide latitude) is accomplished by superimposing three emulsions each differing in sensitivity on one film base. This resulted in an exposure ratio of 10^8. The film may be used in either of two ways:

1. If the fast layer is exposed first (the normal way), the full latitude of 10^8 is realized.
2. If the film is reversed so that only the slow layer is exposed (through the film base), a lower background results, with a latitude of 10^4.

The latter was found to be more useful.

Measurements were performed both on rotating-reflector, Q-switched lasers, and on Kerr-cell-switched lasers. Divergence angles were seen to vary significantly when flash lamps were changed; this was thought to be due to changes in the pump-energy distribution within the ruby rod, rather than to changes in the optical alignment. The method of Q-switching (rotating-mirror or stationary-Kerr-cell) also affected the shape of the pattern obtained.

3.2.3.2 Semiconductor Lasers Semiconductor lasers emit from a narrow strip. The model for the analysis of the laser-beam shape is Fraunhofer diffraction from a single slit. A normally incident uniform plane wave will generate a single-slit diffraction pattern with a slit of height a given by

$$\frac{I(u)}{I_0} = \frac{(\sin u)^2}{u^2} \tag{3.69}$$

where $u = \pi a (\sin \theta)/\lambda$ and θ is the angle between the normal to the slit and the direction of the ray. When the aperture is not uniformly illuminated (the usual situation in lasers) the results will differ. For a beam of uniform phase over the aperture the intensity at the edges of the emitting strip is so small that it may be neglected. The model of the beam becomes a Gaussian with a main lobe somewhat wider than that obtained when a slit emits with uniform amplitude and phase.

Measurements made with properly designed optics indicate that the single-slit diffraction model is good and that the output beam of a semiconductor laser may be treated in a conventional manner except that cylindrical optics are required.

3.3 DETERMINATION OF THE NUMBER OF MODES

In Section 3.1 formulas are given for the spacing of various (TEM_{mnq} or TEM_{plq}) modes for the most commonly encountered laser-resonator structures. The number of cavity resonances contained within the line width of the laser transition is less than the maximum number of oscillating modes of a laser because of inevitable losses. In typical lasers variation of the m, n indices by one unit will produce frequency differences of several MHz. The line width of a given solid-state laser transition is likely to vary over a tremendous range, from less than 0.01 nm to more than 10 nm. In gas lasers line widths are quite regular and are predicted by the Doppler line width of the gas. In a wavelength interval of 0.1 nm at 1000 nm, depending upon the cavity geometry, there will usually be several hundred cavity modes within the line width. The experimental task of determining so large a number of modes would be formidable were it not for the fact that only a few modes have low enough losses to oscillate.

The theory of laser oscillation in multimode cavities indicates that in the ideal case of the homogeneously broadened line, and for pumping rates considerably in excess of threshold (about 50 percent), most of the emitted power would be concentrated in one or, at most, a few of the most favored modes (those with the highest Q, oscillating near the line center where the gain is highest). Multimode operation is almost always encountered in solid state lasers. Several factors contribute to this situation. First, most of the laser transitions are either inhomogeneously broadened, or both homogeneously and inhomogeneously broadened. In the case of a homogeneously broadened line the gain is uniformly reduced across the entire line; in the inhomogeneously broadened line, the gain is reduced in a narrow region in the immediate vicinity of the incident frequency, and very slightly at frequencies removed from ν_i. This phenomenon is known as "hole burning."

A second mechanism that favors multimode operation in solid-state lasers is the large Fresnel number and the consequent low diffraction losses for all modes. Another factor to be considered is the fact that the standing waves in the resonator produce a nonuniform distribution in the density of excited state atoms in accordance with the field variations within the cavity. A similar argument applies to the transverse variations of the field. Thus, the population inversion is high at the nodes and correspondingly low at the antinodes of the field distribution within the cavity. The inhomogeneity of excited states engenders modes that will utilize the anisotropes of inverted population. In gases this situation is partly alleviated by the motion of the gas atoms.

For the different modes to oscillate, it is necessary that the width of the cavity resonances be less than the frequency separation of these resonances. Figure 3.12 illustrates typical values for the line width, resonator width, and laser-emission width of gas lasers. The width of the cavity resonances, taking into account the losses due to the mirrors (usually, the absorption and transmission through the mirrors), is given very approximately by

$$Q = \frac{f}{2\Delta f} = 2\pi f \tau_c = \frac{nfd}{c(1 - R)} \tag{3.70}$$

where R = reflection coefficient of the mirrors, f = resonant frequency of the cavity, Δf = half-width of the cavity resonance, τ_c = lifetime of a photon in the resonator, n = index of refraction of the laser medium, and b is the mirror separation. In solid-state lasers scattering losses are often the determining factor in evaluating the cavity Q because the Fresnel number is usually of the order of 100 or more.

In most sophisticated applications operation in the diffraction-limited lowest-order mode (TEM_{00q}) is required. A number of techniques have been developed to force solid-state lasers to operate in this mode. Use has been made of circular limiting apertures, wires, and special mirror

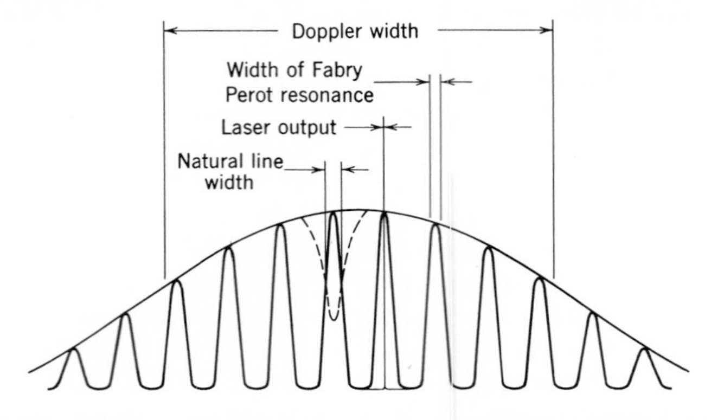

Figure 3.12. Spectral-line width factors in a gas laser. (Reproduced with permission from *Applied Optics*.)

adjustments to obtain single transverse-mode operation. In a helium-neon laser operation in a single axial mode has been obtained at very low power by using a third mirror. The spacing of the third mirror, together with the spacing of the laser mirrors, results in constraints on the cavity that enable only one or a few axial modes to have sufficient gain to oscillate. This method also depends upon low-power operation to increase gain discrimination for nearby modes. Modern single-frequency lasers either use a cavity so short that it will only permit one axial mode or use interference or frequency modulation combined with mode locking to extract high single-frequency power.

Many techniques must be used to study the number of oscillating modes in lasers. In many cases the instruments are the standard tools of the spectroscopists. In addition to the large wavelength range that must be spanned, there are the problems imposed by the various types of lasers.

3.3.1 Use of a Spectrograph to Study Mode Structure of Solid-state and Semiconductor Lasers

In solid-state and semiconductor lasers the mirror separation is usually small enough to permit the resolution of axial modes by spectrographs of moderate resolving power. An example of the technique employed at a wavelength range for which photographic materials are not available is found in the case of the $CaF_2{:}U^{3+}$ laser at 3984 cm^{-3}. The spectra were obtained by using a precision micrometer to scan a lead sulfide detector across the focal plane of a high-dispersion spectrometer into which the laser beam was directed. The resolution was slit-limited to about 0.05 cm^{-1} The axial modes present in the laser output were clearly resolved[45].

A study has also been made of the spectral emission characteristics of CW GaAs lasers[46]. The ends of the diode (the mirror surfaces) were cleaved flat and parallel, and the sides were roughened by virtue of having been string cut. The diodes (11.8 mils long by 2 mils wide) were immersed in liquid He maintained well below the lambda point. The output radiation was directed into the slit (set at 10 μm) of a 1-m Ebert scanning spectrometer. An infrared photomultiplier tube was used to detect the radiation at the exit slit. The data was recorded on a chart while the spectrometer automatically scanned the required wavelength interval.

The frequency spacing observed between the various satellites (TEM_{10q}–TEM_{00q}) was larger by a factor of 10 or 20 than the frequency difference given by (3.4) for the case of a rectangular cavity with a transverse dimension of 2 mils, filled with a material of index of refraction $\approx$ 3.6. The explanation of this phenomenon was believed to be that the GaAs diode did not oscillate uniformly but, rather, in filaments because of variations in the index of refraction and imperfections in the cleaved-end mirror surfaces. Thus the lowest order cavity modes corresponded to plane waves with a cross-sectional area much smaller than that determined by the width of the *p-n* junction. If the width of a filament is taken to be approximately 1/4 to 1/5 of the width of the resonator, agreement with the data can be obtained[46].

The line widths and emitted spectra were studied at two power levels. The smaller width at the higher power level is required by the theory of line narrowing as a result of stimulated emission. The half-width of the spectral line as a function of the increased emission is given by

$$B = \Delta\nu\left(\frac{1 - N\tau_c}{p_t\tau_{21}}\right)^{-1/2}, \tag{3.71}$$

where $\Delta\nu$ = spontaneous emission half-width, N = number of excited ions (electron-hole pairs in the case of semiconductors), τ_c = photon lifetime in the cavity, p_t = number of cavity modes within linewidth, and τ_{21} = half-life for spontaneous emission. The reduction in linewidth with power was found to be in approximate agreement with (3.70). For a mode power output of about 25μW, however, the half-width of an oscillating component, given by (3.70), was approximately 30 MHz. As this was an order of magnitude smaller than the best resolution of the scanning spectrometer the results could not be checked in this way.

3.3.2 Fabry-Perot Interferometers

Fabry-Perot interferometers have been used to study the spectral output of helium-neon gas lasers at 633 nm. The number of dominant modes oscillating simultaneously, and the power distribution among the modes, have been studied. As this work is discussed at length in Chapter 8, the reader is referred to that portion of the book for details.

3.3.3 Electronic Techniques

In the microwave and radio-frequency range the use of square-law detectors for the observation and production of beat frequencies is well known. With the advent of strong, coherent optical sources that lasers provide, these methods can be applied to the detection of optical beats with great sensitivity and efficiency, using photoemitters and photoconductors. The current produced in these detectors is proportional to the light intensity, which in the case of two frequencies is given by

$$\begin{aligned} N(x) &\propto [E_1 \cos \omega_1 t + E_2 \cos \omega_2 t]^2 \\ &= \tfrac{1}{2}E_1{}^2 + \tfrac{1}{2}E_2{}^2 + E_1E_2 \cos(\omega_1 - \omega_2)t + E_1E_2 \cos(\omega_1 + \omega_2)t \\ &\quad + \tfrac{1}{2}E_1 \cos 2\,\omega_1 t + \tfrac{1}{2}E_2 \cos 2\,\omega_2 t \end{aligned} \tag{3.72}$$

Because ω_1, ω_2, and $(\omega_1+\omega_2)$ are infrared or optical frequencies, photoconductive and photoemittive detectors will not respond to these frequencies. The difference frequency, however, will often fall well within the pass band of these detectors and can thus be measured with photomixer diodes[47]. Many configurations of the photodetector diodes have been used to observe optical beats[48] [49]. A traveling-wave phototube, useful in observing optical beats at microwave frequencies, is discussed in Chapter 9.

The high-frequency response of diode photodetectors is limited by the equivalent *RC* time constant of the detector and circuit. Table 3.2 lists semiconductors useful in observing photobeats with a variety of laser-output wavelengths [47]. The most suitable diode structure is listed, that is, *p–n* or *p-i-n*. The critical dimension that determines response is the quantity W_p which is approximately the thickness of the junction layer. The highest beat-frequency response of the detector is given by ω.

The miximum sensitivity of these detectors is given by

$$\frac{S}{N}=\frac{\eta P}{h\nu B_n},\tag{3.73}$$

where S is the power at the signal frequency; N is the total noise power, which consists principally of dark currents in the photodetector, and Johnson noise in the load resistor; η is the quantum efficiency, that is, the number of photoelectrons per incident light quantum; $P =$ the rms power in the incident signal; $h\nu =$ the energy of the incident quanta; and $B_n =$ the noise band-width. This expression for the signal-to-noise ratio (3.73) was derived on the basis that $E_2 > > E_1$, a condition which is not always satisfied, in the detection of beats between laser modes.

The characteristics of the most useful photon counting and thermal detectors are discussed in more detail in Chapter 4.

3.3.3.1 Mode-beating Measurements Before discussing the techniques of mode-beating measurements the relevant mode equation will be characterized in a more useful form. The resonance condition for a wide class of laser resonators having spherical mirrors with radii of curvature b_1 and b_2, separated by a distance d is

$$f=\frac{c}{2d}q+\frac{c}{2\pi d}(1+m+n)\cos^{-1}\left[\left(1-\frac{d}{b_1}\right)\left(1-\frac{d}{b_2}\right)\right]^{1/2}\tag{3.74}$$

here $c =$ speed of light in the laser medium. m and n are transverse mode indices, and q is the longitudinal mode number. This expression is valid so long as

$$0\leq\left(\frac{d}{b_1}-1\right)\left(\frac{d}{b_2}-1\right)\leq 1\tag{3.75}$$

TABLE 3.2 CHARACTERISTICS OF SEMICONDUCTORS USEFUL IN OBSERVING PHOTOBEATS FROM LASERS[a]

Laser System	Emission Line	Gallium Arsenide (300°K)	Silicon (300°K)	Germanium (300°K)	Indium Arsenide (196°K)	Indium Antimonide (77°K)
Ion:Host	Microns			Structure Critical Dimension 3-dB Frequency		
$CR^{+3}:Al_2O_2$ $Sm^{+2}:CaF^2$	0.6943 0.7083	*p-n* $W_p = 2 \times 10^{-5}$cm $\omega = 4.7 \times 10^{10}$	*p-i-n* $W_p = 3 \times 10^{-4}$cm $\omega = 2.9 \times 10^{10}$	*p-n* $W_p = 2.5 \times 10^{-4}$cm $\omega = 3.1 \times 10^{11}$	*p-n* $W_p = 10^{-4}$cm $\omega = 1.1 \times 10^{11}$	*p-n* $W_p = 10^{-4}$cm $\omega = 4.6 \times 10^{11}$
$Nd^{+3}:CaF_2$ $Nd^{+3}:CaWO_4$ $Nd^{+3}:SrMoO_4$ Ne:He	1.046 1.063 1.064 1.15			*p-i-n* $W_p = 10^{-4}$ $\omega = 1.4 \times 10^{11}$	*p-n* $W_p = 10^{-4}$ $\omega = 1.1 \times 10^{11}$	*p-n* $W_p = 10^{-4}$ $\omega = 4.6 \times 10^{11}$
$Tm^{+3}:CaWO_4$ $Ho^{+3}:CaWO_4$	1.91 2.046				*p-n* $W_p = 10^{-4}$ $\omega = 1.1 \times 10^{11}$	*p-n* $W_p = 10^{-4}$ $\omega = 4.6 \times 10^{11}$
$U^{+3}:CaF_2$	2.49				*p-n* $W_p = 10^{-4}$ $\omega = 1.1 \times 10^{11}$	*p-n* $W_p = 10^{-4}$ $\omega = 4.6 \times 10^{11}$

[a] Reproduced with permission from Institute of Electrical and Electronics Engineers.

and the mode size is somewhat smaller than the mirror diameter. In the case in which $b' = b_1 = b_2 >> d$ the latter requirement for the lowest order mode reduces to

$$\frac{b'}{d} < 2\pi^2 N^2 = 2\pi^2 \left(\frac{a^2}{d\lambda}\right)^2 \tag{3.76}$$

This condition is still satisfied by "plane-parallel" solid lasers; that is their deviation from flatness is almost always large enough to satisfy this condition. Denoting the departure from parallelism by λ/Q at the edge of the mirror, we can restate (3.76) as

$$Q < 2\pi^2 N. \tag{3.77}$$

Frequency differences between modes are given by

$$\Delta f = \frac{c}{2d}\,\Delta q + \frac{c}{2\pi d}\cdot\Delta(m+n)\cdot\cos^{-1}\left(1 - \frac{d}{b'}\right). \tag{3.78}$$

The first term on the right-hand side is an axial-mode beat frequency $[\Delta(m+n) = 0]$, the second, a transverse-mode beat frequency ($\Delta q = 0$). For resonators in which $b' >> d$ (such as nominally plane-parallel ruby lasers), the transverse-mode beat frequency can be written as

$$\Delta f_t = \frac{c}{2\pi d}\left(\frac{2d\lambda}{Qa^2}\right)\Delta(m+n). \tag{3.79}$$

For the condition $d = b'$ (a confocal resonator)

$$\Delta f_t = \frac{c}{2d}\,\Delta(m+n)\left(1 - \frac{2}{\pi}\frac{b'-d}{b'}\right), \tag{3.80}$$

and it becomes difficult to distinguish between these beats and beats between axial modes.

3.3.3.2 Gas Lasers Historically, the method of optical beats was first used to study the mode spectrum of a helium-neon laser. A schematic of the experimental apparatus and a drawing of the radio-frequency spectrum analyzer presentation [50] is shown in Figure 3.13. The swept oscillator of the spectrum analyzer provides a plot of the frequency spectrum of the input (hence the name spectrum analyzer) as well as a measure of the relative amplitudes of the oscillating components. According to Figure 3.13 at least three separate longitudinal modes were oscillating. The peaks at 1.3 MHz could be accounted for on the basis that the first unsymmetrical mode on the low-frequency side of each longitudinal mode was oscillating. The mode spacing was in agreement with values computed on the basis of (3.4).

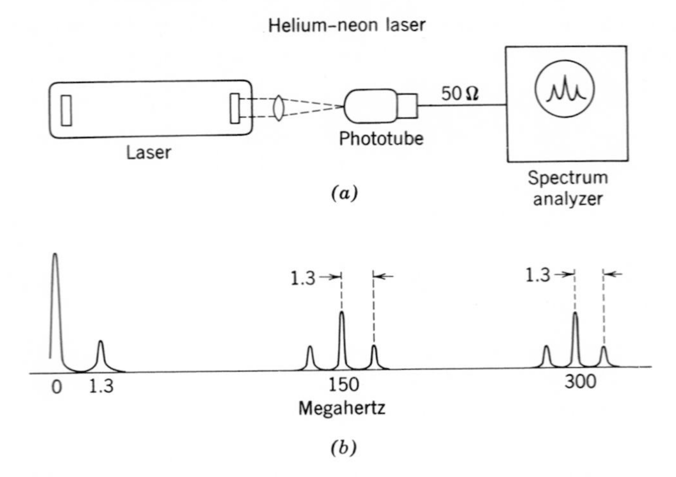

Figure 3.13. Schematic of apparatus used to observe photobeats in the output of a gas laser: (*a*) experimental arrangement; (*b*) beat pattern observed on the oscilloscope. (Reproduced with permission from *Applied Optics*.)

To observe all of the oscillating modes present, two precautions are necessary. First, the E vectors of the two light beams must have a common component of polarization; second, the beams must fall on identical areas of the photodetector. If the two components are linearly polarized at right angles, the detector will contain no output at the BFO (beat-frequency output). In this case, a linear polarizer inserted in front of the photodetector at an angle of 45° to the beam polarization will produce a maximum BFO. Thus it is necessary to use a polarizer and to rotate it so that the BFO is a maximum. To ensure spatial overlap of the laser components, a lens is sometimes employed to focus the components onto a small area of the photodetector.

From the width of the pulses on the spectrum-analyzer display, the half-width of the beating components was inferred to be 7 kHz. If both components are assumed to have a Lorentzian line shape $g(\nu) = (\pi\Delta\nu)^{-1}\cdot\{1 + (\nu - \nu_0)^2/\Delta\nu^2\}^{-1}$ and are of equal width, the width of the BFO is just twice that of a single component. More recent measurements have shown widths as low as a few cycles for a few seconds. The method of optical beats is the only technique available for the measurement of line widths as narrow as those encountered in gas lasers.

A difficulty in the application of the method of optical beats occurs when many modes are oscillating simultaneously. The BFO spectrum may be so complex that unambiguous identification of the oscillating components becomes impossible; for example if there are N components, the number of different combinations taken two-at-a-time is $N(N - 1)/2$. Thus for ten nondegenerate modes (nonidentical frequency differences), the BFO would contain 45 components.

3.3.3.3 Ruby Lasers (Crystal Lasers) The method of optical beats has been extensively used to study the mode spectrum in solids, where the pulsed-spiking type of operation is often encountered. When the mode spectrum may fluctuate rapidly in time, it is not convenient to use a spectrum analyzer. A schematic diagram of the apparatus that has been used to detect beat frequencies too high for direct observation is shown in Figure 3.14. The IF system includes a detector. Using this system, the envelope of the beats between longitudinal modes in a ruby laser (in the gigahertz range) has been displayed. By noting the local oscillator frequencies which maximize the envelope, the beat frequencies and amplitudes are deduced[50]. The BFO amplitude varied widely for a given signal amplitude, indicating fluctuations in the number and amplitude of the simultaneously oscillating axial modes.

Beats between transverse modes in ruby lasers, which are lower in frequency, have been directly observed using a wideband oscilloscope. The oscilloscope trace indicated a ringing sinusoidal wave form in time coincidence with the ruby laser spike, and the frequency was obtained by counting the number of cycles in a known time interval[8].

The mode spectra observed in solid-state lasers has been found to be in agreement with the theoretical predictions of Section 3.3.3.1 for the PPFP and the confocal configurations. In general, the observation of transverse

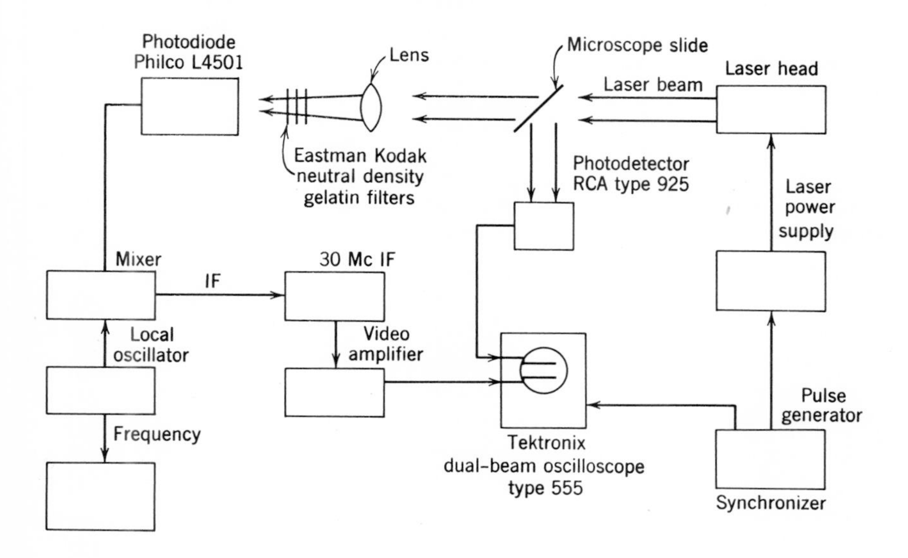

Figure 3.14. Block diagram of apparatus employed in observing optical beats.

modes in solid-state lasers has been difficult, and definitive data has been obtained only in a few cases.

3.3.3.4 Measurement of Transverse-mode Beats. A very advantageous method of measuring transverse-mode beats involves the use of an appropriate photomultiplier as the optical detector.

3.3.3.4.1 Unstablized Lasers This method applies primarily to any laser whose modes are difficult to control continuously, although it can also be used for stable CW lasers if necessary. Mode hopping occurs in pulsed lasers of practically all types. In addition, the prediction of transverse-mode beat frequencies (TMBF's) for solid lasers is difficult, for the cavity shape does not generally correspond to one that can be easily analyzed. Consequently, a wideband system which can record a short-duration beat is necessary.

The laser beam is directed onto the cathode of a photomultiplier tube. Normal incidence is not required, and the beam may be focussed or unfocussed. The detector load resistor should match the characteristic impedance of the remainder of the system. A recommended photomultiplier for 6943 Å and shorter wavelengths is the RCA 7326 or equivalent (S-20 photosurface); for near-infrared lasers, the RCA 7102 or equivalent (S-1 photosurface) is appropriate.

If the laser spikes, a high-pass filter should be added across the load resistor to prevent the Fourier components of the spikes from saturating the remainder of the system. A suitable filter is a constant-k high-pass type: two T sections in cascade, three capacitors in series with two inductors to ground. The first and third capacitors have a value $2C$, with a middle capacitor $= C$, and the inductors $= L$. The filter is described by

$$Z_0 = \left(\frac{L}{C}\right)^{1/2}; \qquad f_0 = \frac{1}{4\pi(LC)^{1/2}}$$

where Z_0 is the characteristic impedance of the system and f_0 is the low-frequency cutoff of the filter; for example if $Z_0 = 200\ \Omega$ and $f_0 = 7$ MHz, $C = 56$ pf and $L = 1.2$ uh.

The signal is then amplified by a wide-band video amplifier. An appropriate amplifier is an HP 460-A distributed amplifier or equivalent, which has a maximum gain of 20 dB, a frequency range of 200 kHz to 160 MHz, and a maximum output signal of 8 V. Two or three units may be necessary, depending on the oscilloscope amplifier and probe.

The amplifier output is then connected to an oscilloscope having a high accelerating voltage and large band-width; due to the change of TMBF's from spike to spike, the oscilloscope must be operated in a single sweep mode. An appropriate instrument is the Tektronix 517A or equivalent with a 24 kV accelerating voltage and 70 MHz band-width. Using the pulse from the laser firing switch as a synchronization, with a pulse generator with an internal delay of 2 μsec or more, the photographed traces will permit the calculation of the TMBF's at any time during the laser pulses. The horizontal and vertical calibration may be achieved simply by substituting a standard signal generator for the photomultiplier, with an output, for example, at 15 MHz.

The technique described above enables an accuracy of about ± 3 percent, being limited only by our ability to read the scales involved in measurement and calibration. Care must be taken to reduce stray capacitance and lead inductance to a minimum when the high-frequency load resistor and coaxial connector are connected to the photomultiplier, for this can easily be the primary band-width determining factor. Also, ground loops in the trigger circuit for the flashlamp must be avoided because radiation from this can be easily picked up and amplified by the sensitive detection system; the ringing that is induced can easily be confused with a TMBF.

3.3.3.4.2 Stabilized Lasers This method applies to any CW-gas, solid, or semiconductor laser the modes and corresponding TMBF's of which are stable in frequency and time. The general scheme is the same as that of Section 3.3.3.4.1, the exception being the method used for the display of data. The photomultiplier, filter (to be used if the CW laser spikes), and amplifier (to be used if the signal level is below the minimum detectable level of the display system) are the same as those used in the method described in Section 3.3.3.4.1.

Three methods of data display are possible. The first simply employs an rf voltmeter. This is useful if a single beat of known frequency is to be measured or monitored. The photomultiplier load resistor is replaced by a parallel LC circuit which is resonant at the frequency of interest, and the rf voltmeter is connected directly across this resonant circuit. A suitable meter would be a Millivac vacuum tube millivolt meter Model MV-18B, which is useful from 1 to 200 MHz and for signal levels as low as 1 mV. The accuracy of the measurement will be mainly determined by the circuit Q, which should be made as high as possible.

The second method of data display employs a radio receiver, which will typically have a range from 0.54 MHz to 31 MHz. Sensitivities are orders of magnitude greater (as good as -118 dBM) than the rf voltmeter method, and the calibration and accuracy are determined by the receiver characteristics and are also superior to the voltmeter method. The photomultiplier load resistor and rf cable should be chosen to match the antenna input of the receiver (generally 70 to 300 Ω); this is not critical, however, because frequencies below 31 MHz are involved, and an antenna "trimmer" is usually provided. Receiver frequency scales are typically accurate to 1 percent.

In the third method of data display a spectrum analyzer is used. This is preferred over the two methods described above because the instrument displays a plot of signal power versus frequency directly on a CRT. The horizontal sweep can range from 10 to 420 MHz at a 60-Hz rate, and the vertical scale can cover a range of 60 dB. This type of operation permits the monitoring of a wide range of beats simultaneously and retains the sensitivity and frequency scale accuracy (approximately) of a receiver. The photomultiplier load resistor should match the input impedance of the spectrum analyzer (usually 50 Ω), with the connection via a 50-ohm cable. The resolution of a spectrum analyzer is generally variable from 1 to 8 kHz; the frequency scale accuracy is 1 MHz or ± 1 percent (whichever is larger), and the sensitivity can be as good as − 110 dBM.

Regardless of the method used, some precautions are necessary. For lasers in which the transverse modes are well described by orthogonal functions it may be necessary to cut half the beam off, outside the cavity, by using a knife edge. This will remove the spatial orthogonality of the modes and permit generation of a larger TMBF photocurrent. If positive mode identification is desired, the mode pattern should be photographed simultaneously with the measurement of the beat frequency. No optics are necessary; simply a shutter and film will suffice. Photomultiplier currents should generally be less than 1 ma averaged over any 30-sec interval; neutral-density filters are convenient for this purpose. Stray capacities and inductances should again be minimized.

3.3.3.5 Spurious Harmonic Generation in Optical Heterodyning If the response of the photodetector is given by

$$i = k_0 + k_1 E + k_2 E^2 + k_3 E^3 + \cdots + k_n E^n + \cdots \tag{3.81}$$

instead of (3.72), the higher order k's will give rise to signals at harmonics of the difference frequency. Specifically, the term $k_{2n}E^{2n}$ will generate a component at $n(\omega_1 - \omega_2)$. This effect can lead to a spurious identification of modes particularly during the search for weak components. To test the photodetector response for the higher terms in (3.81), mechanical (or other means) of chopping the laser beam at a slow rate (usually an audio frequency) should be employed. The detected signals are then examined for harmonics of the chopping frequency. The relative magnitude of the fundamental and harmonics can be used to measure the coefficients of the various terms in (3.81).

3.3.3.6 Mode-pulling Effects The measurement of the frequency difference between oscillating modes can be performed with very high precision, which has led to the observation of some very subtle effects in gas lasers. The spectral output of a PPFP, 1-m helium-neon laser, employing magnetostrictive elements for varying the mirror separation, was studied using the BFO method. It was found that the axial mode beats were not separated by $c/2L$, but rather, were less by about 200 kHz, a discrepancy of 1/800. These effects were similar to earlier observations on mode pulling in the ammonia microwave maser. The pulling results from the interaction of the cavity resonance and the atomic resonance; it is given by

$$\nu_0 = \frac{\nu_\upsilon \Delta\nu_c + \nu_c \Delta\nu_\upsilon}{\Delta\nu_\upsilon + \Delta\nu_c}, \tag{3.82}$$

where ν_0 is the oscillator frequency, $\Delta\nu_c$ is the line width at half-maximum intensity and ν_v is the line center. In the usual case $\Delta\nu_c \ll \Delta\nu_v$ (3.80) becomes

$$\nu_0 \cong \nu_c + (\nu_\upsilon - \nu_c)\frac{\Delta\nu_c}{\Delta\nu_\upsilon} \tag{3.83}$$

and the beat frequency between two modes is given by

$$(\nu_0' - \nu_0) \cong (\nu_c' - \nu_c)\left(1 - \frac{\Delta\nu_c}{\Delta\nu_\upsilon}\right). \tag{3.84}$$

A power-dependent splitting effect of approximately 20 kHz was also found and it was shown that the shift in the oscillation frequency would be obtained from a more exact evaluation of the oscillation frequency:

$$\nu_0 = \nu_c - \frac{\Delta\nu_c}{f}\Delta\phi_G(\nu_0), \tag{3.85}$$

where f is the fractional energy loss per transit, and $\Delta\phi_G(\upsilon_o)$ is the change in phase shift at the oscillating frequency. This term must be evaluated taking into account saturation effects in the presence of oscillation.

This "hole repulsion" can be qualitatively understood on the basis that each component feeds to a certain extent on an inverted reservoir of atoms common to both lines. In the competition for additional atoms both components utilize the atoms on the far side of the resonance because they tend to belong more effectively to each of the components. This effect leads to hole repulsion.

3.3.3.7 Zeeman Effects If a helium-neon laser is placed in a longitudinal magnetic field (in the interior of a long solenoid), the transitions will be split by the magnetic field [52]. The energy levels as a function of the magnetic field are given by

$$E_i(H) = E_i(0) + g_i \beta m_i H, \tag{3.86}$$

where $i = 1, 2$ for the lower and upper level, respectively; $m_i =$ the component of angular momentum in the direction of the magnetic field H, $\beta =$

the Bohr magneton, and g_i = the magnetic splitting factor, which can be estimated using the *J-L* coupling approximation. In the case that $g_1 = g_2 = g$ the frequencies of the right and left circularly polarized components ($m = \pm 1$) are given by

$$v_v \pm = v_c \pm \frac{g\beta H}{h}, \quad (3.87)$$

where the $\pm$ signifies right and left circular polarization, respectively. These two modes can oscillate within the same resonator cavity mode because the two distributions that support the modes are independent of each other. The two components add together to give a linearly polarized wave that rotates at half the difference frequency. If, however, the light is observed with a photodetector that responds to the square of the field, the linearly polarized resultant wave will be observed to rotate at the frequency difference given by[52]

$$v_0^+ - v_0^- \simeq \frac{2g\beta H}{h} \frac{\Delta v_c}{\Delta v_v} \quad (3.88)$$

If the output is observed by passing the light through a polarizer in front of the photodetector, it appears to be amplitude-modulated at a frequency given by equation (3.88).

3.3.4 Summary

Spectrographs and Fabry-Perot interferometers can be used to observe the mode spectrum of lasers over the wavelength range of 200 nm to 0.4 mm. Usually only the methods of optical homodyning or heterodyning are capable of providing the resolution required or to observe the transverse modes in the PPFP resonators and the Zeeman components in gas lasers. Most detectors in the infrared (far infrared particularly) have a very poor high-frequency response. This will limit the applicability of these methods to a long wavelength limit of about 40μm, the limit of the Ge:Au, Zn detectors. The response time of these detectors is given as less than 10^{-8} sec, which indicates that BFO's as high as 100 MHz can be detected in the infrared. Detailed infrared measurements, however, have so far been made only on a few systems, and almost no work has been reported at wavelengths beyond about 2.6μm.

3.4 DETERMINATION OF THE POLARIZATION

The determination of the polarization of the light output of lasers is required in a number of applications, of which the following are examples: the production of optical harmonics and other nonlinear effects, optical heterodyning and homodyning, and interference and diffraction experiments. It has already been mentioned that light beats cannot be observed if the polarization directions of the light signals are perpendicular. In interference experiments we find that if light beams that are identical in all respects, except that they are linearly polarized at right angles, they will not interfere. The same is true for opposite circular polarizations. In quantum mechanics this situation is interpreted as indicating that a light beam (photons) has two internal degrees of freedom. The specification of the light beam in terms of the components of linear polarization or of the components of circular polarization are completely equivalent.

3.4.1 Analysis of the Polarization Properties of Light

In the analysis of the polarization composition of an unknown beam, the following possibilities occur:

1. Unpolarized light,
2. Plane-polarized light,
3. Circularly polarized light,
4. Elliptically polarized light,
5. Unpolarized light plus plane polarized light,
6. Unpolarized light plus circularly polarized light,
7. Unpolarized light plus elliptically polarized light.

If the light is propagating in the z direction in a Cartesian coordinate system, the seven possibilities have a simple mathematical characterization as follows:

$$
\begin{array}{lll}
1) & E_x = E_1 \cos(ft + q_1(t)); & E_y = E_2 \cos(ft + q_2(t)) \\
2) & \quad = E_1 \cos(ft + q_1(t)); & \quad 0 \\
3) & \quad = E_1 \cos(ft + q_1(t)); & \quad = E_1 \cos\left[ft + q_1(t) \pm \dfrac{\pi}{2}\right] \\
4) & \quad = E_1 \cos(ft + q_1(t)); & \quad = E_2 \cos\left[ft + q_1(t) \pm \dfrac{\pi}{2}\right] \\
5) & \quad = E_1 \cos(ft + q_1(t)); & \quad = (1) + (2) \\
6) & \quad = E_1 \cos(ft + q_1(t)); & \quad = (1) + (3) \\
7) & \quad = E_1 \cos(ft + q_1(t)); & \quad = (1) + (4)
\end{array}
$$

The systematic procedure outlined below can be used to analyze the polarization components of an unknown source.

First test: insert a Nicol prism or other polarizer in the beam and rotate it. If the light is completely extinquished for one orientation of the polarizer, the light is completely plane polarized. If the light is unaffected for all settings of the polarizer, the light corresponds to (1), (3), or (6). Finally, if the effect is intermediate between complete and zero extinction, then the light is (4), (5), or (7).

Second test: to distinguish between (1), (3), and (6), insert a quarter-wave plate behind which a polarizer is inserted. If the light is completely extinquished for one setting of the polarizer, the original beam is completely circularly polarized. If no effect is observed upon rotation of the polarizer, the original beam is unpolarized. The intermediate case indicates a mixture of circularly polarized and unpolarized light.

Alternate second test: to distinguish between (4), (5) and (7); rotate the quarter-wave plate and the polarizer independently. If the light is completely extinquished for one setting of the components, the original beam is elliptically polarized. The axes of the ellipse are parallel to the principal axes of the quarter-wave plate when the light is extinguished. Partial extinction indicates that the original light is a mixture of elliptically polarized and unpolarized light. If the polarizer is parallel to one of the principal axes of the quarter-wave plate, for maximum extinction, the original beam is a mixture of plane polarized and unpolarized light.

3.4.2 Polarization Measurements

In principle, the output of a single-mode CW laser would be expected to be almost completely plane polarized. In gas lasers, this is found to be the case. However, most measurements are performed with lasers using external mirrors, in which the discharge tube is terminated with Brewster-angle windows. This favors oscillation in the plane of incidence, for this polarization is transmitted without loss through the windows. It is therefore not unexpected that the output of these lasers is found to be nearly completely plane polarized. In Zeeman-effect studies with the helium-neon laser (Sections 3.3 to 3.7), circularly polarized components in the emitted radiation have been observed.

In the case of solid-state and semiconductor lasers both polarized and unpolarized outputs have been observed. In ruby lasers, the output of 0° orientation rods (*C* axis coincident with the rod axis) has been found to be unpolarized, while that of 60 and 90° orientation rods has been found to be completely plane polarized.

In one experiment the output of a ruby laser was passed through a Glan-Thompson prism[53]. Part of the laser output was fed directly into a monitor phototube, while the beam transmitted through the prism was fed into a second phototube. The phototube outputs were displayed on a dual-beam oscilloscope. In the case of the 60 and 90° rods the output was found to be completely plane polarized, with the electric vector in a direction perpendicular to the plane containing the optic axis. All spikes that were observed showed identical polarization. This result is anticipated from the studies of the R_1 fluorescence, in which it is observed that the component with its E vector perpendicular to the C axis is the strongest. Observation of the output of 0° orientation rods showed no polarization, plane or circular, for temperatures ranging from 100 to 300°K, and for power levels near and far above threshold. Identical results were true for every individual spike. These results should be contrasted with [38] the mode studies of Section 3.2, where it was stated that the two lobe patterns of a ruby laser operated near threshold are plane polarized, for 0° rods of ruby.

3.4.2.1 Senarmont Polariscope In determining the parameters of pulsed laser sources, it is often necessary to obtain the polarization analysis of a single pulse of the laser. An instrument that can be used for this purpose is the Senarmont polariscope, shown in Figure 3.15. At some distance in y along the exit face of the device, a null will appear. This will be the locus of all light rays along which the azimuthal polarization of the incident beam is rotated through the angle necessary to cross it with the analyzer. The rotation is a function of position along the face of the active element and is given by

$$P(y) = dL - L(d - \ell)\frac{y}{W} = A + By \tag{3.89}$$

where $P(y)$ is the total rotation in degrees of a ray parallel to the z axis, d and ℓ are the rotatory power of the right-and left-handed quartz in degrees per millimeter; L and W are the length and width respectively of the prism, and A and B are defined by (3.89). The position of the null is measured from the edge of the active element. The nulls observed with 90° ruby possessed rough edges. This was taken to indicate that the entire emitting rod was not acting uniformly.

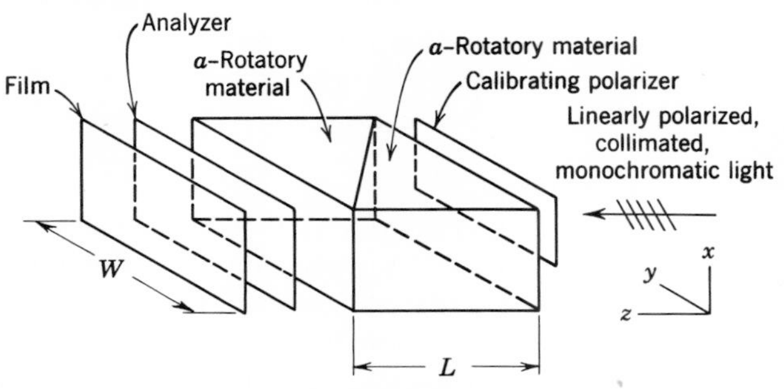

Figure 3.15. Schematic of polariscope for analysis of the polarization of pulsed-laser beams.

Other measurements using a method analogous to that of the Senarmont polariscope indicated that in the case of a 0° rod pumped well above threshold, two planes of polarization at right angles were present. In low angle crystals, although there was a mean angle for the plane of polarization, there was variation about this mean for light from different parts of the end face. Completely polarized light was found to be emitted from a small area of the crystal. At the start of measurements using a 90° crystal, the light was found to be completely plane polarized.

The average value of the angle of polarization of a pulsed ruby laser can be determined to an accuracy of $\pm 2°$ if the laser beam is directed onto the apex of a Brewster-angle cone that rests on a photographic film [55]. The polarizing polariscope operates on the principle that beam components will be reflected from a Brewster angle cone only if the electric vector is perpendicular to the plane of incidence. The photograph must be analyzed to yield the plane of polarization. The value of the technique rests on its ability to record the polarization of Q-switched laser pulses.

3.4.2.2 Semiconductor Lasers The observations of semiconductor lasers showed several features analogous to those derived from observations of ruby lasers, indicating the effects of inhomogeneous laser materials. In the lowest order mode the semiconductor laser light is expected to be completely plane polarized. For some diodes, using the PPFP geometry, the output was found to be almost completely polarized with the electric vector parallel to the junction, and for others the output was found to be nearly completely polarized with the electric vector perpendicular to the junction. In still others neither polarization was observed.

By using diodes with two sets of plane-parallel sides (bouncing-ball configuration) other polarization effects were observed. Close to threshold, both the on-axis radiation and the radiation at 4° were found to be highly polarized, with the E vector perpendicular to the plane of the junction. With increased currents, the on-axis radiation (0°) became less polarized, whereas the radiation emitted at 4° remained highly polarized. This was consistent with the interpretation that the off-axis peaks corresponded to the internally reflected, bouncing-ball modes, which should be plane-polarized because of the internal reflections.

3.4.3 Summary

The polarization of a source may be systematically investigated by the method presented in Section 3.4.1. The precision with which the measurements can be performed is dependent on many circumstances. In the visible and near-visible region it should be possible to attain a precision of about 1 percent. At other wavelengths the major difficulty is the lack of materials that can serve as polarizers, both linear and circular, to permit the analysis of the unknown parameters of the radiation. Materials are available that transmit out to about 80μm, and therefore it appears not unreasonable to anticipate that components can be made. The lack of components, however, tends to make the determination a research job rather than a measurement task that can be simply described.

It is worthwhile to speculate on the possibility of applying homodyning and heterodyning methods to the task of determining the polarization, provided that a fraction of the signal can be polarized and then used for homodyning or that an optical oscillator of the proper frequency for producing a BFO within the bandwith of the detector can be obtained. For these measurements we take advantage of the requirement that the E vectors of the two signals must be parallel for maximum output and zero for the perpendicular alignment. Thus the experiment would consist of observing the BFO as a function of the polarization direction of the optical local oscillator. Implicit in this discussion of heterodyne or homodyne methods is the assumption that the polarization of one of the components can be varied. This method would possess great sensitivity and furthermore could be used for very long infrared wavelengths.

3.5 MEASUREMENT OF BEAM-SPOT SIZE

Two techniques have been used to determine spot size. The laser beam can be made to interact with a material and cause radiation damage. Several examples are outlined in Chapter 4 (Section 4.10). These techniques are especially useful for determining the spot size of pulsed-ruby and infrared lasers. As the beam continuously diverges upon leaving the cavity, the measurement must be correlated with a specific distance from the laser.

A second method of measuring beam-spot size, which is particularly well suited to visible CW gas lasers, is to employ a microscope objective of reasonable aperture (~ 1 cm diameter) and short and accurately known focal length (20 to 30 mm) to enlarge the beam. The distance between the microscope objective and a suitable screen placed a few meters away can be measured with accuracy. The diameter of the suitably enlarged beam can also be measured accurately, particularly if the microscope objective entrance pupil is large compared with the beam. The beam-spot size may then be determined by a simple ratio:

$$\text{Beam-spot size} = \frac{\begin{pmatrix}\text{focal length of}\\ \text{microscope objective}\end{pmatrix}\begin{pmatrix}\text{diameter of}\\ \text{projected spot}\end{pmatrix}}{\begin{pmatrix}\text{distance from entrance}\\ \text{pupil to screen minus the}\\ \text{focal length of}\\ \text{microscope objective}\end{pmatrix}}$$

The principal sources of error in this measurement are the accuracy with which the distances are measured and the accuracy to which the microscope objective focal length is known.

3.6 PRECISION MEASUREMENT OF ATMOSPHERIC DISPERSION

A discussion of the measurement of laser-beam parameters is not complete without the consideration of methods for measuring the dispersion of a medium in which the beam might be propagated. Furthermore, the advent of stabilized gas lasers has made possible more precise measurements of dispersion in cases where such a measurement is an end in itself.

3.6.1 Introduction and Definitions

We define "relative dispersion," $\tilde{\delta}(\lambda, \lambda_0)$, as

$$\tilde{\delta}(\lambda, \lambda_0) = \frac{\delta(\lambda, T, P)}{\delta(\lambda_0, T, P)} = \frac{\eta(\lambda, T, P) - 1}{\eta(\lambda_0, T, P) - 1}, \tag{3.90}$$

where $\delta(\lambda, T, P)$ is the refractivity at temperature T, pressure P, and vacuum wavelength λ. λ_0 is some arbitrary reference wavelength, and n is the refractive index. For gases at normal pressures $\tilde{\delta}(\lambda, \lambda_0)$ is highly independent of T and P [56, 65]; $\tilde{\delta}(\lambda, \lambda_0)$ is usually measured interferometrically, by allowing a change in gas pressure to alter the optical path difference between two interfering beams. If the geometrical path difference remains unchanged, then

$$\tilde{\delta}(\lambda_i, \lambda_0) = \left(\frac{M_i}{M_0}\right)\left(\frac{\lambda_i}{\lambda_0}\right), \tag{3.91}$$

where M_i is the change in the order of interference at wavelength λ_i. λ_i is a variable whose values are restricted to the values of suitably monochromatic spectral lines. Sources are available that provide an accuracy in λ_i/λ_0 of a few parts in 10^8, and stabilized gas lasers will afford even greater accuracy. Measurements of wavelengths will not be discussed here except to note that the method of excitation of the source can cause small but significant shifts in λ_i.

We shall concern ourselves with the matter of obtaining an accuracy in M_i/M_0 approaching that λ_i/λ_0. It is convenient to rewrite (3.91) as

$$\tilde{\delta}(\lambda_i, \lambda_0) = \frac{N_i + \epsilon_i}{N_0 + \epsilon_0} \frac{\lambda_i}{\lambda_0} \tag{3.92}$$

where N_i = integral part of M_i and ϵ_i = fractional part of M_i.

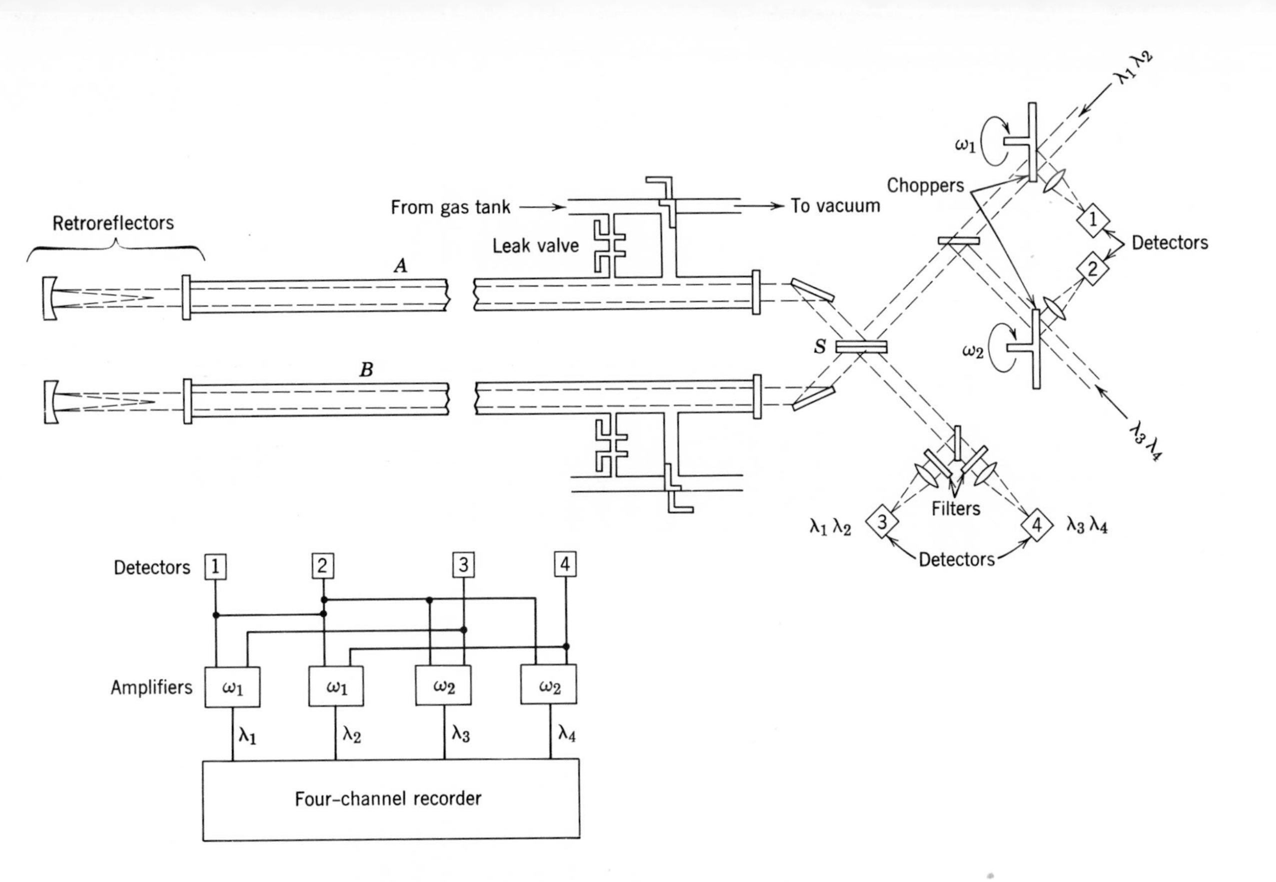

Figure 3.16. System for measuring atmospheric dispersion.

3.6.2 Experimental Arrangement for Measuring Atmospheric Dispersion

For extreme accuracy in measuring atmospheric dispersion it is generally safer to rely on large values N_i than to rely on measurements of ϵ_i to very small fractions of a fringe. This is due to the fact that errors in ϵ_i are largely independent of the magnitude of N_i. The modified Michelson arrangement, shown in Figure 3.16, is therefore recommended. The chambers may be 100 m or longer and the orders of interference N_i may exceed 100,000. This arrangement is essentially that used by Rank et al.[59], except that two-element retroreflectors are used instead of flats. With long chambers this simplifies alignment and reduces the effect of drift. The retroreflectors should be $f/10$ or longer, so that orientation and focus are not critical. The double-passing mirrors should be small to minimize obstruction of the beams.

Separation of spectral lines is carried out partly by means of synchronous detection and partly by means of filters. Many variations are possible. Separation can also be accomplished in a straightforward manner using slits and a prism. This is usually simpler except in the infrared where chopping is required anyhow.

The beam-splitter and retroreflector assemblies should be isolated from vibration and protected from irregular perturbations during measurements (e.g., flexure of the floor or convection past the beams). Steady thermal drift can be tolerated if measurements are made quickly and symmetrically, although at high-order numbers N_i counting fringes takes too long because the fringes vanish if gas is admitted rapidly. Several channels, preferably four, are therefore recorded simultaneously to permit identification of the integers N_i without counting. The gas temperature T and pressure P are measured to help identify N_i, for which thermocouples and mercury manometers are suitable. The fractions ϵ_i are measured from the strip chart. The precision demanded of T, P, and ϵ_i is governed by inequality (3.98).

3.6.3 Alignment Procedure

1. Prefocus the retroreflectors at infinity with an autocollimator.

2. Limit the angular field of each detector to a radius $\theta_i < 1/(N_i)^{1/2}$. Limit the angular field of each source to a somewhat larger value.

3. Position the sources so that they appear superposed on each other. Position the detectors at the images.

4. Tilt the chamber windows just enough to suppress reflected images.

5. Using a telescope, look through beamsplitter S from the detector side and align the optics so that the two small double-passing mirrors appear superposed. Fringes should appear at infinity. Realign the optics until one fringe covers the field of view.

6. Let gas leak into one chamber. Carefully reposition and refocus

retroreflectors and detectors to maximize fringe contrast on the strip chart.

7. Check the isolation of the spectral lines. Regular modulation of the fringes indicates crosstalk. This can cause an error in fringe position of $\Delta\epsilon \approx (1/\pi)$ (percent modulation).

8. For sources of limited coherence equalize the beam paths (and window thicknesses) so that fringe contrast decreases equally, regardless of the chamber filled.

3.6.4 Data-recording Procedure

1. Establish equal and reproducible leak rates into chambers A and B such that the period between fringes is much longer than the response time of the detector system.

2. Adjust the strip chart to a speed preferably over 3 cm/fringe. Adjust the gain of each channel to the maximum usable value.

3. Fill A and evacuate B. Record pressure P and temperature T of A.

4. Start the strip chart. Let gas leak into B until at least 10 fringes have been recorded in each channel. Close the leak valve and stop the strip chart.

5. Repeat Steps 3 and 4 quickly, with the roles of A and B reversed.

6. Repeat Steps 3 and 4 as written, allowing the same time for this repetition as in step 5.

7. If sufficiently accurate values of $\delta(\lambda_1, T, P)$ are not available for identification of N_i repetition of Steps 3 through 6 is required at lower pressures. If a direct count of N_i is desired, do not close the leak valve in Step 4. If sufficiently exact values of $\delta(\lambda_i, T_0, P_0)$ are available from the literature, the "bootstrapping" operation may be begun by calculating R_i from (3.92) and N_0 from (3.93), instead of by counting N_i from the strip chart

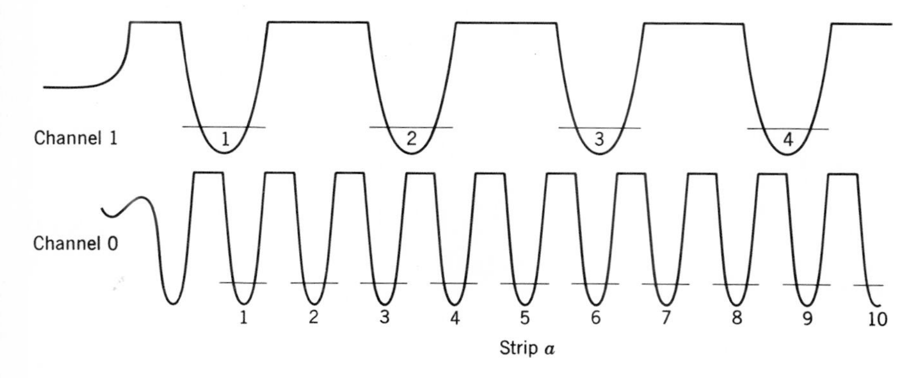

Figure 3.17. Fringe-interpolation procedure.

3.6.5 Fringe Interpolation Procedure

1. Label the three strip recordings "a," "b," "c," in the order taken. Use corresponding subscripts for T and P.

2. Label the four channels on each strip "0," "1," "2," "3," where "0" denotes the one chosen as a reference (preferably one with a short wavelength).

3. Number the fringe minima in each channel of each strip, "$1, 2, \ldots, k_{ij}$" in the sequence recorded. Ignore incomplete fringes at the ends of the strips. Use an odd number of fringes k_{ij} dropping a fringe at one end if necessary.

4. Draw a horizontal line at a convenient fixed distance above the minimum of each numbered fringe, as shown in Figure 3.17. This yields $2k_{ij}$ points of intersection with each trace. Some smoothing of a noisy trace is acceptable.

5. For each channel of strip "a," record the longitudinal coordinates of the $2k_{ij}$ intersection points (the origin is arbitrary; + is toward higher-numbered fringes). Calculate the average values

$$X_{0a}, X_{1a}, X_{2a}, X_{3a}.$$

6. From the coordinates of the first and last fringes in channels 1, 2, 3, of a strip "a," calculate the periods τ_{1a}, τ_{2a}, τ_{3a}.

7. Calculate the fringe displacement ϵ_{ia} between the points X_{ia} and point X_{0a}:

$$\varepsilon_{ia} = \frac{X_{ia} - X_{0a}}{\tau_{ia}}.$$

8. Repeat Steps 5 through 7 for strips "b" and "c."

9. Calculate $\epsilon_i = \epsilon_{ia} + \epsilon_{ib} + \epsilon_{ic}$, $\quad \epsilon_0 = 0$.

$(N_1 + \epsilon_i)$ has been chosen as the number of fringes passing between points X_{0a} and X_{0b} plus the number of fringes passing (in the reverse direction) between points X_{0b} and X_{0c}. Temperature and pressure are measured near the beginning of each strip, and at points X_{0a}, X_{0b}, X_{0c}. A correction K_i must therefore be included in N_i is estimated from the measurement of temperature and pressure:

$$N_i + K_i = 2\delta(K_i, T_0, P_0)\frac{T}{P_0}\left(\frac{P_a L_A}{T_a} + \frac{2P_b L_B}{T_b} + \frac{P_c L_A}{T_c}\right), \tag{3.93}$$

where

$$K_i \approx \tfrac{1}{2}(K_{ia} + 2K_{ib} + K_{ic});$$

L_A and L_B are the lengths of chambers A and B, and T_0, P_0 represent "standard conditions."

3.6.6 Order-number Identification Procedure

Because $\delta(\lambda_i,\lambda_0)$ is highly independent of density it follows from (3.92) that

$$\left(\frac{N_i + \epsilon_i}{N_0 + \epsilon_0}\right)_{\text{high pressure}} = \left(\frac{N_i + \epsilon_i}{N_0 + \epsilon_0}\right)_{\text{low pressure}} \tag{3.94}$$

At the lowest pressure the integers N_i can be identified by direct counting from the strip chart (be careful whenever the count reverses direction). At higher pressures counting is not required. If the pressure increase is not too great there is ordinarily only one set of integers N_i at the higher pressure that is consistent with (3.94). Once these integers have been identified they may be used to identify another set N_i at a still higher pressure, and so on. The N_i at the highest pressure are used to calculate $\tilde{\delta}\ (\lambda_i, \lambda_0)$ from (3.92).

One cycle of "bootstrapping" operation is given here (primes denote values associated with the lower pressure). We have set $\epsilon_0 = 0$; hence (3.94) reduces to

$$N_i = R_1' N_0 - \epsilon_i, \tag{3.95}$$

where

$$R_i' = \left(\frac{N_i + \epsilon_i}{\epsilon_0}\right)'. \tag{3.96}$$

1. Calculate R_i' from (3.96).
2. Determine an approximate value $\tilde{N}_0$ for N_0. According to (3.94).

$$\tilde{N}_0 = (N_0 + K_0)' \frac{(P_a/T_a) + (2P_b/T_b) + (P_c/T_c)}{(P_a/T_a) + (2P_b/T_b) + (P_c'/T_c)} - K_0. \tag{3.97}$$

(We have assumed $L_A = L_B$).

3. Establish the range $\tilde{N}_0 \pm \Delta N_0$ that is consistent with the accuracy of the data on (3.97).
4. Calculate

$$\gamma_i = \frac{\tilde{N}_0}{N_0} \Delta\epsilon_i' + \Delta\epsilon_i,$$

where $\Delta\epsilon_i'$, $\Delta\epsilon_i$ are the maximum possible errors in ϵ_i', ϵ_i, excluding errors in λ_i/λ_0.

5. Check that the measurements are accurate enough that

$$2((2)\gamma_i)^{s-1}\Delta N_0 \ll 1, \tag{3.98}$$

where $s =$ number of channels (preferably four).

6. Using (3.95), calculate N_1 for each trial value of N_0 in the range $\widetilde{N}_0 \pm \Delta N_0$. This can be done rapidly on a desk calculator by repeatedly adding R'_i to the lowest value of N_1. N_0 and N_1 can be displayed on separate registers. Record only those values of N_1 that are within $\pm \gamma_1$ of an integer. Record the corresponding trial values of N_0 also.

7. Repeat Step 6 for $i = 2, 3$. Skip all trial values of N_0 that have been eliminated at lower values of i. If inequality (3.98) holds, only one possibility for N_0 normally remains. This identifies N_1, N_2, N_3 as well as N_0. On rare occasions, the ratios λ_i/λ_0 may be unfortunately chosen. γ_i or ΔN_0 must then be decreased (or s decreased), or some new spectral line must be used.

3.6.7 Applications Remarks

3.6.7.1 Gas Mixtures When a mixture of gases is used, a very slight separation can be made either by selective absorption on chamber walls or dessicants or by differential diffusion through a valve. The amounts of CO_2 and O_2 in natural air are gradually changing[64]. Water vapor may be removed by freezing.

3.6.7.2 Liquids and Solids The procedures described above are easily modified for use on liquids and solids. For liquids short chambers A and B may be equipped with bellows that permit fluid to be squeezed from A to B and vice-versa. For solids plane-parallel samples may be inserted in the beams. A short gas cell in one beam may be used for scanning.

3.6.8. A Possible In-cavity Laser Technique

If the wavelengths λ_i are restricted to those simultaneously present in a single CW laser, it is possible (at least in principle) to obtain high accuracy by such means as that of beating two lasers together while introducing a gas sample into one of the laser cavities. With a gas cell more than a few centimeters long several axial modes will normally be present, thus greatly complicating the identification of the order numbers N_i. Also, when the fractions ϵ_i are measured to extreme accuracy, such effects as mode pulling and lateral refraction at the Brewster-angle windows must be carefully assessed. Undoubtedly, such techniques will be developed.

3.7 REFERENCES

[1] A. Fox and T. Li, *Bell System Tech. J.*, **40**, 453–488 (1961).
[2] G. D. Boyd and J. P. Gordon, *Bell System Tech. J.*, **40**, 489–508 (1961).
[3] G. D. Boyd and H. Kogelnik, *Bell System Tech. J.*, **41**, 1347–1369 (1962).
[4] T. G. Polanyi and W. R. Watson, *J. Opt. Soc. Am.*, **54**, 449 (1964) .
[5] J. Kotik and M. G. Newstein, *J. Appl. Phys.*, **32**, 178 (1961).
[6] S. R. Barone, *J. Appl. Phys.*, **34**, 831 (1963).

[7] T. Li, *Bell System Tech. J.*, **44**, 917 (1965).
[8] C. M. Stickley, Physical Sciences Research Papers, No. 19, AFCRL–64–434, May 1964; Ph. D. Thesis, Northeastern University, Boston, Mass., June 1964.
[9] G. A. Campbell and R. M. Foster, *Fourier Integrals for Practical Applications*, Van Nostrand, Princeton, N.J., 1948, p. 15.
[10] R. E. Collin, *Field Theory of Guided Waves*, McGraw-Hill, New York, 1960, p. 470.
[11] R. J. Collins and J. A. Giordmaine, "New Modes of Optical Oscillations in Closed Resonators," in P. Guivet and N. Bloembergen (eds.), *Quantum Electron.*, Paris 1963 Conference, Columbia University Press, New York, 1964, p. 1239.
[12] E. Snitzer and H. J. Osterberg, *J. Opt. Soc. Am.*, **51**, 502 (1961).
[13] "Plates and Films for Science and Industry," *Kodak*, **P–9**, 4 1962.
[14] G. R. Harrison, R. C. Lord, and J. R. Loofbourow, *Practical Spectroscopy*, Prentice-Hall, Inc., Englewood Cliffs, N. J., 1948, p. 352.
[15] M. C. Adamson, T. P. Hughes, and K. M. Young, "The Effects of Temperature on Ruby Optical Laser Modes Sequences," *Quantum Electron.*, Proc. Third International Conference, Paris, 1964, pp. 1459–1468.
[16] J. C. Borie, M. Durand, and A. Orzag, "Photographie ultrarapide de la face de sortie d'un cristal de rubis fournissant l'emission 'laser'," *Compt. Rend.*, **253**, 2215–2217 (1961).
[17] M. Born and E. Wolf, *Principles of Optics*, Macmillan, New York, 1964, Sec. 8.3.3.
[18] G. L. Clark, R. F. Wuerker, and C. M. York, *A High Speed Photographic Study of the Coherent Radiation from a Ruby Laser*, STL Tech. Paper 9844–0035–RU 000. December 1, 1961.
[19] G. L. Clark, "Light Economics in High Speed Photography," *SPIE J.* **1**, 111, 116 (1963).
[20] W. C. Davis, "Maximizing Exposure-limited Resolution of Practical Rotating Mirror Cameras," *Appl. Opt.*, **3**, 1217–1222 (1964).
[21] W. C. Davis, "Exposure-limited Application of Kerr Cell Cameras," *App. Opt.*, **3**, 1215–1216 (1964).
[22] E. S. Dayhoff and B. V. Kessler, "High Speed Sequence Photography of a Ruby Laser," *App. Opt.*, **1**, 339–341 1962.
[23] E. S. Dayhoff, *The Emission Mode Patterns of Ruby Laser*, Proc. Tenth Colloquium Spectroscopicum International, Spartan Books, Washington, D. C., 1963.
[24] G. R. Hanes and B. P. Stoicheff, 'Time-dependence of the Frequency and Linewidth of the Optical Emission from a Pulsed Ruby Maser," *Nature*, **195**, 587–589 (1962).
[25] M. C. Kurtz, "A New Framing Camera," *J. SMPTE*, **68**, 16–18 (1959).
[26] J. S. Courtney-Pratt, "High Speed Photography and Micrography," *Appl. Opt.*, **3**, 1201–1209 (1964).
[27] W. J. Schenk, S. M. Hauser, and C. Lawler-Wilson, "Kerr Cell Modulation and Thermal Limitations," *EO Items*, **1**, 3–6 (1964).
[28] C. M. Stickley, "A Study of Transverse Modes of Ruby Lasers Using Beat Frequency Detection and Fast Photography," *Appl. Opt.*, **3**, 967–979 (1964).
[29] E. Reisman and G. D. Cummins, *Time-resolved Near and Far Field Patterns for a Rotating-Prism, Giant-Pulse Laser*, Paper FEI3 (post deadline), New York APS Meeting, January 1965.
[30] M. Shimazu, I. Ogura, A. Hashimoto, and H. Sasaki, *Opt. Masers* (Proc. Symp. Optical Masers), **13**, 406 (1963), Polytechnic Press, Brooklyn, N.Y.
[31] "RCA Phototubes and Photocells," *Tech. Man. PT–60*, 1963 p. 25.
[32] D. E. Gray (Coord. Ed.), *American Institute of Physics Handbook*, McGraw-Hill, New York, 1963, 2nd. ed., pp. 6–47.

[33] H. Kogelinik and W. W. Rigrod, *Proc. IRE* (corres.), **50**, 220, (1962).
[34] W. W. Rigrod, *Appl. Phys. Letters*, **2**, 51, 1963.
[35] H. Heinemann and H. W. Redlien, *Proc. IEEE*, **53**, 77 (1965).
[36] G. Kirchhoff, *Berl. Ber.*, **641** (1882); *Ann. Physik*, **18**,663 (1883).
[37] G. Watson, *A Treatise on the Theory of Bessel Functions*, Cambridge University Press, Cambridge, 1956, 2nd ed.
[38] V. Evtuhov and J. K. Neeland, in P. Grivet and N. Bloembergen (eds.), *Quantum Electron.*, Paris 1963 Conference, Columbia University Press, New York 1964, p. 1406.
[39] T. P. Hughes and K. M. Young, *Nature*, **196**, 332 (1962).
[40] A. J. DeMaris and P. Gagosz, *Appl. Opt.*, **2**, 809 (1963).
[41] G. E. Fenner and J. D. Kingsley, *J. Appl. Phys.*, **34**, 3204 (1963).
[42] J. D. Kingsley and G. E. Fenner, in P. Grivet and N. Bloembergen (eds.), *Quantum Electron.* Paris 1963 Conference, Columbia University Press, New York 1964, p. 1885.
[43] R. A. Laff, W. P. Dumke, F. H. Dill, Jr., and G. Burns, *IBM J. Res. Develop.*, **7**, 63–65 (1963).
[44] R. W. Waynant, J. H. Cullom, I. T. Basil and G. D. Baldwin, "Beam Divergence Measurement of Q-Switched Ruby Lasers, Westinghouse Electric Corp., Surface Division, Advanced Development Engineering, Baltimore, Md.
[45] J. P. Wittke, Z. J. Kiss, R. C. Duncan, and J. J. McCormick, *Proc. IEEE*, **51**, 61 (1963).
[46] P. P. Sorokin, J. D. Axe, and J. R. Lankard, *Opt. Masers*, (Proc. Symp. Optical Masers), 13. 487 (1963) Polytechnic Press, Brooklyn, N.Y.
[47] G. Lucovsky, M. E. Lasser, and R. B. Emmons, *Proc. IEEE,* **51**, 166 (1963).
[48] S. Saito, K. Kurokawa, Y. Fujii, and T. Kimura, *Opt. Masers*, (Proc. Symp. Optical Masers), **13** (1963), Polytechnic Press, Brooklyn, N.Y.
[49] Same as 47.
[50] D. R. Herriott, *J. Opt. Soc. Am.*, **52**, 31–37 (1962).
[51] M. Birnbaum and T. L. Stocker, *Appl. Phys. Letters,* **3**, 165 (1963).
[52] R. Paananen, C. L. Tang, and H. Statz, *Proc. IEEE,* **51**, 63 (1963).
[53] D. F. Nelson and R. J. Collins, in J. R. Singer (ed), *Advances in Quantum Electronics*, Columbia University Press, New York, 1961, p. 80.
[54] D. Hellerstein, *Appl. Opt.*, **2**, 801, (1963).
[55] J. D. Boardman, Wright-Patterson AFB, Ohio, private communication.
[56] K. E. Erickson, "Investigation of the Invariance of Atmospheric Dispersion with a Long-Path Refractometer," *J. Opt. Soc. An.* **52**, 777, (1962).
[57] K. E. Erickson, "Precise Determination of Atmospheric Dispersion with Applications to Long-path Interferometry," *ONR Progress Rep.*, Contr. **248**(01), 1961.
[58] Karl-Filip Svensson, "Measurements of the Dispersion of Air for Wavelengths from 2302 to 6907Å," *Arkiv Fysik*, **16**, 361 (1960).
[59] D. H. Rank, G. D. Saksena, and T. D. McCubbin, Jr., "Measurements of the Dispersion of Air from 3651 to 15,300 Å," *Opt. Soc. Am.*, **48**, 455 (1958).
[60] D. J. Schleuter and E. R. Peck, "Refractivity of Air in the Near Infrared," *J. Opt. Soc. Am.*, **48**, 313 (1958).
[61] V. P. Kovonkevich, "Dispersion of Air in the Visible Region of the Spectrum," *Opt. i. Spektrosk opiya*, **1**, 85 (1956) (in Russian).
[62] P. G. Guest, "A Differential Refractometer of High Sensitivity," *Austral. J. Phys.*, **8** (1955).
[63] Bengt Edlen, "The Dispersion of Standard Air," *J. Opt. Soc. Am.*, **43**, 339 (1953).
[64] E. Glueckauf, "Composition of the Atmosphere," *Compendium of Meteorology*, 1951, p. 3.
[65] H. Barrell, and J. E. Sears, Jr., "The Refraction and Dispersion of Air for Visible Spectrum," *Phil. Trans. Roy. Soc. (London)*, **238**, (1939).
[66] J. K. Sharp and E. Q. Vaher, *Investigation of Optical Frequency Translation Techniques*, *Tech. Documentary Rept. AL TDR 64–226*, Wright-Patterson, Air Force Base, Ohio.

3.8 PRINCIPAL SYMBOLS, NOTATIONS, AND ABBREVIATIONS

Symbol	*Meaning*
$A\ (r)$	Amplitude of the electric field at a radial distance r
a, a_1, a_2	Radius of beam, radius of apertures, see text
B_n	Noise bandwidth
b_1, b_2, b'	Mirror radius of curvature
C	Capacitance
c	Velocity of light
D	Axial distance
d	Mirror spacing
d'	Summation of axial length in a composite resonator
E	Electric field
e	Base of natural logarithms
F	Focal length of lens
f	Fractional energy loss per transit
f_g/number	Lens speed for transmitting a Gaussian beam
f/number	Measure of lens speed for Airy disc diffraction limitation
g	Mirror radius
g_i	Magnetic splitting factor
H	Magnetic field intensity
H_m	Hermite polynomial of order m
h	Planck's constant
I, I_i	Intensity, specific value thereof, see text
J_n	Bessel function of nth order
J'	J'_n Derivative of J_n
j	$\sqrt{-1}$
k, k_i	General constant, see text
L	Inductance, length, see text
L^l_p	Associated Laguerre polynomials
m_i	Component of angular momentum in the direction of the magnetic field
N	Fresnel number, total noise power, number of excited ions, number, see text
n, n_1	Index of refraction, specific value, see text
P	rms power in signal
$P\ (y)$	Rotation of the electric vector in a Senarmont polariscope
P_{nm}	mth root of nth-order Bessel function
Q	Resonator quality factor $\Delta v/v$
Q	Mirror concavity factor, departure of mirror from flatness $= \lambda/Q$
R_n	Radius of ring pattern number n

r, s, q	Integers
r, θ, z	Cylindrical coordinates
S	Power in signal frequency
u	Dummy variable, see text
W	Width of a prism
w_s	Radius of beam waist (spot size)
w_1, w_2, w_s	Radius of beam waist at a mirror (spot size)
x, y, z	Rectangular coordinates
x_m	Separation of lobes in TEM_{001} beam pattern
Z_0	Characteristic impedance of an electrical filter
β	Bohr magneton, dummy variable, see text
Δf_t	Transverse beat frequency
$\Delta\lambda$	Increment in λ
$\Delta\phi_G(v_0)$	Phase-shift change at the oscillating frequency
$\delta(\lambda, T, P)$	Refractivity at wavelength λ, temperature T, and pressure P
δ	Relative dispersion
ϵ	Dielectric constant
ϵ_r	Dielectric constant of ruby
η	Quantum efficiency
θ	Plane angle
λ	Wavelength
λ_0	Free-space wavelength
μ	Permeability of free space, see text
v_0	Oscillation frequency
v_c	Line width of half maximum
v_v	Live center
ξ	Dummy variable, see text
ρ, θ	Polar coordinate radius and angle
τ	Spontaneous emission lifetime
τ_c	Photon lifetime in resonator
ω	Normalized dummy variable, see text
ω, ω_i	Angular frequency, particular frequency, see text
ψ, ψ_n	Phase

4

Measurement of Energy and Power

4.0 LASER RADIOMETRY

Radiometry has always been a field of semiprecision measurements. Scientific personnel, skilled in microwave-power measurements, who transfer their attentions to laser measurements soon find that measurements of energy and power in the ultraviolet through infrared portions of the spectrum that exceed 5 percent accuracy are considered good. Accuracies of 1 percent are seldom achieved.

Measurement of radiant energy and power in the ultraviolet, visible, and infrared portions of the spectrum has been a subject of constant study and development for a great many years[1] to [6]. The advent of lasers has introduced some additional problems as well as some simplifications into the subject. Simplifications have been provided by the nearly monochromatic radiation produced by most lasers. As the response characteristics of most detectors do not change appreciably over narrow spectral ranges, the data-reduction problem is made simpler. Furthermore, because only narrow-band radiation is to be measured, it may be possible to use narrow-band filters with some types of detectors, thereby reducing some sources of error and external noise, and reducing re-radiation losses. Some complications are, of course, also evident. Compared with most thermal light sources, much higher energy and power densities can be obtained with lasers; careful use of many detectors is therefore necessary if one is to avoid saturation or radiation damage. The extremely short pulses that can be produced with lasers require correspondingly fast detectors and associated fast pulse instrumentation to measure instantaneous power. A good deal of effort has been spent on perfecting reliable techniques to overcome these complications, much of which will be described below. In addition to the materials presented herein, the reader is referred to several surveys of the various techniques in common use that have already been published[7] to [9].

Although energy (time-integrated measurements) and power (time-resolved measurements) are distinctly different but related quantities, the terms are, unfortunately, sometimes loosely used. This is undoubtedly because some of the techniques that have been traditionally used to measure these quantities from black-body radiators can be used to measure either one. For instance, although most photoelectric detectors and radiation thermopiles are basically designed to measure instantaneous power, they can also be used to measure the total energy in a pulse by integration, provided the instrumental time constants are much shorter than the pulse lengths involved. Conversely, many calorimeters and virtually all photographic methods are basically total-energy measuring techniques, but they can also be used for measuring power if the time history of the radiation is known or measured independently. Although both power and energy measurements are treated in this chapter, the distinction between them will be specific, and the reader is cautioned to keep this difference constantly in mind.

This chapter cannot treat in detail all of the technology that has been developed for infrared, optical, and ultraviolet energy and power measurements. Outlined herein are various techniques available, their advantages and limitations, the emphasis being placed on those that have been found to be particularly useful for measuring laser radiation. The first portions of this chapter will provide the reader a broad entré into the extensive literature on the sensors or radiation transducers. Later portions of the chapter will treat specific device characteristics that affect the precision and accuracy of measurement. Included will be sensitivity, spectral response, power and power-density limits imposed by saturation or radiation damage, and accuracy of calibration and capability of being calibrated. Treatment of the measurements problem is preceded, for clarity, with unit and term definitions.

Laser radiation is detected as a manifestation of qualitative or quantitative effects produced by the interaction of electromagnetic radiation with a material body. A detector or transducer converts, for display and measurement, the electromagnetic energy into a different form, usually electrical, thermal, or mechanical. In general, the electromagnetic flux from a laser is converted into an electrical current, which is converted to an observable quantity through electronic apparatus. There are, however, many power and energy detectors that depend upon producing observable results with no electronic circuitry whatsoever.

Because electromagnetic radiation may be considered as discrete quanta, we are led to the concept of a perfect detector as one in which each quantum of energy produces a measurable effect. At high-energy fluxes the discrete nature of quanta averages out to a continuous signal. An ideal quantum counter with 100 percent quantum efficiency produces a measurable output for each quantum and produces no output in the absence of quanta. Practical detectors of this type are basically quantum counters. Their performance deviates from the ideal, in that detection efficiencies which vary with wavelength are less than 100 percent efficient. Even in the absence of quanta spurious counts are detected from various noise sources. Noise pulses are seldom distinguishable from the real signal, even when the statistical characteristics of the signal and noise are considered.

4.1 DEFINITIONS AND UNITS

The metric system of units is used to express all quantities, using either the Systeme Internationale (MKSA) or the centimeter-gram-second (CGS) system of units, in accordance with the latest recommendations of the Committee on Fundamental Constants of the National Academy of Sciences—National Research Council[10], [11] and is consistent in nearly all respects with the recommendations of the International Organization for Standardization (ISO) and the International Electrotechnical Commission (IEC). The units commonly used in photometry, that is, candles, lumens, and lamberts, are adequate for the purpose intended[12] (namely illumination for the average human eye). Terms more suitable for radiometry include the description of radiation in absolute units. The radiometric quantities more appropriate for the physical description of laser radiation will be defined below.

4.1.1 Radiant Energy

The standard unit of radiant energy in the MKSA system is the Joule, and in the CGS system it is the erg. Conversion factors between these and other commonly employed units of energy are give below:

1 joule (J) $= 10^7$ ergs

1 thermochemical calorie (cal_{th}) $= 4.1840$ J $= 4.1840 \times 10^7$ ergs

1 international steam table calorie (cal_{IT}) $= 4.1868$ J $= 4.1868 \times 10^7$ erg

1 electron volt (eV) $= 1.6021 \times 10^{-19}$ J $= 1.6021 \times 10^{-12}$ erg

(4.1)

4.1.2 Radiant Power

Radiant power is defined as the energy flux or the rate of change of energy with time. The basic unit of power P in the MKSA system is the Joule per second, or watt, and in the CGS system it is the erg per second. Various conversion factors are given below:

$$1\ J/\text{sec} = 1\ \text{W} = 10^7\ \text{erg/sec} = 685\ \text{lm} \quad \text{at } 5550\text{Å}$$
$$= 107/\pi \quad \text{spherical cp at } 5550\text{Å}. \tag{4.2}$$

4.1.3 Radiance and Spectral Radiance

Radiance N is a measure of the "brightness" of a radiating surface; it is defined as the radiant energy emitted in a specified direction per unit time, per unit projected area of surface, per unit solid angle. Radiance is measured in watts per steradian per square centimeter (or square meter). The radiance of a source may vary with wavelength and the "spectral radiance" N_λ is defined as the radiance per unit wavelength interval.

4.1.4 Irradiance and Spectral Irradiance

Irradiance H is a measure of radiant power incident upon a surface; it is defined as the radiant energy per unit time per unit area incident upon the surface. Irradiance is measured in watts per square centimeter (or square meter). The irradiance on a surface may also vary with wavelength, and the "spectral irradiance" H_λ is defined as the irradiance per unit wavelength interval.

4.1.5 Emissivity and Spectral Emissivity

The emissivity e of a surface is defined as the ratio of the power radiated by a hot surface to that radiated by a black body at the same temperature. The emissivity of a surface may vary with wavelength, and e_λ, the "spectral emissivity," is defined as the emissivity per unit wavelength interval.

4.1.6 Relationship of Radiometric to Photometric Quantities

Photometric (visual) quantities are extricably related to radiometric (physical) quantities through standards based upon the spectral response of the eye. Photometric quantities can be derived from radiometric data by integrating the product of the spectral distribution of the radiation and the spectral response of the eye; for example, the photometric equivalent of radiometric spectral irradiance H is illuminance E the effective incident radiation-flux density required to produce a given visual response

$$E = K \int H_\lambda Y_\lambda \, d\lambda \ \ \text{lm/cm}^2, \tag{4.3}$$

where K = 685 (to yield E in lm/cm^2)
= 635,000 (to yield E in lm/ft^2)

and Y_λ is the spectral-response function of the eye. The total luminous flux F is similarly related to the total radiant power P through the eye-response function as

$$F = K \int P_\lambda Y_\lambda \, d\lambda \ \text{lm}. \tag{4.4}$$

Similarly the total radiant intensity of a point source I_r (watts per steradian) is related to the total luminous intensity I_L (lumens per steradian or candlepower) as

$$I_r = K \int (I_L) Y_\lambda \, d\lambda \ \text{W/s}. \tag{4.5}$$

Finally, for a broad source, the total radiance B_r (Watts per square centimeter per steradian) in radiometric units is related to the total luminance or brightness in photometric units (candle power per square centimeter, lumen per square centimeter per steradian) as

$$B_r = K \int (B_{L\lambda}) Y_\lambda \, d\lambda \ \text{W/cm}^2\text{/s}. \tag{4.6}$$

4.1.7 Source Classification and Interrelation

Two kinds of sources are of interest in radiometric measurements: point sources and extended sources. A point source emits with uniform radiant intensity independent of angle. Because ideal point sources do not exist, a satisfactory approximate model is required. We consider a point source as one whose dimensions are less than 10 percent of the distance between the source and the detector and as one that radiates the same radiant flux in all directions.

If a detector of area A is located a distance L from a point source of intensity I_r(W/sr), the power received by the detector will be

$$P = \frac{I_r A}{L^2}, \tag{4.7}$$

and the irradiance at an angle θ with respect to the common axis will be

$$H = \frac{I_r}{L^2} \cos \theta. \tag{4.8}$$

The total flux output or power will be just

$$P = 4\pi I_r. \tag{4.9}$$

Extended sources radiate from one side into a hemisphere. If it is assumed that the source is Lambertian (though few extended sources are), the flux in a given direction is proportional to the cosine of the angle with respect to the source normal. If the extended source is of radius R and the detector is located L units away, the radiometric source intensity B_r will cause a flux density at the detector of

$$H = B_r\left(\frac{\pi a^2 R^2}{R^2 + L^2}\right), \tag{4.10}$$

where the total output flux of the source is

$$P = \pi^2 B_r L^2. \tag{4.11}$$

4.2 CALORIMETRIC METHODS

4.2.1 General Principles

Calorimetric methods of measuring energy and power are defined as those in which radiant energy is absorbed, converted to heat, that creates either a temperature rise in the absorber or change in phase in a part of the measuring instrument[13], [14]. Temperature rise will result in a change in some measure, which can be sensed directly as a temperature change or ascertained indirectly by monitoring changes in the volume or pressure, or some other characteristic, of the absorber. Phase changes can be measured by monitoring the relative amounts of each phase in a two-phase system. Meaningful calorimetric techniques are reversible in the sense that no permanent changes should occur in the absorber, and they can all return to their initial condition in the calorimeter's equilibration time.

Many calorimeters can measure either energy or power; for instance, continuous-flow calorimeters are normally designed to measure the average power from continuous sources or pulsed sources having a high duty factor (pulse-width, pulse-rate product approaching unity). Conversely, other calorimeters are designed to determine the total energy in a pulse by measuring the temperature rise resulting from the absorption of the radiant energy in a known thermal mass. As in all calorimetry, heat losses by conduction, reflection, radiation, and convection must be minimized, or carefully controlled and calibrated, and the time constants involved for thermal equilibrium must be known.

Calorimeters primarily designed for pulse energy measurements may also be used for continuous (CW) power measurements in some cases[15], [16]. If the equilibration time of the calorimeter is short compared with the time required to establish thermal equilibrium between the calorimeter and its environment, the calorimeter may be used directly to measure power. A measure is made of the initial rate of the temperature rise when the CW source is instantaneously started. The rate of temperature rise will be proportional to the rate of energy from the source, or to the CW power. More rigorously, a knowledge of the transient response of a calorimeter to an impulse function may be used to predict the response to a step function and hence to determine the response when suddenly exposed to a CW power source. A caution is in order. The transient response of some calorimetric devices, such as radiation thermopiles, is not uniform over the absorbing surface[17]. This can lead to large errors if the distribution of illumination of the calibration source over the absorber is not identical to that of the source being measured.

4.2.2 Temperature Sensing

The energy absorbed by a calorimeter can be measured by monitoring a temperature change, a volume change, a pressure change, or a change in some other characteristic, such as the electrical resistance of the absorber. The latter may be solid, liquid, or gaseous. All techniques have been successfully used, although temperature sensing is most common.[18], [19]. Calorimeters may be designed to measure kilowatts of average power in continuous-flow types, to the microwatts per gram required for monitoring radiation dosage[20], or the nanowatts of power measured by sensitive radiation thermopiles or Golay cells[6].

There are many factors that must be considered for precision calorimetric measurements. These include changes in the heat capacity, coefficient of expansion, resistance or emissivity with temperature, or changes in the thermal losses with changes in surroundings. Because these are common to all calorimetric measurements, not just to laser measurements, and are thoroughly covered elsewhere, they will not be treated here[13], [14], [19].

There are special problems that arise in the measurement of extremely short pulses that can be achieved with Q-switched lasers. For example many calorimeters are designed so that the energy absorbed at the surface is transferred to the rest of the mass by conduction. If the energy is applied in a time that is short compared to this thermal relaxation time, the instantaneous surface temperature will be much higher than it would be if the same energy were applied in a longer pulse. Because the reradiation losses are proportional to $(T^4 - T_0{}^4)$, which T is the temperature of the surface, and T_0 is the equilibrium temperature in degrees Kelvin, it is apparent that in precision measurement the calorimeter must be designed to have a small temperature rise and fast equilibration time to ensure extremely small reradiation losses if energy measurements are to be essentially independent of pulse length.

It is important that no permanent or irreversible damage be done to the detector as the result of the rapid surface heating. In general, irreversible changes absorb energy that does not necessarily manifest itself as a temperature rise of the absorbing mass. However, some calorimeters have been proposed in which the heat of fusion of the thin layer on the surface will limit the surface temperature rise, and this heat will be liberated to the rest of the thermal mass as it resolidifies. Most of the calorimetric devices for laser measurements that have been reported in the literature[8], [21] to [27] and that are in common use in many laboratories have been designed to operate with surface absorption, though they all have upper limits on the peak powers they can withstand without permanent damage. Some calorimeters utilizing surface absorption of the energy do not measure the change in temperature of the absorber with an independent sensor, but measure the change in some temperature-sensitive characteristics of the absorber, such as resistance [23] or pyroelectric current[28].

In liquids and solids that are most commonly employed in calorimeters, the amount of heat energy ΔQ required to raise the temperature of an absorbing mass m by an amount ΔT is given by

$$\Delta Q = c_p m \, \Delta T \tag{4.12}$$

where the proportionality constant c_p is defined as the specific heat of the absorbing material, at constant pressure, and is often expressed in calories per gram degrees or calories per mole degrees. The specific heat of most materials varies with temperature and pressure. In most cases, the changes with pressure are negligible as long as the measurements are conducted in normal laboratory environments. (This, of course, is not true in gases.) It is apparent that small values of the specific heat c_p and small absorbing masses m are required to maximize the sensitivity of the calorimeter.

A variety of techniques may be used to measure the temperature rise of an absorbing body in a calorimeter. Thermocouples and thermopiles, bolometers, or resistance thermometers are frequently used. The techniques required for the proper use of such devices are thoroughly covered in the literature[13], [18].

The risk of irreversible surface damage is avoided if a liquid calorimeter is used, for the laser energy is absorbed in the liquid bulk. Absorbing solutions must be chosen to avoid local boiling, corrosion, and changes of absorption coefficient (and thus calibration) with time. Small liquid calorimeters have been used that employ temperature sensing[29]. A temperature rise of the order of a degree is detected in the liquid with a thermistor. Calibration is achieved by dissipating a known amount of electrical energy in an immersion heater. Sensitivities of the order of 5 J/mV are achieved[29] with 10 kΩ thermistors using liquid volumes of the order of 1 cc.

A basic disadvantage of using the liquid calorimeter for thermal sensing is the long equilibration time required. The *e*-folding time of the order of two minutes is typical, for convection currents are required to establish temperature uniformity. Great care must be exercised to account for reradiation losses. A large calorimeter, required to cope with laser energies of more than a few joules, will require a longer delay before thermal equilibrium is reached; the cooling corrections will be greater. As shown below, volume sensing provides a more elegant method of energy measurement, for a virtually instantaneous measure of the bulk temperature rise is obtainable.

4.2.3 Volume Sensing

The attractive characteristics of the liquid calorimeter for high-energy, pulsed-laser work are retained when volume changes; they are exactly analogous to the common mercury or alcohol thermometer when they are used as a measure of total pulse energy. The change in volume ΔV of the absorber can be used to measure the absorbed energy, because $\Delta V = V\beta\Delta T$, where V is the volume of the absorber and β is the cubical thermal-expansion coefficient. Therefore, the equation above for heat energy becomes

$$\Delta Q = \frac{m}{V} \cdot \frac{c_p}{\beta} \Delta V = \rho \frac{c_p}{\beta} \Delta V \tag{4.13}$$

where ρ is the density of the absorber. Note that this result is *independent* of the mass or the volume of the absorber. A figure of merit by which to evaluate volume-sensed calorimeter liquids is the ratio of $\beta/\rho c_p$. Thus low densities, small specific heats, and large thermal expansion coefficients are required to maximize the sensitivity of the calorimeter. All of the equations above are approximations, but they are valid for small changes in temperature ΔT or volume ΔV.

An example of a liquid calorimeter is a container that is filled with copper nitrate dissolved in acetonitrile. The concentration of copper nitrate is adjusted so that the transmission through a 3-in. long cell is 10^{-4} of the incident energy at the ruby laser wavelength (0.694μ). Attached to the container is a capillary that is 0.004 in. in diameter and into which the liquid expands. Approximately 2.6 J will raise the liquid level by 1 in.[35]. A piston attachment allows the liquid level to be adjusted for ambient temperature changes. This technique is readily scaled to larger volumes for higher energy measurements.

4.2.4 Pressure Sensing

In contrast to solids and liquids, gases do relatively large amounts of work when heated at constant pressure. That is the specific heat of gas when heated at constant pressure c_p is substantially greater than the specific heat of the gas heated at constant volume c_v. In general

$$c_p - c_v = T\left(\frac{\partial p}{\partial T}\right)_V \left(\frac{\partial V}{\partial T}\right)_p. \tag{4.14}$$

Therefore, the design of a calorimeter that utilizes a gas either to sense the heat from an absorber, as does the Golay cell [6], or actually to absorb the radiation in the gas [2] must be carefully designed to take this into proper account.

4.2.4.1 Golay Cell The Golay pneumatic detector is best suited to detection of far-infrared radiation. The heart of this detector is a cell that contains a confined volume of gas that is heated by the energy absorbed [6*d*], [6*e*]. The absorbed energy raises the temperature of the confined working gas, and the resultant expansion distends a flexible membrane-mirror system that acts as an optical displacement detector. An aperture in the cell compensates for slow drifts in the ambient of the cell.

A typical Golay cell has an aperture of 3/32 in. in diameter and has a sensitivity of 6×10^{-11} W rms equivalent noise input with an instrument time constant of 1.6 sec. The Golay-cell sensitivity is limited by the Brownian motion of the flexible membrane, but the usable sensitivity of a Golay cell extends from the ultraviolet up the microwave range. A special gas filling can be used to reduce the response time to 600 μsec, with a corresponding loss in sensitivity.

The Golay cell, which has been used for nearly two decades for infrared spectroscopy, is being supplanted by liquid-helium cooled (4.2°K) indium-doped germanium photoconductors with response times of the order of 0.1 μ sec[6*f*].

The emissivity of a calorimeter must be high, or known, if accurate measurements are to be made. The design and construction of black-body absorbers with high emissivities have received a great deal of attention in the literature[30] to [32] and need little further evaluation here, except to mention that gold "black" has been found to have a very wide spectral absorption[33]; even the emission of photoelectrons from the gold in the vacuum under ultraviolet radiation does not appreciably affect its efficiency[34]. In general, the technique has been to so design the cavity that any radiation not absorbed on the first reflection within the cavity will be directed so that it will undergo many subsequent reflections within the cavity. Conical cavities appear to provide about the best performance for the least thermal mass, an important consideration in fast-detector design.

4.3 PHOTOELECTRIC METHODS

4.3.1 General Principles

Photoelectric methods for measuring optical energy and power include all radiation detectors in which the absorption of a photon produces an electrically measurable result, such as the emission of an electron or the creation of a hole-electron pair. Such detectors are quantum counters in the sense that the interaction involves a single photon of radiant energy, and the output is proportional to the number of photons rather than to their average energy of power, assuming constant quantum yield.

Any photodetector that gives an output proportional to input power can be used for energy measurement by integration of the detector output. The integration may be performed by a capacitor in which the change in voltage is measured by a high-impedance voltmeter. The capacitor may function either as a source or sink of charge.

4.3.2 Photoemission Devices

One large class of photodetectors includes those photoemissive devices in which an electron is emitted from a photocathode by the absorption of a photon. To escape from the photocathode the electron must have enough energy to overcome the work function of the surface. If the photon has more than the minimum amount of energy, the escaping electron may or may not absorb this excess energy by escaping at a higher velocity. The *maximum* kinetic energy of a photoelectron ejected from a pure metal is given by

$$E_m = \tfrac{1}{2}mv_m{}^2 = h\nu - \phi \tag{4.15}$$

where m = mass of the electron,
v_m = maximum escape velocity,
h = Planck's constant,
ν = frequency of incident photon, and
ϕ = work function of the metallic surface.

This single expression is only true at absolute zero. Actually the electrons in the surface possess thermal energies. Consequently there is an interaction term due to the associated thermal distribution function which causes some spreading out of the sharp cutoff predicted above. Pure metals are seldom used as photoemissive surfaces, primarily because their quantum yields are very low (the order of 0.1 percent), and their work functions are so high that only ultraviolet photons have enough energy to eject an electron.

Some semiconductors have very high quantum yields (up to 30 percent in come cases), and, in addition, the photon energy required to eject an electron

TABLE 4.1 TYPICAL COMBINATIONS OF PHOTOSENSITIVE SURFACES AND WINDOW MATERIALS WHICH CAN PROVIDE THE BASIC SPECTRAL-RESPONSE DESIGNATION STANDARDIZED BY E.I.A[a]

Spectral-response Number	Type of Photodetector	Photosensitive Material	Envelope
S-1	Photocathode	Ag – O – Cs	Lime-glass
S-2[b]		Obsolete	
S-3	Photocathode	Ag – O – Rb	Lime-glass
S-4	Photocathode	Cs – Sb	Lime-glass
S-5	Photocathode	Cs – Sb	UV-transmitting glass
S-6	Photocathode	Na	Unspecified
S-7	Photocathode	Cs – Rb – O – Ag	Pyrex
S-8	Photocathode	Cs – Bi	Lime-glass
S-9	Photocathode	Cs – Sb (semitransparent)	Lime-glass
S-10	Photocathode	Ag – Bi – O – Cs (semitransparent)	Lime-glass
S-11	Photocathode	Cs – Sb (semitransparent)	Lime-glass
S-12	Photoconductor	CdS (crystal with plastic coating)	Lime-glass
S-13	Photocathode	Cs – Sb (semitransparent)	Fused silica
S-14	Photojunction (photocell)	Ge	Lime-glass
S-15	Photoconductor (photocell)	CdS (sintered)	Lime-glass
S-16	Photoconductor (photocell)	CdSe	Lime-glass
S-17	Photocathode	Cs-Sb (reflecting substrate)	Lime-glass
S-18	Photoconductor (vidicon)	Sb_2S	Lime-glass
S-19	Photocathode	Cs – Sb	Fused silica
S-20	Photocathode	Sb – K – Na – Cs (semitransparent)	Lime-glass
S-21	Photocathode	Cs – Sb (semitransparent)	UV-transmitting glass
S-22	Presently unspecified		
S-23	Photocathode	Rb – Te	Fused silica
S-24	Photocathode	Na_2KSb	Lime-glass

[a]Reproduced with permission from Radio Corporation of America.

is often very much less than that required for metallic surfaces. Semiconductors enable the extension of the response of a surface into the visible, and in some cases, into the near infrared. Semiconductors are almost universally used in both photoemissive devices as well as some photoconductive devices. The designations of a number of photosensitive surfaces have been standardized by the E.I.A. and are shown in Table 4.1.

Photoemissive surfaces are used in a variety of widely available detectors, including vacuum phototubes, gas-filled phototubes, and photomultipliers. The techniques developed for using photoemissive detectors have been thoroughly studied and are widely available in the literature[36] to [43]. As a result of the progress in nuclear science, a great deal of effort has gone into the development of photomultiplier tubes for use in scintillation counters[44] where the important characteristics are high speed, high gain, large photocathode areas, and low dark currents.

The use of phototubes for measuring laser radiation had an early beginning[45] and is now very widespread[46] to [48]. For quantitative measurements the usual precautions must be observed; that is using the phototubes within their peak and average current limitations to avoid fatigue and regenerative or degenerative feedback, shielding the tubes against magnetic and electrostatic fields, and using accurately determined voltages from well-regulated power supplies to set phototube gain. The power output of many lasers is so large that suitable beam splitters or attenuators must be employed to reduce the beam intensity to safe levels.

For precise measurements the temperature of many photomultipliers must be controlled or known, for the gain can increase with temperature the order of 0.3 percent/°C for 1P21 tubes[41]. In addition, the quantum yield of photocathodes can increase slightly with temperature near the long-wavelength end of their spectral-response characteristic. The dark current from a photomultiplier increases rapidly with temperature. The illumination of the photocathode by a well-collimated beam with subsequent shadowing by supports[49] can lead to appreciable errors, as well as some possible transient effects on the gain observed in S-1 photocathodes[50], [51].

Some recent developments in photomultipliers promise substantial improvements in their performance for future applications[52] to [56*a*]. The incorporation of photocathodes in traveling-wave tubes has also led to some very fast, wide-band photodetectors[57], [58].

Much work is in progress to improve the quantum yield of photoemissive surfaces, as well as to extend the response into the infrared-wavelength regions over which the yield remains high, or to optimize the yield for a single wavelength. Multilayered, thin-film photoemitters to improve the quantum yield[59] and special techniques to trap the incoming radiation by total internal reflection in the glass face of existing photomultipliers so that the light makes several passes at the photoemissive surface[60] have been proposed. These devices increase the quantum yield from the photocathode, especially toward the long-wavelength end of the spectral-sensitivity curve.

Because the process of photoemission is extremely fast, the current from the photocathode can accurately follow the changes in intensity associated with even the fastest pulsed lasers. The spread in the transit times of the photoelectrons between the photocathode to the anode is the ultimate limit, although the capacity associated with external circuitry usually limits the response before this point is reached (unless great care is taken). The transit-time spread is, of course, reduced as higher voltages are applied across the photodiode tube. This spread is greatly increased by the use of additional stages of secondary-emission amplification, as is done in the use of in photomultipliers. Well-designed photomultipliers can have rise times of the order of a few nanoseconds, and photodiodes a fraction of a nanosecond, which is quite adequate for most requirements.

Because the instantaneous output of the phototube is proportional to the instantaneous intensity or power of the light incident on the photocathode, it is possible to measure the energy in the laser pulse by integrating the phototube output with respect to time[47], [48]. This can be done either by measuring the area under the output-versus-time characteristic or by integrating electrically. The maximum energy in a fixed pulse length that can be measured directly is limited by the power level, incident upon the photosensitive surface, at which response ceases to be linear. The limiting power level of most vacuum photocells is of the order of 1 W; hence the energy limitation is of the order of millijoules for millisecond pulse lengths. Calibration is usually obtained by reference to a calorimeter and absorber of known characteristics.

4.3.3 Photoconductive Devices

Photoconductive materials have found use in the measurement of laser power and energy[61]. In general the spectral-response characteristics of photoconductors can extend further into the infrared than photoemissive devices. The transient response of most commercially available photoconductive cells is much slower than that of photoemissive cells, although with suitable design, the transient response can be comparable to photoemissive devices[62] to [68]. This is fundamentally due to the fact that the production of an electron-hole pair by the absorption of a photon is virtually instantaneous, and the transient response is determined by transit-time effects and the circuit parameters of the photoconductor and associated circuitry. As in photoemissive devices, a dc bias is maintained across the device to collect the carriers freed by the absorption of a photon.

4.3.4 Photovoltaic Devices

Photovoltaic devices have been used for the measurement of laser energy and power[69], [70]. These devices are similar to photoconductive devices except that no external bias is required to separate the electron-hole pairs as they are created by photons. Instead, the junction of two different materials, such as a metal and a semiconductor or two different semiconductors, produces a contact potential. When illuminated, the carriers that are created and flow across the junction constitute the external current produced by the cell. The silicon solar cell is an example of a photovoltaic detector widely used to measure the output power of CW gas lasers in the 300-to-1200-nm wavelength range.

4.3.5 Other Solid-state Devices

The combination of photoconductive junctions with additional transistor-type junctions to produce additional gain has some very attractive possibilities[69], [71]. Additional gain has also been observed in simple junctions under certain conditions, due to an avalanche-like process[64], [64*a*]. The investigation of these devices, for use particularly in the far infrared, has been the subject of intensive study for many years, and a great deal of information is readily available in the literature on their relative merits[72] to [75*a*].

4.3.6 Photoionization in Gases

At sufficiently short wavelengths, where the energy per photon exceeds the energy required to ionize a gas, photoionization of gases has been employed for many years to measure the radiant energy from ultraviolet sources[76], [77]. Nitric oxide is a commonly used gas with broad spectral absorption, a large cross section, and low ionization potential[78, 79]. Other gases that have been employed for this purpose include hydrogen bromide and iodide, oxygen, carbon dioxide, carbon disulfide, acetone, ethylene oxide, and ethyl sulfide[76, 77]. In the vacuum ultraviolet down to about 40 nm, rare gases have been proposed[80]. By comparing the quantum yields of xenon, krypton, and argon, it is found that the quantum yield is unity within 1 or, at most, a few percent. For convenience, the first ionization potentials of the rare gases are listed in Table 4.2. Additional information regarding the photoionization cross sections of some of the rare gases is becoming available[76], [81], [83].

4.3.7 Fluorescence Detectors

When exposed to radiation in the ultraviolet and vacuum ultraviolet, many materials fluoresce, emitting photons often in the visible region where

TABLE 4.2 FIRST-IONIZATION POTENTIAL OF RARE GASES

Gas	Symbol	Electron Volts (eV)	First-Ionization Potential: Wave Number (cm^{-1})	Wavelength (nm)
Helium	He	24.581	198.27×10^3	50.44
Neon	Ne	21.559	173.89×10^3	57.51
Argon	Ar	15.755	127.08×10^3	78.69
Krypton	Kr	13.996	112.89×10^3	88.58
Xenon	Xe	12.127	97.82×10^3	102.23

they can be readily measured with more conventional detectors[84], [85]. By suitable choice of material with a high quantum yield and by careful preparation and handling it is possible to make absolute measurements with high reproducibility. For maximum utility, such phosphors or "wavelength converts" should have a high and constant quantum yield over as wide a spectral region as possible, a yield that should be independent of intensity (within the limits of use), and independent of small changes in temperature. The emission spectrum of the phosphor should also be independent of the exciting wavelength, and the material should be transparent to this emission spectrum. Sodium salicylate[80], [84] to [90] has been widely studied for this purpose because of its wide absorption bands from at least 40 to 360 nm, its durability and reproducibility, and the fact that its fluorescent output is mostly between 370 and 520 nm, a region in which many photoemissive surfaces are particularly sensitive.

Another useful phosphor, dihydroxy-naphthaldiazine, known as liumogen, with an absorption spectrum extending from at least 90 to 460 nm in the visible spectrum, has been studied by Kristianpoller and Dutton[91]. Its emission spectrum extends from about 510 to 620 nm in the yellow-green portion of the visible spectrum.

4.4 PHOTOCHEMICAL METHODS

4.4.1 General Principles

This category of power and energy measuring devices includes photographic and actinometric methods. Both are essentially total-energy measuring techniques, although optical power can be measured in some cases by various auxiliary streaking or timing devices. The output of a CW laser can be measured, for example, by exposing the photochemical detector to the laser output for a known time interval. The power-versus-time characteristics can be measured by sweeping a line or point image of the source across a film, a practice commonly called "streak" photography.

4.4.2 Photographic methods

The use of photosensitive materials as a semiprecision method of recording optical energy is perhaps the most widespread technique today. Because a wealth of readily available information is normally kept in any library or well-equipped photographic laboratory[92], the brief outline of the characteristics of photographic materials given here is hardly sufficient to do more than define the common terminology. (A useful discussion of the evolution of performance criteria for photographic as well as other types of photodetectors was recently published by Zweigu[93]).

The mechanism by which photosensitive emulsions respond to optical energy is generally characterized by a single curve that depicts the optical density of the emulsion versus the logarithm of the exposure, or the amount of radiant energy falling on the film. A typical characteristic curve is illustrated in Figure 4.1. Many variations of this curve are possible, for the curves are a function not only of the particular emulsion, but also of the developer, the development time, and the spectral characteristics of the light source. The optical density of the emulsion is defined as the logarithm of 1 over the fraction of light transmitted through the emulsion. Note that for small exposures, no effect is observed, and the slope of the characteristic curve is zero. With increasing exposure, the slope gradually increases until it reached a maximum value defined as the gamma γ of the film. In general, the slope of gamma of an emulsion increases with development time. The gamma remains approximately constant over a wide range of exposure that determines the "exposure latitude" of the emulsion. Film gamma saturates or flattens out with increasing exposure. In practice, the useful upper exposure limit is often set by increased graininess and loss of definition well before the upper limit of film gamma is reached.

It has been assumed in the discussion of characteristic curves that the exposure along the abscissa of the figure is only a function of the total energy

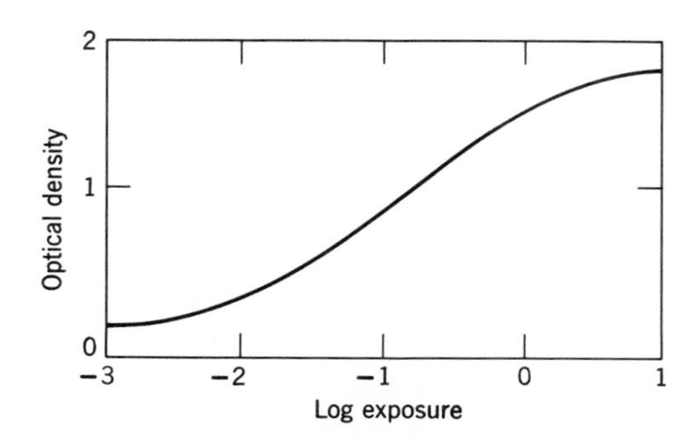

Figure 4.1. Characteristic curve for photosensitive emulsions.

of the light incident on the emulsion. Actually, this is not quite true, and many emulsions show a reduction in sensitivity to either very long, or to very short, exposures containing the same amount of total energy. This reduction in sensitivity is called "reciprocity failure"[94] and may be particularly important for emulsions used to measure or monitor the energy in pulsed laser beams. Many emulsions show a very broad maximum in sensitivity for exposure times in the vicinity of 0.01 to 1.0 sec.

In addition to the characteristics defined above, emulsions differ in their spectral sensitivity characteristics. There are three principal classes:

1. Noncolor-sensitized or "blue-sensitive" emulsions that respond to the ultraviolet and blue-violet portions of the spectrum characteristic of the silver halide.
2. Orthochromatic emulsions that are also sensitive to greens.
3. Panchromatic emulsions that are sensitive to the entire visible spectrum in addition to the ultraviolet.

In addition, there are a number of specially sensitized emulsions, such as infrared films, that extend the response out to beyond 900 nm.

4.4.3 Actinometric Methods

Chemical actinometers have been used by photochemists for many years to measure total energies in the ultraviolet portion of the spectrum. The techniques for using the various actinometers have been highly developed, and energy measurements with precisions of 1, or a few, percent are reported. (Although no very high-power or high-energy lasers have yet been reported in the ultraviolet, it is probably only a matter of time until they are.) In all actinometers the absorption of a quantum of radiation results in a specific reaction with a known quantum yield. The number of quanta absorbed can then be inferred from measurements of the reaction products or of the remainder of the unreacted absorbing material. The characteristics of a useful actinometer include the following:

1. Small variation of quantum yield and high absorption over the wavelength band of interest (for most purposes, the wider the absorption band the better).
2. High sensitivity and precision.
3. Small variation of quantum yield with temperature.
4. Reaction rate linear with intensity.
5. Simple reaction with nonabsorbing, stable photolysis products.

One of the most sensitive and widely used actinometers is based on the photolysis of acidified solutions of potassium ferrioxalate[95–98]; it is useful through the ultraviolet out to wavelengths as long as about 510 nm. Another widely used actinometer, particularly for large energies, is a solution of uranyl oxalate, useful from about 200 to 440 nm [99], [100]. The concentrations are critical and the temperature of the solution must be known or controlled. For intensities below about 10^{13} quanta sec and reasonable exposure times, the malachite green leucocyanide actinometer is useful[101], [102].

A number of gaseous systems have been used as actinometers[100] by photochemists, including hydrogen bromide, hydrogen iodide, oxygen, and carbon dioxide. These are discussed in Section 4.3 under photoelectric methods.

4.5 MECHANICAL METHODS

4.5.1 General Principles

Mechanical techniques for measuring laser energy or power are based upon the fact, as originally predicted by Maxwell, that light exerts pressure on any surface that absorbs or reflects the light. The first experimental measurements of this pressure were performed more than 60 years ago by Nichols and Hull[103] and also by Lebedew[104]. Mechanical measurements are based upon the fact that the energy in a photon is given by

$$h\nu = Mc^2$$

where h = Planck's constant = 6.626×10^{-34} J sec;
ν = frequency Hz;
c = velocity of light = 2.9979×10^8 m/sec;
M = apparent equivalent mass of photon kg.

Therefore [105] a momentum of $Mc = h\nu/c = h/\lambda$ can be associated with each photon, where λ is the wavelength.

4.5.2 Radiation Pressure Devices

When a photon is absorbed at a surface, the momentum of the photon is transferred to the surface. If the photon is perfectly reflected from the surface, the net momentum transferred to the mirror is transferred to the surface. If the photon is perfectly reflected from the surface, the net momentum transferred to the mirror is given by $(2h\nu/c)\cos\theta$ where θ is the angle of incidence measured with respect to the normal to the mirror. The average force F on the mirror is normal to the surface and is given by $F = n(2h\nu/c)\cos\theta$ where n is the average number of photons per second incident of the mirror. Because the power in the optical beam P is given by $P = nh\nu$, this average force can also be expressed as

$$F = \frac{2P}{c}\cos\theta, \tag{4.17}$$

which is independent of the wavelength. We can see that the fluctuations in this force are exactly identical to the statistical fluctuations associated with the "steady" stream of photons. To appreciate the order of magnitude of these forces, it is interesting to note that it takes a total beam power of 1.5 MW to exert a force on the mirror of equal to the weight of 1 g when the beam is normal to the mirror.

For a pulse of photons of length Δt, the total number of photons in the pulse is given by

$$N = n\Delta t, \tag{4.18}$$

and the total momentum transferred to the mirror is given by

$$F\Delta t = \frac{2P\Delta t}{c}\cos\theta = \left(\frac{2h\nu N}{c}\right)\cos\theta = \frac{2E}{c}\cos\theta, \tag{4.19}$$

where E is the total energy in the pulse.

For the case of partial reflection and partial absorption the situation is not quite as simple, for the net force on the imperfect mirror is not normal to the surface for $\theta = 0$. This is due to the fact that the momentum transferred from absorbed photons is in the direction of propagation. If part of the radiation is also scattered by the imperfect mirror, still further complications arise.

To measure the output of continuous-light sources, steady deflections of a torsion pendulum *in vacuo* of the order of 15°/mW have been achieved for beams smaller than about 1 cm² in cross section[106]. By chopping the light source at the resonant frequency of the torsion pendulum, much higher sensitivities could be achieved in principle. This technique is used in some simple experiments to measure radiation pressure in air[107]. To measure the pulse output of lasers it is customary to use the torsion pendulum as a ballistic pendulum; that is the pulse length must be much smaller than the natural period of oscillation of the pendulum [108] to [110].

The impulse produced in the photon ballistic pendulum by the change of photon momentum on reflection can be measured by means of a torsion pendulum with a mirror mounted on one or both ends of the crossarm. The deflection of the crossarm γ for light of energy E incident normally on a single mirror is given by

$$\gamma = \frac{2RE}{c(IK)^{1/2}} \tag{4.20}$$

where R is the radius of the cross arm,
K is the torsional stiffness,
I is the moment of inertia of the system.

Angular deflections of the order of 1 to 2 cm/J have been obtained in ballistic pendulums when a 200 cm optical level arm is used to amplify the deflection. The ballistic-pendulum energy limitation is determined by the power threshold for mirror damage.

Ballistic-pendulum techniques have the advantage that radiation can, in principle, be measured in terms of the fundamental mechanical and physical characteristics of the torsion pendulum. Absolute calibration is possible because all quantities can be measured and controlled quite precisely. These techniques have the additional advantage that only a small fraction of the laser beam is absorbed in the process, the remainder being available for other uses. The many experimental difficulties associated with most reported instruments, however, and the skill required of the experimenter, makes it seem unlikely that these techniques will be developed for routine laboratory measurements in the very near future.

4.6 NONLINEAR FIELD METHODS

4.6.1 General Principles

At the low field strengths or photon densities normally associated with conventional incoherent light sources, the dielectric constant or index of refraction of most dielectric materials remains essentially constant, independent of field strength. At the very high field strengths or photon densities that can be achieved with high-power lasers, however, this approximation is no longer valid, and higher order terms in the polarizability of the medium must be included. These nonlinear effects have stimulated a great deal of interest and activity in the scientific community, both theoretically and experimentally, and the published literature on the subject is growing at an enormous pace[111] to [115]. These effects have rapidly expanded from the first weak detection of the second harmonic of a ruby laser in 1961 to the very efficient (10 to 30 percent) conversion to second harmonic frequencies, the detection of third harmonics and the dc component (optical rectification), Raman scattering and now Raman lasers, and a variety of multiple-frequency parametric effects[116].

It is therefore not at all surprising that some of these effects might prove to be useful for measuring the energy or power of a laser beam. In particular, the dc component that is generated in crystalline materials without centers of symmetry (noncentrosymmetric) have been proposed as a method for monitoring the power in a laser beam passing through the crystal[117]. This effect is normally associated with the generation of even-order harmonics in a nonlinear material, but there is also a close connection between the dc component and the linear electro-optic effect (Pockel's effect) in the crystal[115], [118], [119].

When an electric field is applied to a crystal that does not have inversion symmetry, the resulting polarization can be represented by a polynomial that comprises first-, second-, and higher-order terms of the electric-field strength. When an electromagnetic wave is passed through such a crystal the expression for the polarization includes a dc term that is proportional to the square of the applied-field strength and that is therefore proportional to the power in the transmitted wave.

If a plane-polarized laser beam is passed through a quartz crystal in the direction of the z axis, and the direction of the electric field vector is parallel to the x axis, polarization will occur parallel to the y axis, and measurement of the voltage produced across the crystal in this direction will be an indication of the power in the laser beam[117]. There is no dc polarization of the crystal if the laser beam is circularly polarized, but if the beam is randomly polarized, power measurement can be achieved by separating the beam into two spatially orthogonal components and measuring each separately.

As long as the crystal embraces the whole of the laser beam the output voltage does not depend on the degree of focusing the crystal. For a given crystal size therefore the upper limit for power measurement will be determined by the dielectric strength of the quartz, namely about 10^{12} W/cm^2. At the lower limit powers of the order of 10^3 W can readily be measured. Calibration is necessary if conventional calorimetry techniques are used. By integration of the output signal from the quartz detector a measurement of the laser energy can, of course, be obtained.

4.7 ENERGY-DENSITY MEASUREMENTS

4.7.1 General Principles

In some cases, it is not simply a measurement of the total energy that is required, but a measurement of the energy distribution over the target area, that is, energy per unit area. This may mean the measurement of the energy distribution across the exit aperture of a laser in the near zone, the measurement of the energy density in the focal plane of a lens, or the measurement of the far-field pattern of a laser beam.

Measurements are made in three regions, the *near, Fresnel*, and *far or Fraunhofer* zones of a radiating aperture of diameter D. The *far-field region* is defined as that region in which the angular distribution remains essentially independent of distance from the aperture, beginning at a distance variously taken from $D^2/2\lambda$ to $2D^2/2\lambda$, depending on the accuracy required, and extending to infinity. This inner boundary line is not, of course, a sharp boundary, and the fields are described by neglecting all but the most slowly decaying terms in range in the diffraction integral. At a distance $D^2/2\lambda$, sometimes called the *Rayleigh range*, the maximum phase error from a point in space to different points across the aperture is $\lambda/4$, whereas at $2D^2/\lambda$ the maximum phase error is $\lambda/16$. This far-field region is often called the *Fraunhofer region*[120] to [123] because Fraunhofer diffraction effects are observed in this region and calculated with the same mathematical approximations.

The region nearer the aperture is most frequently called the *Fresnel zone*, or *Fresnel region*. Fewer approximations can be made in describing the fields in this region, although several simplifications can still be made in the analyses. This region extends from the far-field boundary defined above towards the aperture to another boundary defined approximately at the range $D/2\ (D/\lambda)^{1/3}$, where, again, phase errors of the order of $\lambda/16$ are encountered[122]. This inner boundary varies from about one aperture diameter for apertures 10 wavelengths in diameter to about 11 aperture diameters for apertures 10^4 wavelengths across. Table 4.3 summarizes the energy distribution in the various regions.

Between the inner boundary of the Fresnel region and the aperture, essentially no approximations can be made in describing the fields, and the region is normally termed the *near zone*. In this region, the fields can be very complex, and rapidly changing, with sharp shadow boundaries.

Because it is possible to observe far-field or Fraunhofer diffraction phenomena in the Fresnel zone[123], some confusion in terminology has resulted[120]. Because it is often desirable to measure far-field patterns without being in the far field, it is important to understand the application of this extremely useful method. Essentially, the technique involves placing a positive lens with focal length f so that it receives all of the laser radiation and observing the image in the focal plane of the lens; for instance, if the diameter of the half-intensity contour on the spot in the focal plane is d, the half-intensity, far-field angular beamwidth is given by $\theta = d/f$ in radians. For

TABLE 4.3 DISTRIBUTION OF ENERGY FROM A DIFFRACTION LIMITED, SPATIALLY COHERENT SOURCE

Zone of Measurement	Distance L from Aperture D	Phase Error Measured at L from Different Points on D
Near field	$L \sim D$	No approximations
Fresnel zone (near zone)	$L \sim \frac{D}{2}\left(\frac{D}{\lambda}\right)^{\frac{1}{3}}$	$\Delta\phi \sim \lambda/16$
Rayleigh zone	$L \sim \frac{D^2}{2\lambda}$	$\Delta\phi \sim \frac{\lambda}{4}$ Region of parallel beam
Fraunhofer zone (far zone)	$L \sim \frac{2D^2}{\lambda}$	$\Delta\phi \sim \frac{\lambda}{16}$ Region of diverging beam

lasers that are close to diffraction limited it is apparent that long focal-length lenses are required to produce adequately sized images in the focal plane and that the lens introduce negligible errors into the measurement. This last requirement is easily met because very simple lenses should be used to measure the far-field patterns from the best single-mode gas lasers.

The technique described above has been successfully used by a number of authors to study the angular far-field distributions from lasers[124], [125]. Far-field patterns can, of course, also be measured in the far field with no additional optics required, but this is only practical for relatively small apertures[126]. Fortunately, this is true for most gas lasers.

The technique for measuring the far field, described above, is also useful to measure the energy distribution in the focal region of a lens. It is, of course, possible to do this in principle simply by placing a suitable recording medium, such as film, in the focal plane and exposing it with greatly attenuated laser energy. The image is, however, in most cases too small to be very useful, and it is often necessary to monitor this energy distribution continuously. This can be done by sampling a portion of the beam without introducing appreciable errors, such as with an optically flat beam splitter, and recording the image at the focal plane of a long focal-length lens while focusing the remainder of the beam on the target with a shorter focal-length lens. The two images are then nearly identical, assuming that neither lens introduces appreciable errors, but the sampled image size is larger than the main beam spot size by the ratio of the focal lengths of the two lenses.

The most suitable medium for recording the energy distribution will depend, of course, on the wavelength of the laser, the total energy, and the time duration of the recording. Film is a very convenient medium, and a great deal of information is available concerning its proper use. The characteristics of photosensitive emulsions are discussed in Section 4. 4.2 under photochemical techniques of energy measurements. Suitable precautions must be exercised to allow for reciprocity failure (intensity multiplied by exposure time not producing a film density that is independent of the exposure time) and the limited dynamic range of the film. Special types of film, such as Polaroid, Polaroid Infrared, and EG&G's XR extended-range film, are very useful in special cases.

For qualitative or rough quantitative work with moderate- to high-energy lasers the thermal damage occurring to various absorbing surfaces can be used to study the beam pattern. Such surfaces commonly used include exposed Polaroid film and carbon paper[127].

Thermal effects can also be observed and recorded far out into the infrared by visually observing the differential evaporation or condensation of oil on a thin membrane in an evaporograph[128]to[130].

Another technique useful in the infrared utilizes a liquid crystal supported on a thin membrane[131]. The heat pattern from a thick radiation-absorbing layer is transferred to the liquid crystal, which changes color with temperature when illuminated with white light[132].

Suitably sensitized phosphors[133] can also be used to visualize infrared laser images, although they are currently limited to wavelengths shorter than about 1.3 μm.

Photoemissive image converters[134], [135] can also be used successfully for observing laser patterns, but again they are limited by the spectral response of the photocathode, that is, wavelengths below about 1.2 μm for the S-1 photocathode. Vidicon-type pick-up tubes using semiconducting layers sensitive to infrared radiation have also been reported[136], as well as various combinations of electroluminescent layers with photoconductive layers to form imaging panels suitable for use in the infared out to about 1.6 μm[137], some of which can provide very long storage times for the image[138].

4.8 STANDARDS AND CALIBRATION

4.8.1 General Principles

Radiometric standards are based on the Stefan-Boltzmann and Planck laws of black-body radiation, and the development of standard sources that are both reasonably accurate, as well as reasonably easy to use in the average laboratory, has been a long and difficult task, with still much to be done. The fundamental law of black-body radiation, the Stefan-Boltzmann law, is given by

$$W = n^2 \sigma T^4, \tag{4.21}$$

where W is the total radiant flux per unit area of the black-body, n is the index of refraction of the median, and T is the absolute temperature. The proportionality constant σ called the Stefan-Boltzmann constant, can be derived from other physical constants and from

$$\sigma = \frac{2\pi^5 k^4}{15c^2h^3} = 5.6697 \times 10^{-12}\ \text{W/(cm}^2\ \text{deg}^4 K), \tag{4.22}$$

where k = Boltzmann constant = 1.38054×10^{-16} erg deg^{-1}K,
c = velocity of light = 2.997925×10^{10} cm/sec,
h = Planck's constant = 6.6256×10^{-27} erg sec.

This value agrees well with carefully determined radiometric measurements; it is entirely adequate for all currently envisioned radiometric work.

Planck's law specifies the spectral radiance of a black body:

$$N_\lambda = \frac{c_1}{\lambda^5}(\exp c_2/\lambda T - 1)^{-1} \quad \text{W cm}^2/\text{sr/cm}. \tag{4.23}$$

where λ = wavelength in cm,

$c_1 = 2c^2h = 1.10988 \times 10^{-12}$ W cm²/sr,

$c_2 = 1.4380$ cm deg K (by definition on the practical, international temperature scale).

Extensive tables of these functions have been published[139].

The development of standard black-body sources requires the development of radiators with very high emissivities over the spectral region of interest as well as accurate control of the temperature of the source. For moderate black-body temperatures (below 2000°C) often required for absolute calibrations in the infrared, standard black-body sources have been developed that are merely well-designed, carefully controlled furnaces [140]. A number of these are commercially available[141]. They can be used as standards of radiance or irradiance, because the geometry is simple and the brightness over the apertures is quite uniform.

The National Bureau of Standards has set up a number of very carefully designed black-body sources[142] to [144] that are used to calibrate secondary standards that are also commercially available.

Tungsten ribbon filament lamps are available from the Bureau of Standards as secondary standards of spectral radiance over the spectral region from ≈ 0.2 to 2.6 μm[142], [143]. Recent improvements in calibration procedures permit comparisons within about 1 percent in the near ultraviolet, and about 1/2 percent in the visible[145].

Quartz iodine lamps have also been made available recently by the Bureau of Standards as secondary standards of spectral irradiance, which are good to one or a few percent if properly used[143]. Because of their small size and high operating temperatures, relatively high spectral irradiances may be obtained with these lamps. Low-intensity standards of spectral irradiance are also available to cover the same spectral region from 0.25 to 2.6 μm[144] to [146]. These, in general, are nothing more than a standard of spectral radiance combined with a small aperture that becomes a point source of known spectral intensity of the order of 10^{-16} W cm²/nm wavelength interval with an uncertainty of about 10 percent when used in the usual manner[146], [147].

Over limited spectral regions, especially in the ultraviolet where the intensities of the usual secondary standards are low because of their limited operating temperatures, it has been found possible to use mercury-arc lamps as very bright spectral irradiance standards[148].

Very high temperatures can also be maintained in the craters of carbon arcs, and black-body temperatures of around 3800°K with high emissivities have been measured over the spectral range from 0.3 to 4.2μm[149]. These arcs have also been found to be sufficiently reproducible ($\pm$10°K) to be useful as standards of spectral radiance.

Single crystals of tungsten with a plane radiating surface have also been proposed for radiation standards over the range from 1500 to perhaps 3000°K[150].

4.9 ATTENUATORS

4.9.1 General Principles

It is often necessary to attenuate a laser beam by an arbitrary amount to bring the energy or power of the beam within the dynamic range of a particular detector. Alternatively, if a well-calibrated attenuator is available, it is possible to measure the range over which a detector follows some prescribed response law or the deviations from that law.

At the relatively low intensities ordinarily encountered in optics, neutral-density filters are commonly used attenuators. They are simply suitable concentrations of absorbing material in a transparent medium such as gelatin or glass. The spectral-absorption characteristics of the material are relatively independent of wavelength, at least within the visible spectrum, with the result that they are gray or black in appearance, depending on their total absorption. Care should be exercised in using these filters over narrow wavelength bands because the transmission characteristics of typical filters may vary as much as 2:1 over the visual spectrum. For this reason, as well as the fact that the surface reflections from inside stacks of such filters can interact to make the total insertion loss of the array not equal to the sum of the individual insertion losses, each filter or stack of filter must be calibrated at the particular wavelength being studied if reasonably precise measurements are required.

The insertion loss of neutral-density filters does not remain constant at higher powers due to reversible or even irreversible changes within the filter itself. The nature of these changes depends, of course, on the exact mechanism of absorption, but it may involve such factors as depopulation of absorbing levels, the absorption spectra and lifetimes of any excited states produced by the initial absorption of a photon, as well as simple thermal damage to the filter.

A useful technique for detecting nonlinearities in filter-absorption characteristics has been described by Glick[157]. The technique uses a beam splitter to divide a laser beam into two known fractions. Each beam is then sent through filters to two photodetectors. If nonlinearities exist in one of the filters, it will show if the two outputs of the photodetectors are compared, if they are subtracted from each other in a differential amplifier, or, perhaps best, if they are displayed simultaneously on identical orthogonal axes of an oscilloscope. The technique requires that the photodetectors be operated well below levels at which they saturate or become nonlinear and that the beam-splitter characteristics are independent of power. The polarization vector of the laser beam must be lined up either perpendicular or parallel to the plane of incidence on the beam splitter, for most beam splitters have different splitting ratios for the components of polarization in these two planes.

Beam splitters, by themselves, provide a useful technique for attenuating a laser beam, and if the surfaces are optically flat and the material homogeneous, the spatial coherence of both portions of the beam are unaffected by the beam splitter.

One of the simplest as well as one of the more reliable attenuator techniques takes advantage of the $1/R^2$ decay of radiation scattered from a diffuse Lambertian surface[152]. In this case, the laser beam is incident on a diffuse, nonabsorbing target and a small fraction of the scattered energy is collected and measured. If the photodetector has an active receiving area A and the scattering surface is normal to the line between the detector and the surface at a distance R, the instantaneous received power P_s is given by

$$P_s = P_0 \frac{A}{2\pi R^2}, \tag{4.24}$$

where P_0 is the instantaneous laser-beam power incident on a portion of the scattering surface. It is assumed that the dimensions of the illuminated spot are small compared to the distance R. The geometry should be such that the detector area A is large enough to average over a number of lobes in the "speckle" pattern of the illuminated area.

The diffuse, white scattering target can be made of a variety of materials, many of which were originally developed and tested for use in integrating spheres. These include barium sulfate[153], magnesium oxide [153], [154], and magnesium carbonate coatings, although opal glasses and white pyroceramics[155] may withstand higher power densities with more durable mechanical properties and have high reflectivities in the near infrared.

Integrating spheres[156] to[159] can be used in a similar way, the input beam being thoroughly scrambled and diffused so that the exit aperture acts as though it were a Lambertian scatterer. The entire volume of an integrating sphere can be filled with scatterers to reduce the optical loading on any given surface, and the results are still amenable to analysis[159].

Another technique for attenuating a laser beam uses fine wire grids or screens to reflect or scatter a known or variable portion of the energy[160]. Normally, they are based on a simple, geometrical optics design, the attenuation being proportional to the fractional area obstructed by the wires. These grids can be rotated to provide a continuously variable attenuation over ranges of approximately 2:1 or 4:1, and a pair of grids at right angles to each other can further increase the range. The predicted performance of these grids neglects diffraction effects; therefore wire sizes and spacings should be, and usually are, much greater than a wavelength. In addition, there must be a large number of wires within the beam, or erratic results will be obtained. Furthermore it is difficult to array more than a few grids in series with different angular orientations to obtain large attenuations when they are used in well-collimated beams, although large attenuations have been demonstrated for incoherent beams[161]. Thus, although these grids can withstand high peak powers, they are most useful in beams with relatively large cross-sectional areas. Parallel wire grids also have some polarization effects, transmitting slightly more energy polarized perpendicular to the wires than polarized parallel to the wires.

The diffraction effects associated with relatively coarse grids have also been proposed for attenuating laser beams[162]. In this case, the higher diffraction orders are utilized, because in general the energy available decreases as the order increases. The absolute calibration of such a device could be done experimentally, but an optimum design to minimize polarization effects requires the solution of the complete electromagnetic boundary-value problem. Such solutions have recently been obtained[163].

A form of variable attenuator for optical frequencies is analogous to the familiar crossed polaroids used in variable-density goggles. In this case, the first polarizer permits only one linear polarization to pass through. For unpolarized sources, this requires a loss of at least 50 percent of the incident light, but for polarized sources, such as many lasers, the first polarizer can be eliminated. Then, by varying the angle of rotation θ of the analyzer, the transmission of the pair of polarizers will vary as

$$P_t = P_0 \cos^2 \theta, \tag{4.25}$$

where P_0 is the peak transmission with both polarizers aligned. Because polarizers with very low cross-polarized components are easier to achieve for narrow wavelength bands than over large bandwidths, the performance of such attenuators can be very good, although great care must still be exercised if large attenuations are required. For high-power applications, the best-quality crystalline quartz and calcite polarizers without cements or coatings are good choices at the moment. The same effect can be achieved in certain liquids or crystalline solids with strong electric fields to induce a birefringence in the medium. Kerr cells and Pockels cell can be used in this way to achieve an electrically controllable attenuation that can be varied at very high rates.

4.10 QUALITATIVE MEASUREMENT OF LASER-ENERGY DENSITY AND INTENSITY

4.10.1 Carbon Paper and Exposed Polaroid Films

One of the simplest methods of obtaining a qualitative measure of the output-energy density of a laser is to direct the output beam onto a surface that can be damaged. Ordinary carbon paper (carbon side facing the laser beam) and exposed Polaroid film (emulsion side facing the laser) have found wide use with both ruby- and Nd^{3+}-doped materials. At input-energy densities of the order of 1 J/cm^2 carbon paper is clearly marked. Polaroid film yields a yellow-brown spot at the radiation impact spot. Some of the emulsion may be reduced to yield a silver film in the damaged area. The technique is more useful in determining the average shape of the output beam than it is in measuring energy, the effect being virtually independent of energy in the 1-to-100-J range. Perhaps its greatest usefulness is in enabling us to define the shape factor of beams of infrared energy.

The Polaroid-film technique is capable of crude calibration as an energy-measuring tool provided the threshold level for damage is determined.

4.10.2 Estimation of Output Energy with a Razor Blade

A simple qualitative measure of the output energy of a pulsed-ruby or Nd^{3+} laser can be obtained by focussing the output beam upon a stack of double-edged, blue-steel razor blades. When a simple positive double convex lens of $f \approx 10$ cm is used, a hole can be produced in a razor blade at the 1-J level. The number of razor blades damaged in a stack is found to be roughly proportional to energy (in the 1 to 10 J region). At higher energy levels this technique becomes unreliable, particularly when the number of blades extends appreciably beyond the focal point of the lens.

4.10.3 Estimation of Laser-power Output by Burning Paper or Wood

The power output of high-power, CW-gas lasers can be measured qualitatively by noting the threshold of burning, or the time required to burn, through flammable materials. Ion lasers at or near the watt level typically produce raw beam spots 1 to 2 mm in diameter. The unfocussed beam of a

CW-ion laser will burn a hole in a white 3 × 5 in. index card in about 5 secs at the $\frac{1}{4}$- to $\frac{1}{2}$-W level. At the 1-W level a hole will be burned through a dry "Popsicle" stick in less than 30 secs. The unfocussed beam of a 15 to 25 W $He/N_2/CO_2$ laser (with a plasma-tube bore of approximately 1 in.) will cause a red-beam spot to appear on a $\frac{3}{8}$in. thick piece of Transite in about 5 secs.

The techniques outlined herein are simple to apply, although crude "rules of thumb," but are capable of misinterpretation. They are relatively insensitive to wavelength, and though capable of being calibrated, in the loosest sense of the word, they should only be considered as adjuncts to established order of magnitude effects. One of their main values is that they use readily available materials.

4.11 TECHNIQUES OF OPTICAL DETECTION

4.11.1 Definition of Parameters

It is appropriate that the remaining portions of this chapter, which are concerned with the more pragmatic aspect of laser-beam power and energy measurement, be introduced by a section devoted to a definition of terms. In evaluating the characteristics of various detectors it is essential that we start with a good understanding of the various figures of merit and the interpretation of the test condition upon which the definitions are based.

4.11.1.1 Responsivity For detectors with an electrical output the responsivity $\mathscr{R}$ is roughly defined as the ratio of the rms value of the signal voltage to the rms power incident on the detector

$$\mathscr{R} = \frac{\text{rms output voltage}}{\text{rms power incident upon the detector}}. \tag{4.26}$$

This transfer function is usually linear over an appreciable range from threshold to (in sensitive detectors) several mW/cm². When a detector is operated in a nonlinear-response range an incremental responsivity can be defined as a change in the transfer function.

The responsivity of an individual photodetector, wherein one is not free to vary the detector area A or the detective time constant τ_d depends upon five key parameters

$$\mathscr{R} = \mathscr{R}(\lambda, f, P_a, T, g), \tag{4.27}$$

where λ is the wavelength of the source, f is the signal-modulation or chopping frequency (usually a mechanical chopper) used to enable low-drift, high-gain amplification of small signal voltages, P_a is the ambient background-radiation, nonsignal power incident upon the detector, T is the absolute temperature, and g is the gain parameter; the circuit impedance must be specified when the detector is a bolometer.

Most detectors, particularly the semiconductor detectors, have externally adjustable parameters, such as bias voltage, by which the responsibility and detectivity may be varied. The gain parameter g encompasses all these parameters.

Because responsivity is concerned with the magnitude of the detector's output voltage per unit radiant power, we can neglect noise. It serves only to limit the noise voltage; the bandwidth need not be specified. Provided that the signal-frequency time variation is large compared with the detector's response time, the modulation frequency need not be specified. The temperature of the irradiating black-body source and the detector area must be known; black-body responsivity is

$$\mathscr{R}_{BB} = \frac{S}{P_{BB}} = \frac{S}{H_{BB}A} \quad \text{V/W}, \tag{4.28}$$

where S is the rms value of the fundamental component of the signal voltage as measured with the entire detector area exposed, P_{BB} is the incident black-body radiant power, H_{BB} is the black-body irradiance, and A is the detector area.

Responsivity is usually measured when the detector is operated under maximum detectivity conditions while being irradiated with a polychromatic black-body source operated at 500°K. For this reason the measured value of responsivity is of little value in determining the power sensitivity of the detector to monochromatic laser light. The spectral responsivity $\mathscr{R}_\lambda$ of the detector may be obtained by comparison with a radiation thermopile. Whenever a graph or table of the relative spectral responsivity is available, the detector's response at a given wavelength may be obtained as

$$\mathscr{R}_\lambda = \begin{pmatrix}\text{responsivity at the wavelength}\\ \text{of peak response}\end{pmatrix} \times (\text{relative spectral response}).$$

The value of R_{BB} is related to the spectral responsivity and the Planck function as

$$\mathscr{R}_{BB} = \int_0^\infty W(\lambda)\mathscr{R}_\lambda(\lambda)\, d\lambda. \tag{4.29}$$

The detailed shape of the R_λ (λ) curve is reproduced by the D^*_λ discussed below. Plots of R_λ (λ) can usually be obtained from the manufacturer of the detector. If the value of R_λ is determined for the wavelength for which the detector has its greatest responsivity $(R_\lambda)_{max}$, the scale factor can be used with the D^* curve to unfold the R_λ characteristic.

The responsivity is related to other detector parameters defined below:

$$\mathscr{R}_{BB} = \frac{D^* N}{(A \Delta f)^{1/2}}, \tag{4.30}$$

where N is the rms noise voltage, A is the detector area, and Δf is the noise band width.

4.11.1.2. Noise-equivalent Input Noise-equivalent input, NEI is a measure of the minimum irradiance H required to produce a signal-to-noise ratio of unity

$$\text{NEI} = \frac{H}{(\Delta f)^{1/2}} \frac{N}{S} \quad \text{W/cm}^2/\text{Hz}^{1/2}. \tag{4.31}$$

For purposes of specificity, noise band width is normalized to 1 Hz. Noise data are conventionally taken at 4 Hz and the noise voltage divided by 2 (the square root of the bandwidth employed for measurement). NEI depends upon operating conditions, that is

$$\text{NEI} = \text{NEI}(f, T, g, \lambda, P_a). \tag{4.32}$$

4.11.1.3 Noise-equivalent Power Noise-equivalent power NEP is a measure of the minimum power that can be detected. It is the product of NEI by the area of the sensitive surface. NEP is a performance measure of a specific detector rather than a detector material because the definition includes an arbitrary detector area

$$\text{NEP} = \text{NEIA} = \frac{HAN}{S(\Delta f)^{1/2}} \quad \text{W/Hz}^{1/2}. \tag{4.33}$$

The conditions of measurement affect the NEP as they do NEI. Thus we have

$$\text{NEP} = \text{NEP}(T, f, \lambda, g, P_a). \tag{4.34}$$

If manufacturers' values are used for NEP, it is essential that due note be taken of the condition under which the NEP was measured. Standard conditions for its measurement employ a black-body source operated at 500°K.

The radiation used in determining NEP values is normally pulse-modulated by a mechanical chopper to permit low noise amplification of the resultant signal voltages. Modulating frequencies of 90, 400, or 900 Hz are used.

4.11.1.4 Detectivity The reciprocal of NEP is denoted as detectivity D. The term, which is no longer widely used, has value in that a large number for the detectivity corresponds to high detector sensitivity. Detectivity expresses the rms signal-to-noise voltage ratio obtained per watt of incident radiant power.

4.11.1.5 Detector D* and D_λ* For fundamental reasons detector noise is usually proportional to the sensitive area of a detector. The D^* of the detector, conceptually related to detectivity and originally introduced to remove the dependence of NEP on the detector's sensitive area[164], is related to NEP by the expression

$$D^* \quad \frac{A^{1/2}}{\text{NEP}} \quad \text{cm(Hz)}^{1/2}/\text{W}, \tag{4.35}$$

where A is the sensitive area of the detector.

The use of D^* has met wide industry acceptance as a figure of merit, and it is presently used more than NEP to specify the detector's sensitivity. As with NEP, the specified source temperature, chopping frequency, and noice band width must be used to qualify the D^* quoted. Because D^* is spectrally dependent, its value is frequently quoted as a function of wavelength D_λ^*.

For most infrared detectors a constant relates the detector's black-body sensitivity (NEP and D^*) to its spectral sensitivity at the wavelength of peak response. For a given detector, these constants are independent of any characteristic of the specific detector type. Because the shape of the spectral-response curve is independent of the NEP or D^* for a specific type of detector, we can compute the absolute spectral response from the NEP or D^*, the general shape of the spectral-response curve D_λ^*, and scaling constant. Table 4.4 gives the value of this constant for several infrared detectors.

It is common practice to report D^* in the formal D^* (500°K, 900, 1) = a value, or D_λ^* (5μm, 900, 1) = a value. These refer respectively to the value of D^* for a 500°K black-body, or 5 = μm narrow-band source as measured at a 900-Hz chopping frequency, and a 1-Hz noise bandwidth.

Figures 4.2 to 4.6 illustrate the D_λ^* wavelength dependence for typical

TABLE 4.4 TYPICAL INFRARED-DETECTOR CONSTANTS

Detector	Temperature (°K)	Mode	Peak Response Wavelength (μm)	$\frac{D^* \text{(peak)}}{D^* \text{(500°K)}}$ or $\frac{\text{NEP (peak)}}{\text{NEP (500° K)}}$
PbS	300	PC	2.1	200
InAs	300	PV	3.5	35
InAs	195	PV	3.3	50
InSb	77	PV	5.0	5
InSb	77	PC	5.3	5

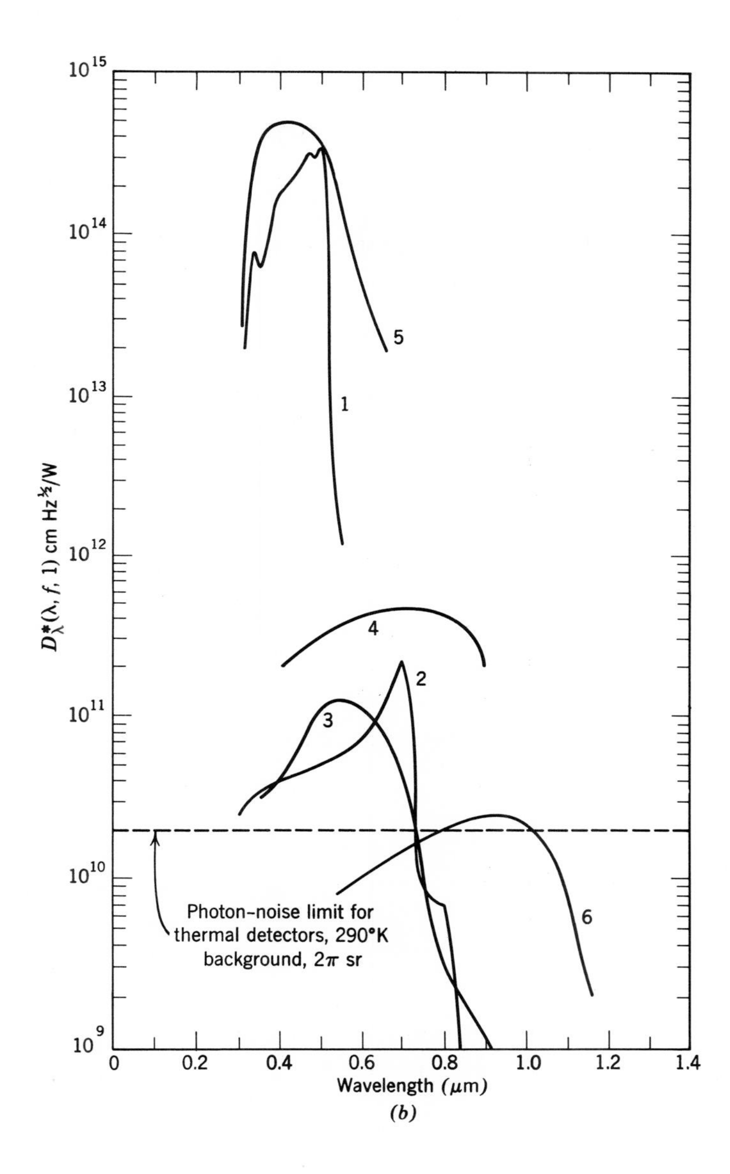

Figure 4.2. Spectral D_λ* of room temperature detectors responding in the visible spectrum. (1) CdS, PC (90 Hz); (2) CdSe, PC (90 Hz); (3) Se-SeO, PV (90 Hz); (4) GaAs, PV (90 Hz); (5) 1P21 photomultiplier (measuring frequency unknown); (6) 1N2175 Si photo-duo-diode, PV (400 Hz).

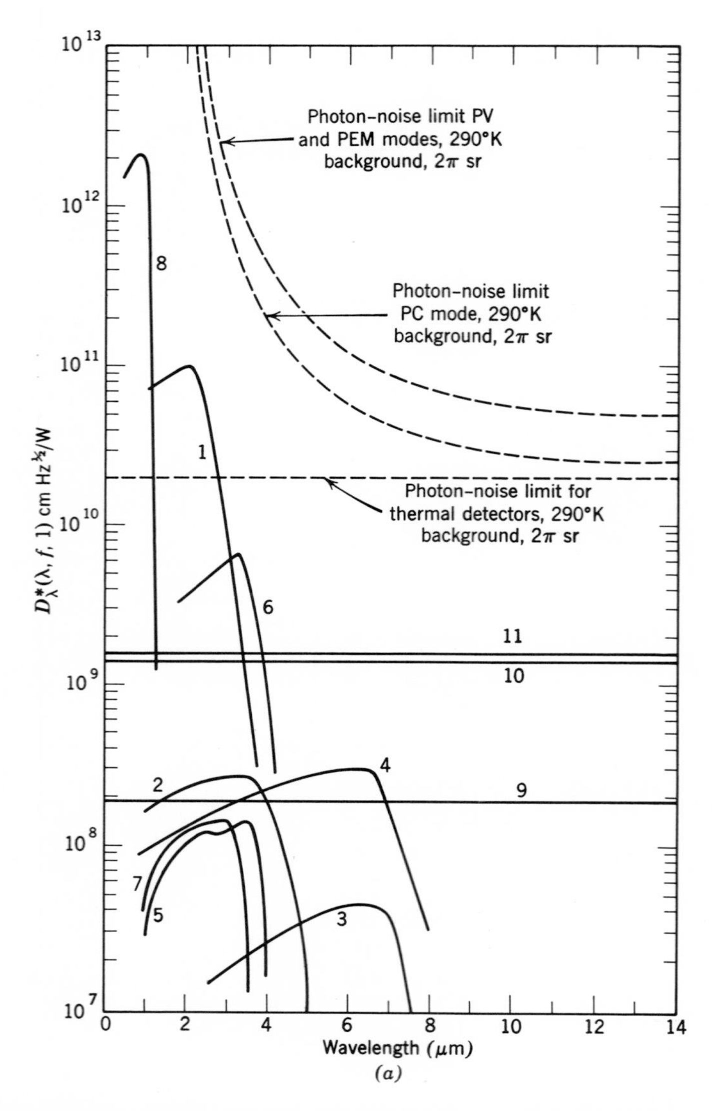

Figure 4.3. Spectral D^*_λ of room temperature detectors. (1) PbS, PC (250 μsec, 90Hz); (2) PbSe, PC (90 Hz); (3) InSB, PC (800 Hz); (4) InSb, PEM (400 Hz); (5) InAs, PC (90 Hz); (6) InAs, PV (frequency unknown, sapphire immersed); (7) InAs, PEM (90 Hz); (8) Tl_2S, PC (90 Hz); (9) Thermistor bolometer (1500 μsec, 10 Hz); (10) Radiation thermocouple (36 msec, Hz); (11) Golay cell (20 msec, 10 Hz).

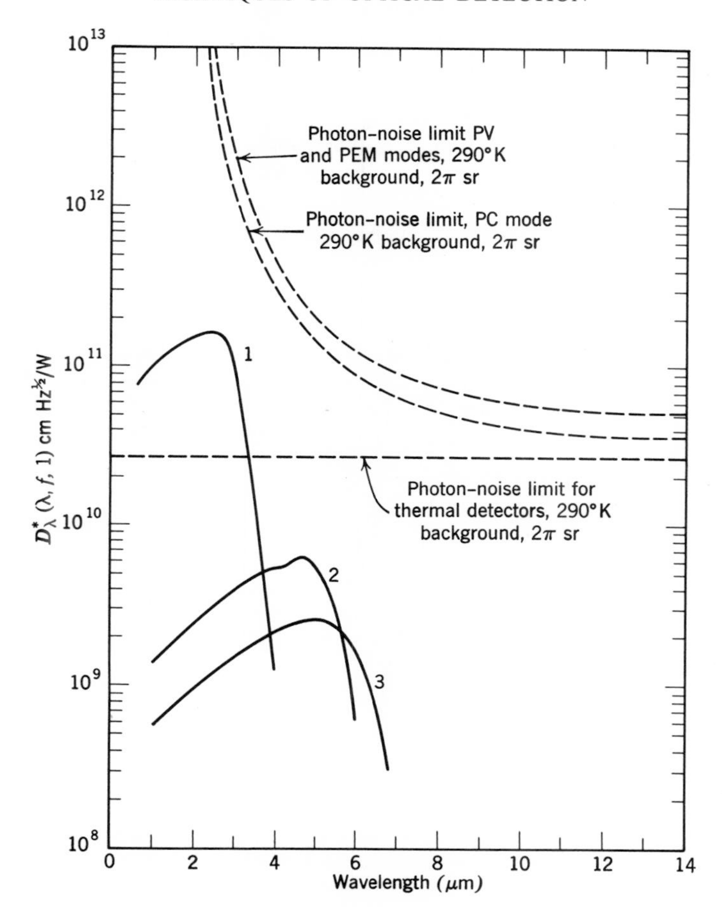

Figure 4.4. Spectral D^*_λ of detectors operating at 195°K: (1) PbS, PC (1000 Hz); (2) PbSe, PC (900 Hz); (3) InSb, PC (900 Hz).

detectors operating in the visible and near infrared at room temperature, 195°K, 77°K, and below[165].

4.11.1.6 BLIP and D** The application of a detector may be *background limited in performance*, BLIP, especially in the infrared, by radiation noise that reaches the detector from the field of view. Clearly the ultimate detectivity is dependent upon the solid angle from which the detector receives photon noise generated by the statistical fluctuations in the number of photons arriving at the detector from the background[166]. To remove

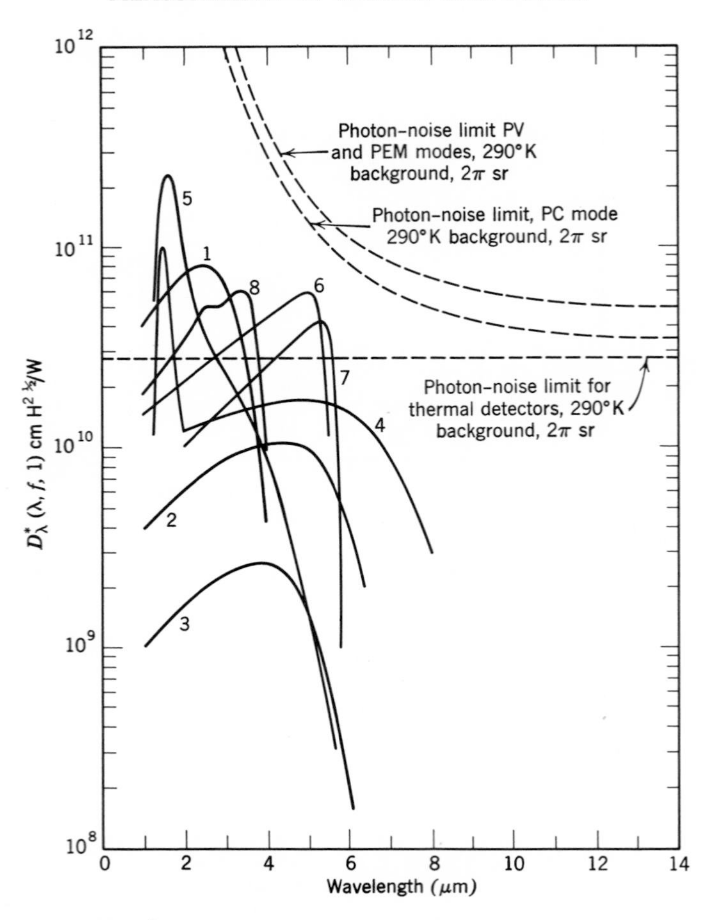

Figure 4.5. Spectral D^*_λ of detectors operating at 77°K: (1) PbS, PC (90 Hz); (2) PbSe, PC (90 Hz); (3) PbTe, PC (90 Hz); Ge:Au, PC (900 Hz); (5) Ge:Au, Sb, PC (90 Hz); (6) InSb, PC (900 Hz, 60° field of view); (7) InSb, PV (900 Hz); (8) Te, PC (900 Hz).

this dependence, the solid angle of irradiation is included in the definition of D^{**}

$$D^{**} = D^*\left(\frac{\Omega}{\pi}\right)^{1/2} \tag{4.36}$$

where Ω is the solid angle of irradiation in steradians.

4.11.1.7 Response- and Detective-time Constants and Techniques for Their Measurement Detector time constants may be divided into two classes: intrinsic and extrinsic. Intrinsic time constants are a function of the basic

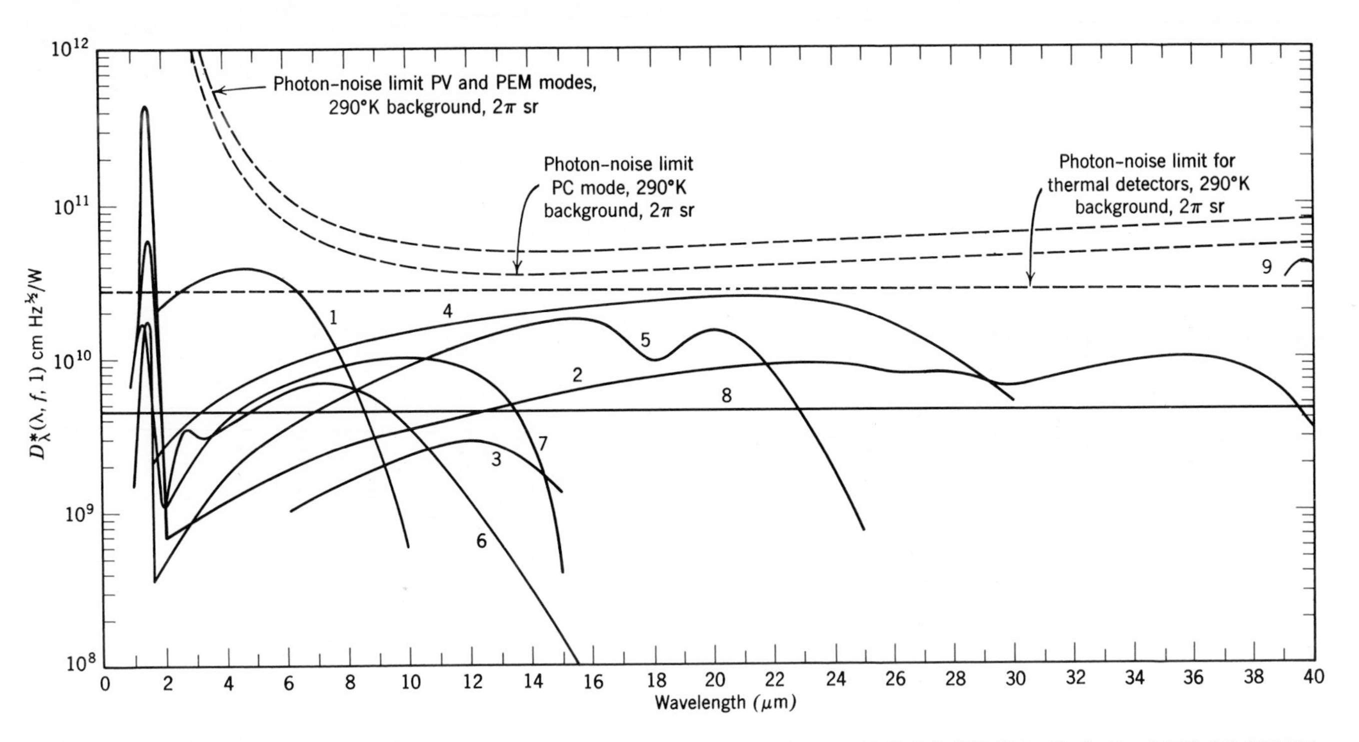

Figure 4.6 Spectral D^*_λ of detectors operating at temperature below 77°K: (1) Ge:Au, 65°K, PC (900 Hz); (2) Ge:Zn, 4.2°K, PC (800 Hz); (3) Ge:Zn, Sb, 50°K, PC (900 Hz); (4) Ge:Cu, 4.2°K, PC (900 Hz, 60° field of view); (5) Ge:Cd, 4.2°K, PC (500 Hz, 60° field of view); (6) Ge-Si:Au, 50°K, PC (90 Hz); (7) Ge-Si:Zn,Sb, 50°K, PC (100 Hz); (8) NbSn superconducting bolometer, 15°K (360 Hz); (9) Carbon bolometer, 2.1°K (13 Hz).

detector type, and, for example, in photoconductors depend upon the detector material. In photoconductors the intrinsic time constant is determined by the majority-carrier lifetime. In photovoltaic and PEM detectors the minority carriers determine the intrinsic time constant. Intrinsic time constants control the detective time constant τ_d. The detective time constant is related fundamentally to D^* as.

$$\tau_d = \frac{\mathscr{R}_m{}^2}{4\int_0^\infty [D^*(f)]^2\, df}, \tag{4.37}$$

where $\mathscr{R}^*_m$ is the peak value of $\mathscr{R}^*(f)$. The detective time constant is an invariant speed-of-response measure of the detector performance, for it cannot be varied by changing the gain-frequency characteristic of the amplifier.

The responsive time constant τ_r of a detector is defined generally as

$$\tau_r = \frac{\mathscr{R}_m{}^2}{4\int_0^\infty [\mathscr{R}(f)]^2\, df} \tag{4.38}$$

or implicitly as

$$\mathscr{R}(f) = \frac{\mathscr{R}_0}{[1 + (2\pi f \tau_r)^2]^{1/2}}. \tag{4.39}$$

In the relation above $\mathscr{R}_0$ is the value of responsivity measured at the zero-chopping frequency (dc), $\mathscr{R}(f)$ is the value dependent on the chopping-frequency, $\mathscr{R}$, $\mathscr{R}_m$ is the maximum value of $\mathscr{R}$, and f is the chopping frequency.

The way in which detector parameters limit the rapidity with which detectors will respond to chopped radiation is illustrated in Figures 4.7 through 4.10. These graphs also demonstrate the temperature dependence of the responsive time constant for several detectors at room temperature, 195°K, 77°K, and below[165]. It is clear from these graphs that most non-thermal detectors are frequency independent below 100 Hz. At chopping frequencies above $f = 1/2\pi\tau_r$, $\mathscr{R} = \mathscr{R}_0/\sqrt{2}$, the responsivity decreases at a rate of 6 dB per octave corresponding to a single time-constant characteristic.

Consider the chopping frequency dependence of $D^*(f)$. Because D^* is the signal-to-noise ratio per unit radiant flux that is independent of area for a defined spectral distribution and unit bandwidth, it can be expressed as a ratio of

$$D^* = \frac{\mathscr{R}(A\Delta f)^{1/2}}{N} \tag{4.40}$$

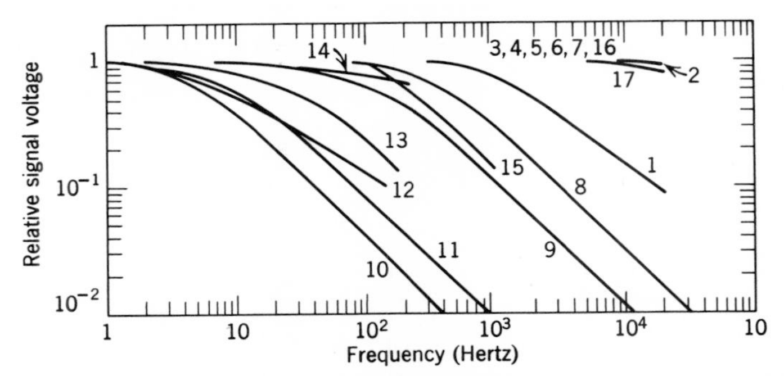

Figure 4.7. Frequency response of room temperature detectors: (1) PbS, PC (250 sec); (2) PbSe, PC; (3) InSb, PC (estimated); (4) InSb, PEM (estimated); (5) InAs, PC (estimated); (6) InAs, PV (estimated); (7) InAs, PEM (estimated); (8) Tl_2S, PC (estimated); (9) Thermistor bolometer (1500 sec, estimated); (10) Radiation thermocouple (36 msec, estimated); (11) Golay cell (20 msec, estimated); (12) CdS, PC; (13) CdSe, PC; (14) Se-SeO, PV; (15) GaAs, PV; (16) 1 P21 photomultiplier; (17) 1 N2175 photo-duo-diode.

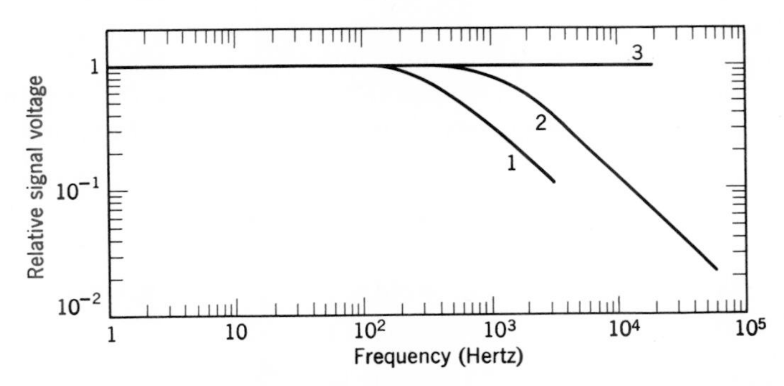

Figure 4.8. Frequency response of detectors operating at 195°K: (1) PbS, PC; (2) PbSe, PC (estimated); (3) InSb, PC (estimated).

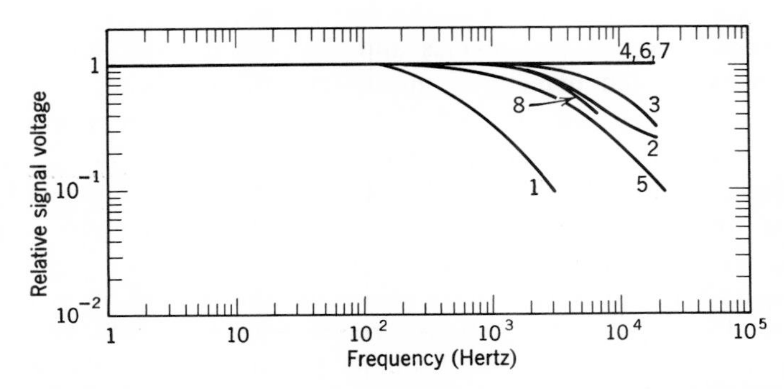

Figure 4.9. Frequency response of detectors operating at 77°K: (1) PbS, PC; (2) PbSe, PC; (3) PbTe, PC; (4) Ge:Au, PC; (5) Ge:Au, Sb, PC; (6) InSb, PC (estimated); (7) InSb, PV; (8) Te, PC.

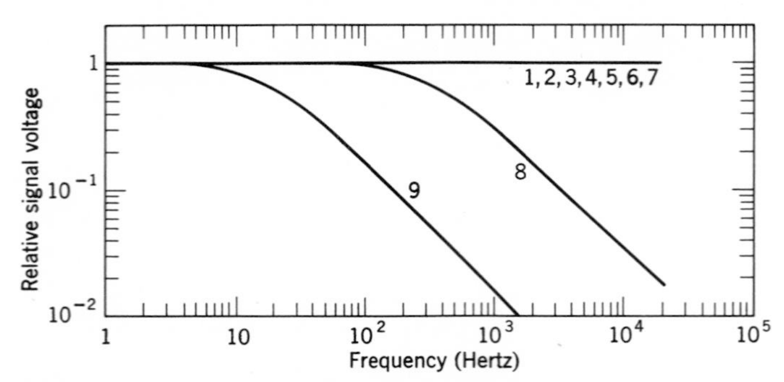

Figure 4.10. Frequency response of detectors operating below 77°K: (1) Ge:Au, 65°K, PC (estimated); (2) Ge:Zn, 4.2°K, PC (estimated); (3) Ge:Zn, Sb, 50°K, PC (unknown); (4) Ge:Cu, 4.2°K, PC (estimated); (5) Ge:Cd, 4.2°K, PC (estimated); (6) Ge-Si:Au, 50°K, PC (estimated); (7) Ge-Si:Zn, Sb, 50°K, PC (estimated); (8) NbSn superconducting bolometer 15°K (estimated); (9) Carbon bolometer, 2.1°K (estimated).

In detectors whose sensitivity is white-noise limited, the frequency dependence of $\mathscr{R}$ and D^* must have the same form, for the noise is not frequency dependent. In semiconductor detectors, in which sensitivity is limited by current noise, the noise voltage exhibits a $(1/f)^{1/2}$ dependence, and D^* has a chopping frequency $1/f$ noise dependence of the form

$$D^*(f) = \frac{Kf^{1/2}}{[1 + (2\pi f \tau_r)^2]^{1/2}}, \tag{4.41}$$

where K is a proportionality constant. D^* exhibits a maximum when $\delta D^*/\delta f = 0$ or

$$f_m = \frac{1}{2\pi\tau_r} \quad \text{and} \quad \mathscr{R} = \frac{\mathscr{R}_0}{\sqrt{2}}, \tag{4.42}$$

where f_m is the chopping frequency of maximum detectivity. From the form of (4.39) and (4.41) it is clear that the maxima in responsivity and D^* occur at different chopping frequencies and that the quoted value for D^* in current-noise-limited detectors must specify the frequency at which the measurement is made. In many fast-response, current-noise-limited detectors the $1/f$ noise-power characteristic reaches the generation-recombination noise plateau at less than 1000 Hz, or long before the response time limit. In these detectors it is found that D^* is virtually independent of f above approximately 1000 Hz.

Response-time measurement of slower detectors may be made with a mechanical chopper. The response versus frequency may be measured directly or may be computed from he signal voltage obtained at two frequencies, S_1 at f_1 on the part of the response curve independent of frequency, and S_2 at f_2 on the portion dependent on frequency. Then equation (4.39) can be solved with S replacing $\mathscr{R}$ to yield

$$\tau_r = \frac{1}{2\pi}\left\{\frac{S_1^{\,2} - S_2^{\,2}}{f_2^{\,2}S_2^{\,2} - f_1^{\,2}S_1^{\,2}}\right\}. \tag{4.43}$$

A more convenient method, based upon the carrier lifetime dependence of τ_r, displays the output voltage of the detector on an oscilloscope when the detector is exposed to a short excitation pulse from an electroluminescent diode (driven by a square-wave generator). The time constant is determined by the $1/\epsilon$ amplitude of the detector decay curve of the detector.

4.11.1.8 Bias Effects in Semiconductor Detectors In general the highest responsivities and detectivities are obtained by operating semiconductor detectors near dc short-circuit conditions. Although the theoretical D^* of a photovoltaic detector is the square root of 2 larger than for a photoconductive cell, state-of-the-art detectors exhibit a much larger improvement factor. Photovoltaic detectors are usually an order of magnitude faster than their photoconductive counterparts. In both detector configurations, however, D^*, $\mathscr{R}$, and N are sensitive to variations in the applied bias voltage or the cancelling bias used to maintain dc short-circuit operation despite developed photocurrent. Measurements of optimum detector-performance characteristics are generally made for optimum bias conditions.

4.11.1.9 Effect of Temperature on Semiconductor Detectors The performance of most detectors used in the intermediate infrared deteriorates rapidly as the temperature is raised above the optimum values the detectors are designed to work at. In general D^*, $\mathscr{R}$, and device impedance decrease approximately three orders of magnitude between 77°K and room temperature. The long-wavelength cutoff tends to increase, whereas the time constant of the detector decreases with device impedance. Above 225°K a considerable decrease in impedance and an increase in responsivity may be obtained by operating some detectors with reverse bias.

4.11.1.10 Field of View The relative D^* can be improved in most detectors by limiting the angular field of view of the sensitive area of the detector. Although a $1/(\sin\theta/2)$ improvement of a factor of 5 is predicted theoretically, if the cone angle θ is decreased to 20°, the actual improvement is only about 2.5. This gain may nonetheless be significant in low-power, laser-gain measurements where the detector is operated D^* BLIP.

4.11.1.11 Detector Noise There are eight kinds of noise to which radiation detectors as a group are subject[167]. These are identified in Table 4.5. Modulation, flicker, and contact noise, though not fundamental, occur in all radiation detectors and are of such practical importance that they must

TABLE 4.5 NOISE IN RADIATION DETECTORS [a]

Kind of Noise	Detectors Concerned	Physical Mechanism
Shot	Thermionic	Random electron emission
Johnson, Nyquist	All	Thermal agitation of current carriers
Radiation	All	Bose-Einstein radiation photon fluctuations
Flicker	Thermionic	Fluctuation in work function
Temperature	Thermal	Wiener-spectrum noise temperature fluctuation of different metals
Generation-recombination		Fermi-Dirac fluctuation of current carriers
Modulation	Junction photocells	Resistance fluctuation in semiconductors
Contact	Bolometers	Contact-resistance fluctuations

[a]Reproduced with permission from Institute of Electrical and Electronics Engineers.

be included in any pragmatic consideration of detection. Shot noise, common to electron-emission-type detectors, has a power spectrum in a temperature-limited diode that may be expressed as

$$P_i(f) = 2e\bar{I}, \tag{4.44}$$

where P_i, is the mean square fluctuation in current per unit of the bandwidth, e is the electronic charge, and I is the mean current. The presence of space charge reduces the noise level without changing the flatness of the spectral distribution. At frequencies high enough to incur electron-transit time effects, electrode capacity casuses the high-frequency portion of the noise spectrum to fall off.

Johnson or thermal-agitation noise appears at the output of every radiation detector. To isolate the Johnson noise component the detector must be as isolated and inactive as a resistor in thermal equilibrium with its surroundings. When there is no current flow through the detector, the Johnson-noise output is of the form

$$P_e(\nu) = 4kTR\frac{h\nu/kT}{\epsilon^{h\nu/kt} - 1}. \tag{4.45}$$

Here R is the dissipative (resistive) part of the electrical impedance. Except at very high frequencies, which $h\nu \gg kT$, the last factor may be omitted.

Temperature noise is the electrical noise in the detector output caused by thermal fluctuations in the active element; it is most significant in thermal detectors wherein radiation is sensed by the generation of an electrical output by virtue of a radiation-induced temperature change. Bolometers, thermocouples, the Golay pneumatic detector, and the thermionic detector are included in this class. The power spectrum of temperature noise is of the form

$$P_e(\nu) = Q^2(\nu)\, W_t(\nu), \tag{4.46}$$

where $Q(\nu)$ is the instrument factor by which a change in temperature produces a change in output voltage, and $W_t(\nu)$ is the Weiner spectrum of the temperature fluctuation, that is the mean-square-temperature fluctuation about its mean value per unit of the frequency bandwidth. The Weiner spectrum may be expressed as

$$W_t(\nu) = \frac{4kT^2 g(\nu)}{g^2 + (h + 2\pi C)^2} \frac{h\nu/kT}{\epsilon^{h\nu/kT} - 1}\cdot \tag{4.47}$$

In this expression C is the thermal capacity of the active element and $g + jn$ is the complex frequency-dependent thermal conductance of the active element to the environment, including conduction caused by radiation exchange.

Radiation or photon noise is due to fluctuations in the radiant flux incident on the detector. The power spectrum of the radiation noise may be expressed as

$$P_e(\nu) = \mathscr{R}^2(\nu) W_p(\nu), \tag{4.48}$$

where $\mathscr{R}$ is the detector responsivity and $W_p(\nu)$ is the Wiener spectrum of the incident power.

In a narrow-frequency band $W_p(\nu)$ may be expressed as[167]

$$W_p(\nu) = 2h\nu\bar{P}(1 - \nu^{-h\nu/kT}), \tag{4.49}$$

where ν is the radiative frequency c/λ and $\bar{P}$ is the mean incident power. For the black-body case (4.49) must be integrated over the radiating spectrum to yield

$$W_p(\nu)_{BB} = 8kT\bar{P} = 8Ak\sigma T^4, \tag{4.50}$$

where A is the detector area, and σ is the Stefan-Boltzmann constant.

Semiconductor detectors are susceptible to two noise sources that have a $1/\nu$ type of frequency dependence. Modulation noise, which has a power spectrum of the form

$$P_i(\nu) = \frac{CI^2}{\nu^\alpha}, \qquad (\alpha \sim 1), \tag{4.51}$$

is due to modulation of the conductivity by causes other than Fermi-Dirac fluctuations in the number of carriers.

Contact noise differs from modulation noise in that its origin can be traced to electrode (ohmic) connections on the semiconductor. The power spectrum of contact noise is of the form

$$P_i(\nu) = \frac{CI^\beta}{\nu^\alpha}, \qquad (\alpha \sim 1, 2 > \beta > 1). \tag{4.52}$$

Flicker noise results from an excess of material-dependent shot noise on the cathode in vacuum tubes. It is observed in shot-noise measurements in temperature-limited vacuum tubes. At low frequencies the flicker noise appears as excess shot noise. Measurements show that flicker noise is least for tungsten, greater for thoriated-tungsten, and greatest for oxide-coated cathodes.

Generation-recombination (GR) noise is developed by semiconductor cells and is due to fluctuations in the number and lifetime of thermally generated carriers. The power spectrum of GR noise may be expressed as

$$P_i(\nu) = \frac{4\bar{I}^2\tau}{n_0[1 + (2\pi\nu\tau)^2]}, \tag{4.53}$$

where n_o is the mean number of carriers in the semiconductor, τ is the mean carrier lifetime, and $\bar{I}$ is the average current through the semiconductor.

The frequency dependence of radiation-detector noise at room temperature, 195°K, 77°K, and below is illustrated in Figures 4.11 through 4.14[165].

4.11.2 Radiation-detector Classification and Summary of Important Performance Characteristics

Within the range of laser frequencies electronic radiation detectors can be subdivided into two broad classes:

1. Photon detectors or particle counters.
2. Energy-integrating or thermal detectors.

These detector classes are distinquished on the basis of the manner in which the detector responds to radiation. Photon detectors essentially measure the rate at which quanta are absorbed. These photon-counting devices require that the incident photons have more than a certain minimum energy before they can detect. Because photon-counting detectors respond to the rate at which photons of a given wavelength are absorbed and the photon rate per watt of incident energy increases directly with wavelength (the number of photons increasing to compensate for the decrease in photon energy with wavelength), the photon-counter response increases with wavelength to the point of wavelength cutoff.

Thermal-radiation detectors, being energy integrators, respond only to the intensity of the absorbed radiant power, regardless of spectral content.

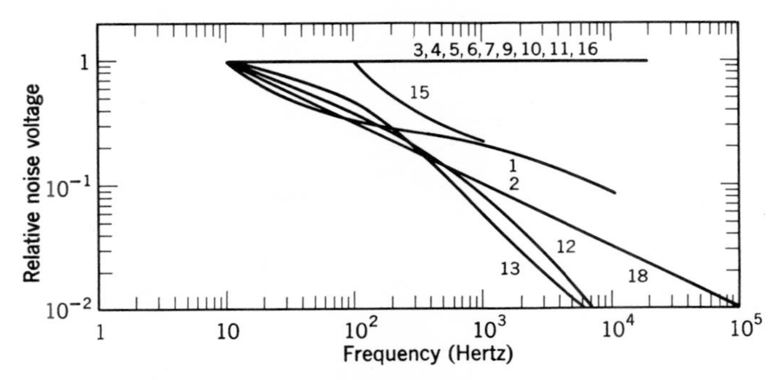

Figure 4.11. Noise spectra of room temperature detectors: (1) PbS, PC(250 μsec); (2) PbSe, PC (assumed). (3) InSb, PC (assumed); (4) InSb, PEM (assumed); (5) InAs, PC (unknown); (6) InAs, PV; (7) InAs, PEM (assumed); (8) Tl_2S, PC (unknown); (9) Thermistor bolometer (assumed); (10) Radiation thermocouple (assumed); (11) Golay cell (assumed); (12) Cds, PC; (13) CdSe, PC; (14) Se-SeO, PV (unknown); (15) GaAs, PV; (16) 1P21 photomultiplier; (17) 1N2175 Si photo-duo-diode, PV (unknown); (18) $1/f$ power law.

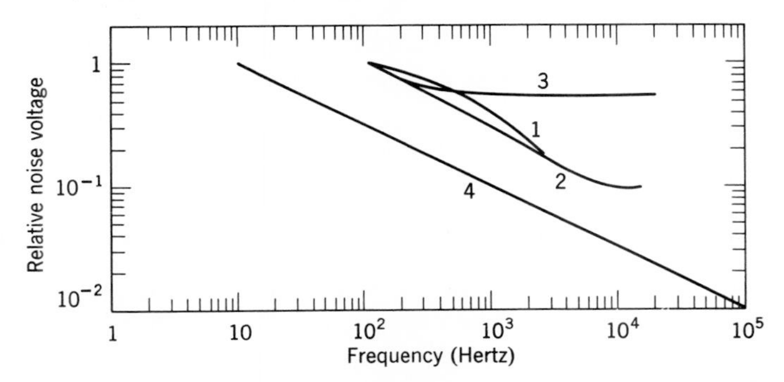

Figure 4.12. Noise spectra of detectors operating at 195°K: (1) PbS, PC; (2) PbSe, PC; (3) InSb, PC; (4) $1/f$ power law.

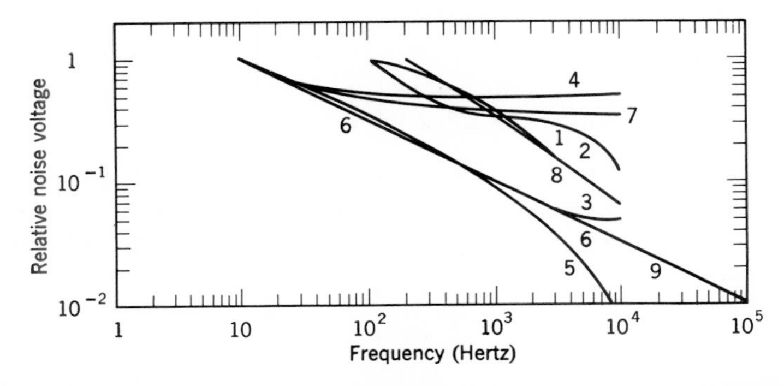

Figure 4.13. Noise spectra of detectors operating at 77°K: (1) PbS, PC; (2) PbSe, PC; (3) PbTe, PC; (4) Ge:Au, PC; (5) Ge:Au, Sb, PC; (6) InSb, PC; (7) InSb, PV; (8) Te, PC; (9) $1/f$ power law.

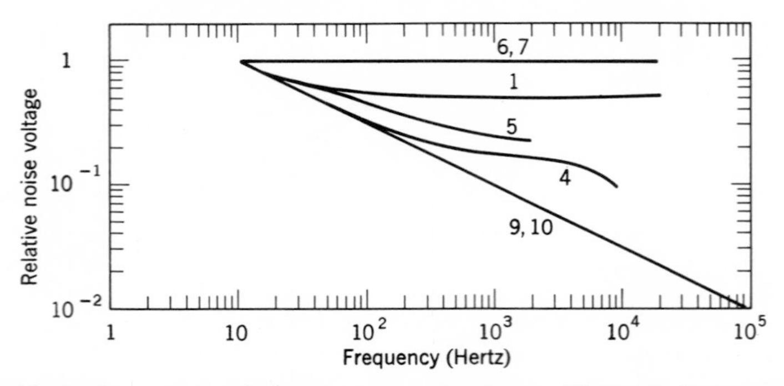

Figure 4.14. Noise spectra of detectors operating below 77°K: (1) Ge:Au, 65°K, PC; (2) Ge:Zn, 4.2°K, PC (unknown); (3) Ge:Zn, Sb, 50°K, PC (unknown); (4) Ge:Cu, 4.2°K, PC; (5) Ge:Cd, 4.2°K, PC; (6) Ge-Si:Au, 50°K, PC (assumed); (7) Ge-Si:Zn, Sb, 50°K, PC (assumed); (8) NbSn superconducting bolometer, 15°K, (unknown); (9) Carbon bolometer, 2.1°K (assumed); (10) $1/f$ power law.

Both thermal and photon detectors have one important common characteristic. They are square-law devices; their output signal is proportional to the input radiant power.

The characteristics of several important radiation detectors are summarized in Table 4.6.

4.11.3 Performance Analysis of Calorimetric Energy-measuring Devices

Calorimetric devices that measure the temperature rise of a known mass of material of known heat capacity have long been used for absolute photometric energy measurements. Application of conventional calorimeters to high-energy, laser-pulse measurements requires that no irreversible damage or change be sustained as the result of a laser pulse. To obtain measurements of absolute energy to an accuracy of 5 percent or better requires careful attention to the details of calorimeter use. With this objective in view, the analysis presented below will be of value in attaining the best performance of a typical calorimeter and will expose the various sources of uncertainty or error.

4.11.3.1 Energy Balance in Calorimeters The energy flow in a typical calorimeter, shown in Figure 4.15, is related to the specific functions of energy collection, absorption, and measurement. The energy balance in simple form is as follows[168, 169]:

$$U_{\text{input}} = U_{\text{calorimeter}} + U_{\text{losses}}. \tag{4.54}$$

The losses shown in Figure 4.15 may be lumped as follows:

$$U_{\text{losses}} = U_{\text{window}} + U_{\text{collector}} + U_{\text{absorber}} + U_{\text{transducer}}.$$

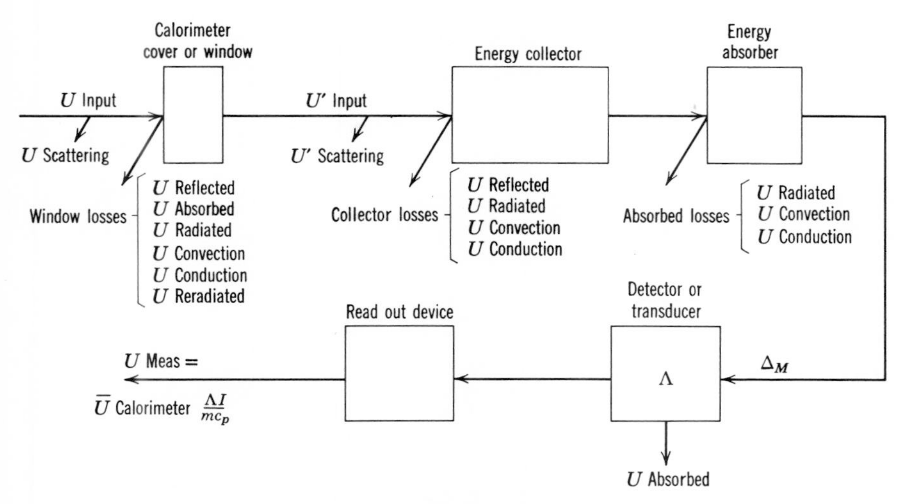

Figure 4.15. Energy flow in a typical calorimeter.

The window losses include conventional reflection and absorption. The reflection losses include Fresnel-mismatch losses as well as surface-scattering losses. The absorption of radiation in the window lends to a number of loss mechanisms. These include a temperature change in the window and the attendant conduction, and convection and radiation losses. At high laser energies, however, some of the energy that is absorbed is lost through excitation of phonon waves and through multiple quantum effects.

Some of the energy that is transmitted through the window is scattered by the medium that separates the window from the energy collector. The remaining energy that impinges on the energy collector that is not absorbed is lost by conduction, convection, and radiation to the environment and by reflection from the collector.

The energy that is transmitted by the collector to the absorber is subject to further convection, conduction, and radiation losses. The remainder reacts with the detector or transducer to produce some change in a measure and ΔM that corresponds to the detector sensitivity Λ (volts per joule). The energy absorbed by the calorimeter causes a temperature rise in the energy collector and absorber that may be expressed as

$$U_{\text{calorimeter}} = mc_p \, \Delta T, \tag{4.55}$$

where ΔT is the temperature change and c_p is the mean equivalent specific heat of the calorimeter of equivalent mass m. The use of equivalence here accounts for the fact that the energy collector and absorber may be made of different materials having different thermal properties.

TABLE 4.6 PERFORMANCE OF ELEMENTAL DETECTORS

Material	Photon or Thermal	Mode of Operation	Film or Single Crystal	N type, P type, or Intrinsic	Operating Temperature (°K)	Wavelength of Peak Response $\lambda_p(\mu m)$	Cutoff Wavelength (50% value) $\lambda_0(\mu m)$	D* (500°K, f, 1) cm Hz½/w (measuring frequency indicated)	$D_{\lambda p}^*$ (λ_p, f, 1) cm Hz½/w (measuring frequency indicated)	Response Time (μsec)	Calculated Optimum Chopping Frequency (Hz)	Resistance per Square (Ω)	Noise Mechanism
[1] PbS	P	PC	F	I	295	2.1	2.5	4.5×10^8 90 Hz	1.0×10^{11} 90 Hz	250	640	1.47 MΩ	Current
[2] PbS	P	PC	F	I	195	2.5	3.0	4.0×10^9 1000 Hz	1.7×10^{11} 1000 Hz	455	350	4 MΩ	Current
[3] PbS	P	PC	F	I	77	2.5	3.3	4.0×10^9 90 Hz	8.0×10^{10} 90 Hz	455	350	5 MΩ	Current
[4] PbSe	P	PC	F	I	295	3.4	4.2	3.0×10^7 90 Hz	2.7×10^8 90 Hz	4	40 kc	50 kΩ	Current
[5] PbSe	P	PC	F	I	195	4.6	5.4	7.5×10^8 900 Hz	6×10^9 900 Hz	125	1270	40 MΩ	Current below 6 kc
[6] PbSe	P	PC	F	I	77	4.5	5.8	2.2×10^9 90 Hz	1.1×10^{10} 90 Hz	48	3300	5 MΩ	Current
[7] PbTe	P	PC	F	I	77	4.0	5.1	3.8×10^8 90 Hz	2.7×10^9 90 Hz	25	6500	32 MΩ	Current
[8] Ge:Au	P	PC	SC	P	77	5.0 (excluding intrinsic peak)	7.1	7.5×10^9 900 Hz	1.75×10^{10} 900 Hz	< 1	Frequency independent above 40 Hz	1.0 MΩ	Current below 40 Hz, *gr* above
[9] Ge:Au	P	PC	SC	P	65	4.7 (excluding intrinsic peak)	6.9	1.7×10^{10} 900 Hz	4×10^{10} 900 Hz	< 1	Frequency independent above 40 Hz		Current below 40 Hz, *gr* above
[10] Ge:Au, Sb	P	PC	SC	N	77	No clearly defined peak exists except for intrinsic excitation		2.9×10^9 90 Hz	2.5×10^{10} at 3μ 90 Hz	110	1500	1.0 MΩ	Current

[11] Ge:Zn (Zip)	P	PC	SC	P	4.2	36	39.5	4.0×10^9 800 Hz	1.0×10^{10} 800 Hz	< 0.01		300 kΩ	Current
[12] Ge:Zn,Sb	P	PC	SC	N	50	12	15	2×10^9 900 Hz	3×10^9 900 Hz				
[13] Ge:Cu	P	PC	SC	P	< 20	20	27	1×10^{10} (60° field of view) 900 Hz	2.5×10^{10} (60° field of view) 900 Hz			0.1 MΩ	Current below 1 kc; *gr* above 1 kc
[14] Ge:Cd	P	PC	SC	P	< 25	16	21.5	7×10^9 (60° field of view) 500 Hz	1.8×10^{10} (60° field of view) 500 Hz				Current below 500 Hz *gr* above 500 Hz
[15] Ge-Si:Au	P	PC	SC	P	50	7.3	10.1	3.1×10^9 90 Hz	7.0×10^9 90 Hz	0.1	Frequency independent to approximately 1 Mc	10 MΩ	*gr*
[16] Ge-Si:Zn, Sb	P	PC	SC		50	10	13.3	4.0×10^9 100 Hz	1.0×10^{10} 100 Hz	0.1	Frequency independent below approximately 1 Mc	20 MΩ	*gr*

Notes (numbers correspond to those in column 1)

[1,2,3]Detectors with time constants ranging from about 1 μsec to 10,000 μsec are available. The detectivity will vary with time constant according to the McAllister relation. The cutoff wavelength may also be shifted to greater values with a sacrifice in detectivity. Detectors operating at 77°K may exhibit double time constants.

[6]May exhibit double time constant.

[7]Resistance may be reduced by grid type of electrodes. Detector is background limited. Performance at 90°K same as at 77°K. May have second time constant for 1.5 μ radiation.

[8]Exhibits long time constant for intrinsic excitation (less than 2 μ). Detectivity improved by cooling to 65°K. See below.

[9]Exhibits long time constant for intrinsic excitation (less than 2 μ).

[10]Detectivity at 90°K equal that at 77°K. Exhibits wavelength-dependent time constant.

[12]Not readily available.

[15]Spectral response can be changed by varying alloy composition. Frequency response may be limited by *RC* time constant.

[16]Spectral response can be changed by varying alloy composition. Frequency response may be limited by time *RC* constant.

TABLE 4.6 (*continued*)

Material	Photon or Thermal	Mode of Operation	Film or Single Crystal	N type, P type, or Intrinsic	Operating Temperature (°K)	Wavelength of Peak Response $\lambda_p(\mu m)$	Cutoff Wavelength (50% value) $\lambda_0(\mu m)$	D* (500°K, f, 1) cm Hz$^{1/2}$/w (measuring frequency indicated)	$D_{\lambda p}^*$ (l_ν, f, 1) cm Hz$^{1/2}$/w (measuring frequency indicated)	Response Time (μsec)	Calculated Optimum Chopping Frequency (Hz)	Resistance per Square (Ω)	Noise Mechanism
[17] InSb	P	PC	SC	I	295	6.5	7.3	1.4×10^7 800 Hz	4.3×10^7 800 Hz	0.2	Frequency independent to 500 kc	20	Thermal
[18] InSb	P	PC	SC	I	195	5.0	6.1	5×10^8 900 Hz	2.5×10^9 900 Hz	< 1	Frequency independent above 500 Hz	60	Current below 400 Hz
[19] InSb	P	PC	SC	P	77	5.0	5.4	1.2×10^{10} (60° field of view) 900 Hz	6×10^{10} (60° field of view) 900 Hz	< 2		10 kΩ	Current
[20] InSb	P	PV	SC	PN	77	5.3	5.6	8.6×10^9 900 Hz	4.3×10^{10} 900 Hz	< 1	Frequency independent above 500 Hz	1 kΩ	Current below 100 Hz *gr* above
[21] InSb	P	PEM	SC	I	295	6.2	7.0	1.0×10^8 400 Hz	3.0×10^8 400 Hz	0.2	Frequency independent below 100 kc	20	Thermal
[22] InAs	P	PC	SC	N	295	3.6	3.8	1.4×10^7 90 Hz	1.4×10^8 90 Hz	0.2	Frequency independent below 100 Hz		
[23] InAs	P	PV	SC	PN	295	3.4	3.7	2.5×10^8 90 Hz	2.5×10^9 750 Hz	< 2		50	Assumed thermal

[24] InAs	P	PEM	SC	N	295	2.5	3.4	1.4×10^7 90 Hz	1.4×10^8 90 Hz	0.2	Frequency independent below 100 Hz		Assumed thermal
[25] Te	P	PC	SC	P	77	3.5	3.8	4.0×10^9 900 Hz	6.0×10^{10} 900 Hz	60	2700	2 kΩ	Current
[26] Tl_2S	P	PC	F	I	295	0.9	1.1		2.2×10^{12} 90 Hz	530	300	5 MΩ	Current
[27] 86% HgTe-14% CdTe	P	PC	SC	I	295	6	6.5	5×10^6	1.5×10^7			≈ 1	
[28] Thermistor bolometer	T	Bolometer			295			1.95×10^8 (1.5 msec) 10 Hz	1.95×10^8 (1.5 msec) 10 Hz	1500	Frequency independent below 30 Hz	2.4 MΩ	Thermal
[29] Radiation thermocouple	T	Thermoelectric effect			295			1.4×10^9 5 Hz	1.4×10^9 5 Hz	3.6×10^4	< 5	5	Thermal
[30] Golay cell	T	Expansion of air			295			1.67×10^9 10 Hz	1.67×10^9 10 Hz	2×10^4	< 5		Temperature
[31] NbSn bolometer	T	Superconducting bolometer			15			4.8×10^9 360 Hz	4.8×10^9 360 Hz	500		0.2	Unknown
[32] Carbon bolometer	T	bolometer			2.1			4.25×10^{10} 13 Hz	4.25×10^{10} 13 Hz	10^4	16	0.12 MΩ	Current

[19] Responsivity is superior to InSb PV, 77°K.
[20] May be either broad-area diffused junction or line type of grown junction. May be operated with or without bias voltage.
[21] Maximum dimensions approx. 2 × 10 mm. Should be transformer coupled to amplifier. Sensitive to magnetic pickup from electrical mains.
[22] Not readily available.
[23] Detector is sapphire immersed.
[24] Not readily available.
[25] Peak detectivity in solar and earth background minimum.
[26] Not readily available.
[27] Not readily available.
[28] Detectors with time constants ranging from about 1 to 50 msec are available.
[29] Widely used in infrared spectroscopy.
[30] Fragile, microphonic.
[31] Not readily available. Noise appears to arise from some unknown mechanism associated with superconductivity.
[32] Not readily available. Use of quartz and paraffin filters cut out response at wavelengths shorter than 40 μ.

TABLE 4.6 (*continued*)

Material	Photon or Thermal	Mode of Operation	Film or Single Crystal	N type, P type, or Intrinsic	Operating Temperature (°K)	Wavelength of Peak Response $\lambda_p(\mu m)$	Cutoff Wavelength (50% value) $\lambda_0(\mu m)$	D* (500°K, f, 1) cm Hz½/w (measuring frequency indicated)	$D_{\lambda p}^*$ (λ_p, f, 1) cm Hz½/w (measuring frequency indicated)	Response Time (μsec)	Calculated Optimum Chopping Frequency (Hz)	Resistance per Square (Ω)	Noise Mechanism
[33] CdS	P	PC	SC F sintered	N	295	0.5	0.51		3.5×10^{14} 90 Hz	5.3×10^4	3	5×10^{11}	Current
[34] CdSe	P	PC	SC sintered		295	0.7	0.72		2.1×10^{11} 90 Hz	1.2×10^4	13	1.5×10^{11}	Current
[35] Se-SeO	P	PV	SC	PN	295	0.55	0.69		1.2×10^{11} 90 Hz	910	160	3 kΩ area dependent	
[36] GaAs	P	PV	SC	PN	295	0.8	0.89		4.5×10^{11} 400 Hz	< 1	160	4.6 MΩ area dependent	Current
[37] IN 2175 photo-duodiode	P	PC	SC	PN	295	0.95	1.07		2.5×10^{10} 400 Hz	8	20 kc	4×10^9	
[38] 1P21 photo-multi-plier	P	PE	F		295	0.40	0.53		5×10^{14} 1000 Hz	< 0.01	Frequency independent to about 100MHz		Shot

[33] Highest responsivity of any photoconductor.
[34] Responds to longer wavelengths and is faster than CdS.
[35] Used in exposure meters.
[36] Useful for star tracking.
[37] Very small overall size.

Because the measurand will exhibit some output that, over a sufficiently small range, is proportional to temperature change through a constant Ψ, as

$$\Delta M = \Psi \, \Delta T = \frac{\Psi U_{\text{calorimeter}}}{mc_p}, \tag{4.56}$$

the measured energy is related to the detector sensitivity

$$U_{\text{measured}} = U_{\text{calorimeter}} \left(\frac{\Lambda \Psi}{mc_p} \right). \tag{4.57}$$

Equation 4.54 is therefore recast as

$$U_{\text{input}} = U_{\text{measured}} \left(\frac{mc_p}{\Lambda \Psi} \right) + U_{\text{losses}}. \tag{4.58}$$

If we could neglect the losses, (4.58) would yield an absolute measure of the laser energy. An indirect measure of laser energy could, of course, be obtained if a suitable calibration could be performed by measuring the output radiation of a source of known spectral emission.

4.11.3.1.1 Significant Energy Losses Fresnel-reflection losses at the input surface of the detector window and from the backscatter inside the calorimeter are both spectrally variable and awkward to account for. In an elementary sense the Fresnel reflection at the input aperture can be accounted for as

$$r(\lambda) = \frac{[n(\lambda) - 1]^2 + v[\lambda]^2}{[n(\lambda) + 1]^2 + v[\lambda]^2}, \tag{4.59}$$

where $n(\lambda)$ is the spectral index of refraction and $v(\lambda)$ is the absorption coefficient of the window. Reflectance calculated from (4.59) will only establish a lower bound, for both n and v become nonlinear at the high electric field and strengths of interest. Low-level calibrations routinely performed to correct for these losses are obviously pointless because the calibrations cannot be extrapolated into nonlinear regions associated with high peak power. Where possible, window-loss effects can be eliminated by removal of the window.

The reflection losses in the collector that result from backscattering are difficult to account for on an analytical basis. They may, of course, be minimized by a calorimeter design that efficiently traps the incoming radiation. The geometry of the calorimeter aperture and the calorimeter surface itself should be designed to minimize backscatter. This in general means that a calorimeter design that maximizes the ratio of the absorbing-surface area to the aperture in the energy collector will enable more accurate measurements.

The relationship between aperture area a and surface area S for the truncated cylinder, the cone, and the sphere are expressed respectively in terms of length L and aperture radius R as

$$\frac{a}{S} = \frac{\pi R^2}{2\pi R^2 + 2\pi RL} = \left[2\left(1 + \frac{L}{R}\right)\right] - 1, \tag{4.60}$$

$$\frac{a}{S} = \frac{\pi R^2}{[\pi R^2 + \pi R\sqrt{R^2 + L^2}]} = \left[1 + \sqrt{1 + \left(\frac{L}{R}\right)^2}\right] - 1, \tag{4.61}$$

$$\frac{a}{S} = 1 - \frac{L^2}{R^2 + L^2} = \left[1 + \left(\frac{L}{R}\right)^2\right] - 1. \tag{4.62}$$

Typical values of a/S are shown for the three geometries in Table 4.7.

The effective emissivity e_0 of the calorimeter can be calculated from a knowledge of the emissivity e of the inner surface from the relation.

$$e_0 = e\,\frac{1 + (1 - e)}{e(1 - a/S) + a/S}. \tag{4.63}$$

Thus a sphere with L/R of 10 makes a substance with an emissivity of 0.5 appear as $e_0 = 0.990$. For the same L/R ratio the effective e_0 for a cylinder and cone are, respectively, 0.973 and 0.953[170].

Radiant-energy losses will occur through the aperture-coupling hole and from the exterior of the energy collector. These losses will follow the Stefan-Boltzmann law

$$U = e_s \sigma S \Phi (T^4 - T_0{}^4) \int_0^\tau dt, \tag{4.64}$$

where e_s is the emissivity of the outer surface of the energy collector and absorber, σ is the Stefan-Boltzmann constant, S is the surface area, Φ is a

TABLE 4.7 EFFECT OF L/R RATIO ON SURFACE-TO-APERTURE AREA OF BLACK-HOLE DETECTORS

		a/S	
L/R	Sphere	Cylinder	Cone
1	0.500	0.250	0.415
2	0.200	0.167	0.309
5	0.0385	0.083	0.164
10	0.010	0.0455	0.0905
20	0.0025	0.025	0.0475
40	0.0005	0.012	0.0245

constant dependent upon the collector-absorber geometry, T, T_0 are respectively the environmental and the collector-absorber temperature, and τ is the time interval of interest.

Obviously e_s should be low and the temperature differential T-T_0 should be made as small as possible consistent with adequate response from the temperature-difference detector. In general the input energy will be deposited over a small fraction of the area of the calorimeter. It is important that the diffusivity of the energy-collector-absorber material be large so that the equilibration time for any local temperature differential will be minimized. Similarly, the heat capacity per unit of mass of the energy absorber should be large so that large temperature gradients are not created for a given energy-sensing capability.

Conduction losses through the calorimeter supports and temperature-sensor leads must be added to the conducting losses of the surroundings. These losses are minimized by making the supports and electrical connectors of small wire and using materials with a low thermal conductivity. The surroundings of the calorimeter can be evacuated or replaced by open foam materials.

4.11.3.1.2 Radiant-energy Collection The energy collector in a laser calorimeter must withstand the localized sample of the beam, disperse the energy quickly over an extended energy-absorbing area, and minimize reradiation or reflection of the incoming beam. If an absorbing calorimeter of the Ulbricht sphere type is used, the inner surface of the sphere should be diffusely reflecting, whereas the outer surface should have low emissivity. The equilibration time of the calorimeter will be less if, in addition, the material of the sphere walls has high diffusivity (thermometric conductivity).

Because the accuracy of the calorimeter deteriorates with increased equilibration time, and thermal equilibration by conduction and convection are determined by the thermal relaxation times of materials (order of seconds), improved performance can be expected through the use of radiation-relaxation time. The incident radiation sample, admitted off axis, may be distributed by multiple reflections from a highly polished inner surface of a silver sphere. The inner surface can be diffusely reflecting; in this event the radiation errors will be smaller. If a specular reflecting surface is used, care must be taken to ensure that negligible energy is lost through imaging on the entrance aperture.

In summary, the energy collector affects the accuracy of the calorimeter if the equilibration time is long; the radiation-loss rate varies as the fourth power of the deviation of the surface temperature of the calories from the surroundings. The time-dependent portion of this loss mechanism depends upon the time that the collector surface or portions thereof remain at elevated temperatures.

4.11.3.1.3 Radiant to Thermal-energy Conversion The absorber, considered as a separable part of the calorimeter, must be able to absorb the impact of the beam and quickly distribute the energy uniformly throughout the calorimeter to minimize radiation losses. Obviously the material of the absorber must have a high enough heat capacity to preclude surface melting or damage. The maximum irradiance that the absorber can withstand in a direction x normal to the surface without melting may be expressed as [168].

$$H_0 = \frac{(c_p \, \Delta T + \Gamma)\rho}{\tau_{\text{pulse}} \alpha e^{-\alpha x}}, \tag{4.65}$$

where c_p is the specific heat at constant pressure, ΔT is the surface-temperature rise during the pulse, Γ is the latent heat of fusion, H_0 is the surface irradiance, τ_{pulse} is the total duration of the radiation pulse, ρ is the mass density, and δ is the penetration depth of the material defined as

$$\delta = \frac{1}{\alpha}.$$

The initial distribution of absorbed radiant power within the absorber is described as

$$P(x) = H_0 \, A e^{-x/\delta},$$

where H_0 is the surface irradiance, A the equivalent spot area, and δ is the penetration depth of the material.

Calorimeters constructed of more refractory materials of low diffusivity for example graphite, where $\kappa \sim 0.003$ instead of silver, where $\kappa \sim 1.71\ \text{cm}^2/\text{Hz}$), will have a longer equilibrium time with less inherent accuracy. Calorimeters of this design are usually capable of a higher energy flux, for thermal diffusivity and conductivity are inversely related.

Liquid calorimeters, which absorb the energy in an extended volume, are inherently capable of higher peak powers but suffer from longer equilibration times, at least in thermally sensed devices, with convective current temperature equilibration times. Provided that an input window is admissible in the calorimeter design, the liquid calorimeter has the capability of absorbing energy in the volume of a liquid chosen to match the spectral range of the laser.

4.11.3.1.4 Shielding and Extraneous Sources of Radiation After the calorimeter has been equilibrated the main sources of energy loss that remain are conduction and radiation losses from external surfaces. Because thermal conductivity determines the rate of heat flow from the calorimeter surface, material with a low thermal conductivity can aid in the reduction of losses (provided that the temperature change induced is not large and the measurement time is sufficiently short).

Radiation and conduction losses can be reduced by using an evacuated wall insulation technique similar to that employed in a thermos bottle.

If an air enclosure is used for the calorimeter, account must be taken of the 680*m*j/°C heat capacity of air. Air will conduct heat from the instrument at a rate which cannot be neglected in calorimeter use and calibration.

4.11.3.2 Temperature-change Detection Energy is detected in most calorimeters through a temperature change. For reasons already outlined, it is desirable that the temperature transducer be capable of detecting a small temperature change with negligible error and high reproducibility. The transducer response should (preferably) be linear and fast compared with the equilibration time of the thermal mass of the absorber. As a practical consideration the transducer should be capable of convenient calibration. Unfortunately no single transducer can satisfy all of these requirements simultaneously. The characteristics of the three commonly used temperature transducers are compared in Table 4.8.

Temperature transducers employ a number of readout instruments, the most common of which are sensitive circuits of the microvoltmeter or Wheatstone-bridge type. The overall accuracy and sensitivity of the temperature-change detector are dependent upon the combined transducer and circuit. Precision instrumentation includes regulating apparatus that essentially frees the instrument from its environment. This includes temperature compensation, regulated sources of supply voltage, isolation from induced sources of noise, and compensation for lead resistance as well as thermal emfs. To achieve accuracies of better than 2 percent in the measurement of temperature change requires expensive instrumentation[171]. The limiting accuracy with which temperature change can be measured is probably not better than 0.001°C corresponding to limiting errors of 1 percent or less. Obviously, temperature fluctuations in the ambient environment are of serious concern.

4.11.3.3 Technique of Energy Measurement of a Pulsed Solid-state Laser Using a Mendenhall-Wedge Ballistic Calorimeter Two configurations of Mendenhall-Wedge calorimeters have been used to measure solid-state laser energy output. In the cone type of configuration, shown in Figure. 4.16, the radiant energy is directed to the entrance of a specularly reflecting (or diffusely reflecting) [172] measuring cone where the energy is almost totally absorbed by the multiple reflections. In planar devices the cross section of the cone is replaced by a large number of sharp-edged wedges (razor blades). The difficulty of machining an accurate polished cone of narrow apex angle is circumvented by the stacked-wedge calorimeter. The

TABLE 4.8 CHARACTERISTICS OF CALORIMETER TEMPERATURE-CHANGE DETECTORS

	Dynamic Range	Sensitivity	Response Time	Linearity	Accuracy	Calibration Stability	Application
Thermistors	-180 to $+100°C$	5×10^{-3} °C	$> 10^{-3}$ Hz	Poor	~2 percent	Poor	Straight-forward
Resistance wire	-180 to $+650°C$	10^{-2}°C	>0.1 Hz	Good	$<0.001°C$	Excellent	Difficult
Thermo-couples	0–1000°C	5×10^{-1}°C	>1 Hz	Adequate with tables	~0.5°C	Good	Easy

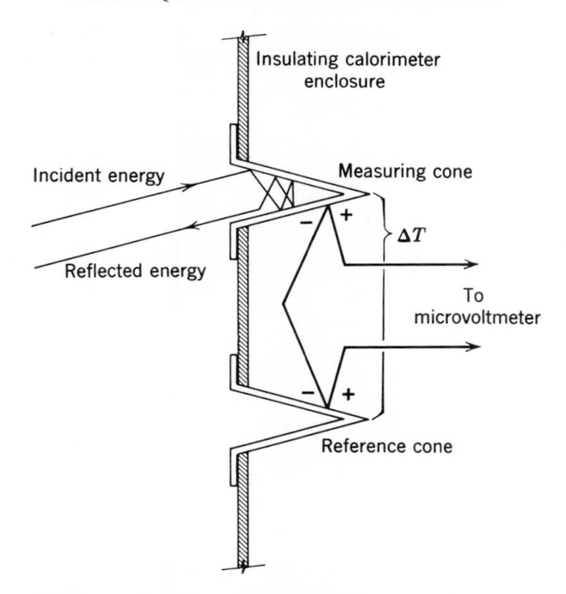

Figure 4.16. Conical Mendenhall wedge laser calorimeter.

entrance aperture of the latter can easily be made large and does not require optics for convenient alignment or reduction of the beam cross section[173].

In both the cone and planar configurations an identical reference sensor is packaged in the same calorimeter so that, to the first order, environmental differences are nulled.

Apparatus for wedge-calorimeter measurements is detailed in Table 4.9. The equipment is arranged in a convenient geometry so that the calorimeter is far enough from the laser to minimize errors due to the leakage of light from the optical pump, spontaneous emission, and electrical transients caused by coupling to the high magnetic fields associated with currents in the optical pump lamp. The beam spot must be smaller than the entrance aperture of the calorimeter. If this requirement cannot be satisfied, a lense must be used to reduce the beam cross section. Care must be taken not to focus the beam to a narrow point on the calorimeter surface or the high-energy beam may damage the calorimeter surface and/or destroy the calibration. The energy measurement must be corrected for the Fresnel reflection and absorption losses of the collimating lense.

Several checks should be made to ensure that the measurements are meaningful. The beam energy should be measured at two locations, both with and without a narrowband filter, to discern the relative errors due to light from spontaneous emission and from the pump lamp. The effect of electrical transients can be determined by intercepting the beam and

TABLE 4.9 APPARATUS FOR MEASURING THE OUTPUT ENERGY OF A PULSED, SOLID-STATE LASER WITH A WEDGE TYPE OF CALORIMETER

Item	Function	Source
Calorimeter	Energy measurement	TRG model 100 Carbon cone calorimeter [a] Razor blade calorimeter [b]
Pulsed solid-state laser	Energy source	Korad Kl or equal
Microvoltmeter or chart recorder	Temperature-transducer readout	HP 425 voltmeter or Varian G-10 chart recorder or equal
Lenses, apertures	Light collimation	Ealing Corp.
Notch filter	Monochromatizer to eliminate pump-lamp radiation	Thin Film Products, Inc. 1.0-nm filter for 694.3 nm, etc.

[a]See [172].
[b]See [173].

repeating the calorimeter measurements. If the laser has external mirrors, it may be possible to insert an absorber between the laser rod and the nonoutput mirror to preclude laser action but retain all the other sources of radiation.

Depending on the calorimeter and the response speed of the readout instrumentation, three characteristics of the output signal may be seen after a laser pulse. The laser output pulse is, of course, very short compared with any thermal time constants of the calorimeter. Its shape will not be resolved. There may, however, be a short, high-amplitude spike indicative of the heat wave that reaches the detector before equilibration. The shape and amplitude of the heat transient will depend on the position of the laser beam with respect to the transducer. Fortunately, the transient peak is not of interest in determining input energy. The initial voltage pulse will decay to a nearly constant value as the energy diffuses throughout the calorimeter. The quasisteady-state output of the temperature transducer is used, with the calorimeter calibration factor, to determine the energy in the laser pulse. After many tens of seconds or minutes the transducer output will decay slowly as the temperature of the calorimeter returns to its quiescent (room) temperature. To minimize deviations in the measurements caused by time-dependent variations in losses it is wise to measure the transducer output after some consistent time interval and to allow the calorimeter to equilibrate an equal period of the order of several minutes between successive measurements. For the most precise work the calorimeter should be permitted to return to the same temperature before each measurement. This is annoying in practice for the time constant is usually very long.

The calorimeter should be shielded from drafts and heat sources such as intense incandescent lamps that fall within the detector's field of view. The energy measuring equipment and experimental arrangement should be static for at least 30 min to minimize thermal drifts before measurements are made.

Particularly when using small-aperture calorimeters it is difficult to be certain that the laser pulse will fall within the aperture. The techniques outlined in Section 4.10 may be helpful, particularly if the laser emits in the infrared. A lense may also be found helpful in reducing the alignment accuracy required by reducing the laser-spot size. Beam refraction obtained by moving the lense yields a convenient means of beam displacement if the position of the laser and the calorimeter are not easily changed.

The output energy of the laser, subject to the errors, corrections, and validity tests outlined above, can be determined by multiplying the measured signal by a calibration factor. The latter, previously established by the calibration of the calorimeter, is usually of the order of microvolts to millivolts per joule.

4.11.3.4 Technique of Energy Measurement with a Resistance-wire Bolometer The change in the resistance value of a wire bolometer, consisting of a bundle of insulated wire, has been used with some success to measure the energy output of pulsed-ruby lasers. Although the equilibration time of a cone type of calorimeter is measured in seconds, the wire bolometer time constant can be as short as 10^{-4} sec[174]. In application, the laser beam is directed into the calorimeter wherein it is reflected, scattered, and absorbed, causing a rapid rise in the resistance of the insulated bundle of wire. The resistance change is directly proportional to laser energy if the wire size is uniform and the temperature coefficient of resistance and specific heat of the wire are independent of temperature.

Although not a true calorimeter, in the sense that it measures radiation independent of wavelength, the resistance-wire bolometer may be conveniently used for measurements in the visible and near infrared at energy levels in the 0.1-to-10 J range under long pulse (10^{-3} sec) conditions. An outstanding feature of this instrument is that the equilibration time is short. The change in resistance of the wire is a measure of the total energy absorbed and, because the change in resistance is independent of the volume distribution of heat in the wire, there is no equilibration time associated with the absorber.

It can be shown that the basic relation between input energy and resistance change is of the form[174]

$$U = \frac{\theta m c_p}{\alpha R_0} \Delta R \quad \therefore \quad U \propto \Delta R, \tag{4.66}$$

where m is the total mass of the resistance wire, c_p is the specific heat, α here is the temperature coefficient of resistance, θ is a constant, and R_0 is the resistance of the wire prior to the impact of the laser-energy pulse. Thus the change in resistance ΔR is related to the input energy in a basic way.

Although the simple relation described above implies that the calorimeter is capable of absolute calibration, it turns out that there are several sources of error that must be considered. These include the effect of the 0.0003 in layer of insulating varnish on the wire, which increases the heat capacity as much as 29 percent, Fresnel reflection as the calorimeter window (≈ 8 percent) and backscattering of energy from the wire bundle (≈ 18 percent).

The resistance-wire bolometer commonly employs a Wheatstone-bridge with a dummy bolometer for energy measurement. A decade box or calibrated multiturn potentiometer may be used to null the bridge circuit before a measurement is made. Rather than measuring the resistance before and after a laser pulse, it is more convenient to determine a ratio N of the meter deflection d that develops as the result of a known change in the decade resistor ΔR. Using the calorimeter calibration factor χ the energy equation becomes

$$U = \frac{\chi\,\Delta d}{N} \quad \text{J}.$$

If the calorimeter is constructed from 1000 ft of No. 40 B&S gage enameled copper wire loosely and randomly packed into a 50 ml beaker silvered on the inside and provided with a flat glass window, the energy sensitivity will be approximately

$$U = 2.38\,\Delta R \quad \text{J}.$$

Application of this calorimeter must be restricted on the high side by the tendency of the insulation to deteriorate (approximately 10 J) and on the low side by the inherent drift of the instrument (10^{-3} J). Its use to measure the output of Q-switched lasers is not recommended.

4.11.3.5 Photoelectric Technique for Measuring Laser-energy Output Within the wavelength range where photoelectric surfaces are useable the minimum detectable laser energy can be reduced about three orders of magnitude if a vacuum photocell is used in place of a calorimeter. A direct measure of laser energy can be obtained using a photocell to place a charge on a high-quality (polystyrene) capacitor in a self-integrating circuit, as is shown schematically in Figure 4.17. The apparatus required is listed in Table 4.10.

The operating principle of the energy meter described above is straightforward. A charge Q is placed upon the capacitor C by power supply B to

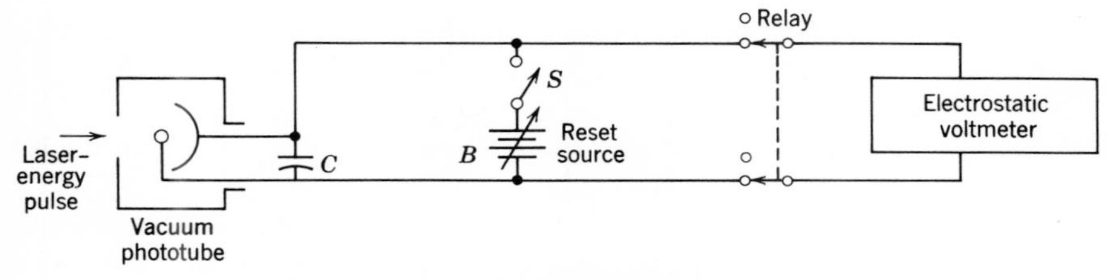

Figure 4.17. Schematic of a photoelectric laser-energy meter.

produce a capacitor voltage V that polarizes the photocell. The laser pulse induces a current pulse through the photocell and then partly discharges the capacitor. It can be shown that as long as the photocell current is linear related to the radiation intensity, the time integral of the photocurrent is linearly related to the time integral of the laser power. Thus we may relate the charge removed from the capacitor to the input energy[175]

$$U = \int_0^\tau P(t)\,dt = \frac{h\nu}{\eta e} = \int_0^\tau I(t)\,dt, \tag{4.67}$$

where I is the photocurrent, τ is the laser pulse length, η is the photocell quantum efficiency, e is the electronic charge, and $h\nu$ is the laser photon energy.

The Voltage decrement ΔV on the capacitor C is

$$\Delta V = \frac{Q}{C} = \frac{1}{C}\int_0^\tau I(t)\,dt. \tag{4.68}$$

The sensitivity ϑ of the photocell may be expressed as

$$\vartheta = \frac{I}{P} = \frac{\eta e}{h\nu} \quad \text{A/W}. \tag{4.69}$$

$$Q = C\,\Delta V = \int_0^\tau I(t)\,dt.$$

TABLE 4.10 APPARATUS FOR PHOTOELECTRIC MEASUREMENT OF LASER-BEAM ENERGY

Item	Function	Source
Photocell	Energy measurement	RCA 6570 or equal (S-1 surface)
Electrostatic voltmeter	Charge measurement	Sensitive research instrument model ESD
Integrating capacitor	Charge accumulator	J. Q. Fast

Using (4.67) and (4.68) the energy sensitivity of the photocell may be expressed in terms of the capacitor voltage decrement as

$$U = \frac{Q}{\vartheta} = \frac{C\,\Delta V}{\vartheta} = \left(\frac{h\nu}{\eta e}\right) C\,\Delta V \tag{4.70}$$

Direct application of the integrated-photocurrent technique of measuring laser-output energy is limited to the energy range of 10^{-2} to 10^{-4} J without employing attenuators or using high-sensitivity, high-impedance amplifiers.

The energy range of the photoelectric calorimeter can be most accurately extended if a scattering attenuator is used. The radiance of the laser beam may be decreased by using inverse-square-low attenuation, as is illustrated in Figure 4.18. Diffusely reflected energy from the scattering surface is examined at near-normal incidence by locating the photocell a distance R from the scatterer. The photocell cathode is defined by an aperture of area s. The energy received by the photocell is reduced by the ratio

$$\frac{U_{\text{photocell}}}{U_{\text{laser}}} = \frac{s\rho}{2\pi R^2}, \tag{4.71}$$

where ρ is the reflectivity of the scatterer at the laser wavelength. Attenuation ratios of the order of 10^8 are feasible. The errors in this technique are associated with the aperture and photocell-to-scatterer length measurements. A correction factor is required if radiance measurements are made at large angles.

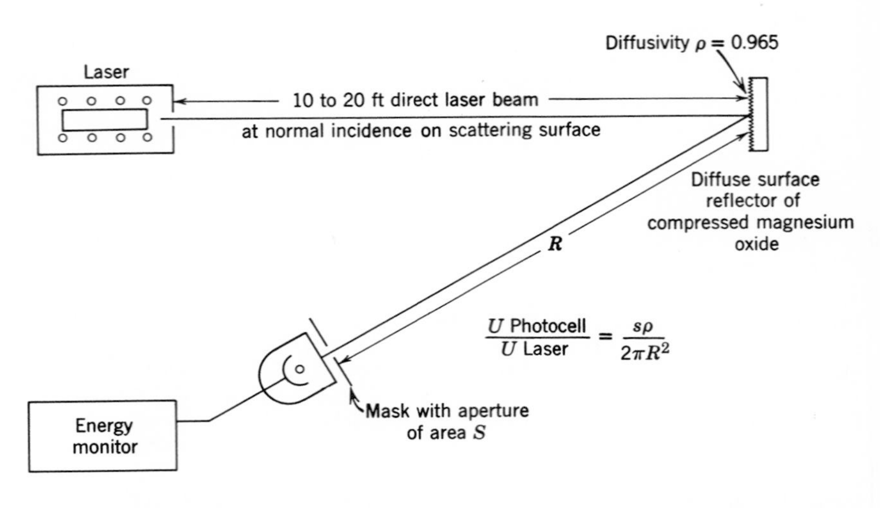

Figure 4.18. Reduction of laser radiance with a scattering attenuator.

Several precautions should be observed in applying the photoelectric technique of energy measurement. Due to the nonuniform sensitivity of the photocathode, it is necessary that a large fraction of the surface be illuminated. Errors will be introduced in the measurements if the anode of the photocell casts a shadow on the cathode. This effect can be eliminated by properly orientating the photocell or using a planar photocell. If a conventional photocell is used and A_i is the illuminated area of incidence on the unoccluded cathode and A_a is the area of the anode, the data may be corrected[176] by multiplying the voltage change by the simple ratio of $A_i/(A_i\text{-}A_a)$.

Although the photoelectric-energy meter is best calibrated with a standard light source, nominal values for the variables and parameters in (4.70) can be used to check the results.

4.11.4 Photo-counting Devices for Time-resolved Measurement of Laser Energy

A detector system for measuring time-resolved, absolute radiant-power requires a sensor that can respond with sufficient speed and dynamic range to perform the transducing function (conversion of radiant energy to electrical energy) with high fidelity. In addition, the electronic circuitry associated with photon-counting equipment with high speed and a wide dynamic range must have an adequate bandwidth to pass all the frequency components of the system without distortion. It is further desired that output impedance of the detector be low enough to enable the detector, for display purposes, to drive a traveling-wave type of oscilloscope with a low impedance.

A number of detection techniques have been evolved for the measurement of light power. When most of these are extended to the high-intensity level of pulsed, solid-state lasers, their size, response times, dynamic range, and saturation properties are incompatible with the task of rendering accurate measures of time-resolved laser intensity. A typical detector system capable of being used for the measurement of the high-intensity of the laser-output power consists of an attenuator for reducing the intensity of the laser radiation to a satisfactory level, a detector to transduce the optical energy into a current or voltage, and the readout instrumentation to display the wave shape (or a peak-reading meter). It is clear from Figure 4.19 that the power input in a photoelectric detection system is simply related to the photocurrent (to the extent that the system is linear) by the simple relation

$$P = \frac{I}{\vartheta\alpha}, \tag{4.72}$$

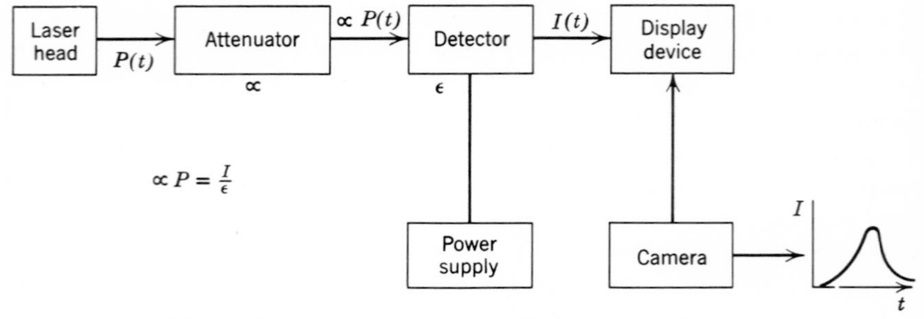

Figure 4.19. Schematic of laser peak-power measurement.

where I is the photocurrent, ϑ is the sensitivity of the detector in amperes per watt and α is the loss factor of the input attenuator.

Taking the logarithmic derivative of expression (4.72)

$$\left|\frac{\Delta P}{P}\right| = \left|\frac{\Delta I}{I}\right| + \left|\frac{\Delta \vartheta}{\vartheta}\right| + \left|\frac{\Delta \alpha}{\alpha}\right|, \tag{4.73}$$

it is clear that the magnitude of the error in power measurement is associated with fractional changes in photocurrent, responsivity, and attenuation constant. Neglecting all the systematic losses in the photoelectric detection system for the moment, such as wavelength-dependent reflections, due to the Fresnel mismatch, from surfaces, sources of background radiation interference caused by large rates of change of exciting current normally associated with flash lamps, and so on, it is clear that there remain three major sources of error (nonlinearities in the photocurrent, the sensitivity, and the attenuation constant).

A stringent requirement must be placed on both the uniformity of response and the linearity of phototubes used for high power measurements of lasers. The response must be linear over a very large dynamic range and must not display space-charge saturation effects. Early experience in measuring the peak-power output of lasers with photomultipliers demonstrated that space-charge saturation limits the available photocurrent in photomultipliers to the extent that the response even at low dynode voltages becomes intolerably nonlinear. Special high-speed phototubes are, fortunately, now available in configurations that enable high-speed operation and linear intensity-to-photocurrent conversion characteristics, with reasonable quantum efficiencies, at peak currents up to several amperes. Table 4.11 lists a number of high-speed photoelectric detectors commercially available today. Careful note should be taken of the fact that these are power detectors, adequate for high-power laser measurements, and are not suitable as signal detectors for communications systems wherein detectivity is of paramount importance.

TABLE 4.11 HIGH-SPEED PLANAR PHOTOCELLS [a]

	FW 162	FW 128	FW 114	FW 114A	F 4000	FW 127
EIA cathode type	*S*4	*S*4	*S*4	*S*20	*S*1	*S*4
Peak current (A)	0.1	0.5	5	5	5	30
Rise/fall time (nsec)	0.5/0.4	0.5/0.5	0.5/0.8	0.5/0.8	0.5/0.8	1/2
Capacitance (pF)	1.5	3.6	7	7	7	26
Operating voltage (kV)	1	1	2.5	2.5	2.5	2.5
Linear dynamic range	10^9	2.5×10^8	10^9	10^9	10^8	3×10^9
Design operating impedance (Ω)	–	100	125	125	125	50

[a]Products of the Farnsworth Division of ITT Corp.

The choice of optical attenuators is of great concern in high-power laser measurements for the following reasons.

1 Tendency for the attenuation factor to change during a laser pulse as a result of quantum bleaching.
2 Tendency for the attenuators to fluoresce and produce radiation at wavelengths far removed from the wavelength of the incident radiation (but falling within the response characteristics of the photodetector).
3 Tendency for dielectric surfaces to shatter under high peak-power, short-pulse conditions encountered in *Q*-switched lasers.
4 Susceptibility of film-type attenuators to irreversible change over several pulses or during a single pulse.
5 Tendency for attenuators to be polarization sensitive.
6 Tendency for attenuators to be wavelength sensitive.

The most promising technique available today for obtaining reproducible linear attenuation of high-intensity laser beams uses a magnesium oxide block as a diffuse target upon which the laser beam is allowed to impinge. When properly prepared the target can operate effectively as an attenuator because only a small sample of the diffusely reflected laser energy is transmitted to the detector by the target. This enables us to measure a fraction of the laser power rather than requiring us to expose the detector directly to the full laser output.

For the target to be useful it must be highly diffuse and have a specular reflectance characteristic sufficiently low that negligible error is introduced, if other than diffuse radiation reaches the detector. Even the best optical surfaces do not obey the Lambertian Law for angular dependence of diffusely reflected intensity. Therefore either the target angle must be fixed in all observations or a correction factor must be applied to the result to remove the angular dependence of diffusivity.

Obviously the target should be made of a material that does not evaporate, deteriorate, or change color when repeated high-power pulses impinge on the surface. A material with a high melting point is required. Magnesium oxide, long used as a standard reflecting surface in photometry, although adequate for low laser power (e.g., in the 10-kW range) is so soft that the layers of magnesium oxide tend to flake and peel under repetitive laser pulses. Perhaps the most satisfactory surface available to date is a block of compressed magnesium oxide with a high purity, but even this surface tends to yellow after prolonged exposure to laser radiation and air. These difficulties can generally be overcome if the surface of the block is rubbed with fine sandpaper.

4.11.4.1 Photoelectric Technique of Measuring Peak Power and Energy Output of *Q*-switched Lasers Measurement of power and energy outputs of *Q*-switched lasers, at peak power levels above 10^7 W is distinguished by two phenomena that present unique photometric problems: extremely high radiance and light pulses of nanosecond duration. The peak power is sufficiently large to engender a number of second-order effects, typical of which are stimulated Raman emission and Brillouin and Rayleigh scattering. Nonlinear optical phenomena can cause serious errors unless each step in the measurements process is carefully examined.

4.11.4.1.1 Fast Photodetector To minimize the magnitude of error in the power measurements of *Q*-switched lasers, a fast photocell with a very large dynamic photoresponse is required. High-current vacuum photodiodes of the planar type previously discussed will satisfy this requirement if they are properly applied. Furthermore, these photodiodes offer the advantage of being able to withstand high voltage, thereby minimizing the electron transit time effect; they also supply photocurrents at levels adequate to drive low-impedance circuitry critical to fast pulse measurements. Because these devices can be operated in the constant-current regime, they are capable of linear photoresponse over a very large range. Where the criterion of ± 10 percent nonlinearity between input flux and output current is applied, the input light flux being uniformly distributed over the effective photocathode area, the maximum photocurrent is approximately one-half the space-charge-limited current.

An instrument that contains the planar photocell must provide a coaxial housing that permits a good impedance transformation from the largely lumped capacitance of the photodiode to the resistive impedance ($\sim 125\,\Omega$) of the transmission line which leads to the oscilloscope. High-frequency matching to the coaxial cable is accomplished with a lumped inductance within the housing.

The instrument package must provide a low inductance source of high voltage to ensure that an appropriate reservoir of charge is available to sustain operation of the photocell in the constant-current regime while the latter is exposed to the high radiance pulse. (A schematic mechanical assembly of a typical instrument is shown in Figure 4.20.) This assembly illustrates two additional features that are desirable in an instrument. The package should contain a divergent lense that will enlarge the beam spot to fill the face of the photocell, thereby averaging out the effects of nonuniform sensitivity over the photosurface. Application of the photocell often requires its optical isolation from a stray light source such as a flash lamp. A narrow-band filter can be inserted in front of the photocell to monochromatize the sensitivity to the wavelength range of interest effectively.

The circuitry that provides the high-voltage bias for the planar photocell can conveniently include a diode integrator that will provide a voltage proportional to power at the conclusion of the laser pulse.

4.11.4.1.2 Procedure for Q-Switch Laser Peak-Power Measurement A good arrangement of experimental equipment for laser peak-power measurements is illustrated in Figure 4.21. Also, the instrumentation outlined in Table 4.12 is a convenient reference.

Because the total radiant output of a Q-switched laser is far too large to permit direct exposure of the planar photocell, a means of linear attenuation must be employed to enable a meaningful reduction in the flux intercepted by the detector. As we have already mentioned, the usual means of optical attenuation is unsatisfactory, the beam is scattered by a diffuse target[177], so that the power density is reduced by spreading the energy over a hemisphere of radius R. Although a compressed magnesium oxide block is one of the best presently available scattering targets, it is not prefectly Lambertian. Furthermore, the diffusivity of magnesium oxide is wavelength dependent,

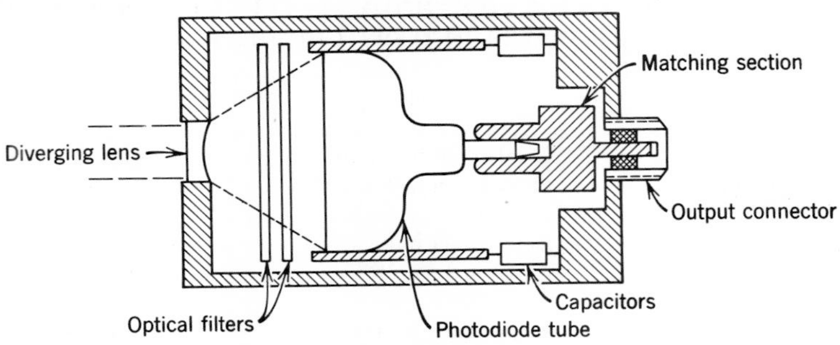

Figure 4.20. Mechanical assembly of a fast-photocell instrument for measuring Q-switched laser pulses.

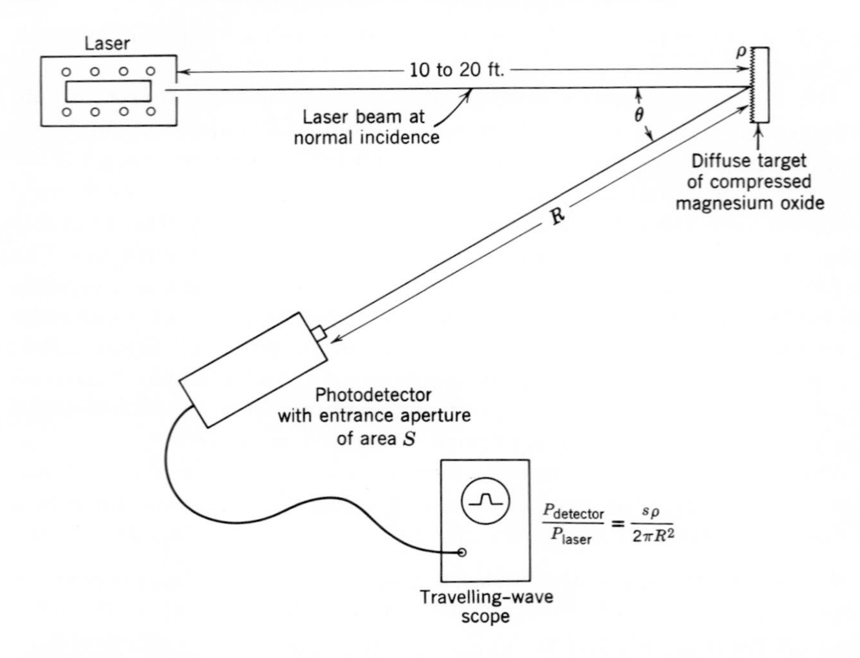

Figure 4.21. Apparatus arrangement for measurement of Q-switched laser power output.

particularly in the infrared[178], as is shown in Figure 4.22. (Preparation of magnesium oxide targets with reproducible characteristics turns out to be more of an art than is desired for precision measurements.)

Provided that the laser beam impinges on the target at normal incidence and the angle between the detector and the beam is kept small, the laser-beam intensity, as measured by the detector of entrance aperture area s,

TABLE 4.12 APPARATUS FOR PHOTOELECTRIC MEASUREMENT OF PEAK POWER IN Q-SWITCHED LASERS

Item	Function	Source
Traveling-wave oscilloscope	Display output of photocell	Tektronix model 519
Oscilloscope camera	Photograph high-speed oscilloscope trace	Tektronix C-12
Fast photoelectric detector	Detect laser pulse	Korad KD-1 (or equal) and associated power supply
Compressed MgO block	Scatter laser beam	Baker Chemical
Q-switched laser	Pulsed light source	Korad K2Q or equal

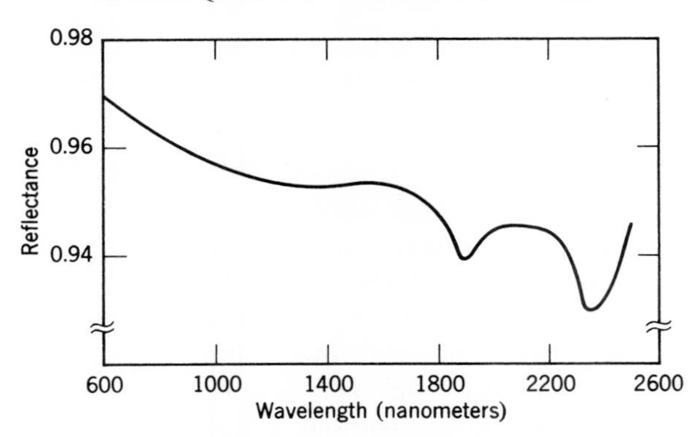

Figure 4.22. Absolute spectral diffuse reflectance of magnesium oxide in the near infrared [178].

is reduced to

$$P_{\text{detector}} = P_{\text{laser}}\left(\frac{s\rho}{2\pi R^2}\right) \tag{4.74}$$

Because R is arbitrary and adjustable over a large range, linear attenuation ratios of nearly 70 to 80 dB can be achieved. The upper limit in the ratio is set by the requirement that the detector aperture area be large enough to average over a number of lobes in the "speckle" spatial-interference pattern of the illuminated area.

If the angle between the detector and the target departs markedly from a small value, two defects will arise. First, the detector result must be cosine corrected for the reduced off-axis reflectance of the Lambertian radiator. Second, the effects of particle size of the magnesium oxide crystals on diffuse reflectance are magnified.

Assuming good geometry, the power output of the laser is measured by the following procedure:

1. Adjust the laser to be measured for consistent performance at the desired power level.
2. Locate the detector, as shown in Figure 4.21, at an appropriate distance from the target to receive an adequate signal for display on the oscilloscope. Determine accurately, measure, and record the distance R.
3. Photograph the output-pulse shape of the detector.

For a freshly sanded target surface $\rho \simeq 0.96$ at 694.3 nm. Using the detector-aperture area, the value of R and ρ, calculate the attenuator ratio (4.74). From the photograph of the output-pulse wave shape and the manufacturer's calibration of the photocell, determine P_{detector} and compute

P_{laser} from the ratio given in (4.74) ratio. Using a planimeter, determine the area under the output-pulse wave shape. From its integral compute the energy content of the pulse.

4.11.4.1.3 Measurement of the Energy Output of a Q-Switched Laser The peak-power output that can be achieved with some *Q*-switched lasers is now so large that many total absorption laser calorimeters satisfactory for low-power operation (about 500 kW) cannot be used because they will be irreversibly damaged in a single pulse. Damage can be precluded if a liquid calorimeter is used that admits the total energy of the *Q*-switched laser pulse to the absorbing liquid through a transparent window capable of transmitting high peak power without sustaining permanent damage. Sapphire, fused quartz, borosilicate, and barium crown glasses, as shown in Table 4.13, have been found to have the greatest resistance to damage from *Q*-switched, ruby-laser pulses.

The fluid used in the calorimeter must absorb the laser radiation totally and in such a way that the absorbed energy is transformed into heat rather than into other forms of energy such as vaporization or chemical-bonding energy. Monotomic metal-ion solutions have been found to

TABLE 4.13 RUBY-LASER DAMAGE THRESHOLD FOR SEVERAL CORNING GLASSES[a]

Corning Code No.	Damage Threshold (GW/cm^2)	Notes
GE-106	2.4	Made from natural quartz
X191-AST	10.0	Experimental high elastic modulus glass
1723	6.0	Tempered aluminosilicate glass
1723	15.4	Annealed aluminosilicate glass
7070	45.0	Borosilicate glass—low elastic modulus
7740	6.0	Borosilicate glass (tempered)
7900	490.0	VYCOR (96% silica glass)
7905	85.0	Less water than 7900
7940	470.0	High-purity silicate glass
8361	500.0	White-crown optical glass
8363	5.6	Extra-dense-flint optical glass
8370	710.0	Borosilicate crown glass
8387	710.0	Dense barium crown glass
8461	2.0	Lanthanum borate optical glass
0311	4.8	Salt-treated
0311	37.0	Not salt-treated

[a]Reproduced with permission from *Applied Optics*.

satisfy these conditions. The fluid container must not react chemically with the contents, or the calibration will change with time. Plated copper has been found to satisfy this condition and the requirement of high thermal conductivity. A simplified mechanical assembly of a liquid calorimeter is illustrated in Figure 4.23.

The energy deposited in the calorimeter may be sensed as a temperature or a volume change. Temperature changes are sensed with either a thermocouple pair, with sensitivities of 0.162 J/μV, or a thermistor with sensitivities of 5 J/mV. Liquid calorimeters typically have absorbing volumes of the order of a few cubic centimeters and are capable of overall accuracies of better than ± 3 percent when carefully and recently calibrated. Their energy range is typically 0.1 to 10 J. They may be calibrated conveniently by supplying a known quantity of heat to an immersed resistance element.

Application of the liquid calorimeter to the measurement of the energy of a Q-switched laser is conveniently combined with the photoelectric measurement of power output. The equipment listed in Table 4.14 will be found convenient for the combined measurement of energy and power. Accurate measurement of high peak powers by photoelectric techniques may require verification that phototube saturation and nonlinear optical phenomena are not interfering with results. Assuming that the liquid calorimeter used for a reference is accurate, the total energy output of the laser may be measured and compared to the results obtained with the equipment arrangement shown in Figure 4.24.

Two photocells are employed in the measurement of the energy of a Q-switched laser, one to monitor the laser output for multiple pulses, which provides a check through electronic integration of the beam energy, and one to measure the fast laser pulse. Two oscilloscopes are required, one to monitor the fast laser pulse and one to monitor the laser output during the

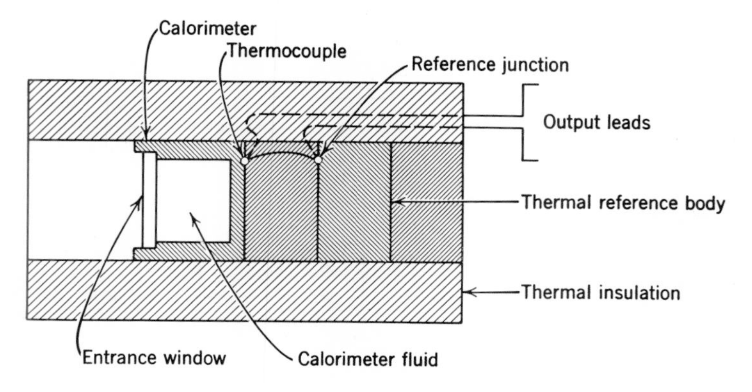

Figure 4.23. Simplified mechanical assembly of a total-absorption calorimeter.

TABLE 4.14 APPARATUS FOR SIMULTANEOUS MEASUREMENT OF ENERGY AND POWER OF A Q-SWITCHED RUBY LASER

Item	Function	Source
Oscilloscope	Display energy and check for multiple pulses	Tektronix 531
Fast oscilloscope	Display peak power	Tektronix 519
Chart recorder	Record energy output on chart	Varian G-10
Fast photocells (2)	Power and energy measurement and beam monitoring	Korad KD-1 or equal with MgO Targets
Liquid calorimeter	Energy measurement	Korad K-J or equal
Cameras (2)	Photograph waveforms on oscilloscopes	Tektronix C-12
Microvoltmeter	Measure calorimeter output	HP 425A
Pulsed Q-switched laser	High peak-power light source	Applied lasers 1014Q4

entire optical pumping pulse and also to display the electronically integrated laser-energy pulse. Two cameras provide a simultaneous photographic record of the events.

The measurement proceeds as follows:

1. Adjust the laser for consistent operation at the peak-power level of interest.

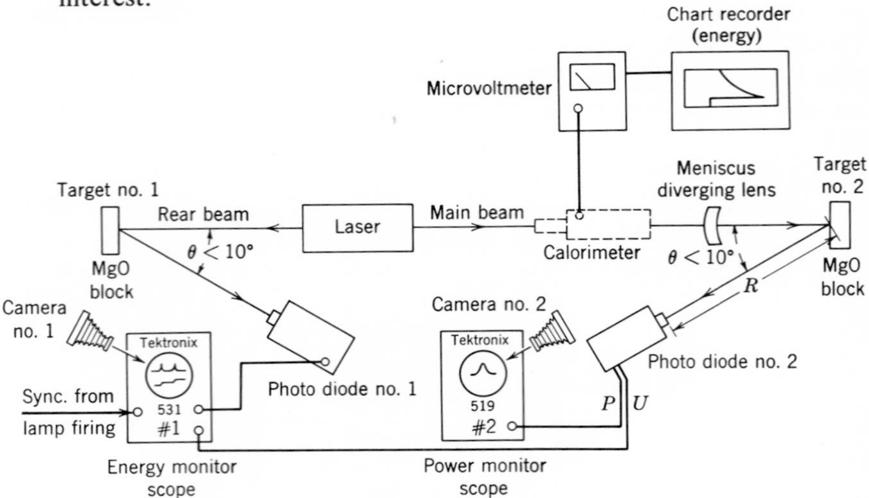

Figure 4.24. Apparatus arrangement for simultaneous measurement of the energy and power of a Q-switched ruby laser.

2. Place the monitor photocell so that it yields a convenient signal level from the diffuse reflector irradiated by the rear beam of the ruby laser. A spacing of a few feet will generally be found adequate. Adjust the oscilloscope to trigger at the start of the flashlamp pulse and set the sweep speed to display the full period of the pump pulse. The sensitivity of the second sweep must be adjusted when the integrated output of photocell No.2 is obtaned.
3. Locate target No. 2 about 10 to 20 ft from the laser and place its face normal to the beam. Place a diffusing lense between the laser and the target to preclude radiation damage. Place photocell No.2. at an angle of 10° or less with respect to the beam and spaced at a distance R adequate for a satisfactory signal display on the fast oscilloscope. Connect the electronically integrated signal output to the dual-beam oscilloscope to obtain an energy record.
4. Locate the calorimeter so that it can be placed accurately in the direct beam in an easily reproduced location. Adjust the sensitivities of the microvoltmeter and chart recorder to yield nearly full-scale values.
5. Make the checks previously described to eliminate electrical and stray light interference.
6. Accurately measure and record the distance between photocell No.2 and target No.2 and the angles with respect to the laser beam. Calculate the attenuation factor using (4.74).
7. Alternately measure the beam energy with the liquid calorimeter and the power and energy with the photocells. The output of the former will appear on the strip-chart recorder. The latter must be derived from the photographs of oscilloscopes 1 and 2 using their respective gain and time settings.

Analyze the data obtained from the procedure described above. The liquid-calorimeter measure of energy should agree both with the integrated energy output of photocell No.1 using its calibration factor and the energy obtained from mechanically integrating the area under the pulse from the photograph obtained on oscilloscope No.2. The appropriate calibration factors must, of course, be used. If the data do not agree, the discrepancy may be traced to electrical or optical interference or to "prelasing" or "afterlasing" caused by multiple pulse output from the laser. The latter can easily be recognized as steps in the integrated output pulse of photocell No.2.

Using the liquid calorimeter as a reference, the MW/V calibration of photocell No.2 can be checked if the pulse-to-pulse reproducibility of the laser is adequate. Using the pulse duration and the total pulse energy, the peak power can be calculated without reference to the oscilloscope calibration. Using the peak power so calculated and the oscilloscope calibration and the attenuation factor from equation 4-74, the MW/V sensitivity of the photocell can be determined and computed with the calibration. Note that due account must be taken in the calculations above, of the insertion loss of the beam diverging lens. This usually amounts to about 7 percent, but it will depend upon the lens material used.

4.11.4.1.4 Power Measurements in Amplifier Systems The power measurements of Q-switched, oscillator-amplifier laser systems sometimes require special techniques. To attain the highest possible power densities (limited by damage to the amplifier crystal), it is desirable to measure simultaneously and thereby adjust the output of the oscillator and the amplifier during system optimization. A very convenient method for accomplishing the measurements with a minimum of equipment is shown in Figure 4.25.

A beam splitter is used to sample the oscillator output. The sampled portion is transmitted through colored-glass attenuating filters and through a second beam splitter to a diffuse reflector. By varying the position of this reflector and the transmission of the filters, the deflection of the oscilloscope trace due to the oscillator output can be adjusted.

The amplifier output is transmitted through a weakly diverging lens to spread the beam sufficiently so that the power density at the target is below the target damage threshold. It is advisable to use a concave-convex (meniscus) lense orientated as shown in Figure 4.25, to prevent focusing of reflected energy back into the laser (thus avoiding damage to the crystal).

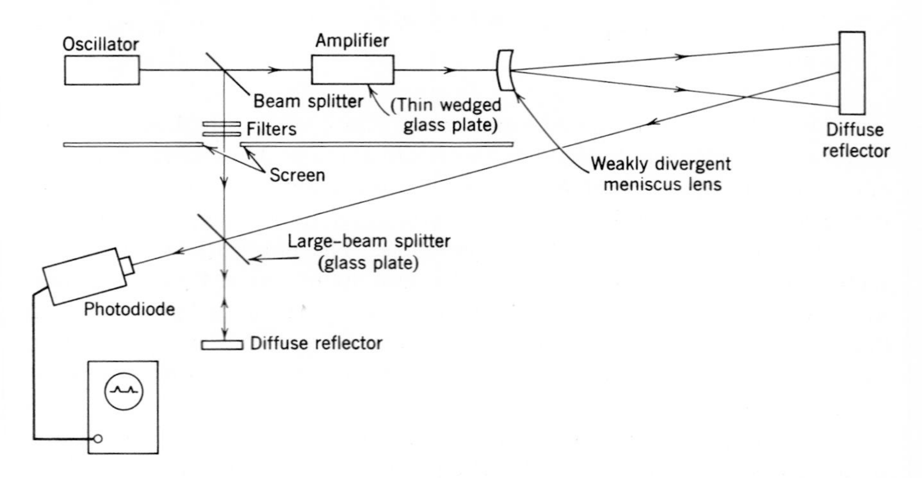

Figure 4.25. Arrangement of apparatus for the measurement of power in an amplifier system.

If the distance from the photodetector to the main target is sufficiently large, two separate signals will be seen on the oscilloscope, separated by the difference in transit times.

Measurement of the gain of the amplifier can be accomplished conveniently by first adjusting the system for equal (or nearly equal) deflections with the amplifier removed. When the amplifier is replaced, one of the attenuating filters is moved from the oscillator sampling path to the photodetector, the attenuation factor being close to the anticipated power gain. This will maintain the calibration of the oscillator pulse but decrease the sensitivity to the amplifier signal by the attenuation factor of the filter.

Using the technique described in the paragraph above, the oscillator and amplifier deflections can both be maintained near the maximum allowable value, thus yielding the maximum accuracy.

A light screen should be so placed as to prevent stray reflections from reaching the photodetector, in particular those from the input of the amplifier stages(s) and from the diverging lens. This technique described above can be extended to a chain of amplifier stages.

4.11.4.1.5 Extension to Other Wavelengths There is nothing special about 694.3 nm in any of the procedures described. In applications to other wavelengths, the most likely being 1.06 μm at this time, the only precautions to be observed are to be sure that the reference calorimeter absorption is not changed, or, if it is changed, that it is changed by a known factor, and that the reflectivity characteristics of the diffuse target are known at the new wavelength. Pressed magnesium oxide is excellent to at least 1.1 μm. At short wavelengths, less than 0.5 μm, we must be concerned more about the aging of the target and clean the block occasionally to provide a new surface.

4.11.4.1.6 Nonlinear Sources of Error in Peak-power Measurements The measurement of high laser powers can be complicated by the nonlinear interactions of light with materials. The properties of optical components, such as spectral filters, attenuators, beam splitters, reflectors, and windows, will, in general, change at sufficiently high flux densities.

The linearity of the reflectivity of the sintered MgO block and its conformity to the cosine law for angles close to normal have been verified to power densities of 5×10^8 W/cm^2. At greater fluxes it becomes markedly sublinear. The onset of this condition is sometimes accompanied by a white luminous display near the surface of the block.

Colored-glass filters, such as those supplied by Corning and Schott, and absorbing filters in general may either become more absorbent at high fluxes, by multiple photon absorptions, or their absorptions may decrease or "bleach." Most of the commonly used optical filters have been checked and found to be linear up to power densities of 5×10^8 W/cm^2. These tests are conveniently performed by commutation of filters or by placing them alternately in the full flux and in the sampled flux. Another convenient method is to vary the flux density with a set of lenses of varying focal lengths, noting any transmission changes. Notable nonlinear filters are the CdS:Se glasses, such as Scott RG-10, which bleach readily and have been used as passive Q-spoilers for generating giant laser pulses.

The stimulated Raman effect and the related stimulated Brillouin scattering phenomenon are nonlinear effects that have significant bearing on the problem of power measurements. Because the solid-state laser may be operating multimode, large statistical fluctuations in the nonlinearprocesses can occur, and no single laser pulse should be considered as typical without verifying reproducibility. Raman light can be distinguised most easily by spectral filtering with a spectrometer, colored glasses, or interference filters. Each substance used in high-power laser applications should be treated as potentially capable of engendering its unique set of stimulated Raman lines at unique wavelengths. Much the same consideration must be given to the stimulated Brillouin scattering, which may be considered a Raman process mediated by acoustical modes. The spectral shifts are generally less than a wave number and higher resolution must be used to detect their presence.

4.11.5 Power and Energy Measurements Using an Optical-polarization Transducer Detector

A high-intensity laser beam propagating through a crystalline medium that lacks inversion symmetry develops a dc polarization in the medium, and the magnitude of the polarization can be shown to vary linearly with the power in the laser beam. This principle can be utilized to build a transmission type of device to measure power in a laser pulse in that the energy in the pulse can be determined by passing the putput of the device through an integrating circuit. This section describes the theoretical considerations and experimental details of the study made into the feasibility of utilizing the nonlinear effects to measure laser power.

4.11.5.1 Theoretical Considerations The phenomenon of dc polarization can be easily explained by assuming a scalar mathematical model for the polarization. Consider an electromagnetic wave. The scalar form of polarization including the nonlinear terms is given by

$$p = a_1 E + a_2 E^2 + a_3 E^3 + \cdots, \tag{4.75}$$

where a_1, a_2, a_3, ... are the first-, second-, third-order polarization coefficients. Substituting for the radiation field $E = E_0 \cos \bar{\omega} t$ in (4.75), we have

$$p = a_1 E_0 \cos \omega t + \frac{a_2}{2} E_0^2(1 + \cos 2\omega t) + \frac{a_3}{4} E_0^3(3 \cos 3\omega t + \cos 3\omega t). \tag{4.76}$$

It can be observed from (4.76) that a dc polarization of magnitude $a_2 E_0^2/2$ is developed in the medium and that it is directly proportional to the power in the radiation field.

In general, the polarization coefficients are tensors. For a crystal that belongs to class 32 (e.g., crystalline quartz) the second-order polarization along the x,y,z axes can be written as [180]

$$\begin{aligned} P_x &= \alpha(E_x^2 - E_y^2) + \beta E_y E_z, \\ P_y &= \beta E_x E_z - 2\alpha E_x E_y, \\ P_z &= 0 \end{aligned} \tag{4.77}$$

where E_x, E_y, and E_z respectively are the components of the electric-field intensity along with x-, y-, and z-axes of the crystal, and α and β are the nonvanishing terms of the second-order polarization coefficient tensor. It has been shown the second-order polarization tensor is the same as the electro-optic tensor[181].

Consider a linearly polarized wave of amplitude E_0 and angular frequency ω, with the plane of polarization making an angle θ with respect to the x axis and propagating in the z direction. From equation 4.77

$$P_x = \frac{\alpha E_0^2}{2} \cos 2\theta, \tag{4.78}$$

$$P_y = \frac{\alpha E_0^2}{2} \sin 2\theta,$$

$$P_z = 0,$$

$$p = \frac{\alpha E_0^2}{2}. \tag{4.79}$$

The orientation of E and p for the case is shown in Figure 4.26 in which xx, yy are the crystal axes and $x'x'$, $y'y'$ are arbitrary choice of coordinate axes. Note that if the incident polarization makes an angle θ with the x axis of the crystal, the dc polarization makes an angle -2θ with it.

4.11.5.2 Theory of the Power Meter It can be observed from (4.79) that by measuring the dc polarization, the power in a linearly polarized beam can be determined. Because any type of polarization can be considered as

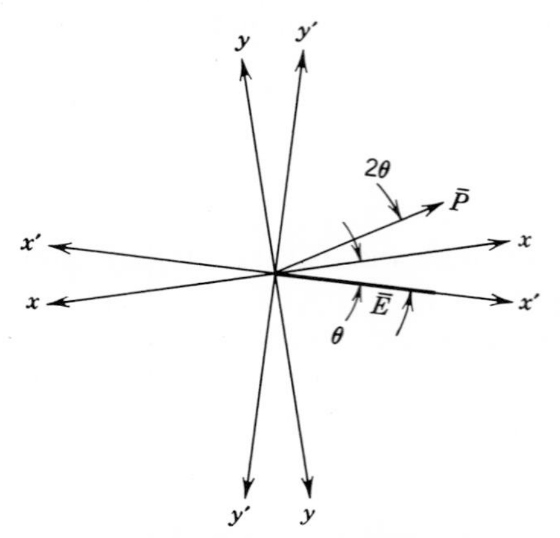

Figure 4.26. Orientation of P and E with respect to the crystal axes xx and yy. (Reproduced with permission from the Polytechnic Press of the Polytechnic Institute of Brooklyn.)

composed of two linearly polarized waves, power in any laser beam can be obtained by using this principle.

Consider a circular cylindrical speciman of a class-32 crystal, which is uniaxial with the laser beam of circular cross section propagating in the z direction and concentric with the crystal. Let the radii of the beam and the crystal be a and b respectively. For a plane polarized laser beam in the x direction, the potential outside the crystal can be shown to be equal to[180]

$$V = \frac{2\alpha\eta}{\pi\epsilon_0(1 + \epsilon_r)} P_L \frac{1}{r} \cos\theta \tag{4.80}$$

where P_L is the power in the laser beam, η the intrinsic impedance of the medium in the transverse direction, ϵ_0 the permitivity of the free space surrounding the crystal, ϵ_r the relative dielectric constant of the crystal medium and r and θ the cylindrical coordinates. Equipotential lines are described by (4.80); contours of $r = k \cos\theta$, where k is an arbitrary constant, describe equipotential surfaces.

The basic construction of the power meter consists of placing a pair of these equipotential lines corresponding to $r = k_1 \cos\theta$, where k_1 is a constant. If the beam is assumed to be polarized along the x axis, the voltage across the electrodes is given by

$$V = kP_L, \tag{4.81}$$

$$k = \frac{4\alpha\eta}{(k_1\pi\epsilon_0 1 + \epsilon_r)}. \tag{4.82}$$

The equivalent circuit for the system is shown in Figure 4.27, where C_Q

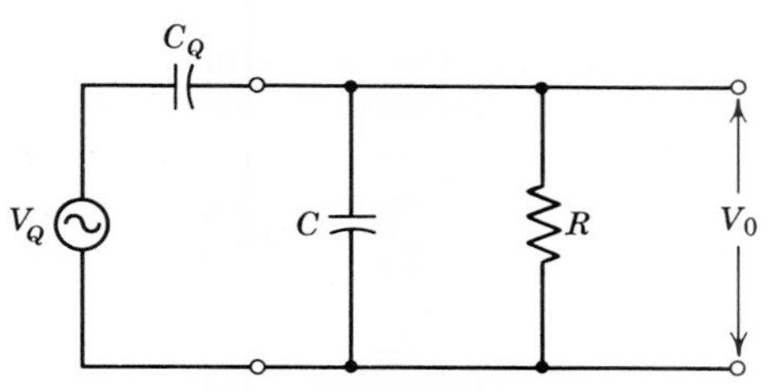

Figure 4.27. Equivalent circuit of the detector. (Reproduced with permission from the Polytechnic Press of the Polytechnic Institute of Brooklyn.)

is the capacitance formed by the two electrodes, C the input capacitance, and R the input resistance of the measuring device. The output response to an idealized square-wave pulse is shown in Figure 4.28. For a faithful reproduction of the pulse, the time constant of the circuit should be large compared to the duration of the pulse.

4.11.5.3 Construction of the Detector A cutaway view of a crystal quartz detector mount is shown in Figure 4.29. The quartz is mounted by supports at both ends. The electrodes are of the shape $r = k \cos \theta$ and are placed

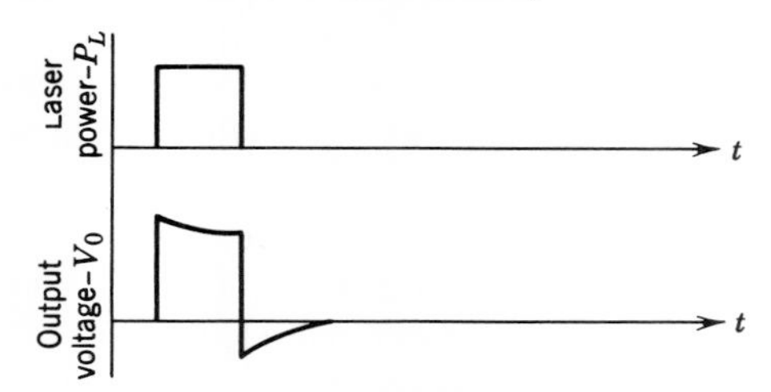

Figure 4.28. Detector output response to a square laser pulse. (Reproduced with permission from the Polytechnic Press of the Polytechnic Institute of Brooklyn.)

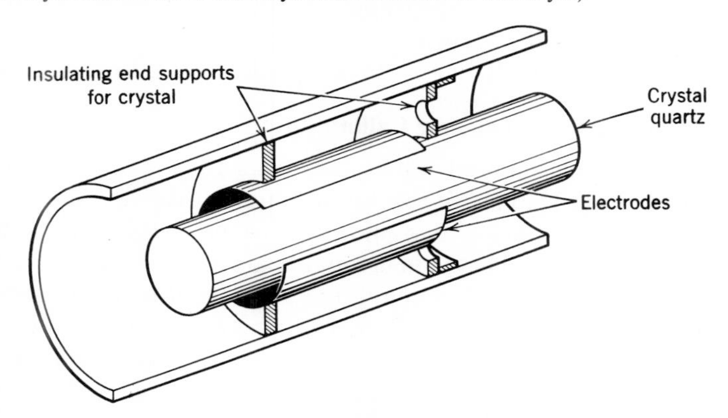

Figure 4.29. Cutaway view of a quartz detector. (Reproduced with permission from the Polytechnic Press of the Polytechnic Institute of Brooklyn.)

perpendicular to the x axis. The output from the two electrodes is fed to a high-imput impedance, balanced-cathode-follower preamplifier. The quartz mount is fitted inside a metal cylinder to provide shielding from external noise pickups. The preamplifier, using a subminiature tube, is built directly inside the mount shield to reduce the capacitance to a minimum. The output of the detector preamplifier is fed to a fast-response recorder.

4.11.5.4 Calibration and Measurements Procedure To calibrate an ideal power meter, it is necessary that the value of the second-order polarization coefficient be known. The practical power meter, however, can be calibrated externally. The area under the amplifier-output pulse corresponds to the total energy in the beam. If we compare the positive integral of the pulse with a reference calorimeter, we obtain the enegy calibration.

The output voltage of the preamplifier has the same shape as that of the laser pulse, that is the instantaneous amplitude of the power meter pulse is directly proportioned to the intensity of the laser beam.

4.11.5.4.1 Performance Characteristics The detector sensitivity depends upon the magnitude of the nonlinear polarization coefficient, the diameter of the crystal, and the electrode geometry. A typical sensitivity value for a crystal 2 cm in diameter and 1 cm in length will be 10 to 50 mV per kilowatt of laser intensity.

The lower limit on the measurement of power is set by the noise of the preamplifier; it is of the order of kilowatts. The upper limit is reached when the electric-field intensity of the laser pulse approaches the breakdown-field strength in the crystal. This, however, can be extended somewhat by defocusing the beam, for the detector output depends only on the total power of the beam and is independent of the beam diamter. To obtain maximum resolution and accuracy the beam cross section must be circular and must be accurately aligned with the longitudinal z axis of the crystal.

4.11.6 Technique for Attenuator Calibration

Filters and attenuators used in high-power laser beams may be calibrated by dividing the laser beam with a semitransparent prism and comparing the amplitudes of the outputs of two photocells wherein the signals obtained are subtracted and displayed on a dual-beam oscilloscope. A series of measurements thus obtained, when filters or attenuators are introduced in the transmitted beam and the oscilloscope gain on the appropriate channel, is adjusted to equalize the signal amplitude, yields the desired data.

The attenuation ratio is obtained directly from the gain ratio provided several precautions are observed. First, the oscilloscope gains must be accurately calibrated for all decades used. Second, a polarizing filter must be placed in front of the main laser beam. Because the main output beam has a time-dependent polarization ratio, errors will result unless linearly polarized light is available to the beam splitter.

The net division of the signal between the two channels can be equalized, thus nulling the oscilloscope differential amplifier output, by varying the angle of the polarizer and the beam splitter. Because the polarization ratio of the laser beam is sensitive to the output-power level, it is necessary to perform the measurements when the laser is producing reproducible pulses.

The photo response of both sensors can be compared over a large dynamic range by the null method described above. When calibrated attenuators are available, this null method can be used to check the linearity of the photocell. The absolute calibration of the photocells may be used if desired[183].

The arrangement of experimental apparatus for attenuator calibration is illustrated in Figure 4.30. The apparatus listed in Table 4.15 will be found helpful in making measurements.

4.11.7 Measurement of Average Power in CW Lasers

Average power measurements of CW lases are usually made with a conventional thermopile or, over a narrow spectral region, with photovoltaic devices. A typical thermopile will detect signals in the power range from $10\,\mu$W to 100 mW.

4.11.7.1 Average Power Measurements with a Thermopile The application of a thermopile to CW-power measurements is straightforward, although several precautions must be observed if accurate measures of power are to be obtained. Because thermopiles are composed of an array of junctions, of fair size, facing the radiation flux, the response of the instrument is highly nonuniform; thermopile responsivity varies as much as 8:1 over the

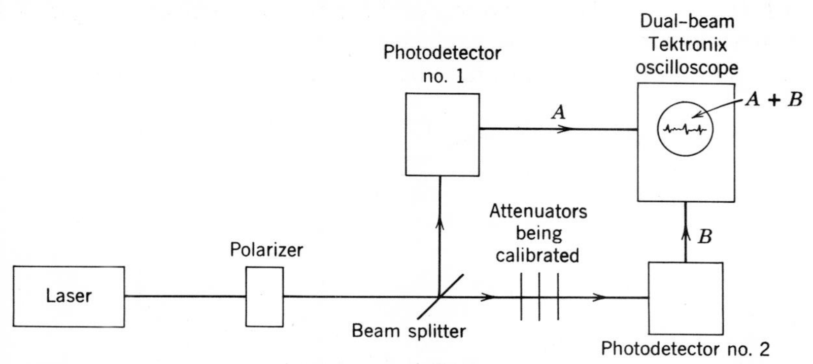

Figure 4.30. Apparatus arrangement for calibration of attenuators and filters with a pulsed-ruby laser.

TABLE 4.15 APPARATUS FOR CALIBRATION OF ATTENUATORS AND FILTERS WITH A PULSED-RUBY LASER

Item	Function	Source
Pulsed-ruby laser	High-intensity light source	Raytheon LR-1 or equal
Calcite, Glan-Thompson polarizer and angle divider with 10 mm aperture	Remove residual beam polarization	Industrial Optics
Fast photodetectors (2) with power supply	Measure beam intensity	Optics technology or equal
Wideband oscilloscope with differential amplifier	Act as null detector with decade attenuators	Tektronix 556 with 1A1 plug-in unit
Beam splitter	Beam division	Ealing
Attenuators	Items to be tested at high intensity	Barnes Engineering or equal

sensitive area[184]. The correct measure of the average power is therefore not attained unless the entire surface of the thermopile is exposed to the laser beams. Furthermore, the laser beam must be free of structure; that is the laser must be operating in the TEM_{00} mode to ensure freedom from a response that is area sensitive. These restrictions and the slow response of the thermopile have fostered the development of photovoltaic devices for more-routine power measurement.

The junction thermopile is commonly believed to be insensitive to wavelength. Careful examination of the spectral sensitivity of thermopiles, using an extremely low-loss, gold-foil, cone-shaped detector, which is believed to be insensitive to wavelength, from the ultraviolet to at least 2 μm, indicates that the thermopile response peaks at near 100 nm and drops monotonically toward 200 nm and 20 μm. A typical calibration curve is given in Figure 4.31. From these data it appears that, exclusive of voltmeter and systematic errors, the calibration of thermopiles may be as much as 13 percent low at 20 μm

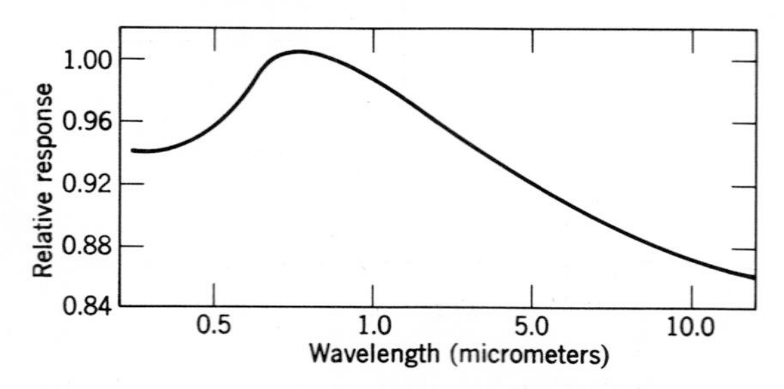

Figure 4.31. Spectral sensitivity of a junction thermopile.

and 6 percent low at 200 nm. For the errors to fall within a $\pm$ 3 percent bound the thermopile should be restricted to the wavelength range of 200 nm to 2 μm. The arrangement of the equipment for power measurements of a CW gas laser is illustrated in Figure 4.32. The apparatus listed in Table 4.16 will be found convenient.

Note that the gas laser must either be placed at a distance sufficient to insure that the TEM_{00} beam spot fills the detector aperture or an auxiliary meniscus lense must be used to diverge the beam. In the latter case, the convex side of the lense should face that gas laser, so that lense reflections will not cause interactions with the cavity mode structure of the gas laser.

Because the thermopile is sensitive to long wavelengths, care must be taken to shield the detector from sources of heat. Without great reduction in sensitivity, the spectral range of the thermopile can be limited by placing the quartz optical flat, furnished with the instrument, in front of the sensitive area. This both filters the incoming radiation and reduces the effects associated with thermal convection currents.

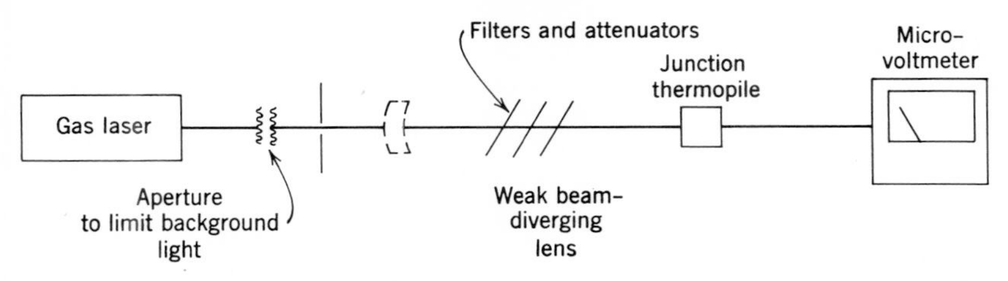

Figure 4.32. Equipment arrangement for the measurement of the power of a CW-gas laser.

TABLE 4.16 APPARATUS FOR MEASURING THE AVERAGE POWER OUTPUT OF A CW-GAS LASER

Item	Function	Source
Gas laser	Intense light source	h nu systems 3410 argon ion laser
12-junction thermopile bismuth-silver	Power measurement	Eppley
Attenuators	Control of beam intensity	Schott-Jena
Notch filters	Reduction of background	Thin-Film Products, Inc. 5-nm bandwidth centered on laser lines
Microvoltmeter	Measure thermopile output voltage	HP 425A
Meniscus lense	Diverge beam	Ealing

If the laser power output is large compared with 100 mW, calibrated attenuators must be placed in front of the thermopile. To ensure that the attenuators are not damaged by the high average-power density, they should be placed directly in front of the thermopile. When attenuators are used, care must be taken to ensure that the refraction of the transmitted beam does not cause the beam spot to deflect away from the central position. It is also necessary to tilt the attenuators so that the reflection from the back surface does not couple into the laser mode structure.

4.11.7.2 Average Power Measurements with a Photovoltaic Cell Routine measurements of the CW-power output of a gas laser can be conveniently made with semiconductor photocells. The responsivity of a selenium cell is well suited to power measurements in the near-ultraviolet and visible range. Silicon cells overlap and cover the wavelength spectrum out to approximately 1.2 μm. Uncooled indium arsenide and indium antimonide may be used respectively to approximately 3 and 6 μm. At longer wavelengths it is more economic to return to the thermopile for routine power measurements.

The application of photovoltaic cells to power measurements is straightforward as long as a few precautions are observed. To obtain the largest photometric intensity range, the cells should be operated into a low impedance, typically less than 100 Ω. Because the circulating currents will be relatively large unless the average radiance is reduced, photovoltaic cells are usually restricted to total power inputs of 100 mW or less.

A basic advantage of the selfgenerating detector is that the power meter can be made compact and highly portable. Within the spectral response range of the detector a correction curve suffices to measure power at wavelengths for which the meter is calibrated. Most instruments are calibrated for a unit response at 632.8 nm. Because manufacturers usually publish curves of broad-band response, it is necessary to convert the results to narrow-band responsivity or calibrate the cell using a thermopile as a reference.

4.11.7.2.1 Conversion of Photo Voltaic-cell-Response Curves From Broad Band to Narrow Band The narrow-band response of a photovoltaic cell may be derived from its broad-band response to a black-body radiator[185]. The black-body calibration must be made with a small load resistor, of about 10 Ω. Then the output current I is directly proportional to light power P over four to five orders of magnitude:

$$I = \vartheta P, \tag{4.83}$$

where ϑ is the detector sensitivity in A/W. The total black body radiant emittance from the source is

$$W = \int W(\lambda)\, D\lambda = n^2 \sigma T^4, \tag{4.84}$$

where σ is the Stefan-Boltzmann constant and n is the dielectric constant of the transmitting medium.

The fraction of the total radiation contained in a wavelength band $\Delta\lambda$ is

$$\frac{\Delta W}{W} = \frac{\Delta\lambda\, \overline{W}(\lambda)}{W}, \tag{4.85}$$

where $\bar{W}(\lambda)$ is the median value of $W(\lambda)$ at the center of the wavelength increment $\Delta\lambda$. The part of the total current due to this part of the total radiation is proportional to

$$\Delta I = \overline{B}(\lambda)\, \overline{W}(\lambda)\, \Delta\lambda, \tag{4.86}$$

where $B(\lambda)$ represents the spectral response of a solar cell in relative units. If $\bar{B}(\lambda)$ is the median value of $B(\lambda)$ at the center of the band and ζ is a constant of proportionality,

$$\Delta I = \zeta B(\lambda) W(\lambda)\, \Delta\lambda. \tag{4.87}$$

Passing to the limit $\Delta\lambda \to 0$,

$$I = \zeta \int B(\lambda)\, W(\lambda)\, d\lambda. \tag{4.88}$$

Because the integral limits cover the entire spectral range, (4.88), is the expression for the current response to white light. The sensitivity ϑ for white light is known: thus ζ can be determined:

$$\zeta = \frac{\vartheta W}{\int B(\lambda) W(\lambda)\, d\lambda}. \tag{4.89}$$

The current response for narrow-band light at a wavelength λ is

$$I_\lambda = \zeta B_\lambda P_\lambda. \tag{4.90}$$

Substituting in (4.89)

$$I_\lambda = \left(\frac{\vartheta W}{\int B(\lambda) W(\lambda)\, d\lambda} \right) B_\lambda P_\lambda \tag{4.91}$$

which becomes

$$P_\lambda = \frac{I_\lambda}{\vartheta}\left(\frac{\int B(\lambda)W(\lambda)\,d\lambda}{B_\lambda W}\right) = \frac{I_\lambda}{\vartheta}C \tag{4.92}$$

where the constant C can be determined by numerical evaluation from the known spectral response $B(\lambda)$ of the solar cell. It is necessary only to evaluate the ratio

$$C = \frac{\int B(\lambda)W(\lambda)\,d\lambda}{B_\lambda W\,d\lambda}. \tag{4.93}$$

As an example, consider measuring the power output of a GaAs diode. At room temperature GaAs emits at $\lambda = 990$ nm. For a silicon solar cell at room temperature and $\lambda = 990$ nm, C becomes 0.213. The peak of the spectral-response curve will shift to shorter wavelength at 77°K ($\lambda = 845$nm) so that a new value of C is required. In this case $C = 0.211$.

Measuring the voltage V devoloped across a resistor R,

$$V_\lambda = I_\lambda R = \frac{P_\lambda \vartheta}{C} R \tag{4.94}$$

or

$$P_\lambda = \frac{CV_\lambda}{R\vartheta} = KV_\lambda \tag{4.95}$$

For $R = 100\ \Omega$, $\vartheta = 0.1$ A/W, $C = 0.231$, and $K = 2.31 \times 10^2$ W/V. Using a microvoltmeter is possible to measure as little as 10^{-8} W.

4.11.8 Radiometric Standards

4.11.8.1 NBS Standard of Total Irradiance The NBS standard of total radiation (total irradiance) consists of a carbon-filament lamp operated at temperatures between 1600 and 2200°K. At these temperatures most of the irradiance falls between the wavelengths of about 1 and 3 μm. This standard is derived through comparisons of the irradiances from a group of lamps with the irradiance from a black body. For this comparison the black-body temperature is usually set at approximately 1400°K and a thermopile, heavily coated with lampblack, is employed as a detector. Because a heavily coated thermopile has been found to be closely uniform in sensitivity with wavelength between the visible and about 3 μm in the infrared, it will give an acceptably accurate evaluation of an 1800°K lamp filament in terms of a 1400°K black body. Recent reviews of detectors substantiate the validity of the earlier measurements in this area that the standard of the carbon-filament lamp for total irrandiance is adequate for use in the calibration of properly blackened thermal detectors over the range from a few microwatts to several hundred microwatts per square centimeter. To cover higher ranges of irradiance[184], work is in progress to devise a secondary standard yielding an irradiance approximating 100 to 150 mW/cm².

4.11.8.2 NBS Standard of Spectral Radiance Since 1960 the NBS standard of spectral radiance in the 250 to 2600 nm spectral range has been a tungsten-ribbon filament lamp manufactured by the General Electric Company as their GE 30A/T24/3 lamp rated at 6V, 30A. This lamp is equipped with a special 1–1/4-in. fused-silica window placed parallel to and located at a distance of three to four inches from the plane of the filament. Typical values of the spectral radiance of tungsten-ribbon lamps in microwatts per square centimeter per 100 nm at 1 for a 1 mm² source[186] are given in Table 4.17.

4.11.8.3 NBS Standard of Spectral Irradiance A new standard of spectral irradiance in the wavelength range from 250 to 2600 nm was established in 1963 in the form of a 200-W quartz-iodine lamp with a coiled tungsten filament operating at 3000°K. The standard is based upon the General Electric Company lamp GE 6.6A/T4Q/1 CL-200-W quartz-iodine lamp. Because of its higher operating temperature, this lamp emits a relatively larger flux in the ultraviolet, thus overcoming objections to previous sources. The spectral irradiance of one of the new standard lamps is given in Table 4.18 in microwatts per square centimeter per nanometer at 42 cm (measured from the axis of the lamp filament and normal to the plane of the lamp press) when operated at 6.5A. The maximum uncertainty in spectral irradiance ranges from 8 percent in the ultraviolet to 3 percent in the visible and infrared. The data are corrected for water-vapor absorption at 1.1, 1.4, 1.9, and 2.6 μm. The inverse-square law should not be applied to these lamps at distances less than 40 cm. Corrections for water-vapor absorption are usually small at 50 cm for most laboratory environments.

4.11.9 Determination of Detector Spectral Sensitivity

The spectral sensitivity of a detector, for example, a silicon solar cell, can either be made by comparison with a detector of known spectral sensitivity and a monochromatic source (such as a gas laser) or by measuring the suitably monochromatized output of a standard source of irradiance, as is discussed above.

TABLE 4.17 SPECTRAL RADIANCE OF GE 30A/T24/3 TUNGSTEN LAMPS[b]

Wavelength	Lamp No. 20 (25 A)		Lamp No. 20 (30 A)		Lamp No. 16 (35 A)	
(μm)	p^a	q^a	p^a	q^a	p^a	q^a
0.250					4.13	5
0.260					9.59	5
0.270					1.88	4
0.280					3.40	4
0.290					5.84	4
0.300					9.72	4
0.320					2.34	3
0.350					7.00	3
0.400					2.76	2
0.450					7.27	2
0.500	8.00	3	4.04	2	1.56	1
0.550	1.88	2	8.14	2	2.72	1
0.600	3.64	2	1.39	1	4.05	1
0.650	6.24	2	2.13	1	5.66	1
0.700	9.56	2	2.99		7.17	1
0.750	1.35	1	3.74	1	8.59	1
0.800	1.66	1	4.38	1		
0.900	2.22	1	5.42	1		
1.000	2.74	1	6.12	1		
1.100	3.12	1	6.38	1		
1.200	3.32	1	6.41	1		
1.300	3.36	1	6.24	1		
1.400	3.30	1	5.94	1		
1.500	3.16	1	5.55	1		
1.600	2.96	1	5.04	1		
1.700	2.74	1	4.52	1		
1.800	2.45	1	3.94	1		
1.900	2.20	1	3.45	1		
2.000	1.97	1	3.03	1		
2.100	1.78	1	2.65	1		
2.200	1.58	1	2.37	1		
2.300	1.40	1	2.13	1		
2.400	1.26	1	1.92	1		
2.500	1.13	1	1.66	1		
2.600	1.02	1	1.52	1		

[a] $N_1 = P \times 10^{-q} \mu W/cm^2$ at 1 meter for $\Delta\lambda = 0.1 \mu m$.

[b] Reproduced with permission from the *Journal of Research of the National Bureau of Standards*.

TABLE 4.18 SPECTRAL IRRADIANCE OF NBS STANDARD 200-W QUARTZ-IODINE, TUNGSTEN-FILAMENT LAMPS[a]

Wavelength (nm)	Spectral Irradiance (μW/cm²/nm at 43 cm)
250	0.0051
260	0.0093
270	0.0158
280	0.0253
290	0.0380
300	0.0545
320	0.104
350	0.237
370	0.366
400	0.643
450	1.26
500	2.04
550	2.93
600	3.88
650	4.79
700	5.54
750	6.11
800	6.51
900	6.72
1000	6.51
1100	6.07
1200	5.53
1300	4.97
1400	4.44
1500	3.93
1600	3.46
1700	3.03
1800	2.63
1900	2.29
2000	1.98
2100	1.73
2200	1.52
2300	1.36
2400	1.22
2500	1.12
2600	1.04

[a] Reproduced with permission from *Applied Optics*.

If the detector is to be used to measure the output of a 633 nm gas laser, the sensitivity may be easily determined as follows:

1. Adjust the gas laser to give stable output on the TEM_{00} mode.
2. Place a 633 nm narrow-band filter in front of the laser to eliminate

extraneous radiation. Alternatively, place a prism in front of the laser and disperse the radiation so that only 633 nm radiation is available.

3. Place a thermopile at a distance from the apparatus such that the aperture of the instrument is filled. The calibrated quartz window should be in place to isolate the detector from infrared radiation as well as convection currents.
4. Using a microvoltmeter and the thermopile calibration, determine the value of beam power.
5. Place the properly terminated solar cell at the same location and repeat the measurements. If the size of the solar cell is very small compared with the thermopile, a lens may be employed to reduce the spot to a convenient size. The insertion loss of the lens must then be accurately determined and its value used to correct the data which require its use.

By comparing the voltage output of the silicon cell with that of the thermopile, and the calibration of the latter, the solar-cell sensitivity is determined for the specific wavelength of interest.

Calibration at other specific wavelengths can be determined with argon-, krypton-, and xenon-ion lasers. A monochrometer can also be used with a blackbody light source to establish a more continuous calibration.

The alternate method of calibrating an unknown detector is to measure the flux from a standard source of irradiance using a monochromator or narrow-band filter. The transmission loss through the filter must be accurately known. Calibration is obtained by operating the standard source of irradiance in accord with NBS recommendations and measuring the detector response (through the monochromatizer) at a known distance. The irradiance of the source and the transmission characteristics dependent on the wavelength of the filter being known, the spectral sensitivity of the detector can be calculated.

It is important to note that standard sources, particularly tungsten-filament lamps, should be operated in an environment that minimizes stray radiation reflection, among other things. Calibrations should always be made under controlled geometric conditions using a stabilized source of supply voltage for lamp excitation.

4.11.10 Evaluation of Infrared-detector Sensitivity

With the exception of spectral response, detector parameters can be measured with the apparatus outlined in Table 4.19, arranged as shown in Figure 4.33. Measurements performed on this black-body test set must be made under a rigidly controlled set of measurement conditions. A typical set of conditions employed is as follows:

$500°K$ black-body flux density at detector	$2\ \mu W\ rms/cm^2$
Chopping frequency	900 Hz
Background temperature	300°K
Noise bandwidth	4 Hz
Bias	as required to set the operating point recommended by the diode manufacturer

The black-body standard source is located at an appropriate distance from the cooled detector to yield the desired reference flux. The detector signal-to-noise ratio is measured under conditions similar to those outlined above. The detector area being known, the NEP and D^* can be computed from relations (4.33) and (4.34):

$$\text{NEP} = \frac{HAN}{(\Delta f)^{1/2} S},$$

$$D^* = \frac{A^{1/2}}{\text{NEP}}.$$

The responsivity is, of course, obtained directly from the value of the signal output and the detector area, as in (4.28):

$$R_{BB} = \frac{S}{HA}.$$

TABLE 4.19 APPARATUS FOR MEASURING INFRARED-DETECTOR CHARACTERISTICS

Item	Function	Source
Black-body standard source	Stable source of 500°K Black-body radiation	Barnes Engineering model 11-200
Oscilloscope	Observe detective time constant	Tektronix 545
Chopper	Modulate beam at 10 to 1000Hz	Infrared Industries model 801-B
Pulsed-infrared light source	Measure time constant broad band of infrared detectors	A-101
Detector	Item to be tested	Philco ISC-363
Bias source	Set detector operator's point	Fabricate
Amplifier	Amplify signal	Tektronix 122 preamplifier
Wave analyzer	Separate and analyze chopped signal	Hewlett-Packard 302A with 297A sweep drive
Microvoltmeter	Responsivity measurements	Kiethly model 601

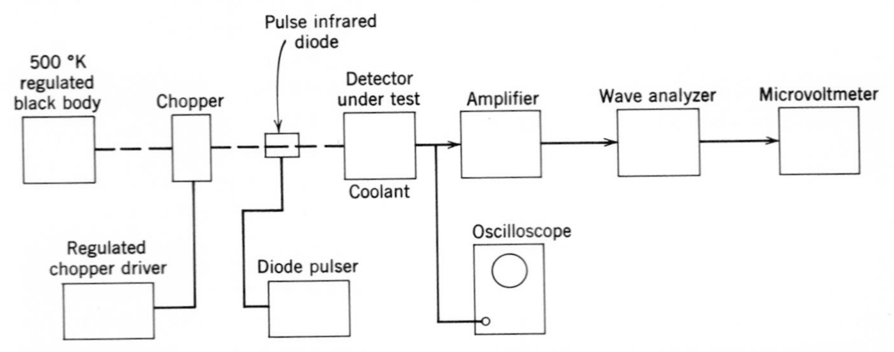

Figure 4.33. Apparatus arrangement for evaluating the performance of infrared detectors.

The detector time constant can be determined by monitoring the detector output as a function of chopping frequency. It will be necessary to use a pulsed infrared diode to determine the time constant of fast detectors.

4.11.11 Technique for Using Total-radiation Standards

NBS-calibrated, standard, tungsten-filament sources of radiance and irradiance can be purchased and used to establish a secondary standard for calibration of thermopiles and other radiation detectors. When these lamps are operated under stipulated conditions the total flux and the incremental flux per unit wavelength interval may be determined by comparing the fluxes with the primary standard to an accuracy of about 0.5 percent. The absolute accuracy of the primary standard is about 1.0 percent[188].

An appropriate radiation calibration facility can be constructed as follows.

1. Place a black cloth-covered shield of 1 × 1 m edge 1 m to the rear of the standard lamp.
2. A similar opaque shield having a central opening 10 cm wide × 15 cm high is placed 25 cm in front of the standard lamp. The front shield should be equipped with an opaque 20 × 20 cm shutter placed between the shield and the lamp.
3. The lamp should be centered vertically and horizontally.
4. The instrument to be calibrated should be placed at a distance 2 m from the lamp and facing the opening.

Several precautions should be observed in making the radiation measurements. As a typical example, we consider the calibration of a radiation thermopile. First, keep in mind that the overall accuracy of the results is somewhat dependent upon temperature and humidity. Highest accuracy will be obtained if the measurement is made for a room temperature of 25°C and a relative humidity of 60 percent. Corrections can be applied to the results if the measurement conditions depart markedly from the above[188].

The best results are obtained when the lamp is operated from a source of regulated power and the lamp current is maintained at recommended values. Tungsten deposits on the lamp envelope change the calibration slowly with time. For this reason a standardized lanp should only be used to calibrate other lamps, the latter being used for radiation-calibration measurements.

Before the calibration lamp is operated the shutter of the radiation-calibration facility should be opened and closed with the thermopile to be tested in place. This enables us to determine the amount of stray thermal radiation falling upon the radiometer. Tests should, of course, be made in an environment free from extraneous sources of electromagnetic radiation. The calibration should be made in a room without windows or provided with opaque drapes to avoid errors from sunlight and overall skyshine (which is cloud-cover dependent). The room should be dimly lighted to avoid errors from intense incandescent sources. Errors may also be caused if the wall temperatures of the room vary markedly and induce convection currents. Air-conditioning equipment may be another source of unwanted convection currents.

The wall and screen to the rear of the lamp may be cooler than the shutter, which will cause a negative deflection on the dc voltmeter that reads the output voltage from the detector. The correction to the deflections observed with the lamp operating are then positive.

The calibration lamp should be turned on and operated for at least 5 min (particularly for high-watt lamps) to permit the filament pins and connections to reach equilibrium. Each source should, of course, be operated under recommended conditions to insure that the original comparison conditions are duplicated as closely as possible.

If the thermopile or detector is provided with a window, it will be desirable to calibrate the unit with and without the window.

Consider calibration of a fluorite-windowed thermopile with a 50-w carbon-filament lamp. With the thermopile exposed at the 2-m point and the calibration standard operated at the recommended 0.350 A, the radiant flux is 84.9×10^{-6} W/cm^2. The fluorite window transmits 0.916 or 77.8×10^{-6} W/cm^2. If the thermopile is read by a galvanometer that deflects 3.15 cm, the sensitivity of the instrument is 24.7×10^{-6} W/cm^2.

If a wavelength-sensitive detector is to be calibrated in this system, the equivalent black-body temperature and the interval of the flux per unit of wavelength must be known to evaluate the information obtained. Data of this type are given in the tables in this chapter.

If an extremely narrow-band source, such as a gas laser, is to be measured, the detector can be calibrated with a notch filter of known transmission properties with due attention given to the large contributions in the transmitted flux represented by the skirts of the transmission curve. In any event, the results of the calibration with a standard source can be compared with a thermopile the calibration of which is checked by the radiation calibration facility.

4.12 REFERENCES

[1] R. A. Smith, F. E. Jones, and R. P. Chasmar, *The Detection and Measurement of Infrared Radiation*, Oxford University Press, London, 1957.
[2] R. DeWaard and E. M. Wormser, *Proc. IRE.* **47**, 1508 (1959).
[3] M. R. Holter, S. Nudelman, G. H. Suits, W. L. Wolfe, and G. J. Zissis, *Fundamentals of Infrared Technology*, Macmillan, New York, 1962.
[4] R. F. Potter and W. L. Eisenman, *Appl. Opt.* **1**, 567 (1962).
[5] J. A. Jamieson, R. H. McFee, G. N. Plass, R. H. Grube, and R. G. Richards, *Infrared Physics and Engineering*, McGraw-Hill, New York, 1963, Chapter 5.
[6*a*] "RCA Phototubes and Photo Cells," *Tech. Man. PT-60*, Radio Corporation of America, Lancaster, Pa., 1963.
[6*b*] J. Scharpe, "Photoelectric Cells and Photomultipliers," *Electronic Technology* (June and July, 1961). Available from EMI/US 1750 N. Vine St., Los Angeles 28, Calif.
[6*c*] *Golay Infra-Red Detector*, Eppley Laboratory, Inc., Newport, R. I.,
[6*d*] M. J. E. Golay, *Rev. Sci. Instr.*, **18**, 347 (1949); *ibid.*, **10**, 357 (1949); *ibid.*, **20**, 816 (1949).
[6*e*] M. J. E. Golay, *Proc. IRE*, **40**, 1161 (1952).
[6*f*] Crocker, Gebbie, Kimmitt, and Mathias, *Nature*, **201**, 250 (1964).
[7] H. Bernstein and L. W. Carrier, *Development of Devices for Measuring Optical Radiation*, Final Report, 11/19/62–4/19/63. AD-420 574, Electro-Optical Systems, Inc., Pasadena, Calif.
[8] R. T. Daly, "Measuring Laser Performance," *Microwaves*, **3**, 50 (1964).
[9] P. J. Bateman, "Measurement of Laser Power and Total Energy," *Proc. Conf. Lasers Applications*, London, 1964.
[10] *Nat. Bur. Std. Tech. News Bull.*, (October 1963).
[11] "New Values of the Physical Constants as Recommended by the NAS-NRC," *Phys. Today*, **17**, 48 (1964).
[12] D. E. Gray, *American Institute of Physics Handbook*, McGraw-Hill, New York, 1963, 2nd ed.
[13] I. Estermann, *Methods of Experimental Physics*, Academic, New York 1959, Vol. I, Ch. 6.
[14] Annual Calorimetry Conferences, *Physics Today*, **17**, 50 (1964).
[15] A. Broido and A. B. Willoughby, *J. Opt. Soc. Am.*, **48**, 344 (1958).
[16] W. F. Flood, *Rev. Sci. Instr.*, **30**, 487 (1959).

[17] R. Stair, W. E. Schneider, W. R. Waters, and J. K. Jackson, *J. Opt. Soc. Am.*, **54**, 1386 (1964).
[18] *Temperature, Its Measurement and Control in Science and Industry*, Reinhold, New York, 1941, 1955, 1962, Vols I, II, and III.
[19] National Bureau of Standards, *Precision Measurement and Calibration*, Handbook 77, Selected Papers on Heat and Mechanics, 1961, Vol. *II*.
[20] E. R. Schleiger and N. Goldstein, *Rev. Sci. Instr.*, **35**, 890 (1964).
[21] S. Koozekanani, P. Debye, A. Krutchkoff, and M. Ciftan, *Proc. IRE*, **50**, 207 (1962).
[22] T. Li and S. D. Sims, *Appl. Opt.*, **1**, 325 (1962).
[23] R. M. Baker, *Electronics*, **36**, 36 (1963).
[24] J. A. Calviello, *Proc. IEEE*, **51**, 611, (1963).
[25] V. S. Zuev and P. G. Kryvkov, *Instruments and Exper. Techn.*, **3**, 563 (1963).
[26] F. Davoine, J. L. Macqueron, and A. Nouilhat, *le J. Phys.*, **24**, 1103 (1963).
[27] J. A. Ackerman, *Appl. Opt.* **3**, 644 (1964).
[28] B. L. Mattes and T. A. Perls, *Rev. Sci. Instr.*, **32**, 332 (1961).
[29] E. K. Damon and J. T. Flynn, *Appl. Opt.*, **2**, 163 (1963).
[30] W. L. Eisenman, R. L. Bates, and J. D. Merriam, *J. Opt. Soc. Am.*, **53**, 729 (1963).
[31] W. L. Eisenman and R. L. Bates, *J. Opt. Soc. Am.*, **54**, 1280 (1964).
[32] J. Dimeff and C. B. Neel, *Reference Black Body Is Compact, Convenient to Use*, NASA Technical Brief 63–10004, Ames Research Center, Moffett Field Calif., 1964.
[33] L. Harris, R. T. McGinnies, and B. M. Siegel, *J. Opt. Soc. Am.*, **38**, 582 (1948).
[34] J. A. R. Samsom, *J. Opt. Soc. Am.*, **54**, 6 (1964).
[35] D. Sell and J. Emmett, Stanford University (Private Communication).
[36] L. Dunkelman, *Typical Absolute Spectral Response Characteristics of Photoemissive Devices*, ITT Industrial Laboratories, 3700 E. Pontiac St., Fort Wayne i, Ind.
[37] V. K. Zworykin and E. G. Ramberg, *Photoelectricity and its Application*, Wiley, New York, 1949.
[38] A. Sommer, *Photoelectric Tubes*, Methuen, London, 1951.
[39] S. Rodda, *Photo-Electric Multipliers*, McDonald, London, 1953.
[40] J. Sharpe, "Photoelectric Cells and Photomultipliers," *Electronic Technology* (June and July 1961). Available from EMI, 1750 N. Vine St., Los Angeles Calif.
[41] *RCA Phototubes and Photocells," Tech. Man. PT-6*, Radio Corporation of America, Lancaster, Pa., 1963.
[42] *Philips Photomultiplier Tubes*, Philips Electron Tube Division, 750 S. Fulton Ave., Mt. Vernon, N. Y., 1963.
[43] *DuMont Multiplier Phototubes*, DuMont Laboratories, Clifton, N. J., 1963.
[44] "Proceedings of the Sixth, Seventh, and Eighth Scintillation Counter Symposia," *IRE Trans. Nuclear Science*, 1958, 1960, 1962.
[45] T. H. Maiman and R. H. Haskins, *Phys. Rev.*, **123**, 1151 (1961).
[46] A. L. Glick, *Proc. IEEE*, **50**, 1835 (1962).
[47] E. Schiel, *Proc. IEEE*, **51**, 365 (1963).
[48] E. Kinoshita and Takeo Suzuki, *J. Appl. Phys.* (Japan), **2**, 311 (1963).
[49] M. Stimler and G. P. Worrell, *Appl. Opt.* **3**, 538 (1964).
[50] R. G. Brewer, *J. Opt. Soc. Am.*, **52**, 832 (1962).
[51] W. P. Ganley, *J. Opt. Soc. Am.*, **53**, 297 (1963).
[52] O. L. Gaddy and D. F. Holshouser, *Proc. IEEE*, **51**, 153 (1963).
[53] R. Hankin and E. Dallafio, *Proc. IEEE*, **53**, 412 (1964).
[54] O. L. Gaddy and D. F. Holshouser, *Proc. IEEE*, **53**, 616 (1964).
[55] N. C. Wittwer, *Fast-Rise Photomultiplier*, Twenty-second Conference on Electron Device Research, Cornell University Ithaca, N. Y., 1964.

[56] I. Dunkelman, *J. Quant. Spec. Radiative Transfer*, **2**, 533 (1962).
[56a] D. Blattner, *et al.*, *RCA Review*, **26**, 22 (1965).
[57] B. J. McMurtry, *IEEE Trans. Electron Devices*, ED-10, **4**, 219 (1963).
[58] B. J. McMurtry, *Appl. Opt.*, **2**, 767 (1963).
[59] R. M. White, *Proc. IEEE*, **51**, 1662 (1963).
[60] W. D. Gunter, Jr., E. F. Erickson, and G. R. Grant, *Appl. Opt.*, **4**, 512 (1965).
[61] G. Lucovsky, M. E. Lasser, and R. B. Emons, *Proc. IEEE*, **51**, 166 (1963).
[62] L. U. Kibler, *Proc. IRE*, **50**, 1834 (1962).
[63] H. Inaba and A. E. Siegman, *Proc. IRE*, **50**, 1823 (1962).
[64] E. Ahlstrom and W. W. Gartner, *J. Appl. Phys.*, **33**, 2602 (1962).
[64a] K. M. Johnson, *IEEE Trans. Electron Devices*, ED-12, **55** (1965).
[65] R. P. Riesz, *Rev. Sci. Instr.*, **33**, 994 (1962).
[66] R. L. Williams, *J. Opt. Soc. Amer.*, **52**, 1237 (1962).
[67] L. K. Anderson, *Proc. IEEE*, **51**, 846 (1963).
[68] R. H. Rediker, F. M. Quist, and B. Lax, *Proc. IEEE*, **51**, 218 (1963).
[69] E. J. Schiel, *Solid State Design*, **4**, 12 (1963).
[70] E. J. Schiel and J. J. Bolmarcich, *Proc. IEEE*, **51**, 1780 (1963).
[71] H. N. Yu, *Proc. IEEE*, **51**, 945 (1963).
[72] R. A. Smith, F. E. Jones, and R. P. Chasmar, *The Detection and Measurement of Infrared Radiation*, Oxford University Press, London, 1957.
[73] R. W. Gelinas and R. H. Genoud, *A Broad Look at the Performance of Infrared Detectors*, Presentation to Conference on Infrared Detectors at Syracuse University, AD-295-044, 1959.
[74] *Infrared Quantum Detectors*, IRIA State-of-the-Art Report No. 2389-50-T, Contract No. NONR 1224(12), Infrared Laboratory, University of Michigan, Ann Arbor, Mich., 1961.
[75] R. F. Potter and W. L. Eisenman, *Appl. Opt.*, **1**, 567 (1962).
[75a] G. A. Morton, *RCA Review.*, **26**, 3 (1965).
[76] G. L. Weissler (ed.) *J. Quant, Spectr. Radiative Transfer*, **92**, (1963), Pergamon, New York.
[77] C. R. Masson, V. Boekelheide, and W. A. Noyes, Jr., *Technique of Organic Chemistry*, Interscience New York, 1956, Vol. II, pp. 298–299.
[78] K. Watanabe, F. F. Marmo, and C. Y. Inn, *Phys. Rev.*, **91**, 1155 (1953).
[79] See Reference 36, pp. 533–544.
[80] J. A. R. Samson, *J. Opt. Soc. Am.*, **54**, 6 (1964).
[81] J. A. R. Samson, *J. Opt. Soc. Am.*, **54**, 842 (1964).
[82] O. P. Rustgi, E. I. Fisher, and C. H. Fuller, *J. Opt. Soc. Am.*, **54**, 745 (1964).
[83] R. E. Huffman, Y. Tanaka, and J. C. Larrabee, *J. Chem. Phys.* **39**, 902 (1963).
[84] F. S. Johnson, K. Watanabe, and R. Tousey, *J. Opt. Soc. Am.*, **41**, 702 (1951).
[85] K. Watanabe and C. Y. Inn. *J. Opt. Soc. Am.*, **43**, 32–35 (1953).
[86] R. A. Knapp, *Appl. Opt.*, **2**, 1334 (1963).
[87] R. A. Knapp and A. M. Smith, *Appl. Opt.*, **3**, 637 (1964).
[88] R. Allison, J. Burns, and A. J. Tuzzolino, *J. Opt. Soc. Am.*, **54**, 747 (1964).
[89] N. Kristianpoller, *J. Opt. Soc. Am.*, **54**, 1285 (1964).
[90] R. Allison, J. Burns, and A. J. Tuzzolino, *J. Opt. Soc. Am.*, **54**, 1381 (1964).
[91] N. Kristianpoller and D. Dutton, *Appl. Opt.*, **3**, (1964).
[92] The following booklets are available from The Sales Service Division, Eastman Kodak Co., Rochester, N. Y. 14650.
P–9: *Kodak Plates and Films for Science and Industry.*
M–3: *Infrared and Ultraviolet Photography.*

E–7: *Kodak Color Films.*

G–1: *Kodak Photographic Papers.*

F–13: *Kodak Films in Rolls.*

[93] H. J. Zweig, *Phot. Sci. Eng.*, **8**, 305 (1964).

[94] *Handbook of High-speed Photography*, General Radio Co., West Concord, Mass., 1963.

[95] C. A. Parker, *Proc. Royal Soc.* (London), **A, 220**, 1140 (1953).

[96] C. G. Hatchard and C. A. Parker, *Proc. Royal Soc.* (London), **A, 253**, 1203 (1956).

[97] R. Livingston, *J. Phys. Chem.*, **44**, 601 (1940).

[98] C. A. Parker, *Trans. Faraday Soc.*, **50**, 383 (1954).

[99] G. S. Forbes and W. G. Leighton, *J. Am. Chem. Soc.*, **52**, 3139 (1930).

[100] C. R. Masson, V. Boekelheide, and W. A. Noyes, Jr., *Technique of Organic Chemistry*, Interscience, New York, 1956, Vol. II.

[101] J. G. Calvert and H. J. L. Rechen, *J. Am. Chem. Soc.*, **74**, 2101 (1952).

[102] L. Harris, J. Kaminsky and R. G. Simard, *J. Am. Chem. Soc.*, **57**, 1151, 1154 (1935).

[103] E. F. Nichole and G. F. Hull, *Phys. Rev.* **13**, 307 (1901). *ibid* **17**, 26 (1903).

[104] P. Lebedew, "Investigations of the Pressure of Light," *Ann. Physik*, **6**, 433 (1901).

[105] R. W. Ditchburn, *Light*, Interscience, New York, 1963, 2nd ed.

[106] H. V. Neher, *Am. J. Phys.*, **29**, 666 (1961).

[107] R. E. Pollack, *Am. J. Phys.*, **31**, 901 (1963).

[108] J. J. Cook, W. L. Flowers, and C. B. Arnold, *Proc. IRE*, **50**, 1693 (1962).

[109] M. Stimler, *Determination of the Energy of a Pulsed Ruby Laser Beam by Transfer of the Photon Momentum to a Ballistic Torsional Pendulum*, NOLTR 63–82, U. S. Naval Ordnance Laboratory, White Oak, Md., 1963. Available from the Office of Technical Services, U.S. Department of Commerce, Washington, D.C., 20230, as AD 420 469.

[110] M. Stimler, Z. I. Slawsky, and R. E. Grantham, "Torsion Pendulum Photometer," *Rev. Sci. Instr.*, **35**, 311 (1964).

[111] J. A. Giordmaine, *Scientific American*, (1964).

[112] R. W. Terhune, *Inter. Sci. Tech.* (1964).

[113] P. A. Franken and J. F. Ward, *Rev. Mod. Phys.*, **35**, 23 (1963).

[114] N. Bloembergen, *Proc. IEEE*, **51**, 124 (1963).

[115] P. Grivet and N. Bloembergen (eds.) *Quantum Electron.*, Paris 1963 Conference, Columbia University Press, New York, 1964.

[116] N. Bloembergen and Y. R. Shen, *Phys. Rev.*, **133**, A-37 (1964).

[117] A. K. Kamal and M. Subramanian, *Opt. Masers*, (Proc. Symp. Optical Masers) **13**, 601, (1963), Polytechnic Press Brooklyn, N. Y.

[118] J. A. Armstrong, H. Bloembergen, J. Ducuing, and P. S. Pershan, *Phys. Rev.*, **127**, 1918 (1962).

[119] J. F. Ward and P. A. Franken, *Phys. Rev.*, **133**, A183 (1964).

[120] R. C. Hansen, "Fraunhofer and Fresnel," *Microwave J.*, **25**, 13 (1962).

[121] S. Silver, *Microwave Antenna Theory and Design* McGraw-Hill, New York, 1949, Ch. 6.

[122] J. W. Sherman, III, *IRE Trans. on Antennas and Propagation*, V, AP–10, N. 4, p. 399–408, 1962.

[123] M. Born and E. Wolf, *Principles of Optics*, Macmillan, New York, 1964, 2nd ed., Ch. 8.

[124] D. B. Bortfeld, R. S. Congleton, M. Geller, R. S. McComas, L. D. Riley, W. R. Sooy, and M. L. Stitch, "Influence of Optical Quality on Ruby Laser Oscillators and Amplifiers," *J. Appl. Phys. (Comm)*, 35, N. 7, p. 2267 (July 1964).

[125] R. W. Waynant, J. H. Cullom, I. T. Basil, and G. D. Baldwin, *J. Opt. Soc. Am.*, **54** 1390 (1964).
[126] T. V. George, L. Slama, and M. Yokoyama, *Appl. Opt.*, **2**, 1198 (1963).
[127] R. Jeanes, *Appl. Opt.*, **3**, 318 (1964).
[128] M. Terai, *J. Appl. Phys.* (Japan), **3**, 421 (1964).
[129] P. J. Ovrebo, R. R. Sawyer, R. H. Ostergren, R. W. Powell, and E. L. Woodcock, *Proc. IRE*, **47**, 1629 (1959).
[130] S. N. Bobo, *Research/Development*, **14**, 22 (1963).
[131] J. R. Hansen, J. L. Ferguson, and A. Okaya, *Appl. Opt.*, **3**, 987 (1964).
[132] G. W. Gray, *Molecular Structure and the Properties of Liquid Crystals*, Academic, New York, 1962.
[133] *Kodak IR Phosphor Screens*, Eastman Kodak Co., Apparatus and Optical Division, Rochester, N. Y. 14650.
[134] *Second Symposium on Photoelectronic Image Devices*, Imperial College London, September 1961. (Volume XVI of *Advances in Electronic and Electron Physics*, Academic, New York, 1962.
[135] *Proceedings of the Image Intensifier Symposium*, NASA SP 2, Ft. Belvoir, Va., October 1961. (Available from Supt. of Documents, U. S. Government Printing Office, Washington D. C.)
[136] W. Heiman and C. Kunze, (Ref. 15) p 217.
[137] F. H. Nicoll, *Rev. Sci. Instr.* **30**, 9, (1959).
[138] *Thorn Image Retaining Panel*, Thorn House, Upper Saint Martin's Lane, London, W C 2, England.
[139] M. Pivovonsky and M. R. Nagel, *Tables of Blackbody Radiation Functions*, Macmillan, New York, 1961, See also D. Hahn, *et al.*, *Seven-place Tables of the Planck Function for the Visible Spectrum*, Academic, New York, 1964.
[140] P. E. Schumacher, *J. Opt. Soc. Am.*, **54**, 1386 (1964).
[141] *Optical Industry and Systems Directory*, Optical Publishing Co., Pittsfield, Mass.
[142] R. Stair, R. G. Johnston, and E. W. Halback, *J. Res. Natl. Bur. Std. (U. S.)*, **64A**, 291 (1960).
[143] R. Stair, W. E. Schneider, and J. K. Jackson, *Appl. Opt.*, **2**, 1151 (1962).
[144] R. Stair and W. E. Schneider, *Standards, Sources, and Detectors, of Solids*, Session III, Measurement Techniques, NASA, San Francisco, March 1964, pp. 1–20.
[145] H. J. Kostkowski, D. E. Erminy, and A. T. Hattenburg, *J. Opt. Soc. Am.*, **54**, 1386 (1964).
[146] R. Stair, W. B. Fussell, and W. E. Schneider, *Appl. Opt.*, **4**, 85 (1965).
[147] E. W. Engstrom, *Rev. Sci. Instr.*, **26**, 622 (1955).
[148] F. J. Studer, *J. Opt. Soc. Am.*, **54**, 1386 (1964).
[149] M. R. Null and W. W. Lozier, *Temperature, Its Measurement and Control in Science and Industry*, Reinhold, New York, 1962, Vol. III, Pt. I, Article 55.
[150] H. H. Blau, Jr., W. S. Martin, and E. Chaffee, *Temperature, Its Measurement and Control in Science and Industry*, Reinhold, New York, 1962, Vol. III, Pt. 2, Article 99.
[151] A. L. Glick, *Proc. IRE*, **50**, 1835 (1962). See also A. L. Glick, "Comment on 'A Method for Calibration of Laser Energy Output,'" *Proc. IEEE*, **51**, 1360 (1963).
[152] R. C. C. Leite and S. P. S. Porto, *Proc. IEEE*, **51**, 606 (1963).
[153] W. E. K. Middleton and C. L. Sanders, *Illum. Eng.*, **1**, 254 (1953).
[154] P. A. Tellex and J. R. Waldron, *J. Opt. Soc. Am.*, **45**, 19 (1955). See also H. K. Hammond, III, *J. Opt. Soc. Am.*, **45**, 904, (1955).
[155] N. A. Voishvilloo and M. N. Smolkin, *Opt. Spectr.*, **16**, 491 (1964).
[156] J. W. T. Walsh, *Photometry*, Constable, London, 1953, 2nd ed.
[157] J. A. Jacquez and H. F. Kuppenheim, *J. Opt. Soc. Am.*, **45**, 460 (1955).

[158] O. E. Miller and A. J. Sant, *J. Opt. Soc. Am.*, **48**, 828 (1958).
[159] P. J. Richetta, *J. Opt. Soc. Am.*, **55**, 21 (1965).
[160] F. R. Bryan, *Appl. Spectroscopy*, **17**, 19 (1963).
[161] P. A. Newman and R. Binder, *Rev. Sci. Instr.*, **32**, 351 (1961).
[162] R. Gerharz, *Proc. IEEE*, **52**, 438 (1964). See also *Proc. IEEE*, **53**, 105, (1965).
[163] G. W. Stroke, *J. Opt. Soc. Am.*, **54**, 846 (1964).
[164] R. Jones, *Proc. IRIS*, **2**, 9 (1957).
[165] P. W. Kruse, L. D. McGlauchlin, and R. B. McQuistan, *Elements of Infrared Technology*, Wiley, New York, 1962.
[166] R. Jones, *Proc IRIS*, **5**, 35 (1959), See also *J. Opt. Soc. Am.*, **50**. 1058 (1960).
[167] R. Jones, *Proc. IRE*, **47**, 1481 (1959).
[168] H. Bernstein and L. W. Carrier, *Development of Devices for Measuring Optical Radiation*, Dept. No. RTD TDR 63–3015, July, 1963 AFSWC, Kirtland Air Force Base, New Mexico.
[169] J. Strong, *Proceedings in Experimental Physics*, Prentice-Hall, Englewood Cliffs, N. J., 1938, Chapter 12, pp. 493–530.
[170] A. Gouffé, *Revue d'Optique*, **24**, 1 (1945).
[171] W. J . Cooper, *et al.*, *Rev. Sci. Inst.*, **30**, 557 (1959).
[172] T. Li and S. D. Sims, *Appl. Optics*, **1**, 325 (1962).
[173] J. Ackerman, *Appl. Optics*, **3**, 644 (1964).
[174] R. M. Baker, *Electronics*, **36**, 36 (1963).
[175] E. J. Schiel, *Proc. IEEE*, **51**, 365 (1963).
[176] M. Stimler and G. P. Worrell, *Appl. Optics*, **3**, 538. (1964).
[177] R. C. C. Leite and S. P. S. Porto, *Proc. IEEE*, **51**, 606 (1963).
[178] C. L. Sanders and W. E. Knowles Middleton, *J. Opt. Soc. Am.*, **43**, 58 (1953).
[179] J. H. Cullom and R. W. Wagnant, *Appl. Optics*, **3**, 989 (1964).
[180] A. K. Kamal and M. Subramanian *Opt. Masers, (Proc. Symp on Optical Masers)*, (1963), Polytechnic Press, Brooklyn, N. Y.
[181] J. A. Armstrong, *et al.*, *Phys. Rev.* **127**, 1918 (1962).
[182] A. L. Glick, *Proc. IRE*, **50**, 1835 (1962).
[183] R. W. Engstron, *RCA Rev.*, **21**, 184 (1960).
[184] R. Stair and W. E. Schneider, *Proc. Symp Thermal Radiation Solids*, WPASD, NASA, NBS, 1964.
[185] E. J. Schiel and J. J. Bolmarcich, *Proc. IRE*, **51**, 1780 (1963).
[186] R. Stair, R. G. Johnston, and E. W. Halbach, *J. Res. Natl. Bur. Std.* (U. S.), **64A**, 291 (1960).
[187] R. Stair, W. E. Schneider, and J. K. Jackson, *Appl. Optics*, **2**, 1151 (1963).
[188] Bureau of Stds. Bulletin No. 227, **11**, 87 (1914). and *J. Res. Natl. Bur. Std.* (U. S.), **11**, 79, (1933) and *ibid.*, **53**, 211 (1954).

4.13 PRINCIPAL SYMBOLS, NOTATIONS, AND ABBREVIATIONS

Symbol	*Meaning*
A	Detector active receiving area
a	Aperture-surface area
B	Spectral response of a detector in relative units
B_r	Total radiance or radiometric source intensity

BLIP	Acronym for *B*ackground *L*imited *I*n *P*erformance
C	Constant, capacity, thermal capacity of active detector element; see text
c	Velocity of light
c_p, c_v	Specific heat at constant pressure or constant volume
D	Diameter of aperture
$D, D^*, D_\lambda^*, D^{**}$	Measures of detectivity; see text
d	Meter deflection
E, E_i	Electric field, specific values thereof
E, E_m	Illuminance, total energy, maximum kinetic energy; see text
e, e_λ	Emissivity, spectral emissivity, electronic charges, see text
F	Luminous flux, force; see text
f, f_m	Signal-chopping frequency, chopping frequency of maximum detectivity
g	Detector-gain parameter
$g + jh$	Complex thermal conductance
H, H_λ	Irradiance, spectral irradiance
H_0	Surface irradiance
h	Planck's constant
I	Current, moment of inertia; see text
$\bar{I}$	Mean current
I_L	Luminous intensity
I_r	Radiant intensity of a point source
K	Torsional stiffness
k	Boltzmann constant
L	Length, distance
L/R	Length L to aperture radius R
M	Apparent mass of proton
m	Mass
N	Ratio, rms noise voltage, total number of photons, radiance; see text
N_λ	Spectral radiance
NEI	Noise equivalent input
NEP	Noise equivalent power
n	Average photon rate
$n(\lambda)$	Spectral index of refraction
n_0	Mean number of carriers in semiconductor
P	Radiant power
P, P_o, P_s, P_t, P_a	Power, specific values thereof; see text
P_e	Radiation noise power

P_x, P_y, P_z	Second order polarization
P_i	Mean square fluctuation in current per unit bandwidth
P_{BB}	Incident black-body radiant power
PEM	Photoelectromagnetic
PV	Photovoltaic
PC	Photoconductive
P	Polarization
Q	Instrument factor, charge; see text
ΔQ	Change in heat energy
R, R_0	Radius vector, resistance, specific value thereof; see text
R, R_{BB}	Responsivity, black-body responsivity
S	Rms value of signal voltage, surface area of calorimeter; see text
T, T_i	Absolute temperature, specific values thereof, see text
U, U_i	Energy, specific values thereof
V	Volume, voltage; see text
v_m	Maximum velocity
W	Total radiant flux of a black body
W_p	Wiener spectrum of incident power
W_t	Wiener spectrum of temperature fluctuations
X, Y, Z	Orthogonal axes
$Y\lambda$	Spectral respnse of the eye
α	Reciprocal penetration depth, numeric, loss factor, second order pollarization tensor; see text
β	Cubical coefficient of expansion, second order polarization tensor; see text
Γ	Latent heat of fusion
φ	Angular deflection
δ	Penetration depth
ϵ_0	Permitivity of free space
ϵ_r	Relative dielectric constant of a crystal
ζ	Fractional relative spectral response of a detector at a given wavelength
η	Intrinsic transverse impedance of a medium, quantum efficiency; see text
ϑ	Detector sensitivity
θ	Angle, constant of proportionality; see text
κ	Diffusivity
λ	Wavelength
r	Frequency
ρ	Mass density, diffusivity; see text
σ	Stefan-Bollzmann constant

τ	Mean carrier lifetime, time interval; see text
τ_d	Detective time constant
τ_r	Responsive time constant
υ	Absorption coeficient of calorimeter window
ϕ	Work function
Φ	Geometrical constant
Ψ	Constant
X	Calorimeter calibration factor
Ω	Solid angle in steradians
ω	Angular frequency

5

Measurement of Gain Parameters

5.0 INTRODUCTION

Chapter 5 treats parameter measurements of some of the less obvious properties of a laser cavity and the quantum-electronic medium that determine laser performance. Several techniques are described for the measurement of single-pass gain. One section augments the gain measurements described in Chapter 7 (Sections 7.3 and 7.4). Methods of mode matching are treated, and one section, on lifetime measurement, includes several ways for determining these characteristics in solids, liquids, and gases. Methods of measuring the electron energy and energy density in gas-laser plasmas are included. Techniques are treated for measuring mirror transparencies in the limit of very high reflectance as well as experimental methods for determining values for reflectance that optimize laser power output. A technique for determining the degree of population inversion in a Q-switched laser is given. Concluding sections treat the problem of cavity losses and techniques for their measurement. The chapter begins with a review of pertinent laser parameters.

5.0.1 Spontaneous and Induced Transitions

A laser usually consists of a medium of negative electromagnetic absorption bounded by a suitable resonant cavity. One role of the resonator is to reduce radiation losses in a low-gain amplifying medium which it does by recirculating the narrow-band electromagnetic energy, thereby enabling stimulated emission to replenish losses. The resonant cavity is not essential to the production of electromagnetic energy having the radiation characteristics of a laser, namely, spectral narrowing, high degree of temporal and spatial coherence, and a high degree of collimation. Radiation having these properties has repeatedly been demonstrated in the infrared, visible, and ultraviolet using high-gain plasma tubes in mirrorless-laser configurations[1, 2, 3].

The stable resonator of a conventional laser (stable in the discrete electromagnetic-mode sense) is essential to the production of standing waves in the medium and the reduction of losses. A significant improvement is obtained in the laser properties of the light source if the coherence brightened or superradiant light source is enclosed in a cavity.

While propagating in a resonant cavity, an electromagnetic wave is subject to a number of cavity-engendered losses, the most important of which are:

1. Transmission, absorption, and scattering in the terminal windows.
2. Walkoff losses and mode conversion due to imperfect imaging by the mirrors, windows of the plasma tube, and inhomogeneities in the index of refraction of the laser medium (particularly in solid-state and semiconductor lasers).
3. Scattering by optical inhomogeneities[4] and surface imperfections—the former in solid-state and semiconductor lasers, the latter by plasma-tube windows in gas lasers.
4. Diffraction at the extremes of the laser bore.
5. Absorption in the amplifying medium.

The cavity laser, like its electronic counterpart, may be regarded as a high-gain amplifier of internally generated noise. For sustained oscillation to occur, it is necessary that the gain of the medium, which increases with population inversion, be sufficient to overcome losses. Negative attenuation or gain at the laser wavelength is produced by induced transitions or stimulated emission. Spontaneous emission, which occurs on the same transition, competes with induced emission for the energy available in the excited states of the laser medium. Because spontaneous emission randomly depletes stored energy, it is a serious source of loss not directly associated with the cavity.

A number of significant laser parameters may be derived from a consideration of the threshold for oscillation. The justification for this treatment is the convenient introduction of a number of spectroscopic and electromagnetic notions and relationships of value in this and other chapters of the book.

Consider an atomic system in thermal equilibrium at a temperature T. The density of the excited states of a pair of related energy levels obey Boltzmann statistics:

$$\frac{n_j}{g_j} = \frac{n_i}{g_i} \exp[-h\nu_{ji}(kT)], \tag{5.1}$$

where n_j and g_j are the population density in the upper energy level U_j, and n_i, g_i relate to the lower energy level U_i. Under the influence of the thermal radiation field, equilibrium demands that transitions $i \rightarrow j$ and

$j \to i$ occur with equal probability. The spontaneous emission lifetime for a dipole transition $j \to i$ is just the reciprocal of the probability per unit time that an atom in the upper state U_j will spontaneously emit a photon in a random direction and change to the lower energy state U_i. It can be shown from quantum theory that the spontaneous lifetime may be expressed in terms of the square of the dipole moment matrix element $[\mu_{ji}]^2$ summed over all degenerate initial and final states, g_i[6]. Thus

$$A_{ji} = \frac{64\pi^4 \nu^3 |\mu_{ji}|^2}{3hv^3 g_j} = \frac{1}{\tau_{ji}}, \tag{5.2}$$

where $v = c/n$ is the velocity of light in a medium with an index of refraction n and the other constants have their usual values.

For allowed (electric-dipole) transitions τ_{ji} is of the order of 10^{-8} sec. Magnetic-dipole transition times are of the order of 10^{-3} sec, and quadrupole transitions are of the order of 1 sec. Magnetic dipole and quadrupole transitions correspond to metastable states.

An electron in a given energy level can either gain or lose energy upon interacting with an externally applied electric field. It can be shown classically that there is a time constant τ associated with this interaction[5]:

$$\tau = 2\pi m v r_0 / e\epsilon_0 \cos\phi, \tag{5.3}$$

where m is the electron mass, $2\pi v$ is the angular frequency of rotation in an orbit of radius r_0, e is the charge on the electron, and ϵ_0 is the externally applied electric field. Depending upon the sign of the phase factor $\cos\phi$, the electron can either gain energy from or add energy to the external electric field[6]. The possibility of directional, field induced energy exchange distinguishes the stimulated emission process from the random spontaneous emission process. Spontaneous emission described by (5.2) is independent of forcing radiation fields.

The black-body radiation density per unit frequency interval $\rho(v)$ can be expressed as a product of the average free-space energy density of electromagnetic modes ($\times$ 2 for random polarization) and the average energy of a harmonic oscillator

$$\rho(\nu) = \frac{8\pi h\nu^3}{v^3} \frac{1}{\exp(h\nu/kT) - 1}. \tag{5.4}$$

It may also be expressed as a product of the average energy per cavity mode $\bar{U}$ and the mode density per unit frequency interval $p(\nu)$. Thus

$$\rho(\nu) = p(\nu)\bar{U} = \frac{8\pi h\nu^2}{v^3} \cdot \frac{h\nu}{\exp(h\nu/kT) - 1}. \tag{5.5}$$

The corresponding excitation number of the radiation field $\bar{n}$, that is the average quanta per mode, is

$$\bar{n} = \frac{\overline{U}}{h\nu} = \frac{1}{\exp(h\nu/kT) - 1}. \tag{5.6}$$

It can be shown that a dynamic equilibrium exists in a system described by (5.1) if the induced transition rate is proportional to the black-body energy density per unit frequency interval. The equilibrium, which must hold at any temperature, is expressed as

$$\rho(\nu)B_{ij}n_i = A_{ji}n_j + \rho(\nu)B_{ji}n_j. \tag{5.7}$$

$$\begin{pmatrix}\text{Thermal radiation}\\ \text{field-induced}\\ \text{absorption } i \rightarrow j\end{pmatrix} = \begin{pmatrix}\text{Spontaneous}\\ \text{emission}\\ j \rightarrow i\end{pmatrix} + \begin{pmatrix}\text{Thermal radiation}\\ \text{field-induced}\\ \text{emission } j \rightarrow i\end{pmatrix}$$

where

$$B_{ij} = \frac{8\pi^3|\mu_{ij}|^2}{3h^2 g_i} \tag{5.8}$$

and

$$B_{ji} = \frac{8\pi^3|\mu_{ji}|^2}{3h^2 g_j}. \tag{5.9}$$

B_{ij} is the transition probability for radiation-field-induced absorption, and B_{ji} is the corresponding transition probability for induced emission. The first term on the right side of (5.7) is a consequence of the principle of detailed balancing, for (5.7) must hold for any temperature.

In the high-temperature limit $p(\nu) \rightarrow \infty$ as $T \rightarrow \infty$ in (5.6), whereas an incomplete statement of temperature equilibrium, namely

$$n_i B_{ij} \rho \overset{?}{=} n_j A_{ji},$$

contradicts the hypothesis of the attainment of equilibrium.

Evidently, from (5.8) and (5.9), if $|\mu_{ij}|^2 = |\mu_{ji}|^2$, a consequence of isotropy,

$$\frac{B_{ji}}{B_{ij}} = \frac{g_i}{g_j}, \tag{5.10}$$

where g_i and g_j are the statistical weights of the lower and upper states. Equation 5.10 implies the existence of equilibrium for if $B_{ij} \lessgtr B_{ij}$ energy would be emitted or absorbed.

Evidently, from (5.2) and (5.9).

$$\frac{A_{ji}}{B_{ji}} = \frac{8\pi h v^3}{v^3} \tag{5.11}$$

or

$$A_{ji} = \left(\frac{8\pi v^2}{v^3}\right) h v B_{ji}. \tag{5.12}$$

The spontaneous-transition coefficient is proportional to the induced-transition coefficient. The coefficient of proportionality is the product of the mode density per unit-frequency interval and the energy of the transition. Thus, although the induced-transition rate is independent of wavelength, the spontaneous-transition rate increases at shorter wavelengths.

The induced-transition rate in the black-body radiation field is obtained from (5.7).

$$\rho(v) = \frac{A_{ji} n_j}{B_{ij} n_i - B_{ji} h_j} = \frac{A}{B}\left(\frac{n^2}{n_i - n_j}\right),$$

or

$$\rho(v) = \frac{A}{B}\, \bar{n}. \tag{5.13}$$

The induced-transition rate W'_i is proportional to the average number of quanta per mode

$$W'_i = \frac{B}{A}\, \rho(v) = \frac{v^3}{8\pi h v^3}\, \rho(v). \tag{5.14}$$

W'_i can be expressed in terms of (5.2) as

$$W'_i = \frac{8\pi^3 |\mu_{ji}|^2}{3h^2 g_j}. \tag{5.15}$$

Implicit in the derivation of (5.15) is the assumption that $\rho\,(v)$ is a constant over the interaction-frequency range of the atomic system for which emission and absorption occurs. This limits the utility of (5.15) because the monochromaticity of the laser signal is so large that the energy density per unit-frequency interval is no longer correctly specified. If we introduce the concept of atomic-line shape and replace $\rho\,(v)$ by the intensity expressed as a Dirac delta function, the induced emission expression will be correct.

Consider the normalized absorption line shape function $g(\nu)$ defined as

$$\int_0^\infty g(\nu)\, d\nu = 1. \tag{5.16}$$

The shape of $g(\nu)$ versus ν may arise as the result of a spread of energy in a solid so that the atomic resonance has a Lorentz shape:

$$g_L(\nu) = \frac{\Delta\nu}{2\pi\left[(\nu - \nu_c)^2 + \left(\frac{\Delta\nu}{2}\right)^2\right]} \tag{5.17}$$

where ν_c is the center of the atomic resonance line and $\Delta\nu$ is the full width at half-maximum.

Considering a gas at low pressure, where the line shape is perturbed by the Doppler broadening from thermal motion, the atomic resonance will be Gaussian in shape:

$$g_G(\nu) = \frac{2\sqrt{\pi \ln 2}}{\pi \Delta\nu_D} \exp\left\{-\left[\left(\frac{\nu - \nu_c}{\Delta\nu_c}\right)\sqrt{\ln 2}\right]^2\right\} \tag{5.18}$$

and

$$\Delta\nu_D = \frac{\nu_c}{c}\left(\frac{2kT \ln 2}{M}\right)^{1/2} \tag{5.19}$$

is the full width at half-maximum.

Without regard to the detailed shape of $g(\nu)$, we may write an expression for the spontaneous-emission rate into the frequency interval $d\nu$ as $Ag(\nu)\, d\nu$ The corresponding induced-emission rate into $d\nu$, due to a radiation density $p(\nu)$ per unit-frequency interval becomes

$$W_i' g(\nu)\, d\nu = \rho(\nu) A g(\nu)\, d\nu \frac{\upsilon^3}{8\pi h \nu^3}. \tag{5.20}$$

The energy flux $I(\nu)\, d\nu$ in the frequency interval $d\nu$ is just $\upsilon\rho(\nu)\, d\nu$. Therefore, writing (5.20) in terms of the energy flux

$$W_i' = \frac{\upsilon^2}{8\pi h \nu^3 \tau_{ji}} g(\nu) I(\nu)\, d\nu, \tag{5.21}$$

and replacing $I(\nu)$ with the Dirac function corresponding to a monochromatic wave of frequency ν,

$$I(\nu) = I_\nu \delta(\nu' - \nu),$$

and integrating over all frequencies, the induced-transition rate reduces to

$$W_i(\nu) = \frac{\upsilon^2}{8\pi h \nu^3 \tau_{ji}} g(\nu) I_\nu. \tag{5.22}$$

Because $W_i^{\prime} = \bar{n}A_{ji}$ at any portion of the atomic line width, the induced-transition rate per mode is equal to the product of the spontaneous emission rate and the number of quanta per mode.

Consider the condition for oscillation threshold. The rate of increase of mode intensity due to induced emission is

$$\left(\frac{dI}{dt}\right)_1 = h\nu v W_i\left(n_j - n_i\frac{g_j}{g_i}\right). \tag{5.23}$$

The rate of power loss from the cavity can be expressed in terms of a photon lifetime in the cavity if we lump all of the losses into one parameter, $Q = \omega\tau_c$. The power loss may be expressed as

$$\frac{dW}{dt} = -\frac{\omega W}{Q} = -\frac{W}{\tau_c}. \tag{5.24}$$

The rate of decrease in the mode intensity becomes

$$\left(\frac{dI}{dt}\right)_2 = -\frac{I}{\tau_c}. \tag{5.25}$$

Oscillation can exist when

$$\left(\frac{dI}{dt}\right)_1 \geq \left(\frac{dI}{dt}\right)_2 \tag{5.26}$$

$$h\nu v W_i\left(n_j - n_i\frac{g_j}{g_i}\right) \geq \frac{I}{\tau_c}.$$

Critical inversion or threshold may be expressed in terms of (5.22) as a constant times the ratio of spontaneous lifetime to photon lifetime

$$\left(n_j - n_i\frac{g_j}{g_i}\right) \geq \frac{8\pi\nu^2}{v^3 g(\nu)}\left(\frac{\tau_{ji}}{\tau_c}\right). \tag{5.27}$$

At line center the normalized line-shape function for a Lorentzian line becomes

$$g_L(\nu)\Big|_{\nu=\nu_c} = \frac{\Delta\nu}{2\pi\left[(\nu-\nu_c)^2 + \left(\frac{\Delta\nu}{2}\right)^2\right]} \tag{5.28}$$

Then the excess population density required to sustain oscillations becomes

$$\left(n_j - n_i \frac{g_j}{g_i}\right)_L = \frac{4\pi^2 v^2 \Delta\nu}{v^3}\left(\frac{\tau_{ji}}{\tau_c}\right). \tag{5.29}$$

Similarly, at line center, the normalized line-shape function for a Gaussian line becomes

$$g_G(\nu)\bigg|_{\nu=\nu_c} = \frac{2\sqrt{\pi \ln 2}}{\pi \Delta\nu_D}. \tag{5.30}$$

For a Gaussian line the population excess required to sustain oscillations becomes

$$\left(n_j - n_i \frac{g_j}{g_i}\right)_G = \frac{4\pi^2 v^2 \Delta\nu_D}{v^3 \sqrt{\pi \ln 2}}\left(\frac{\tau_{ji}}{\tau_c}\right). \tag{5.31}$$

An interesting property of the Gaussian line shape may be observed if the generalized expression for the entire line width is used, namely (5.18). Thus the population excess becomes

$$\left(n_j - n_i \frac{g_i}{g_j}\right) = \frac{3hg_j}{16\pi^3 \tau_c |\mu_{ij}|^2} \sqrt{\frac{\pi}{\ln 2}} \left(\frac{\Delta\nu_D}{\nu_c}\right), \tag{5.32}$$

but

$$\Delta\nu_D = \frac{\nu_c}{c}\sqrt{\frac{2kT \ln 2}{M}}. \tag{5.33}$$

As a result, for a Gaussian shape-line, the critical population density becomes independent of wavelength:

$$\left(n_j - n_i \frac{g_j}{g_i}\right) = \frac{3hg_j\sqrt{\pi}}{16\pi^3 \nu \tau_c |\mu_{ij}|^2}\left(\frac{2kT}{M}\right)^{1/2}. \tag{5.34}$$

The equivalent expression for a Lorentzian line is

$$\left(n_j - n_i \frac{g_j}{g_i}\right) = \frac{3hg_j \Delta\nu}{16\pi^2 \nu_c \tau_c |\mu_{ij}|^2}. \tag{5.35}$$

The incremental gain, γ in the z direction, may be expressed as

$$\gamma = \frac{1}{I}\frac{dI}{dz} = \frac{W_i h\nu}{I}\left(n_j - n_i \frac{g_j}{g_i}\right). \tag{5.36}$$

At line center, for a Gaussian line, (5.32) becomes

$$\gamma = \frac{v^2\sqrt{\pi \ln 2}}{4\pi^2 \nu^2 \Delta\nu_D \tau_{ji}}\left(n_j - n_i \frac{g_j}{g_i}\right), \tag{5.37}$$

and for a Lorentzian line at $v = v_c$

$$\gamma = \frac{v^2}{4\pi^2 \nu^2 \Delta\nu\, \tau_{ji}} \left(n_j - n_i \frac{g_j}{g_i} \right). \tag{5.38}$$

Many transition probabilities have been experimentally determined for a number of elements. The information is usually tabulated in terms of oscillator strengths of f values[7]. The oscillator strength may be regarded as a measure of the degree to which the ability of the atom to absorb or emit a given spectral line resembles such an ability on the part of a classical oscillating electron. It is defined as[8]

$$f = \frac{Mv}{\pi\left(n_i \dfrac{g_j}{g_i} - n_j \right) e^2} \int_0^\infty k_\nu \, d\nu \tag{5.39}$$

where

$$k_\nu = -\frac{1}{I}\frac{dI}{dz} = \frac{h\nu}{v} I(\nu)(n_i B_{ij} - n_j B_{ji}), \tag{5.40}$$

and k_ν is the attenuation coefficient of the median in the z direction.

An important relation exists between the integrated absorption, the upper-state lifetime, and the population density

$$\begin{aligned} \int_0^\infty k_\nu \, d\nu &= \frac{h\nu}{v} n_i B_{ij} \left(1 - \frac{g_i n_j}{g_j n_i} \right) \\ &\cong \frac{h\nu}{v} n_i B_{ij}. \end{aligned} \tag{5.41}$$

Thus, once the integrated absorption has been evaluated, the B value may be obtained if the populations are known.

Using (5.10) and (5.12), (5.41) may also be expressed in terms of the spontaneous emission lifetime:

$$B_{ij} = \frac{g_j}{g_i} B_{ji},$$

and

$$B_{ji} = \frac{v^3}{8\pi h \nu^3 \tau_{ji}} \tag{5.42}$$

Therefore

$$\int_0^\infty k_\nu \, d_\nu = \frac{v^2}{8\pi \nu^2 \tau_{ji}} \frac{g_j}{g_i} \left(1 - \frac{g_i n_j}{g_j g_i} \right). \tag{5.43}$$

Using (5.39) and $\lambda\nu = v = c/n$,

$$f\tau_{ji} = \frac{Mv}{8\pi^2 e^2} \frac{g_j}{g_i} \lambda_{ji}^2 = \frac{Mv^3}{8\pi^2 e^2 \nu^2} \frac{g_j}{g_i} \tag{5.44}$$

where λ_{ji} is the wavelength in the median. Equation 5.39 enables us to calculate the lifetime once the f value has been measured, or vice versa. Values of $f\tau_{ji}$ and g_j/g_i are tabulated[8]. Table 5.1 gives some typical $f\tau_{ji}$ and g_j/g_i values for several elements[9].

The incremental gain may be expressed in terms of the f values rather than the spontaneous-emission time. Considering a Gaussian line,

$$\gamma = \frac{2e^2(\pi \ln 2)^{1/2}}{Mv\Delta\nu_D} \left(n_j - \frac{n_i g_j}{g_i}\right) f. \tag{5.45}$$

At threshold the net loss per pass of the entire ensemble α must equal the net gain. Let α be the normalized net loss per pass (in nepers) divided by the resonator length. Then $\gamma = \alpha$ expresses the threshold condition. The critical population density for a Gaussian line may be expressed conveniently in terms of f values as

$$\left(n_j - n_i \frac{g_j}{g_i}\right) = \frac{Mc\Delta\nu_D \alpha}{2e^2 f(\pi \ln 2)^{1/2}}. \tag{5.46}$$

Using (5.46), known values of α, and a table of f values, we can determine the required critical population density for oscillation.

It will be instructive to compute the order of magnitude of τ_c. Consider a PPFP resonator with flat mirrors having a power-reflection coefficient R spaced at a distance d. The rate of energy loss (5.24) becomes

$$\frac{\omega W}{Q} = \frac{W}{\tau_c} \cong \frac{v(1-R)W}{d},$$

so that the photon lifetime in the cavity is approximately[10]

$$\tau_c = \frac{d}{v(1-R)}. \tag{5.47}$$

In a typical argon-ion laser $d = 100$ cm. $v = c/n = 3 \times 10^{10}$ cm/sec, $R = 0.85$, so that

$$\tau_c = \frac{100}{(3 \times 10^{10})(0.15)} = 2.22 \times 10^{-8} \text{ sec}.$$

TABLE 5.1 PARTIAL TABLE OF OSCILLATOR STRENGTHS[a]

Element	Resonance Line	g_2/g_1	λ_{21} (nm)	$f_{ji} = 1.51\,(g_j/g_i)$ $\lambda_0^2 \times 10^9$
Li	$2^2S_{1/2}$-$2^2P_{1/2}$	1	670	6.80
Li	$2^2S_{1/2}$-$2^2P_{3/2}$	2	670	13.6
Na	$3^2S_{1/2}$-$3^2P_{1/2}$	1	589	5.24
Na	$3^2S_{1/2}$-$3^2P_{3/2}$	2	589	10.46
K	$4^2S_{1/2}$-$4^2P_{1/2}$	1	769	8.94
K	$4^2S_{1/2}$-$4^2P_{3/2}$	2	766	17.8
Cs	$6^2S_{1/2}$-$6^2P_{1/2}$	1	894	12.1
Cs	$6^2S_{1/2}$-$6^2P_{3/2}$	2	852	21.9
Mg	3^1S_0 -3^3P_1	3	457	9.48
Mg	3^1S_0 -3^1P_1	3	285	3.68
Ca	4^1S_0 -4^3P_1	3	657	19.5
Ca	4^1S_0 -4^3P_1	3	422	8.09
Zn	4^1S_0 -4^3P_1	3	307	4.28
Zn	4^1S_0 -4^1P_1	3	213	2.07
Sr	5^1S_0 -5^3P_1	3	689	21.5
Sr	5^1S_0 -5^1P_1	3	460	9.64
Cd	5^1S_0 -5^3P_1	3	326	4.80
Cd	5^1S_0 -5^1P_1	3	228	2.37
Ba	6^1S_0 -6^3P_1	3	791	28.3
Ba	6^1S_0 -6^1P_1	3	553	13.9
Hg	6^1S_0 -6^3P_1	3	253	2.91
Hg	6^1S_0 -6^1P_1	3	185	1.55
Tl	$6^2P_{1/2}$-$7^2S_{1/2}$	1	377	$f/A = 2.15$
Tl	$6^2P_{1/2}$-$6^2D_{3/2}$	2	276	$f/A = 2.31$
Tl	$6^2P_{3/2}$-$7^2S_{1/2}$	1/2	535	$f/A = 2.20$

[a]Reproduced with permission from Cambridge University Press.

Using $Q = \omega\tau_c$ for $\lambda = 488$ nm,

$$Q = (2\pi)\left(\frac{3 \times 10^{10}}{4.88 \times 10^{-5}}\right)(2.22 \times 10^{-8}) \cong 8.6 \times 10^7.$$

The calculation made above will, of course, yield an unrealistically large value for Q and τ_c, for they neglect many of the cavity-loss factors, such as diffraction, scattering, and absorption, among others. They are, however, good order-of-magnitude guides.

5.0.2 Spectral Line-Shape Phenomena

The treatment of spectral line shape is restricted to gases because the analytical results and phenomenology are adequate to form the foundations for measurement techniques, and more than a qualitative treatment is required to illucidate all the variables, which are, at present, still a subject of active research[11, 12, 13].

Spectral lines in gases are subject to five types of broadening:

1. Natural broadening due to the finite lifetime of the excited state.
2. Doppler broadening due to thermal motion of the atom.
3. Lorentz broadening due to collisions with different kinds of atoms or foreign gases.
4. Holtzmark or collision broadening due to impacts with other gas atoms of the same type.
5. Stark broadening due to collisions with electrons or ions.

Both items 3 and 4 are sometimes lumped under the same heading and referred to as "pressure broadening."

The natural breadth of a spectral line $\Delta\nu_N$ may be shown[14] to be the Einstein coefficient (5.2) divided by 2π:

$$\Delta\nu_N = \frac{A_{ji}}{2\pi} = \frac{1}{2\pi\tau_{ji}}. \tag{5.48}$$

$$\tau \sim 10^8,\ \Delta\nu_N \sim 10 \text{ MHz}.$$

Because $\tau \sim 10^{-8}$, $\Delta\nu_N \sim 10\text{MHz}$.

In principle the natural width of a spectral line could be measured as the absorption line produced in the output of a continuous-spectrum light source. The beam of light from the continuous-spectrum source is transmitted through an atomic beam of the material the natural line width of which is to be measured. The description is only of academic interest, first, because the resolution required is beyond that available with spectrographic instruments and, second, because the thermal motion of the atoms, that is the Doppler velocities, would tend to mask the effect.

Almost the entire contribution to line broadening in gases is due to the Doppler motion of the atoms. This is a circumstance that results from the fact that, in general, lasers operate in the pressure range from 1 – 1000 mTorr. We shall, therefore, not treat items 3 and 5 listed above but will refer the interested reader to the literature[9].

It is well established that the absorption coefficient of a gas, subject to Doppler broadening, can be expressed as

$$k_\nu = \frac{I}{I}\frac{dI}{dz} = -k_0 \exp\left(-\left[\frac{2(\nu - \nu_c)}{\Delta\nu_D}\sqrt{\ln 2}\right]^2\right)$$

Integrating the absorption coefficient over the line and using (5.43),

$$\int_0^\infty k_\nu \, d\nu = \frac{1}{2}\sqrt{\frac{\pi}{\ln 2}}\, k_0 \, \Delta\nu_D$$

$$= \frac{v^2}{8\pi\nu^2\tau_{ji}}\left(\frac{g_j}{g_i}\right)\left(1 - \frac{g_i n_j}{g_j n_i}\right).$$

Using (5.43), the value of k_0 becomes

$$k_0 = \left(\frac{2}{\Delta\nu_D}\sqrt{\frac{\ln 2}{\pi}}\right)\left(\frac{\lambda_{ji}^2 g_j}{8\pi g_i}\right)\left(\frac{n_i}{\tau_{ji}}\right) \tag{5.49}$$

Because the Doppler effect and natural broadening arise from entirely independent processes, the combined absorption coefficient may be calculated by considering each portion of the Doppler line width to be broadened by natural damping, or vice versa.

In some instances the effect of collision broadening (item 3) cannot be neglected, and a folded Gaussian-Lorentz line shape best approximates the data[15]. Because Doppler and collision-broadening effects are independent, the line shape may be calculated on the same basis as described above, namely, by considering every infinitesimal portion of the pure Doppler-broadened line as being subject to Lorentz broadening. The normalized line-shape function then becomes[9].

$$g(\nu) = \frac{\sqrt{\ln 2}}{\pi^{3/2}\Delta\nu_D}\int_{-\infty}^{\infty} \frac{\Delta\nu\left\{\exp\left[-\left(\nu_D \dfrac{\sqrt{\ln 2}}{\Delta\nu_D}\right)^2\right]\right\}}{\Delta\nu^2 + (\nu - \nu_c - \nu_D)^2}\, d\nu_D \tag{5.50}$$

where v_D is a running variable in the band $v - v_c$.

The response of the atomic system to intense monochromatic light is determined by the nature of the line-broadening mechanism. For a homogeneously broadened line the same atoms are responsible for gain over different regions of the line, a homogeneously broadened line being defined in the limit as a line with no Doppler broadening:

$$\frac{\Delta\nu_D}{\Delta\nu_N} = 0 \tag{5.51}$$

For an inhomogeneously broadened line, the Doppler width is large compared with the natural width. In the limit

$$\frac{\Delta\nu_N}{\Delta\nu_D} = 0 \tag{5.52}$$

In an inhomogeneously broadened line the atoms responsible for absorption and gain tend to act independently. In most gas lasers the line width is the inhomogeneously broadened Doppler line width of a gas.

A phenomenon that characterizes an inhomogeneously broadened line is "hole burning" or gain saturation. A hole results when the saturating power is confined to a frequency range small compared with the inhomogeneous line width[16].

5.1 METHODS OF MEASURING OPTICAL GAIN

Two methods have been developed for measuring single-pass gain. In high-gain lasers, of which a helium-xenon plasma at 3.5 μm is typical, it is convenient to use an attenuator method, in which lossy elements are inserted inside the laser, directly in the beam path, and the gain is determined from the number and absorption calibration of the attenuators used. This method has the advantage of requiring only one plasma tube, the laser whose properties are to be studied. The attenuation method tends to give a more optimistic gain in that it does not take into account narrowing of the spectral line and consequent increase in gain as the cavity power level is decreased.

Single-pass gain can be measured in a conventional way using an amplifier technique. A stable signal-laser oscillator is required, the frequency of which can be tuned to match the center of the inhomogeneously broadened line width of the laser amplifier for which the gain is to be measured. The amplifier method of gain measurement requires careful alignment and mode matching. Considerable care must be taken, especially with high-gain lasers, to ensure that aperture scattering does not lead to regeneration and cause errors. Also a large degree of isolation is required between the oscillator laser and the amplifier under test to prevent interaction between the stages. The amplifier method of measuring gain results in a value of gain averaged over the cross section of the plasma tube and also over a small portion of the Doppler line width. Gains measured by this method are always smaller than those measured by the attenuator method.

The amplifier method of measuring Doppler line width and gain narrowing in high-gain, CW-gas lasers is described in Chapter 7 (Sections 7.3 and 7.4), wherein the subject of laser bandwidth is treated. The gain of CW amplifier can be determined using these measuring techniques. The attenuation method of gain measurement of a CW xenon lases is described below. After the discussion of CW measurements a section is devoted to gain measurements in pulsed-gas lasers.

5.1.1 Gain Measurement in a CW-gas Laser by the Attenuator Method

When the gain, and not the details of the line shape of the laser transition, is of interest and, furthermore, when the particular laser transition

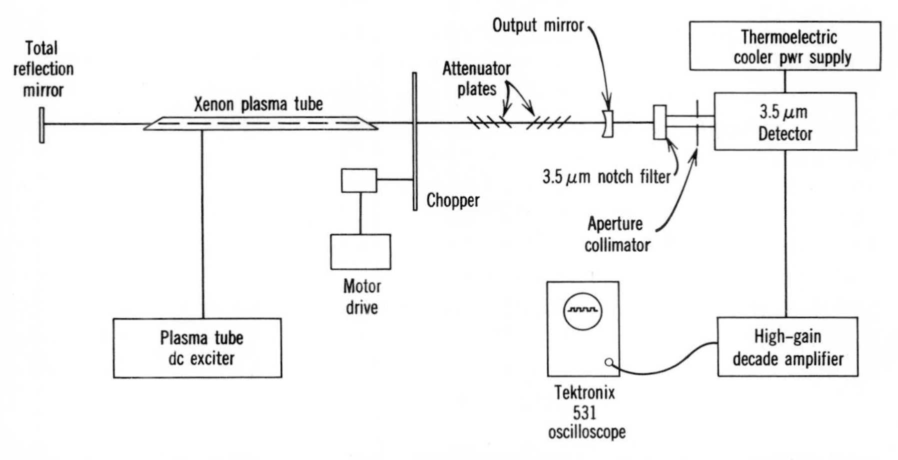

Figure 5.1. Apparatus arrangement for measuring CW-laser gain by the attenuator method.

has high gain, as, for example, the 3.39 μm transition in neon or the 3.508 μm transition in xenon, the attenuator method may be used to obtain a measure of over all gain. The apparatus arrangement shown in Figure 5.1, as listed in Table 5.2, will be found convenient for making gain measurements on the 3.5 μm line in xenon.

TABLE 5.2 APPARATUS FOR ATTENUATOR MEASUREMENT OF SINGLE-PASS GAIN

Item	Function	Source
CW Xenon Laser	Laser under test	*hν* systems Model 3367
Detector	Measure 3.5 μm power	Philco IAT 704 and power supply
Flat-glass wedge	Alternate beam	Industrial Optics
Chopper	Interrupt 3.5 μm beam	Infrared Industries 801-B
Amplifier	Amplify chopped signal of detector	HP-450A
Filter	Isolate 3.5 μm signal	Spectrum Industries notch filter 50°A half-width
Aperture	Limit background noise	Fabricate from black cardboard
Oscilloscope	Measure signal	Tektronix 531 or equal

Several precautions should be observed when the gain measurements are made to ensure that the results will be meaningful. The 3.5 μm notch filter should be tilted with respect to the beam axis to preclude the return of 3.5 μm energy to the laser cavity. The detection system should be a few meters away from the laser to reduce the amplitude of the spontaneous emission 3.5 μm noise generated by the laser[17]. The chopper motor should be mounted on a suspension independent of the laser cavity to minimize vibration-induced changes in laser gain due to mirror motion. The aperture collimator in front of the detector should be of a large enough size to prevent the occlusion of the beam spot, while shielding the infrared detector from background noise caused by laboratory instruments and incandescent lights. The aperture in the chopper plate must be large compared with beam-spot size to preclude apodization of the beam. The loss of the individual attenuator plates should be checked by beam insertion (techniques discussed in Chapter 4, Section 4.11.6). Care must be taken that the attenuators are clean, for oil films may markedly change the absorption characteristics. Because the gain of a xenon laser depends strongly on excitation power it is vital that the current be highly regulated.

The procedure for gain measurement is as follows:

1. Adjust the xenon laser for operation at the desired power level and allow the unit to stabilize. Turn the mirror positions for the highest power output on the lowest-order TEM_{00} mode.
2. Set the chopper to rotate at a convenient speed that does not match with the power-line frequency.
3. Turn on all electronics including the diode thermoelectric cooler and allow the laser to equilibrate.
4. Using the oscilloscope, detector, amplifier, and chopper, align the laser beam with the detector for maximum signal.
5. While observing the chopped signal on the oscilloscope, install the aperture collimator and adjust for maximum reduction in noise for minimum change in signal. Insert the 3.5 μm notch filter and check that it does not complete an optical path to the cavity. This can be most easily tested by aligning the filter front surface with the cavity (as observed by an increase in the detector reading) and subsequently changing the alignment, while watching the oscilloscope signal, until the signal enhancement disappears.
6. Record the signal-to-noise ratio as successive attenuators are inserted into the laser cavity in pairs to minimize beam displacement.
7. Continue to insert attenuators until oscillation ceases. Without disturbing the orientation of the attenuator, remove the assembly from the cavity and place it outside the laser cavity between the laser and the detector to check attenuators. Repeat steps 4 and 5 as necessary to ensure that the attenuator assembly has not displaced the beam with respect to the detector. Measure the net single-pass beam loss through the attenuator assembly. Compare these results with the two-way attenuation observed in step 5.

Note that the attenuator method of gain measurement determines the average of the folded two-pass gain. Note that the placement of the attenuators at the nonoutput end of the cavity is not recommended because the detector must then distinguish between an attenuated signal and the 3.5 μm laser noise that has been amplified by the full single-pass gain.

For a laser operating in steady-state conditions the single-pass gain in the active medium must equal the single-pass loss in the laser and cavity[18]. Therefore the attenuation method yields

$$\exp[2(\gamma_c - \alpha)d] = 0, \tag{5.53}$$

where d is the length of the discharge, α is the average single-pass incremental loss including the transmittance of the mirror and γ_c is the single-pass incremental gain defined at line center as

$$\gamma_c = \frac{2}{\Delta\nu_D}\left(\frac{\ln 2}{\pi}\right)^{1/2}\frac{\pi e^2}{Mv}\left(\frac{g_i}{g_j}n_j - n_i\right)f. \tag{5.54}$$

where the symbols have the previously defined meanings.

The justification for using (5.24) is as follows. If the laser is operating well above threshold, the gain in an inhomogeneously broadened line is reduced by the nonlinearity represented by hole burning to a value that just balances the cavity losses[18]. If, in contrast, the laser is operating exactly at threshold, the gain is just sufficient to equal the losses, and no appreciable hole burning results. An accurate measure is therefore obtained of the maximum small signal gain.

It is instructive to compare the results of two gain measurements on the same laser using first the attenuation and second the conventional amplification method. The data presented in Table 5.3 illustrate the higher gain measured by the attenuation method[19, 20].

TABLE 5.3 COMPARISON OF CW-LASER GAIN MEASURED BY ATTENUATOR AND AMPLIFIER METHODS
(7.0 mm bore laser tube)[19]

Method	Incremental Gain dB/m
Attenuator	4.45
Amplifier	3.25

5.1.2 Single-pass Gain Measurements in Pulsed Lasers

Techniques for measuring gain in pulsed solid-state lasers are treated in Chapter 4 (Section 4.11.4.1.4). This section considers the problems encountered in measuring pulsed-gas lasers.

The two general methods for measuring laser gain outlined above can be applied to pulsed-gas lasers with, however, some important differences in technique.

As before the gain can be inferred using the attenuator method, from measurements of the insertion loss that will just extinguish oscillation. In pulsed-gas lasers, however, the gain, being simultaneously current-density and pressure dependent may not be the same throughout the pulse (even if the pulse-current amplitude is constant; it seldom is). Thus, although it is easy to determine the value of attenuation which extinguishes laser action, the dynamical characteristic of the discharge presents much more information to the investigator. In those pulsed-gas lasers that develop inversion during the current pulse, as contrasted with those that only exhibit pulsed laser-action in the afterglow (that is after the current pulse is extinguished) or the self-terminating type of gas lasers which generate nanosecond pulses, a measurement can be made of the attenuation required to extinguish the beam at some time during the current or optical pumping pulse[21, 22, 23]. An oscilloscope that displays the detector output versus current or optical pump time enables the determination of the current or value of excitation that corresponds to the attenuation and therefore the threshold of gain. From these data we can determine the gain-versus-excitation characteristic. The apparatus outlined in Table 5.4 and arranged as illustrated in Figure 5.2 will be found convenient for measuring gain of pulsed ionized gas lasers by the attenuation method.

Neither the self-terminating nor the afterglow type of pulsed-gas lasers lend themselves to convenient measurement of other than maximum gain [24, 25, 26]. The pulse duration of self-terminating or cyclic pulsed-gas lasers depends upon establishing inversion in a gas during the initial portion of an excitation pulse by preferential population of some upper

TABLE 5.4 APPARATUS FOR GAIN MEASUREMENT IN PULSED-GAS LASERS BY THE ATTENUATOR METHOD

Item	Function	Source
Pulsed-ion laser	Laser under test	$h\nu$ systems Model 3355 and pulse modulator Model 2030 or 2080
Notch filter	Line selection	Thin Film Products, Inc.
Photomultiplier and Collimator	ream intensity measurement	$h\nu$ systems Model 3227
Neutral-density attenuators	Attenuate beam intensity	Optics Technology
Oscilloscope	Measure intensity-current dependence	Tektronix 536 with (2) type G and (1) type T plug in units

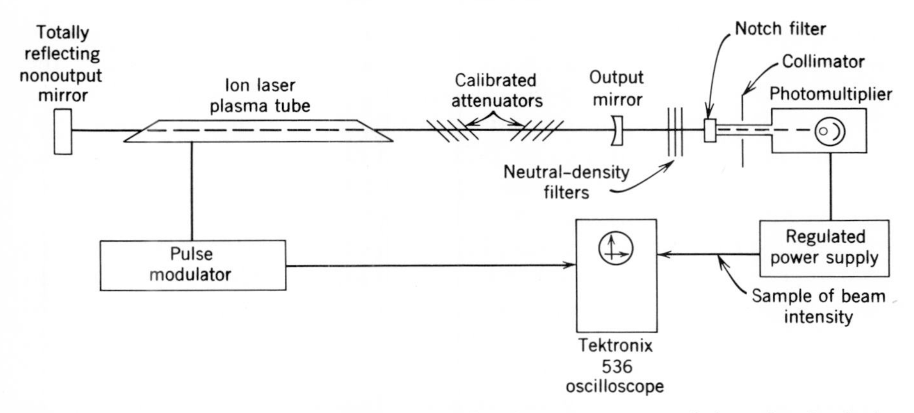

Figure 5.2. Arrangement of instrumentation for the measurement of the gain of pulsed, ionized gas lasers by the attenuator method.

laser level (either directly or through cascade processes). Laser action is usually terminated when the population of the associated lower laser level, which is characteristically limited by a metastable lifetime, becomes equal to the upper state level. Obviously the detailed relationship between the input impulse and the subsequent laser output is difficult to examine directly. Although the mechanism that produces inversion is different in the afterglow laser, the time-separation phenomena associated with laser action similarly negates measurements of direct-gain excitation.

Because gain measurement by the attenuator method only requires one laser and is simpler and more easily performed, it is the method most frequently used. It is difficult to apply the attenuator method to low-gain lasers because low insertion-loss attenuators are not available. A pulsed variation of the technique of resonant-cavity spectroscopy can be applied if the plasma tube can be conveniently segmented for separate excitation. This technique will be described in a subsequent section.

Measurement of gain by the attenuator method in pulsed-gas lasers is entirely analogous to that described for the CW laser. The insertion loss, which causes laser action to cease entirely or at some predetermined value of the excitation time, is determined from the value of attenuation used. As the double-pass gain is measured, the single-pass gain may be unfolded if it is clear that saturation does not predominate.

Gain measurements by the amplifier method require techniques that extend the CW method to the time-dependent domain. When the amplifier method is used, and input and output pulses are displayed on an oscilloscope. The suitably attenuated output pulse can be optically delayed by reflecting it over a 20 to 30 m path (for short-pulse lasers) and displayed on the same sweep with the input pulse. A dynamic-null method can be used for pulses of microsecond or greater duration. A suitably attenuated sample of the amplifier-output beam is subtracted from the sample of the oscillator-output beam, and the difference is displayed on an oscilloscope. The time-dependent, gain-saturation phenomna are then observed as time-dependent deviations from a null. The sum of the value of the optical and electronic attenuation required to produce a null measures the single-pass gain averaged over the diameter of the plasma tube.

Several problems arise in applying the steady-state techniques of the amplifier method to pulsed-gain measurements. The oscillator- and amplifier-excitation times must be coincident. This is usually accomplished by exciting both plasma tubes with the same modulator and connecting the plasma tubes in series. Noninductive shunt resistors enable the adustment of current density in the oscillator to match the higher current requirements of an amplifier of larger bore.

Because pulsed-gas lasers generally have higher gains than their CW counterparts, much greater care must be taken to ensure isolation between the oscillator and amplifier exit and entrance mirrors. If this precaution is not taken, regeneration will occur in the oscillator and cause errors in the gain measurements, particularly at low values of attenuation.

Spontaneous emission contributes to the output of the amplifier. The effect can be identified as follows. The output intensity I_1 is observed with the oscillator on and projecting through the amplifier. The amplifier is then pulsed simultaneously and a new value of intensity I_2 determined. The value of I_2 includes spontaneous emission from the amplifier (and the oscillator if the latter is not suitably separated from the amplifier by an adequate distance). Next, the oscillator laser is turned off and a third intensity measurement I_3 is made. The net time-dependent gain becomes $(I_2 - I_3)/I_1$. The net attenuation of the amplifier is determined by a fourth intensity measurement I_4 in which the oscillator is turned on but the amplifier is physically removed. The amplifier loss is then $I_4 - I_1$.

Even though the null method of gain measurement gives more precise results it is also worthwhile to observe the time-dependent output of the laser. The output-versus-excitation dependence should also be monitored to guard against errors of interpretation and hysteresis in the gain profile. The latter is inevitable in pulsed-gas lasers, for the gain frequently not only depends upon the value of the excitation at the instant but upon the past history of the pulse. If the wavelengths of a pulsed-gas laser match those of a CW-gas laser, a CW-gas laser can be used as a convenient source oscillator. This simplifies the problems of synchronization and gain-saturation monitoring.

Electrical noise is a common difficulty encountered in high-power, pulse-modulated circuitry. Proper shielding, the prevention of ground

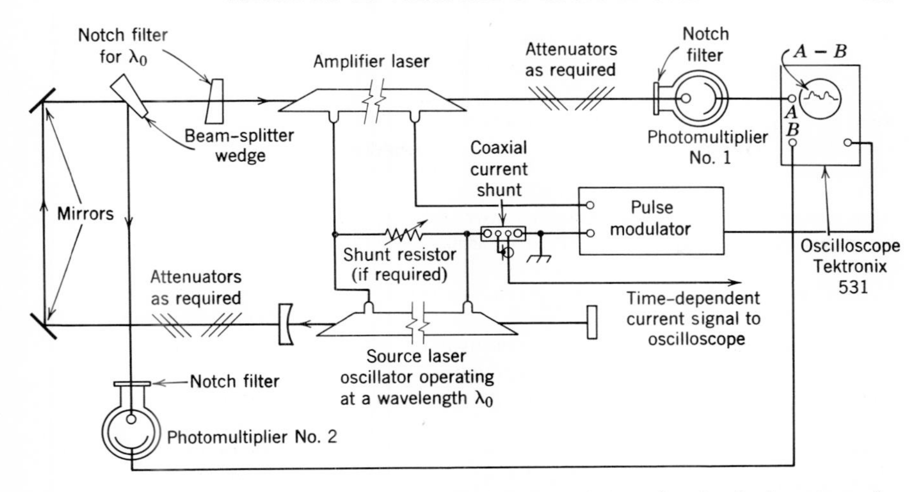

Figure 5.3. Apparatus arrangement for measurement of the gain of pulsed-gas lasers by the amplifier method.

loops, and the use of low-inductance (wide-strap) returns for high-pulsed currents will assist in minimizing the interference problem. A careful check of the noise environment is required before pulsed-gas laser measurements are made. Frequently differential amplifier techniques can be used to cancel out unwanted signals.

A typical experimental arrangement for measuring the gain of a pulsed-gas laser by the amplifier method is shown in Figure 5.3. The equipment listed in Table 5.5 will be found helpful in making measurements. Application of the instrumentation will be obvious from the foregoing discussion.

5.1.3 Measurements of Gain Variation Over the Laser Diameter

The gain of any laser generally varies over the diameter of the active medium, being highest in the core and tapering off to the edges. In gas lasers this gain characteristic is more uniform and repeatable, whereas in optically pumped solid-state lasers the gain is strongly influenced by the optical pump geometry and can change markedly upon replacement of flash tubes or through motion of the laser rod itself. The gain variation of solid-state lasers is best studied during a single pulse using photometric techniques. We consider here only the radial-gain variation in a gas laser. The extention of techniques to CW-solid-state lasers will be obvious.

The radial variation of gain is best explored with a laser beam of very small diameter. Either the beam of the test laser must be apodized or a plasma tube of small bore can be employed for the beam probe. To achieve the best alignment, it is customary to place the probe laser and the detector on a Y-shaped support that will enable the probe laser and

TABLE 5.5 APPARATUS FOR MEASUREMENT OF THE GAIN OF PULSED-GAS LASER BY THE AMPLIFIER METHOD

Item	Function	Source
Source laser	Pulsed laser to illuminate amplifier	$h\nu$ systems Model 3800
Amplifier laser	Amplifier laser under test	$h\nu$ systems Model 3355
Noninductive resistors	Oscillator plasma current shunt	Ohmweave
Notch filter (3)	Monochromatize beam	Thin Film Products, Inc. 5.0 nm bandwidth
Attenuators (2 sets)	Reduce sensitivity of Photomultipliers	Optics Technology
Photomultipliers (2)	Measure time-dependent laser intensity	$h\nu$ systems Model 3227
Coaxial current shunt	Measure current at low impedance	T & M Industries .005 ohm coaxial shunt
Oscilloscope	Measure differential gain	Tektronix 531

detector to be translated as a unit across the aperture of the amplifier tube. If the assembly is translated by a gear-driven table, such that a voltage proportional to position can be derived from a suitably located potentiometer, the output of the detector and the position indicator can be used to drive an x–y recorder. The latter will plot gain as a function of radius. The effect of the tube walls and reflections will be measured if the data are taken first with the amplifier on and second with it turned off. Care should be taken to align the axis of the oscillator accurately with the amplifier.

5.1.4 Resonant-cavity Absorption Spectroscopy for Measuring Low Values of Gain in CW Lasers

The measurement of relative populations in optically connected energy states of atomic ensembles in the gaseous state has been of great interest for many years[9]. The time-honored method of measuring relative populations between optically connected states involves the use of single- or dual-plasma tube, single-pass absorption spectrophotometry. The gain or loss in intensity of an irradiating beam of the appropriate wavelength yields the absorption characteristics of the medium. This technique has the disadvantage that, for small photon cross sections, long paths are necessary for measurement of the resulting small gain or attenuation. This objection is partially overcome by using optical delay-line techniques inside a White-type absorption cell. The latter technique is limited to the investigation of transitions that are not aperture dependent and requires mirrors of extremely high reflectivity to minimize losses. A modification of the multiple-reflection technique has been used to study low-gain laser transitions[27]. Essentially the method places the electronically excited absorption cell inside a Fabry-Perot cavity. Elimination of the illuminating light source ensures that only those transitions that exhibit negative absorption will yield an appreciable output signal. The technique has been referred to as resonant-cavity spectroscopy [27, 28, 29].

The single-pass gain of a medium under investigation is obtained by comparison of the resonant intensity I_R with the nonresonant intensity I_{NR} from one end of the optical cavity containing the fluorescing atoms. The nonresonant intensity is directly proportional to the spontaneous emission intensity in the medium I_0. If the medium has gain, a nonvanishing number Γ exists such that

$$\Gamma = \frac{\bar{I}_R - \bar{I}_{NR}}{\bar{I}_R} = R \exp \xi d \tag{5.55}$$

where the bar over I_R and I_{NR} indicate a mean value over a phase period.

If the power gain $\exp \xi d$ differs only slightly from unity, the power reflectivity R must be accurately known to prevent masking of the difference. If two measurements of Γ are made, each for different lengths of active medium, this difficulty can be circumvented. It is easily seen that

$$\frac{\Gamma_2}{\Gamma_1} = \exp \xi(d_2 - d_1) \tag{5.56}$$

or

$$\frac{1}{d_2 - d_1} \ln \frac{\Gamma_2}{\Gamma_1} = \xi.$$

The accuracy with which d_1 and d_2 can be determined limits this technique unless d_1 and d_2 are several tens of centimeters long. Note that this technique lumps all losses into one parameter R. The difference between the actual reflectivities of the mirrors and the value of R represents another non-negligible source of error which must be determined if the gain is to be measured with desired precision. Nevertheless, the utilization of an active Fabry-Perot resonator enhances and enables the measurement of small values of gain.

5.1.4.1 Properties of an Active Fabry-Perot Cavity To simplify the derivation of the gain equations for the active Fabry-Perot cavity, consider a cavity of infinite aperture containing excited atoms. Figure 5.4 illustrates a real Fabry-Perot cavity on the right side of the figure and an infinite set of virtual cavities in image space on the left. The virtual cavities are shown for book-keeping purposes to account for the emitted radiation in multiple

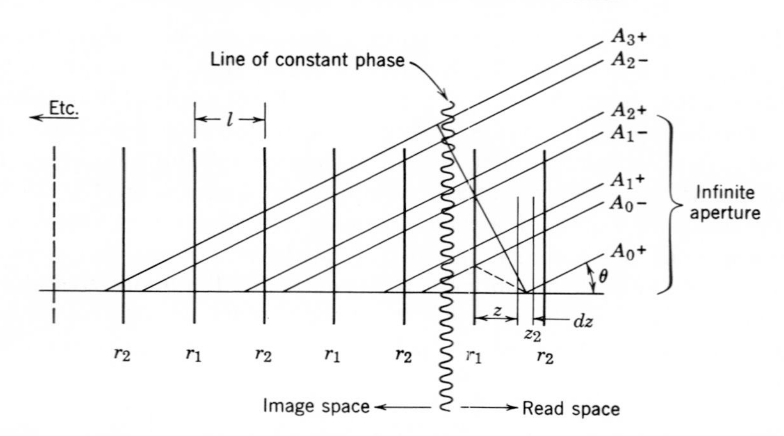

Figure 5.4. Schematic diagram for calculation of external intensity from plane-parallel Fabry-Perot cavity containing fluorescent atoms. (Reproduced with permission from the Polytechnic Press of the Polytechnic Institute of Brooklyn.)

reflections. The emitted radiation can be divided into two plane waves, one wave having a velocity component in the positive direction and one in the negative direction. The amplitude of the *n*th-reflected wave, having a positive z propagation component, is A_n^+, whereas the opposite is A_n^-:

$$A_n^+ = t_1(r_1r_2)^n \exp\left\{\frac{-\gamma[(2n+1)d - z]}{\cos\theta}\right\} \cdot \exp j\phi \exp(2j\beta nd \cos\theta),$$

$$A_n^- = r_1t_2(r_1r_2)^n \exp\left\{\frac{-\gamma[(2n+1)d + z]}{\cos\theta}\right\} \cdot \exp j\phi \exp(2j\beta(nl + z)\cos\theta), \tag{5.57}$$

various A_n's are coherent and may be summed

$$A_T = \sum_{n=0}^{\infty} A_n.$$

An algebraic theorem permits the summation of a geometric progression to obtain

$$A_t = \frac{t_2 \exp j\phi \exp(-\gamma d/\cos\theta)}{[1 - r_1r_2 \exp(-2\gamma d/\cos\theta)\exp(2j\beta d\cos\theta)]} \cdot \left[\exp\left(\frac{\gamma z}{\cos\theta}\right) + r_1 \exp(-\gamma z\cos\theta)\exp(2\beta z\cos\phi)\right]. \tag{5.58}$$

The intensity I is obtained from the product $I = A * A$ where $A *$ is the complex conjugate of A, and
$I = t_2 \exp(-\gamma d/\cos\phi)$.

$$\left[\frac{\exp(2\gamma z/\cos\theta) + r_1 \exp(-2\gamma z/\cos\theta) + 2r_1 \cos(\beta d \cos\theta)}{1 + r_1{}^2 r_2{}^2 \exp(-4\gamma d/\cos\theta) - 2r_1 r_2 \exp(-2\gamma d/\cos\theta)\cos 2\beta d \cos\theta)}\right]. \tag{5.59}$$

The incremental power per unit solid angle associated with the inverted states in the axial increment dz between z and $z + dz$ may be expressed as

$$dP(z) = \frac{n_2 h\nu_{21}}{4\pi\tau_{21}} A_t(z)A_t^*(z)\, dV = I_0 A_t A_t^*\, dz, \tag{5.60}$$

where n_2 is the population of the excited state, τ_{21} is the spontaneous lifetime associated with the state, and ν_{21} is the frequency associated with the transition $2 \rightarrow 1$. Let

$$I_0 = \frac{n_2 h\nu_{21} d}{4\pi\tau_{21}}$$

$$dI(z) = I_0 A_t A_t^*\, dz/d. \tag{5.61}$$

$\xi = t^2 \exp(-2\gamma d/\cos\theta)$.

$$[1 + r_1{}^2 r_2{}^2 \exp(-4\gamma d/\cos\theta) - 2r_1{}^2 r_2{}^2 \exp(-2\gamma d \cos\theta)\cos(2\beta d \cos\theta)]^{-1}. \tag{5.62}$$

Then 5.61 reduces to

$$dI(z) = I_0 \xi[\exp(2\gamma z/\cos\theta) + r_1{}^2 \exp(-2\gamma z/\cos\theta) + 2r_1 \cos(2\beta z \cos\theta)]\, dz. \tag{5.63}$$

To simplify further, let $\vartheta = 2\gamma/\cos\theta$, $\beta = \omega/c$, and

$$\delta = 2\beta d \cos\theta = \frac{2\,\omega d \cos\theta}{c}.$$

Upon integration from $z = 0$ to $z = d$ and further simplification,

$$I(\delta) = I_0 t^2 \left(\frac{1 - \exp(-\vartheta d)}{\vartheta d}\right)\left[\frac{1 - r_1{}^2 \exp(-\vartheta d)}{(1 - r_1 r_2 \exp \vartheta d)^2 + 4r_1 r_2 \exp(\vartheta d)(\sin^2 \vartheta/2)}\right]. \tag{5.64}$$

For definiteness consider the observation of the cavity emission at the diffraction-limited angle $\theta = 0 \pm \lambda/4\pi a$ for λ = wavelength of emission; $2a$ is the aperture diameter, and as $\cos\theta$ is slowly varying about $\theta = 0$, δ is only a function of angular frequency ω. Thus

$$I(\omega) = (I_0 t_2^2)\frac{1 - \exp(-\vartheta d)}{\vartheta d} \cdot \left[\frac{1 + r_1^2 \exp(-\vartheta d)}{[1 - r_1 r_2 \exp(-\vartheta d)]^2 + 4 r_1 r_2 \exp(-\vartheta d)\sin^2 \dfrac{\omega}{2F_\sigma}}\right], \tag{5.65}$$

where

$$F_\sigma = \frac{c}{2d\cos\theta} \quad \text{is the free spectral range,} \tag{5.66}$$

$I(\omega)$ is periodic in ω becoming maximum when

$$\frac{\omega_0}{2F\sigma} = \pi m, \tag{5.67}$$

or

$$\omega_0 = 2\pi m F_\sigma, \qquad m = \text{an integer}. \tag{5.68}$$

A plot of (5.65) appears in Figure 5.5, superposed on a Doppler profile the width of which is assumed large compared to F_σ.

The ordinary monochromator used to single out the spectral line of interest is seldom capable of sufficient wavelength resolution to observe bandwidths less than the free spectral range of a one-meter laser cavity. Thus the observed intensity will be an average over one or more of these

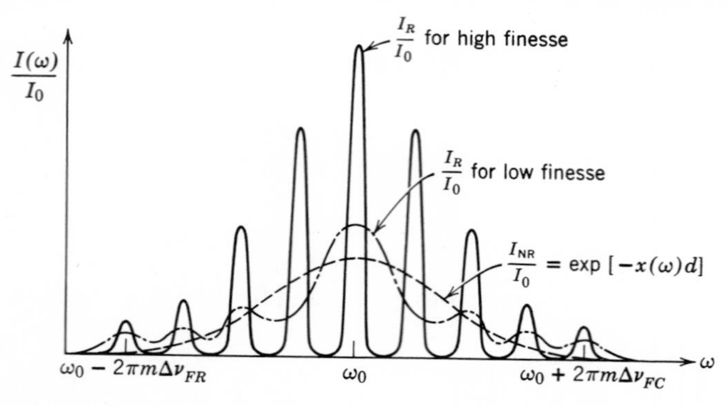

Figure 5.5. Plot of instantaneous intensity versus phase. (Reproduced with permission from the Polytechnic Press of the Polytechnic Institute of Brooklyn.)

modes. On the assumption that ϑ is essentially constant over a period in δ (or over one free-spectral range F_σ), the integral of (5.65) yields the average

$$\bar{I}_R = \frac{1}{2\pi}\int_0^{2\pi} I(\delta)\, d(\delta), \tag{5.69}$$

which becomes

$$\bar{I}_R = I_0\left(\frac{1-\exp(-\vartheta d)}{\vartheta d}\right)T_2\,\frac{[1+R_1\exp(-\vartheta d)]}{[1-R_1R_2\exp(-\vartheta d)]} \tag{5.70}$$

$$R_1 = r_1{}^2, \text{ and } R_2 = r_2{}^2.$$

where $R_1 = r_1{}^2$, and $R_2 = r_2{}^2$.

Similarly it may be shown that

$$\bar{I}_{NR} = I_0\left(\frac{1-\exp(-\vartheta d)}{\vartheta d}\right)T_2 \tag{5.71}$$

when $R_1 = 0$, and $T_2 = t_2{}^2$.

Thus (5.56) is directly deducible from the equations above with the additional simplification that $R_1 = R_2 = R$.

5.1.4.2 Experimental Technique for Resonant-cavity-absorption Spectroscopy A schematic drawing of the experimental arrangement for resonant-cavity spectroscopy is given in Figure 5.6. External mirrors from the optical cavity: the active medium is contained in a plasma tube with Brewster-angle ends. Two choppers are employed, one in the cavity, making it alternately

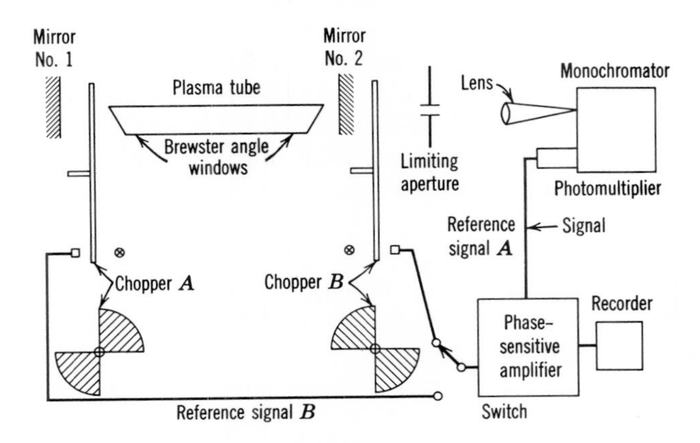

Figure 5.6. Experimental layout for low-gain measurements and resonant-cavity absorbtion spectroscopy. (Reproduced with permission from the Polytechnic Press of the Polytechnic Institute of Brooklyn.)

resonant and nonresonant, and one in front of the cavity. The spectral line to be studied is isolated by a monochromator the entrance slit of which is placed at the focus of the lens and aperture used to limit the field of view. The photoelectrically detected intensity on the output of the monochromator is fed into a recorder attached to a phase-sensitive amplifier that produces signals proportional to the difference in the signals appearing alternately at the input. If the blade of the chopper in the cavity is alternately in and out of the beam path at the same time that the external chopper blade is out of the beam path, the laser cavity is alternately resonant and nonresonant, and a signal is obtained that is proportional to $\bar{I}_R$-$\bar{I}_{NR}$. If the blade of the internal chopper is kept open and the external chopper is operated, the laser cavity is always resonant and either the background signal or $\bar{I}_R$ is measured. Neglecting background signals, the ratio of these signals is Γ.

Depending on the nature of the excitation technique, care must be taken to ensure identical excitation per unit path length of the active medium when plasma tubes of different lengths are used (to eliminate the dependence on R losses). Although the method is independent of the cavity configuration employed a flat-mirror configuration is more useful because it eliminates any lens effect when the cavity is nonresonant.

5.1.5 Technique for Measuring the Amplification Coefficient in Neodymium-doped Glass at 1.06 μm

A knowlege of the amplification coefficient of a laser material is basic to any calculation of its potential performance as a laser amplifier or oscillator. To gain such knowledge by computation alone would require a detailed understanding of the ligan-field environment of the active ion and the influence of the field on the transition probabilities involved. Although computation might be feasible for single-crystal hosts, a computation for glass would be quite difficult because of the variation in the field from ion to ion. An experimental approach is therefore indicated to determine the amplification coefficient, at 1.06 μm in neodymium-doped glass.

The amplification coefficient is an indication of the gain in intensity per unit path experienced by radiation being transmitted through a medium. It is completely analogous to its negative counterpart, the absorption coefficient. The equation

$$\frac{1}{I_0}\frac{dI}{dz} = \gamma \tag{5.72}$$

relates the radiation intensity I at a distance z from point zero to the

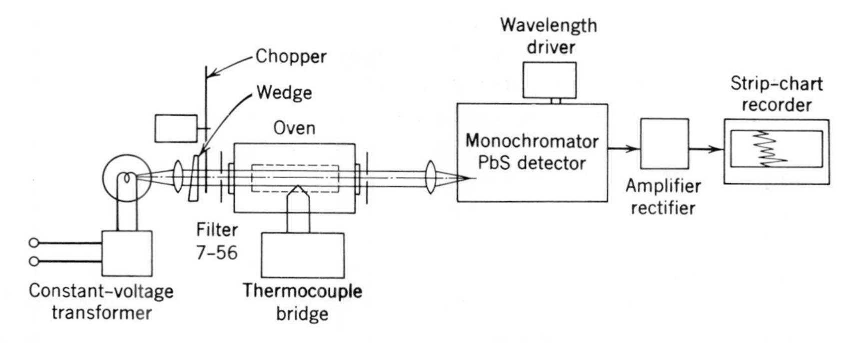

Figure 5.7. Schematic layout of components of a single-beam transmittance recording system.

intensity I_0 at point zero, z being taken along the direction of propagation of the energy and γ being the amplification coefficient.

To measure the amplification coefficient γ directly for neodymium-doped glass, the gain in intensity of 1.06 μm radiation could in principle, be observed after the radiation had traversed a known path in the glass. The problem with this approach is that difficulty in determining the number of neodymium ions active in producing the observed amplification. That is only those ions residing in the $^4F_{3/2}$ energy level are effective in amplifying 1.06 μm radiation, but only a small fraction of the total ions present in the glass, which have been excited by the pumping radiation, will contribute to amplifier gain. An accurate determination of this fraction could be made from detailed knowledge of the spectral and temporal output of the flash lamp, the absorption capability of the neodymium for this output, the lifetime of the excited neodymium ions, and the efficiency of the optical coupling between the flash lamp and the absorbing neodymium ions.

To avoid the complexity of the measurements listed above, an indirect method has been developed[30], based on the established theoretical relationship between the absorption coefficient and the amplification coefficient that produces radiation transitions between two energy states of a material. This method involves only the measurement of the absorption coefficient of neodymium-doped glass at 1.06 μm as a function of temperature and a simple computation based on this measurement.

The essential device for measuring absorption as a function of temperature in the methods described above is a spectrophotometer equipped with a furnace in the sample compartment.

Apparatus for a single-beam system could be arranged as illustrated schematically in Figure 5.7. The various parts of this system may be described as follows.

1. Source: a small, low-voltage lamp, such as the G.E. 1630, driven by a regulated transformer, a collimator lens with a 1 to 4 in. focal length, followed by a Corning 7-56 filter, or an equivalent, to remove the energy from short wavelengths that would cause the neodymium glass to fluoresce; and a chopper to modulate the light at a frequency to which the detector will respond.
2. Oven: a tube furnace, equipped with windows, with a minimum temperature gradient along its length that is equipped with apertures to confine the measurement beam to the interior of the glass rod under test; and a thermocouple located in physical contact with the rod to measure its temperature.
3. Monochromator: an apparatus equipped with a collector lens with a 1 to 4 in. focal length to focus the source lamp on the input slit, the exit slit being arranged to focus the spectral line being studied on a lead sulfide detector for relatively uniform response in 1 μm region that is motor driven to scan the wavelength region from 0.9 to 1.2 μm smoothly.
4. Output system: an amplifier, preferably tuned to the chopper frequency, a suitable linear rectifier, and a strip-chart recorder.

The resultant plot of transmittance as a function of wavelength from either a double-beam spectrophotometer or the single-beam system described will in general not be a straight line but will be influenced by the wavelength variation of the various elements of the system. Fortunately these effects are slow functions of the wavelength compared to the absorption band of the neodymium at 1.06 μm, for the effects may be represented as a slope or gradient in the 100 percent line. This line can be identified as a dip in the transmittance curve that occurs as the 1.06 μm wavelength region is scanned that yields a reference value with which the 1.06 μm and transmission can be compared. The comparison, in turn automatically compensates for reflection losses, scatter, and so on which might influence the absolute magnitude of the reading. (The zero of the instrument should be well adjusted to make this technique accurate.) An example of this technique is shown in Figure 5.8.

After a number of points have been recorded at temperatures in the 300° to 600° Kelvin range, following the technique described above, the absorption coefficient ζ for each point may be calculated from the formula

$$\zeta = \frac{1}{d} \ln \frac{100}{T_0}, \tag{5.73}$$

where T_0 is the amount of the transmittance calculated in Figure 5.8 and d is the length of the glass sample in centimeters. These values of ζ as a

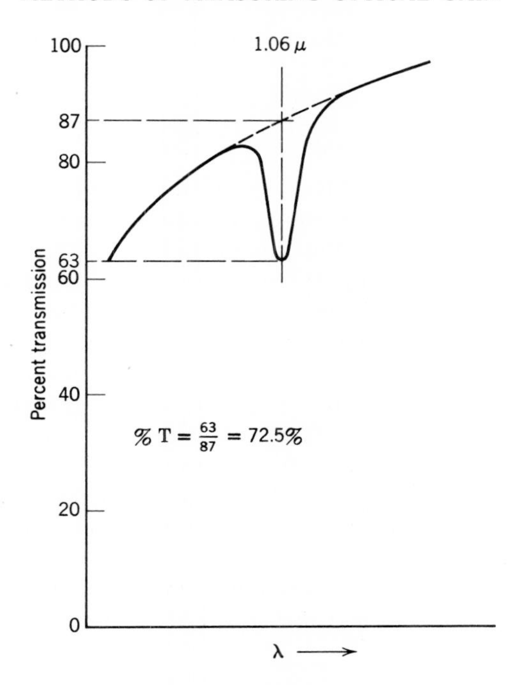

Figure 5.8. Interpretation of typical recorder trace to determine 1.06 μ transmittance.

function of absolute temperature are the essential data for computing the amplification coefficient in neodymium doped glass at 1.06 μm.

To understand the principle of the computation involved in deriving the amplification coefficient from the absorption data, consider the three energy levels of neodymium given in Table 5.6. Only those neodymium ions in the $^4I_{11/2}$ state can contribute to the absorption at 1.06 μm, and the number

TABLE 5.6 PROPERTIES OF THE ENERGY LEVELS OF NEODYMIUM IONS INVOLVED IN AMPLIFICATION-COEFFICIENT MEASUREMENT

Number	Energy	Level		Degeneracy
2	1.42eV	$^4F_{3/2}$		$g_2 = 4$
		↑	1.06μ absorption	
1	0.25eV	$^4I_{11/2}$		$g_1 = 12$
0	0eV	$^4I_{9/2}$		$g_0 = 10$

of ions in the $^4I_{11/2}$ state is proportional to temperature in the following way[28]:

$$N_i = \left(\frac{g_j}{g_i}\right) N_0 \exp\left(\frac{-\Delta U}{kT}\right), \tag{5.74}$$

where the g's are the degeneracies of the states involved, ΔU is the energy difference between the states, k is the Boltzmann constant, $(0.86 \times 10^{-4} \text{eV}/°\text{K})$, and T is the absolute temperature.

Because N_i is directly proportional to ζ, a plot of $\ln \zeta$ as a function of $1/T$ will yield a straight line the slope of which will be $-\Delta U/k$. From this slope, ΔU may be determined. (A representative plot of this type is shown in Figure 5.9).

Next, N_i is computed at some temperature such as 400° K, for which ζ is known, from (5.73). N_0 is taken as the number of ions of neodymium present in the glass per cubic centimeter and is easily determined from the weight percent of neodymium oxide present in the glass and the physical density of the glass. N_0 will not vary significantly in the temperature range 300° to 600° K (< 1 percent).

The absorption cross section per ion is obtained from the value for ζ by N_i. The relation between this value of the absorption cross section per ion

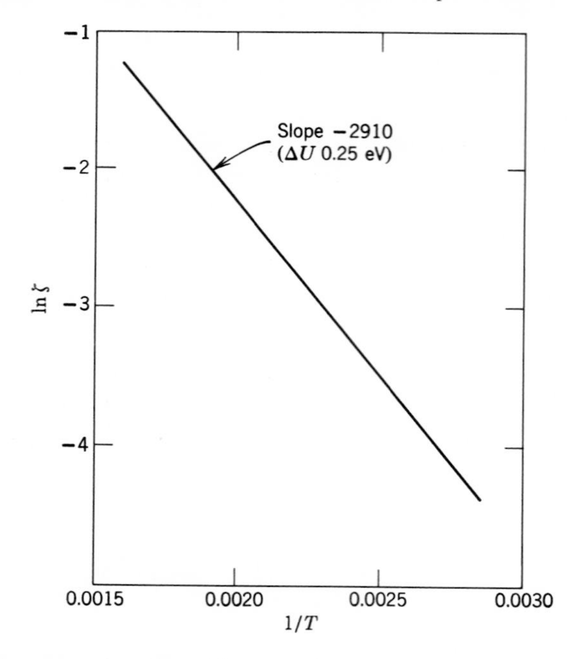

Figure 5.9. Representative plot of $\ln\zeta$ as a function of the reciprocal temperature used in determining ΔU.

at 1.06 μm and the corresponding amplification cross section simply involves the degeneracies of the two levels as follows[31]:

$$\sigma_{21} = \left(\frac{g_1}{g_2}\right)\sigma_{12}, \tag{5.75}$$

where σ_{12} is the cross section for absorption and σ_{21} is the cross section for amplification.

Once σ_{21} has been determined, it is of interest to relate it to the stored energy available in the 1.06 μm transition, 1.07 eV or 1.875×10^{-19} J per excited ion.

Because γ, the amplification coefficient, is equal to the product of the amplification cross section and the number of excited ions, and each excited ion has 1.875×10^{-19} J of stored energy,

$$\gamma = 5.34 \times 10^{18}\sigma_{21}, \tag{5.76}$$

where γ is now the amplification coefficient per centimeter per available joule of stored energy.

It will be necessary to provide a glass sample with a minimum of about 7×10^{21} neodymium ions/cm^2 for the technique described above to have sufficient sensitivity. This corresponds to a rod 7.6 cm long with 6 percent (by weight) doping. With lower doping percentages correspondingly longer rods will be required.

The wavelength resolution required from the monochromator is 0.005 μm or better to read the peak of the 1.06 μm absorption band with photometric accuracy. This will require slit widths of 50 μm or less for a typical monochromator with a rock salt, lithium fluoride, or crystal quartz prism.

5.1.6 Mode Matching

In many applications and measurements it is necessary to transform a Gaussian beam into a Gaussian beam; for example it may be necessary to transform the emergent beam from an oscillator to provide optimum injection into a resonant amplifier or a passive Fabry-Perot cavity designed to study the axial mode structure of the beam[32]. In these cases the oscillator beam may be matched, by inserting lenses, to the natural mode of the terminal system. (The analysis of the lens-matching system is straightforward but too complex to warrant detailed treatment here; the interested reader is referred to the literature for details[33]). The results of the detailed analysis lead to simple matching formulae.

The laser beam in a laser cavity is characterized by its minimum spot size w (beam radius) and the location of the beam waist. To transform the beam into a new waist with a minimum radius w_2 requires consideration of a matching length f, defined in terms of $w_1 w_2$ as

$$f_0 = \frac{\pi}{\lambda} w_1 w_2 \tag{5.77}$$

where λ is the wavelength. The original beam of radius w_1 is transformed into w_2 if a lens of focal length f is used, where $f > f_0$. If the lens is located at an axial distance d_1 from the point where w_1 is measured, at a distance d_2 from the lens the waist radius will be w_2. The distances d_1 and d_2 between the lens and the beam minima satisfy the matching relations[34]

$$\frac{d_1}{f} = 1 \pm \frac{w_1}{w_2}\left[1 - \left(\frac{f_0}{f}\right)^2\right]^{1/2}, \tag{5.78}$$

$$\frac{d_2}{f} = 1 \pm \frac{w_2}{w_1}\left[1 - \left(\frac{f_0}{f}\right)^2\right]^{1/2}, \tag{5.79}$$

where the same sign must be chosen for both equations for the results to have physical significance. Matching is not possible if $f < f_0$. If $f = f_0$ $d_1 = d_2 = f_0 = f$, and the beams are equal and located in the focal planes of the lens.

When more than one lens is used in a system of beam transformation the formulae still apply as long as f is the net focal length of the lens combination and d_1 and d_2 are the principal planes. If w_1 and w_2 are known system parameters f_0, w_1/w_2, d_1 and d_2, all functions of λ through equation (5.77) can be expressed in terms of system parameters alone. Beam transformations between sets of curved mirrors may be calculated by some way considering the mirror in terms of its equivalent lens characteristic.

As an example of mode matching, consider the injection of a beam into an optical delay line[35]. A gas laser operating at 633 nm with a mirror spacing of 1.7 m is used. The cavity geometry is approximately hemispherical and consists of a mirror with a radius of 1 m and a flat output mirror. The minimum beam radius is computed to be $w_1 = 0.37$ mm[36]. The laser output beam is to be injected into a pair of lenses with a 12.5 m focal lens spaced 50 cm apart. The minimum beam radius of the optical delay line is computed to be $w_2 = 0.7$ mm. Using (5.77) we obtain a matching focal length $f_0 = 1.3$ m. Using (5.78) and (5.79), a lens of arbitrary focal length $f \geq f_0$, can be used, and d_1 and d_2 can be computed. If $f = f_0$, $d_1 = d_2 = f = 1.3$ m.

It can be shown that a single, formally identical equation governs the three properties of Gaussian beam propagation:

1. The phase-front curvature and the beam radius in terms of the distance from the beam waist and the minimum beam radius.
2. The propagation of a Gaussian beam in free space.
3. The transformation of a Gaussian beam through a lens.

These characteristics may be presented graphically on an impedance chart, such as a Smith chart, on which the complex mismatch coefficient representing the power coupling coefficient between two Gaussian modes can be determined[37]. The interested reader is referred to several papers that treat the graphical solutions of Gaussian mode problems[38, 39, 40, 41].

5.1.7 Gain Saturation and Pulse Distortion in Laser-pulse Amplifiers

The amplification of a laser pulse is subject to time variation in its transit through a laser amplifier whenever the amplitude of the pulse is sufficient to cause the photon density in the signal wave front to approach the inverted photon density in the active medium[42, 43]. A concise analysis follows.

Let $P_{\mathrm{IN}}(t)$ be the instantaneous power in the input pulse at the entrance to an amplifier, $P_{\mathrm{OUT}}(t)$ be the amplifier output pulse power, G_0 be the small-signal, unsaturated, singlepass gain of the amplifier just before the pulse arrives, and τ_a be the optical transit time through the amplifier. Also let W_0 be the total energy stored in the inverted states of the amplifier before the pulse arrives; that is, $W_0 = h\nu\, \Delta N/2$, where ΔN is the population inversion. Assume the pulses are short compared with the pumping period and neglect any contribution to inversion during the passage of the pulse. The integrated energies of the input and output pulses up to a time τ may be represented as

$$W_{\mathrm{IN}}(t) = \int_{-\infty}^{t} P_{\mathrm{IN}}(t)dt,$$

$$W_{\mathrm{OUT}}(t) = \int_{-\infty}^{t} P_{\mathrm{OUT}}(t)dt.$$

Then, from analysis, the output power may be given by

$$P_{\mathrm{OUT}}(t+\tau) = P_{\mathrm{IN}}(t)\left[\frac{G_0}{G_0 - (G_0 - 1)G_0 - W_{\mathrm{IN}}(t)/W_0}\right]. \tag{5.80}$$

The gain experienced by an increment of input energy $P_{\mathrm{IN}}(t)\, dt$ arriving at a time t depends only on the initial gain G_0 and on the integrated input-pulse energy $W_{\mathrm{IN}}(t)$ up to that time compared with the available energy W_0.

Thus, prior to gain saturation, the gain is independent of the input pulse shape.

Equation 5.80 may be manipulated into the form

$$P_{\mathrm{IN}}(t-\tau) = P_{\mathrm{OUT}}(t)\left[\frac{1}{1+(G_0-1)G_0 - W_{\mathrm{OUT}}(t)/W_0}\right], \tag{5.81}$$

which permits direct calculation of the input-pulse shape required to obtain a specified output-pulse shape. If W_{IN} and W_{OUT}, respectively, represent the integrated values of energy in the pulses at the input and output, (5.80) and (5.81) can be combined to yield

$$G_0^{(W_{\mathrm{IN}}/W_0+1)} = G_0^{(W_{\mathrm{OUT}}/W_{\mathrm{IN}})} + G_0 - 1. \tag{5.82}$$

Equation (5.82) gives a unique relation between the input- and output-pulse energies, the amplifier stored energy, and the initial amplifier gain, independent of the details of the amplifier configuration or the pulse shape used.

One important parameter in the design of a high-energy pulse amplifier is the total energy gain from the amplifier during a pulse $G_p = W_{\mathrm{OUT}}/W_{\mathrm{IN}}$. Another important parameter is the energy efficiency of the amplifier, that is the ability of the amplifier to transfer a large fraction of its stored energy to the pulse during transit. The efficiency of the pulse-energy conversion is given by the ratio

$$\eta = \frac{W_{\mathrm{OUT}} - W_{\mathrm{IN}}}{W_0}.$$

There is a unique relation between η and G_p for each value of G_0. From (5.82) it is apparent that we can only obtain a high-energy conversion efficiency and a large pulse-energy gain by operating with a large initial gain G_0. Then a pulse energy gain of $G_p = 5$ corresponds to $G_0 = 10$ and $\eta = 0.5$. If $G_0 = 100$, the same value of G_p corresponds to $\eta = 0.9$. The disadvantage of operating at a high initial gain is that the amplifier is more difficult to stabilize and the losses due to the amplified spontaneous emission (not included in the analysis above) become much greater. Typical amplifiers are designed for an initial gain of 15 to 17 dB to preclude instabilities caused by reflections from interstage isolation elements among other things[44].

5.1.8 Gain Saturation in CW-gas Lasers

Gain saturation, which in an inhomogeneously broadened laser transition results in hole burning when the saturating power is confined to a frequency range that is small compared to the inhomogeneous linewidth[45], has been subject to extensive experimental investigation in the 3.392 μm line of a helium-neon laser[46]. It has been found that the $3s_2 \rightarrow 3p_4$ neon transition at 3.392 μm[47] and that this transition originates on the same upper level as the 633 nm transition. The observed 50 dB meter gain is of the order of 65 times greater than the corresponding gain in the 633 nm transition. The higher gain results from a Doppler width 5.4 times as small, an inversion population about 2.5 to 3 times as large, and an oscillator strength about 4 to 5 times as large as that associated with the 633 nm transition.

The treatment of the gain saturation, presented below, follows closely the work of Gordon, White, and Rigden[46] and will be restricted to consideration of a single Doppler-broadened line centered at some frequency ν_c and an incident wave at frequency ν. The case for which $\Delta\nu_N/\Delta\nu_D >> 0$ is of interest here. The rate equations for a four-level laser, for those excited atoms whose Doppler-shifted center frequency is ν' can be written

$$\dot{n}_3(\nu', z) = S_3(\nu') - n_3(\nu', z)[A_3 + B'_{32}(\nu', \nu)I(\nu, z)/4\pi] + n_2(\nu', z)B'_{23}(\nu', \nu)I(\nu, z)/4\pi, \tag{5.83}$$

$$\dot{n}_2(\nu', z) = S_2(\nu') + n_3(\nu', z)[A_{32} + B'_{32}(\nu', \nu)I(\nu, z)/4\pi] - n_2(\nu', z)[A_2 + B'_{23}(\nu', \nu)I(\nu, z)/4\pi]$$

where n_3 (ν', z) and n_2 (ν', z) are the volume density of atoms in the upper and lower laser levels in the unit frequency range at ν' on plane z along the axis of the discharge. The intensity at plane z is given by I (ν, z), and the pumping rates to levels 3 and 2 are denoted S_3 (ν') and S_2 (ν'). The effective total spontaneous decay rates corresponding to the Einstein A coefficients from levels 3 and 2, including the radiation trapped decay to the ground state, are denoted by A_3 and A_2. When radiation to the ground state is trapped, the effective total A coefficient is A_0 $(1 + g)$, in which A_0 is the A coefficient for radiation to the ground state, A is the total coefficient, and $g << 1$ is a trapping parameter. The rate for spontaneous emission from level $3 \rightarrow 2$ is denoted by A_{32}.

The rate of stimulated emission for atoms with a Doppler frequency of ν', when the stimulating frequency is ν, is given by

$$B'_{32}(\nu', \nu) = \frac{g_2}{g_3} B'_{23}(\nu', \nu) = B_{32} \frac{(2/\pi\Delta\nu_N)}{1 + [2(\nu - \nu')/\Delta\nu_N]^2} \tag{5.84}$$

in which the factors g_2 and g_3 represent the degeneracy or statistical weights. The stimulated emission probability has a Lorentzian line shape corresponding to a natural linewidth $\Delta\nu_N$ and is normalized so that

$$\int_0^\infty B'_{ij}(\nu', \nu)d\nu = B_{ij}, \tag{5.85}$$

in which B_{ij} is the appropriate Einstein B coefficient[9].

The growth rate for the intensity in the maser medium can be written

$$\frac{dI'(\nu, z)}{dz} = \frac{h\nu I'(\nu, z)}{4\pi} \int_0^\infty [B'_{32}(\nu', \nu)n_3(\nu', z) - B'_{23} \cdot (\nu'\nu)n_2(\nu', z)]d\nu' \tag{5.86}$$

in which $h\nu$ is the photon energy. An absorption term independent of the maser medium, of the form

$$\text{constant } I'(\nu,z),$$

has been omitted because diffraction loss, the only significant possibility, can be made negligible by design.

The integrand in (5.86) can be determined by simultaneous solution of (5.83) for the steady-state case, $\dot{n}_3 = \dot{n}_2 = 0$. Performing the necessary algebra yields

$$B'_{32}n_3 - B'_{23}n_2 = \frac{B'_{32}[S_3/A_3 - (g_3/g_2)(S_3A_{32} + S_2A_3)/A_2A_3]}{1 + [(g_2/g_3)(A_3 - A_{32})/A_2A_3 + 1/A_3]B'_{32}I/4\pi}. \tag{5.87}$$

The pumping terms S_3 and S_2 have a Doppler profile of the form $S_i = S_{i0} \exp - [2(\nu' - \nu_0)/\Delta\nu_D]^2 \ln 2$, corresponding to the thermal velocity distribution of the atoms. Substituting (5.84) and (5.87) into (5.86) yields

$$\frac{1}{I'(\nu, z)}\frac{dI'(\nu, z)}{dz} = k_0 \int_0^\infty d\nu' \left(\frac{2}{\pi\Delta\nu_N}\right)$$
$$\left(\frac{\{1 + [2(\nu - \nu')/\Delta\nu_n]^2\}^{-1} \exp\{-[2(\nu' - \nu_c)/\Delta\nu_d]^2\}\ln 2}{1 + \eta I(\nu, z)(2/\pi\Delta\nu_N)/\{1 + [2(\nu - \nu)\Delta\nu_N]^2\}^{-1}}\right), \tag{5.88}$$

in which the parameters k_0 and η are given by

$$k_0 = h\nu \frac{B_{32}}{4\pi}\left[\frac{S_{30}}{A_3} - \frac{g_3}{g_2}\frac{(S_{30}A_{32} + S_{20}A_3)}{A_2A_3}\right] \tag{5.89}$$

and

$$\eta = \left[\frac{g_3}{g_2}\frac{(A_3 - A_{32})}{A_2 A_3} + \frac{1}{A_3}\right]\frac{B_{32}}{4\pi}. \tag{5.90}$$

In the derivation given below the incident signal frequency is at the line center so that $\nu = \nu_c$. Making this simplification, (5.88) can be written as the incremental gain

$$\gamma = \frac{1}{I}\frac{dI}{dz} = \frac{k_0}{\pi}\int_{-\infty}^{+\infty}\frac{dx \exp -(\epsilon x)^2}{(1 + x^2 + I)} \tag{5.91}$$

in which the new integration variable is defined as $x = 2(\nu' - \nu_c)\Delta\nu_N$ and I is a normalized intensity defined by

$$I = \eta I(\nu_c, z)\left(\frac{2}{\pi\Delta\nu_N}\right). \tag{5.92}$$

The parameter

$$\epsilon = \left(\frac{\Delta\nu_N}{\Delta\nu_N}\right)\sqrt{\ln 2} \tag{5.93}$$

measures the degree of inhomogeneous broadening.

The integral in (5.91) can be written (making the substitution $y = x^2$)

$$2\int_0^\infty \frac{\exp[-(\epsilon x)^2]}{1 + I + x^2} = \int_0^\infty \frac{\exp[-(\epsilon y)^2]}{(1 + I + y)y^{1/2}}$$
$$= \pi(1 + I)^{-1/2}\exp(1 + I)\epsilon^2\{1 - \mathrm{Erf}[(1 + I)^{1/2}\epsilon]\}. \tag{5.94}$$

The integral on the right is tabulated[48].

The incremental gain, (5.91), can now be written

$$\frac{1}{I}\frac{dI}{dz} = \frac{k_0}{(1 + I)^{1/2}}e^{(1+I)\epsilon^2}\{1 - \mathrm{Erf}[(1 + I)^{1/2}\epsilon]\}. \tag{5.95}$$

In the limit $\epsilon = (\Delta\nu_N/\Delta\nu_D)\sqrt{\ln 2} \to 0$ $\mathrm{Erf}\,[(1 + I)^{1/2}\epsilon] \to 0$,

and equation 5.95 becomes

$$\gamma = \frac{1}{I}\frac{dI}{dz} = \frac{k_0}{(1 + I)^{1/2}}, \tag{5.96}$$

which can be recognized as the appropriate limit for inhomogeneous broadening.

In the opposite limit, $\epsilon \to \infty$

$$\mathrm{Erf}[(1 + I)^{1/2}\epsilon] \to 1 - [\exp -(1 + I)\epsilon^2]/(1 + I)^{1/2}\epsilon\pi^{1/2}.$$

Equation 5.95 reduces to

$$\gamma = \frac{1}{I}\frac{dI}{dx} = \frac{(k_0/\epsilon\pi^{1/2})}{1 + I} = \frac{k_0\,\Delta v_D/\Delta v_N\sqrt{\pi \ln 2}}{(1 + I)}, \tag{5.97}$$

which is the appropriate limit for pure homogeneous broadening[49]. Reference to (5.89) indicates that $k_0\Delta v_D$ is independent of Δv_D, for the source terms S_{30} and S_{20} relate to a rate per-unit-frequency internal and consequently are proportional to Δv_D^{-1} for a constant total-inversion population.

For the 3.392 μm transition, $\Delta v_D \approx$ 350 MHz and Δv_N is estimated from the spontaneous-emission coefficient to be of the order of 20 MHz. As a result $\epsilon \approx 0.05$, and it is possibly appropriate to consider (5.95) in the limit where terms of order ϵ^2, but not ϵ, are neglected. Writing Erf $[(1 + I)^{1/2}\epsilon] \approx 2\,(1 + I)^{1/2}\,\epsilon/\pi^{1/2}$, (5.95) reduces to

$$\frac{1}{I}\frac{dI}{dz} = \frac{k_0}{(1 + I)^{1/2}}\,[1 - 2(1 + I)^{1/2}\epsilon/\pi^{1/2}]. \tag{5.98}$$

Assuming an incident intensity I_0 at $z = 0$, the intensity at $z = d$ is defined by the equation

$$\int_{I_0}^{1} I^{-1}(1 + I)^{1/2}[1 + 2(1 + I)^{1/2}\epsilon/\pi^{1/2}]\,dI = k_0\,d. \tag{5.99}$$

In the approximation $\epsilon^2 << 1$. Equation (5.99) can also be written

$$\int_{1+I_0}^{1+GI_0} (t - 1)^{-1}(t^{1/2} + 2t\epsilon/\pi^{1/2})\,dt = k_0\,d \tag{5.100}$$

in which the substitution $t = 1 + I$ has been made, and the gain has been defined as $G = I/I_0$.

Equation 5.100 can be integrated to yield

$$2(1 + GI_0)^{1/2} - 2(1 + I_0)^{1/2} + \ln\left[\frac{(1 + I_0)^{1/2} + 1}{(1 + I_0)^{1/2} - 1}\right]\left[\frac{(1 + GI_0)^{1/2} - 1}{(1 + GI_0)^{1/2} + 1}\right] + \frac{2\epsilon}{\pi^{1/2}}\,[(G - 1)I_0 + \ln G] = k_0\,d. \tag{5.101}$$

In the limit $I_0 = 0$, (5.100) becomes

$$(1 + 2\epsilon/\pi^{1/2}) \ln G_0 = k_0 d \tag{5.102}$$

in which $G_0 = G\ (I_0 = 0)$ is the small signal gain. The small signal-gain parameter is defined by $k_0/\ (1 + 2\epsilon/\pi^{1/2})$ and is somewhat reduced from its value k_0 at the line center. This implies that even in the absence of any saturation effects the signal searches the population away from the line center. Because the inversion is lower in the wings, the actual gain parameter is smaller than its value at line center. Only in the limit of a very broad inhomogeneous line, $\epsilon \to 0$, does the gain parameter have its value at the line center. Writing

$$\ln\left[\frac{(1 + I_0)^{1/2} + 1}{(1 + I_0)^{1/2} - 1}\right]\left[\frac{(1 + GI_0)^{1/2} - 1}{(1 + GI_0)^{1/2} + 1}\right] = \ln G + \ln\left[\frac{(1 + I_0)^{1/2} + 1}{(1 + GI_0)^{1/2} + 1}\right]^2 \tag{5.103}$$

and substituting, we can write

$$2(1 + GI_0)^{1/2} - 2(1 + I_0)^{1/2} + (2\epsilon/\pi^{1/2})(G - 1)I_0$$

or

$$= \ln(G_0/G)(1 + 2\epsilon/\pi^{1/2})\left[\frac{(1 + GI_0)^{1/2} + 1}{(1 + I_0)^{1/2} + 1}\right]^2, \tag{5.104}$$

$$G_0 = G\left\{\left[\frac{(1 + I_0)^{1/2} + 1}{(1 + GI_0)^{1/2} + 1}\right]\exp[(1 + GI_0)^{1/2} - (1 + I_0)^{1/2}]\right\}^{2/(1 + 2\epsilon/\pi^{1/2})}$$

$$\cdot\ [\exp(2\epsilon/\pi^{1/2})(G - I)I_0/(1 + 2\epsilon/\pi^{1/2})] \tag{5.105}$$

In the form as it is represented by (5.105) it is relatively simple to determine the functional dependence of $G(I_0)$ for a given G_0 and ϵ by first plotting a family of curves of G_0 versus G for a given I_0. From this set we can determine the curve $G(I_0)$ for any desired G_0. (It should be noted that the slope of G versus I_0 is not zero at $I_0 = 0$.) However a plot of G versus I_0 in decibels has a zero slope for I_0 (dB) $= -\infty$.

5.1.8.1. Gain Narrowing in a CW-gas-laser Amplifier The theory obtained above treats the 3.39μm laser as a four-level device in terms of level populations. By carrying the frequency dependence through the calculations and treating the line as nearly inhomogeneous[50], expression can be derived for the incremental gain γ:

$$\gamma = \gamma_0\left\{\frac{\exp(-\omega^2)}{\sqrt{1 + I/I_0}} - \frac{1 - 2\omega F(\omega)}{R}\right\} = \frac{1}{I}\frac{dI}{dz} \tag{5.106}$$

where $1/R$ is small and is a measure of the departure from an inhomogeneous line

$$\left(R = \frac{\sqrt{\pi}}{2} \times \frac{\Delta\nu_D}{\Delta\nu_N\sqrt{\ln 2}}\right). \tag{5.107}$$

If we integrate over a length d, we can drive an unsaturated gain at the line center G_0 (O) as follows: $\ln G_0$ (O) $= (1 - 1/R)\,\gamma_0 d$; the gain $G(\omega)$ varies across the line depending on ω, where

$$\omega = 2(\ln 2)^{1/2}\,\frac{\nu - \nu_c}{\Delta\nu_D}. \tag{5.108}$$

At $\omega = \mu$, $G(\mu) = \frac{1}{2}G(0)$. If we solve the resulting transcendental equation graphically for μ, assuming a symmetrical line of half-gain width Δv, we find the result of Δv $(1/R \neq 0)$ of 125 MHz.

If the line is treated as being inhomogeneous, an incremental gain is produced of

$$\gamma = \gamma_0\,\frac{\exp(-\omega^2)}{\sqrt{1 + I/I_0}} = \frac{1}{I}\frac{dI}{dz}.$$

Carrying through the steps as above yields a half-gain width of Δv $(1/R = 0)$ $= 119$ MHz. The data indicate that Δv_N is negligible compared with Δv_D within the experimental error[50]. Therefore measurement of an amplifier's gain line width does not necessarily yield the Doppler width of the line unless the amplifier is unsaturated and the net gain is sufficiently low. It should also be clear that the gain halfwidth measured in the saturated regime will equal the Doppler width only accidentally! (See Figure 5.10).

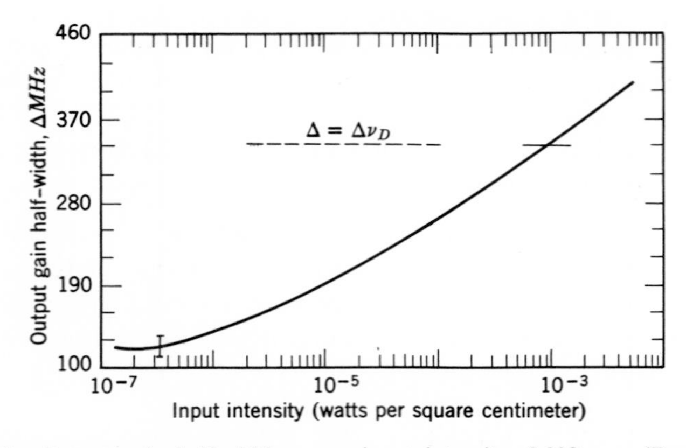

Figure 5.10. Output gain half-width versus input intensity, 3.392 μ m. (Reproduced with permission from *Applied Optics*.)

Following the treatment given above[46], we can derive a fractional gain per unit length γ:

$$\gamma = \frac{1}{I(v, z)} \frac{dI}{dz} = \gamma_0 \int_{-\infty}^{\infty} \frac{g(v, v')\exp(-\Omega^2)}{1 + \eta I g(v, v')} d\Omega \tag{5.109}$$

where

$$g(v, v') = \frac{2/\pi \Delta v_N}{1 + [2(v - v')/\Delta v_N]^2}, \tag{5.110}$$

I is intensity, Δv_N is the radiative line width in Hz for the transition, η is constant for a particular amplifier and transition, and γ_0 equals v times another constant of the amplifier. Taking

$$\Omega = \frac{2\sqrt{\ln 2}}{\Delta v_D}(v' - v_c),$$

the integral for γ can be transformed to the following:

$$\frac{\gamma}{\gamma_0} = \frac{a}{\pi} \int_{-\infty}^{\infty} \frac{\exp(-\Omega^2)\, d\Omega}{\alpha^2 + (\Omega - \omega)^2} \tag{5.111}$$

where

$$\omega = \frac{2\sqrt{\ln 2}}{\Delta v_D}(v - v_c),$$

$$a = \frac{\Delta v_N}{\Delta v_D}\sqrt{\ln 2} \tag{5.112}$$

$$\alpha = a\left[1 + \frac{2\eta}{\pi \Delta v_N} I(v, z)\right]^{1/2},$$

and Δv_D is the Doppler-broadened line width in Hz, and v_c refers to the line-center frequency. Thus

$$\frac{\pi\alpha}{a}\gamma_0 = K(\omega) = \int_{-\infty}^{\infty} \frac{\alpha \exp(-\Omega^2)}{\alpha^2 + (\omega - \Omega)^2} d\Omega.$$

This integration has been performed[9] and leads to

$$K(\omega) = \sqrt{\pi} \int_{-\infty}^{\infty} \exp(-\alpha x) \exp\left(-\frac{x^2}{4}\right) \cos \omega x \, dx. \tag{5.113}$$

Using the definite integral

$$\frac{\sqrt{\pi}}{2} \exp(-\omega^2) = \int_{-0}^{\infty} \exp(-t^2) \cos 2\omega t \, dt$$

and introducing the tabulated function $F(w)$

$$F(\omega) = \exp(-\omega^2) \int_0^{\omega} \exp x^2 \, dx = \int_0^{\infty} \exp(-t^2) \sin 2\omega t \, dt, \tag{5.114}$$

we find $K(\omega)$ for small α values by approximating $\exp(-\alpha x)$ as $1 - \alpha x$.

$$\frac{\gamma}{\gamma_0} = \left[\frac{\exp(-\omega^2)}{\sqrt{1 + 2\eta/\pi\Delta\nu_N}}\right] \frac{2a}{\sqrt{\pi}} [1 - 2\omega F(\omega)]. \tag{5.115}$$

Now let

$$R = \frac{\sqrt{\pi}}{2a} \qquad \phi = 1\sqrt{+\frac{2\eta}{\pi\Delta\nu_N}} I \tag{5.116}$$

and

$$h(\omega) = R \frac{\exp -\omega^2}{1 - 2\omega F}.$$

Then γ simplifies to

$$\gamma = \frac{2\phi \, d\phi}{(\phi^2 - 1) \, dz} = \gamma_0 \frac{\exp(-\omega^2)}{h(\omega)} [h(\omega) - \phi]. \tag{5.117}$$

If we integrate over length d,

$$\frac{\gamma_0 d}{h} \exp(-\omega^2) = J(\omega) = \int_{\phi 1}^{\phi 2} \frac{2\phi^2 \, d\phi}{(\phi^2 - 1)(h - \phi)}. \tag{5.118}$$

$J(\omega)$ can be integrated by parts to yield

$$J(\omega) = \frac{1}{h^2 - 1} \ln\left(\frac{\phi_2^2 - 1}{\phi_1^2 - 1}\right) + \frac{h}{h^2 - 1}$$

$$\times \left[\ln\left(\frac{\phi_2 - 1}{\phi_2 + 1} \times \frac{\phi_1 + 1}{\phi_1 - 1}\right) - 2h \ln\left(\frac{h - \phi_2}{h - \phi_1}\right)\right]. \quad (5.119)$$

Hence

$$\gamma_0 d \exp(-\omega^2) = \frac{h}{h^2 - 1} \ln\left(\frac{\phi_2^2 - 1}{\phi_1^2 - 1}\right) + \frac{h^2}{h^2 - 1}$$

$$\times \left[\ln\left(\frac{\phi_2 - 1}{\phi_2 + 1} \times \frac{\phi_1 + 1}{\phi_1 - 1}\right) - 2h \ln\left(\frac{h - \phi_2}{h - \phi_1}\right)\right]. \quad (5.120)$$

Let

$$\beta = \frac{I(v, z)}{I_0} = \frac{2\eta}{\pi \Delta v_N} I(v, z)$$

and note that the single-pass gain $G(\omega) = I_2/I_1 = \beta_2/\beta_1$. Then, in the small signal-gain region ($I_2 - I_1 << 1$), we find for the unsaturated gain $G_0(\omega)$

$$\ln G_0(\omega) = \frac{h(\omega) - 1}{h(\omega)} \gamma_0 d \exp(-\omega^2), \quad (5.121)$$

which becomes at line center ($\omega = 0$)

$$\ln G_0(0) = \left(1 - \frac{1}{R}\right)\gamma d. \quad (5.122)$$

Results are given above for $\omega = 0$, where $(1 + 1/R)$ in $G_0(0) = \gamma_0 d$, which, when expanded for $1/R << 1$, yields the result above. At some frequency $\omega - \mu$, $G(\omega)$ drops to $\frac{1}{2} G(0)$, and

$$\ln \tfrac{1}{2} G_0(0) = \left(1 - \frac{1}{h}\right)\gamma_0 d \exp(-\mu^2)$$

$$= \left[\exp(-\mu^2) \frac{1 - 2\mu F(\mu)}{R}\right]\gamma_0 d = \ln G_0(\mu). \quad (5.123)$$

Because

$$\gamma_0 d = \frac{\ln G_0(0)}{1 - 1/R},$$

$$\exp(-\mu^2) - \frac{1 - 2\mu F(\mu)}{R} = \left(1 - \frac{1}{R}\right) \frac{\ln \frac{1}{2} G_0(0)}{\ln G_0(0)} = f(\mu). \tag{5.124}$$

The transcendental equation above must be solved to find μ. Taking $\Delta v_N \sim 25$ MHz, $\Delta v_D = 340$ MHz.

$$R = \frac{\pi^{1/2}}{2} \frac{\Delta v_D}{\Delta v_N (\ln 2)^{1/2}} = 14.45. \tag{5.125}$$

Assuming a symmetrical line of half-gain with Δv, $\omega = \Delta v / \Delta v_D \sqrt{\ln 2}$ or $\Delta v = 1.2 \Delta v_D \omega$. Using these values, the equation above is solved graphically for μ. Results Δv $(1/R \neq 0)$ are as follows:

$G_0(0)$	Δv observed MHz	$\Delta v(1/R \neq 0)$ MHz	$\Delta v(1/R = 0)$ MHz
5000	126 ± 12	125	119

A simpler treatment than that given above and that yields line-width values Δv within the experimental error treats the line as inhomogeneous. This amounts to neglecting the radiative width compared with Doppler width ($\Delta v_N << \Delta v_D$) and, accordingly, $1/R \sim 0$. The preceding analysis then reduces to

$$\gamma = \frac{1}{I(v, z)} \frac{dI}{dz} = \frac{\gamma_0}{\sqrt{1 + I/I_0}} \exp(-\omega^2), \tag{5.126}$$

which, when integrated, yields ($\beta = I/I_0$)

$$\gamma_0 d \exp(-\omega^2) = 2\sqrt{1 + \beta} \ln \frac{\sqrt{1 + \beta} - 1}{\sqrt{1 + \beta} + 1} \Bigg|_{\beta_1}^{\beta_2} \tag{5.127}$$

For a small signal gain[51] we find in the limit $\beta_2 - \beta_1 << 1$ that

$$\gamma_0 d \exp(-\omega^2) = \ln\left(\frac{\beta_2}{\beta_1}\right) = \ln G_0(\omega). \tag{5.128}$$

The half-gain frequency μ satisfies the equation

$$\exp(\mu^2) = \frac{\ln G_0(0)}{\ln \frac{1}{2}G_0(0)} \tag{5.129}$$

from which μ can be easily found.

5.2 MEASUREMENTS OF ELECTRON ENERGY, ENERGY DENSITY, AND TEMPERATURE IN GAS-LASER PLASMAS

5.2.1 Microwave Perturbation Techniques for Measuring Electron Energy and Energy Density in a Laser-plasma Tube

A study has been made of the helium-neon discharge in small-bore ($2\ a \leqq 6$ mm) cylindrical tubing using cavity-perturbation techniques to obtain measurements of electron-density measurements[51]. It is possible to measure the electron density and average electron energy in a gas-laser plasma tube as a function of discharge current, tube diameter, and gas pressure by making measurements on the positive column of the discharge[51a].

A square base TE_{101}-mode, rectangular, microwave cavity with a base dimension of w and resonant at the s band was used. The plasma tube was placed at the center of the cavity with its axis perpendicular to the base to keep the microwave electric field parallel to the discharge boundary.

In the perturbation calculation an integral of the form $\int n(r)E^2\,dV$ must be evaluated over the volume of the discharge V_p, where $n(r)$ is the electron-density distribution in the cross section, and E is the unperturbed electric field of the cavity. Assuming that $n(r)$ is determined by the lowest-order diffusion mode, that is $n(r) = n_0 J_0\ (\chi r/a)$, where a is the radius of the discharge tube and $\chi = 2.405$, we can evaluate the integral neglecting only terms of order $(a/w)^4$, where w is the dimension of the base. As long as the collision frequency for momentum transfer is much less than ω_0 (which is true for all discharge parameters considered) the perturbation calculations yield the following relation for the frequency perturbation of the cavity $\Delta\omega$:

$$\frac{\Delta\omega}{\omega_0} = 4\ \frac{J_1(\chi)}{\chi} \left(\frac{\omega_{po}}{\omega_o}\right)^2 \frac{V_p}{V_c} \left[1 - \pi\left(\frac{V_p}{V_c}\right)\left(1 - \frac{4}{\chi^2}\right)\right] \tag{5.130}$$

where ω_0 is the unperturbed resonant frequency of the cavity, V_c is the volume of the cavity, and $\omega_{p0} = (n_0 e^2/\epsilon_0 M)^{1/2}$ is the plasma frequency corresponding to an electron density n_0.

Frequency-perturbation measurements were obtained and the electron density was calculated from (5.130) for tubes of various diameters filled with 5:1 helium-neon mixtures to pressures such that *pd*, the pressure times the diameter, had the value 3.6 Torr mm. In all cases the electron density was found to be a linear function of the discharge current for the range of currents considered. This is in agreement with the results of the spontaneous intensity measurements.

Because many of the rates and production terms pertinent to the laser action in the helium-neon plasma are functions of the average electron energy, it is important to know how this energy depends upon the various discharge parameters. If the rf noise power emitted by the discharge in a small frequency interval is measured, the dependence may be ascertained because the noise power can be related to the average electron energy[29, 30].

The noise power leaving the cavity due to the discharge was compared to the noise power emitted by a noise lamp using a standard noise comparison scheme with an attenuator placed in front of the cavity and the noise lamp. The cavity design was such that the band width of the i.f. amplifier would be much smaller than the band width of the cavity. Then if we adjust the attenuator (attenuation -α_c) in front of the cavity and the attenuator (attenuation -α_{nl}) in front of the noise lamp so that the noise power from the cavity and the noise power from the noise lamp become identical, we can determine the equivalent noise temperature of the discharge.

It has been shown that the equivalent noise temperature of the discharge T_n is given by

$$T_n = \left[\frac{\alpha_{nl}/\alpha_c}{(1-\Gamma^2)(1-\psi_0'/\psi_0)}\frac{T_{nl}-T_0}{T_0}+1\right]T_0, \qquad (5.131)$$

where T_{nl} is the equivalent noise temperature of the noise lamp, T_0 is the room temperature, and Γ is the cavity reflection coefficient on resonance with the discharge on. The quantities ψ'_0 and ψ_0 are the cavity VSWR's on resonance (or their reciprocal depending upon whether the cavity is overcoupled or undercoupled) when the discharge is on and off respectively.

Placing the dicharge tube in a cavity for the noise measurements has two advantages. Both the noise and density measurements may be performed without physically moving the tube, and, more important, an unambiguous relationship is obtained between the equivalent noise temperature of the discharge and the measured data.

The noise temperature of the discharge or equivalently the average electron energy has been found to be independent of the discharge current over the range 5 to 100 ma and for values of *pd* between 2 and 5

Torr mm. This result is consistent with the electron density being a linear function of the discharge current and shows that in this region the electron density may be varied without varying the average electron energy. The electron temperature for a given tube diameter varied approximately as the reciprocal of the gas pressure. A factor of 3 increase in pressure caused an expected factor of 2 reduction in electron temperature. The reduction in electron temperature with increasing pressure is due to the reduced loss of ions by diffusion to the walls. For values of $pd > 5$ Torr mm the noise temperature decreased at the higher currents. This effect may be ascribed to the onset of ionization by cumulative electron collisions.

The scaling law for helium-neon tubes is based on the assumption that the average electron energy, as a function of *pd*, should be independent of the tube diameter. The results of measurements taken on four tubes with different diameters (each filled with a 5:1 mixture of helium-neon) indicate the validity of the assumption.

5.2.2 Measuring the Effect of Binary Gases on the Electron Temperature and Electron Density in Gas-laser Plasmas

The effect on electron temperature and electron density of adding helium to the plasmas of gas lasers can be studied by using a triple-probe technique[54]. This technique avoids the limitations of the double-probe method by tracing out the complete probe characteristic[56]. It has the disadvantage, however, of producing a larger plasma disturbance because a greater electron current is sampled by the probe.

During the measurements fine wire probe is located near the center of the plasma tube to measure the probe characteristic. A second, large-area probe is located as remotely as possible from the first probe to extract a compensatory current from the plasma. A third small monitor probe located near the first is used to measure the change in the plasma potential caused by the first probe.

Two methods must be used to reduce errors due to rectification of the rf used to excite the electrodeless discharge in the plasma.

1. The measuring probes must be located in the plasma near the rf ground potential.
2. Large rf impedances must be used in series with the measuring leads.

The source of these errors is, of course, the nonlinear probe characteristic. The probe characteristic is measured by applying a 2 ms linear ramp to the electrode and measuring the probe electron current extracted from the plasma. The characteristic is recorded on an oscilloscope and is photographed as the binary gas is added to the plasma. From an analysis of the probe volt-ampere characteristic we can infer the electron temperature and electron density for each new set of conditions[55]. The electron-current density to the probe J_e is related to the electron-current density in the plasma J_p as

$$J_e = J_p\, e\, p\, (-eV_0/kT_e) \tag{5.132}$$

where V_0 is the probe potential and T_e is the absolute temperature of the plasma electrons. From the slope the probe characteristic, at a voltage above the floating potential V_t of the plasma, we can determine e/kT_e and thus the electron temperature of the plasma[57]. Values of $T_e \sim 10^5$ are typical.

5.3 TECHNIQUES FOR MEASUREMENT OF LIFETIME

5.3.1 Measurement of Lifetime in Gases

In recent years the problems associated with the storage and release of energy by metastables, recombining atoms, and recombing ions in active and afterglow gaseous plasmas have received increasing attention. The situation with regard to the long-lived neutral excited particles, which constitute an impurity, is particularly challenging in plasma physics for the following reasons:

1. The impurities may be present in the plasma in large concentration by virtue of their long lifetime and neutrality.
2. The means for detecting impurities is as yet semiquantitative and awkward to apply.
3. The particles constituting the impurity may release their energy selectively to other constituents of the plasma in a resonance process favored by appropriate energy balance or quantum selection rules.

As a partical matter, the operation of certain gaseous lasers depends on metastable-atom collisions for a partial contribution to the population inversion of the radiating state.

Metastable states of an atom or a molecule have an abnormally long lifetime due to the fact that the radiative decay by electric dipole transition is forbidden. A sizable accumulation of metastables may, therefore, be present in laboratory discharge plasmas, and they constitute a reservoir of energy which may be released to atoms, or the walls of the environment. The resonance transfer of energy from the metastable of helium to the impurity of neon in the first gas laser is an example of this.

When metastables are contained in a low pressure gas having convenient dimensions, the lifetime is ordinarily not determined by forbidden modes of radiation. Important modes of deactivation are the following:

1. Diffusion to the walls (low pressure or vacuum).
2. Three-body collision with two neutral gas atoms (high pressure).
3. Excitation or ionization (Penning effect) of an impurity.
4. Radiation induced by two-body collision.
5. Transition to a radiating level by (a) external dc field-induced state mixing, (b) absorption of radiation and/or impact with an energetic electron or atom.
6. Collision with a thermal electron.
7. Formation of a metastable molecule by three-body collision.
8. Collision between pairs of metastables, resulting in the ionization of one.

Experimental conditions may be frequently arranged such that one of (1) to (8) dominates.

In contrast to the conditions listed above, metastables are produced as a result of electronic bombardment directly or indirectly by radiative cascade and often as a product of charge recombination. Because these excitations are very difficult to account for quantitatively, the determination of metastable lifetime is never made during or in the region of excitation.

All experiments used to determine lifetimes of atomic states incorporate a method of excitation, a method for photodetection, and a scheme for recording or displaying data. The three basic techniques employed in direct lifetime measurement are the following.

1. Modulated excitation with phase-shift analysis.
2. High-impact-current method employing direct display of the decay curve on an oscilloscope.
3. The coincidence or sampling technique (which has been automated to give a chart record of the decay curve).

The modulated excitation technique is the oldest and probably still the best adapted to the study of extremely short-lived states ($< 10^{-9}$ sec), although the only recent work using this technique is being done in Germany[58]. Because the analysis of phase-shift data requires a pure exponential decay, cascading cannot be tolerated. Thus modulated photoexcitation is employed for its selective excitation of specific states. In this technique a comparison between the phase of the source modulation and the scattered-light modulation as a function of modulation frequency can yield lifetime data. However it is clear that photoexcitation is limited to resonance levels, which greatly restricts the applicability of this system.

The high-impact current technique for measuring the lifetimes, pioneered by Holzberlein[59], consists of apparatus designed to display a graph of a single excitation-decay sequence on an oscilloscope. This requires far greater light output than it required by any other technique because the statistics must yield a curve rather than individual photon bursts.

The sampling technique is the most sophisticated technique for measuring atomic lifetimes. This was pioneered by Heron, McWhirter, and Rhoderick[60] and was automated and brought to its present state of development by Bennett and Dalby[61]. This technique employs bursts of electrons to excite the gas under study and a photomultiplier light detector. By gating the photomultiplier with a gate much shorter than the lifetime being measured it is possible to sample the light intensity over a large number of decay repetitions. By using a high repetition rate on the excitation pulse and slowly delaying the sampling rate, the integrated output of a very large number of decay repetitions may be graphed automatically with a chart recorder.

The integration of a large number of excitation decay events makes it possible to display the average decay curve without requiring high light intensity; in fact, an individual photon count is possible. This permits the use of weaker electron-beam densities for excitation and the use of a monochromator with a moderate-signal aperture to select spectral lines. Smaller current densities give a better control of electron impact energies, and the use of a monochromator adds flexibility to the system. Photomultiplier dark current becomes important in this experiment, however, and requires special consideration.

The relative advantages of each system described above are not mutually exclusive, and some hybrid system may be particularly adapted to very weak light or very fast decays. At present it seems that the phase-shift technique is applicable to the shortest lifetimes but is limited to resonance states. The oscilloscope display of a single decay curves allows a decay-curve-repetition rate such that cascade effects can be observed and optimized visually while excitation energies and burst width can be adjusted. It serves as an excellent search tool for side effects such as cascading. The sampling technique seems to provide the most flexibility and probably the greatest precision within the resolution time of the photomultiplier and electronic pulse techniques.

5.3.2 Direct Measurement of Atomic Lifetime by an Inverted-triode-excitation, High-impact-current Method

Before the experimental equipment needed to measure the atomic lifetime is described, the theoretical basis for the measurement technique will be developed.

An equation can be written governing particle density of the *i*th excited state in a gas.

$$\frac{dN_i}{dt} = -\left[\sum_j A_{ij} N_i + N_i \Lambda(P, T)\right] + \left[\sum_k A_{ki} N_k + J_p N_o \sigma_c + I N_o \sigma_p\right]. \quad (5.133)$$

Here $\Sigma_j A_{ij}$ is the transition probability per unit time that governs the loss process due to spontaneous emission. The Λ (P, T) is a lumped-loss term including all loss processes in which N_i appears as a simple factor. These may include stimulated emission, collisions of the second kind and diffusion to the walls, among others. Λ (P, T) is usually very small compared to $\Sigma A_{ij} N_i$. The second bracket above includes the important gain terms $\Sigma_k A_{ki} N_k$ = (cascading from upper states), $J_p N_o \sigma_c$ (electron-impact excitation), and $I N_o \sigma_p$ from photoexcitation.

Assuming a step function of excitation power, the state-density equation yields a build up of N_i with the following form:

$$N_i(t) = \frac{\text{excitation rate}}{\sum A_{ij} + \Lambda} \left[1 - \exp(\sum A_{ij} + \Lambda)t\right]. \quad (5.134)$$

This implies that a duration of excitation power longer than a mean lifetime $\tau_{ij} = 1/(\Sigma A_{ij} + \Lambda)$ does not result in an appreciable increase in state density. If, however, there is cascading from higher energy states, a mechanism that alters the simple exponential form of the state-density curve, the relative magnitudes of the longer-lived cascade components can be reduced by using a short pulse of excitation.

The differential equation may be solved assuming no further excitation beyond some time zero. The state density is found to be

$$N_i(t) = N_i(0) \exp - (\sum A_{ij} + \Lambda)t. \quad (5.135)$$

Most loss processes increase with pressure, yielding a decrease in apparent lifetime. A lack of lifetime dependence on pressure indicates that a true spontaneous life is being obtained.

In the case of resonance levels (those which can radiate to the ground state via dipole radiation) radiation imprisonment tends to yield increasing lifetime with pressure, approaching

$$\bar{\tau}_{ij} = \frac{1}{\sum_j A_{ij} + \Lambda} \quad \text{where } E_j < E_i \text{ but } j \neq 0. \quad (5.136)$$

Spontaneous emission yields light-intensity dependence as follows:

$$I_{ij}^{*} = N_i A_{ij} \tag{5.137}$$

where I_{ij}^{*} is the photon-source intensity for photons created by spontaneous transitions between the i and j states. It is clear that the time dependence of the excited atom density N_i is accurately followed by the light-output intensity. Thus light intensity may be analyzed to determine the time dependence of the excited-state density.

The heart of the measuring system is a cylindrical triode excitation tube, built to produce a beam of impact electrons radiating inwardly from a cathode capable of producing 2 to 4 A. The massive cathode is heated by magnetic induction through the pyrex envelope. Within the cathode there are two grids, a space-charge grid and an inner grid to control the electron energy entering the central excitation zone.

The cathode was machined of manganese-free nickel and painted with a commercial barium carbonate cathode preparation thinned to the consistency of milk with acetone and brushed onto the inside surface with a clean brush. The nonemitting surfaces were painted with a filament-insulating compound which greatly reduced darkening of the windows by nickel vapors from the cathode. A tantalum iris, spot welded to the cathode end, provided further protection for the window. The inner grids were made of platinum mesh formed and spot welded over accurately sized aluminum mandrels. Alumina ceramic tubing was used for grid spacers. Coaxial glass seals were used to propagate a 5×10^{-9} sec cutoff bias signal into the grid region with minimum reflection at the vacuum seal.

Operation of the system is more easily visualized if you refer to Figure 5.11. During the measurements the space-charge gird is grounded, and the cathode is driven negative for approximately 10 μsec. Concurrently the inner grid is driven negative to give the desired electron energy within the excitation zone. Before the cathode pulse is terminated, a sharp negative cutoff step is applied to the inner grid using a 2D21 thyraton operated at around 2000V. The step is voltage divided down to around 100 V at 19 Ω (the characteristic impedance of the coaxial grid sytem). Radial penetration of electrons into the center of the excitation tube requires space-charge neutralization; therefore the excitation of the central zone is delayed until a plasma could be built up. This delay ranges from 1 to 6 μsec depending upon gas pressure.

The light output is observed through optical interference filters with a 931A photomultiplier. The photomultiplier voltage divider is pulsed with a flat-topped 20 amp 3000 V pulse 2 μsec in duration. Even though the dynode multiplication is only moderately voltage sensitive, the pulse

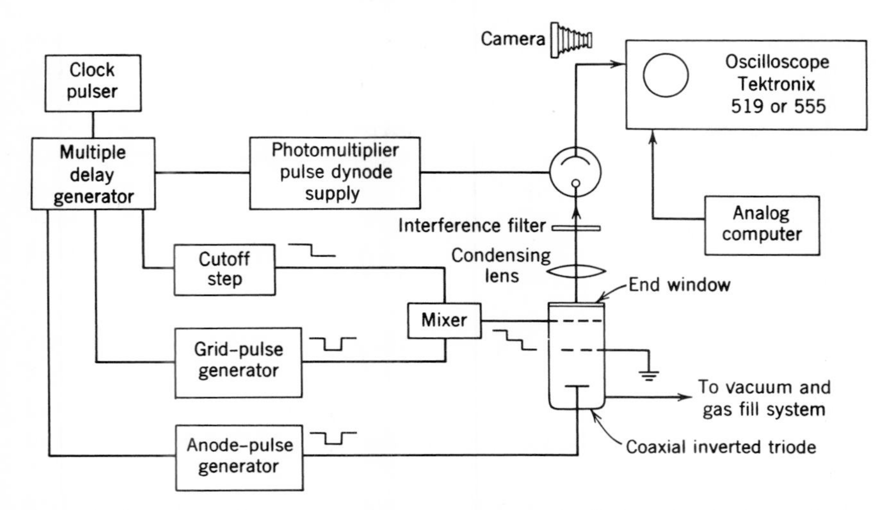

Figure 5.11. Electronic apparatus for measuring atomic lifetimes by the inverted triode method [59].

voltage must be held accurately constant because the amount of the change is multiplied by the number of dynodes contributing to the overall gain.

Space-charge saturation of the photomultipliers was observed for the brightest lines being studied. By using the inverse-square law and a regulated lamp it is possible to discover at what output current the saturation effects set in. (It should be mentioned that the overvoltage-pulse technique improves the amplification, the resolution time, and reduces space-charge saturation for a given output.) The photomultiplier output was displayed on an oscilloscope, photographed, and analyzed by comparing the signal with an analog computer that superimposed several adjustable decay curves for comparison with the experimental curves.

5.3.3 Metastable-lifetime Measurement by Absorption in Gases Using an Afterglow Probe

The method of lifetime determination described below consists of observing the decay of metastable density in the afterglow, which follows a pulsed gaseous discharge. Absorption of resonance radiation by the discharge is measured by a probing signal. This technique is more adaptable to laser technology than experiments that detect the metastables by secondary electron emission, resonance radiation absorption, and fluorescence among others.

Let a beam of radiation (generally in the visible or near-visible spectrum) from an external lamp traverse a medium containing N metastable particles that can absorb a wavelength λ_i while undergoing transition to the ith upper state. A satisfactory and convenient source is a low-pressure capillary gas discharge occurring in the same type of gas as is in the test cell. The transmitted radiation may be detected by a spectrometer photocell-oscilloscope combination, the spectrometer dispersion being sufficient to resolve the discrete spectrum of the lamp. For a given spectral line, let $I(\nu_i, z)\,d\nu$ be the beam intensity at distance z in the cell, at frequency ν_i with band width $d\nu$, and let B_{ji}, B_{ij} be the Einstein "B" coefficients for beam-induced absorption ($j \rightarrow i$) or emission ($i \rightarrow j$). The intensity is diminished according to

$$d[I(\nu_i, z)d\nu]/dz = -[\Delta n(\nu_i)h\nu_i B_{ji} I(\nu_i, z) - \Delta n_i(\nu_i)h\nu_i B_{ij} I(\nu_i, z)], \quad (5.138)$$

where $\Delta n(\nu_i)$, $\Delta n_i(\nu_i)$ are the concentration in the metastable or ith excited states capable of absorbing or emitting radiation ν_i in bandwidth $d\nu$. The absorption coefficient is defined as before: $k = (1/I)(dI/dz)$. Because the broadening of a spectral line is small compared with the center frequency (ν_{ci}) or wavelength (λ_{ci}), an integration over the observed line and substitution for the "B" coefficient yields

$$\int_{i\text{th line}} k(\nu_i)d\nu = h\nu_{ci}(B_{ji}n - B_{ij}n_i)$$

$$= \frac{\lambda_{ci}^2}{8\pi}\frac{g_i}{g_j}\frac{n}{\tau_{ij}}\left(1 - \frac{g_j}{g_i}\frac{n_i}{n}\right) \quad (5.139)$$

where g_i, g_j are the degeneracy factors of the jth and ith levels, and τ_{ij} is the lifetime of state i for spontaneous decay to state j. Unless the probing radiation is too intense or the medium is highly excited, $n_i << n$, and we retain only the first term in the final form of equation (5.139). The lifetime (τ_{ij}), oscillator strength (f_{ij}), and spontaneous-emission coefficient (A_{ij}), of the transition are related as previously developed by

$$f_{ij}\tau_{ij} = \frac{f_{ij}}{A_{ij}} = \frac{mc}{8\pi^2 e^2}\frac{g_j}{g_i}\lambda_{ci}^2. \quad (5.140)$$

Precise determination of the oscillator strength of a transition, either theoretically or experimentally, is difficult. An absolute determination of n is fortunately not required for a lifetime measurement, although consultation of relative f values from the literature[62, 63] may be helpful in the selection of the probing spectral lines to be used in the experiment.

The absorption lines are predominantly Doppler-broadened:

$$k(\nu) = k_0 \exp - [2(\nu - \nu_c)\sqrt{\ln 2}/\Delta\nu_D] \equiv k_0 \exp(-\Omega^2) \quad (5.141)$$

where $\Delta\nu_D$ is the Doppler-frequency width

$$\Delta\nu_D = \frac{\nu_c}{c}(8kT \ln 2/M)^{1/2}, \qquad (5.142)$$

M is the mass of the atom, and T is the gas temperature.

Using (5.141), k_0 may be obtained from (5.139):

$$k_{ci} = \frac{2}{\Delta\nu_D}\left(\frac{\ln 2}{\pi}\right)^{1/2} \frac{\lambda_{ci}^2}{8\pi} \frac{g_i}{g_j} \frac{n}{\tau_{ij}}. \qquad (5.143)$$

The method used for determining n_j is discussed by Mitchell and Zemansky[9].

Low-intensity spectral radiation centered at λ_0 is produced by an external capillary lamp and beamed through the medium, where n_j is to be determined from the beam absorption. Because the radiator and absorber are identical particles, the lamp emissivity may be represented by

$$E(\nu) = E_0 \exp[-(\Delta\alpha)^2] \qquad (5.144)$$

where α is the ratio of the emission line width to the absorption line width.

High-resolution examination of spectral lines emitted by Cenco capillary lamps in general show the lines have a Doppler profile, but because the emitting atoms are in a plasma where fields may perturb the radiator, we cannot expect $\alpha = 1$ a priori.

The ratio α can be measured by artificially doubling the beam in the absorption cell[64]. The value of α is $1.5 < \alpha < 2$ for many lines, which suggests that a minor self-absorption in the source is responsible for $\alpha \neq 1$.

The integrated-beam absorption $A(\alpha, d)$ over a path d is

$$A(\alpha, l) = \frac{\int_{-\infty}^{\infty} \exp - (\Omega/\alpha)^2[1 - \exp - (k_0 d \exp(-\Omega^2))]d\Omega}{\int_{-\infty}^{\infty} \exp - \left(\frac{\Omega}{\alpha}\right)^2 d\Omega} \qquad (5.145)$$

Values of $A(\alpha, l)$ are tabulated in the literature[9]. Changing spectral lines is particularly useful when the absorption on a given line becomes high (approximately 50 percent) because the central region of the spectral line is most sensitive to small changes in n_i and is the first region to be saturated. When diffusion alone controls the loss of metastables, which absorb the probing radiation, the profile of metastable density should be $J_0(2.4r/R)$. However when a binary gas is introduced in quantity, the major loss mechanism may be volume destruction, and a uniform metastable density profile must result. When the absorption tube is long and the light beam is not collimated to a cone angle smaller than the tube solid angle subtended at the source, the experimental absorption will yield the average of metastable density across the tube diameter. The absorption technique permits measurement of k_0 and $n(t)$, from which is obtained the lifetime.

From the integrated absorption coefficient, k_0, the path length d, a measured or assumed value of α, and the curve of Figure 5.12, the value of $A(\alpha, d)$ can be obtained; from the reciprocal of $A(\alpha, d)$ we can compute the spontaneous-emission lifetime of the metastable state.

The disposition of typical experimental apparatus used for the measurements described above is given in Figure 5.13. Measurements of the total cross section for de-excitation can be made by observing the rate of decay of metastable concentration and/or the decay of spectral-light intensity following the termination of a current pulse in the discharge tube.

The plasma tube is typically a glass tube of $2a \leq 1$ cm bore, and the current pulse required to create the plasma ranges from 10 to 50 ma for 50 - 150 μsec; the gas pressure is usually about one mm Hg. Cold cathode electrodes made of tungsten wire are typically used to preclude interaction between the plasma and cathode materials. A conventional outgassed

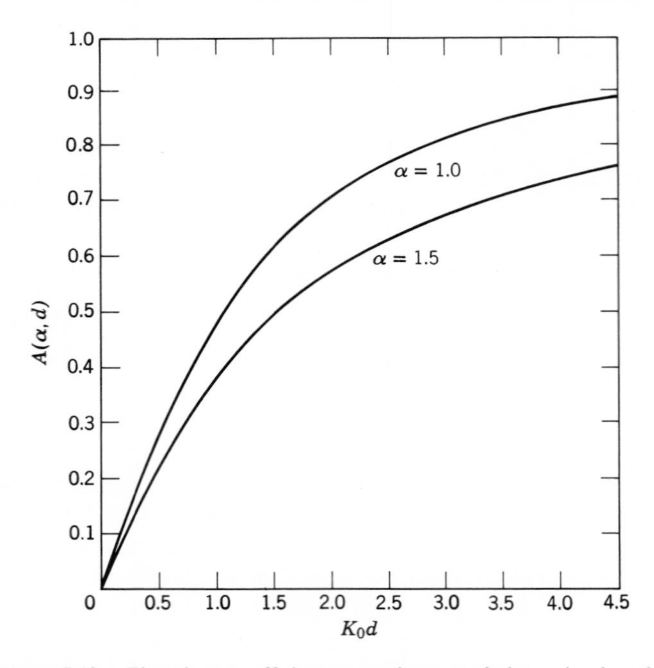

Figure 5.12. Einstein A coefficient versus integrated absorption length.

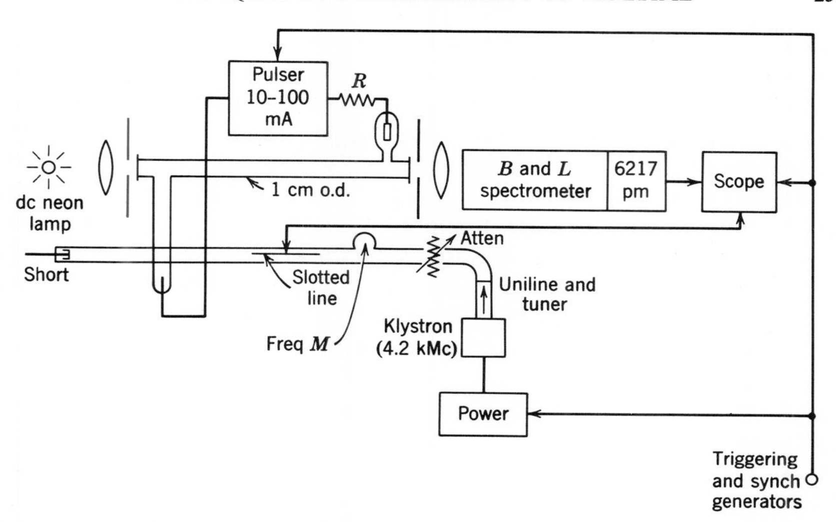

Figure 5.13. Schematic diagram of apparatus used to measure metastable lifetimes and electron temperatures in a plasma discharge.

vacuum system, where the background gas pressure was about 10^{-6} Torr, is used for gas filling and mixing. Spectroscopically pure gases must be used to avoid uncontrolled contamination.

Optical observations are made along an axial path in the afterglow about 17 cm long with a Bausch & Lomb f/5 grating spectrometer and 6217 photomultiplier detector. The electron density is usually low ($n_e \leqq 3 \times 10^{10}$ electron/cc at $t \leqq 50$ μsec in afterglow), thereby negating the possibility of electron-ion recombination light even if collected along the long optical path.

Many oscillator strengths, f, for spectral lines were measured years ago using the method of anomalous dispersion[65]. This gave accurate relative f numbers for the lines; however absolute values were obtained only after a lengthy and less rigorous argument was constructed based on gas-discharge equilibria and the f-sum rule. It is frequently convenient to assume some f value, for example $f = 0.5$, and use it as a reference value throughout measurements. Corrections can be made at a later time if more accurate data become available.

Measurements of electron density, used in the determination of σ_i, are based upon calculations[66] for a dielectric post in a waveguide. If the ratio of the collision frequency to the signal frequency $v_m/\omega << 1$,

$$(\omega_p a)^2 = \frac{b^3 \omega^2 \delta}{2\pi \lambda_{g0}^{\ 2}} \tag{5.146}$$

where a is the effective radius of the plasma tube, b is the dimension of the rectangular guide in which a TE_{01} mode propagates, λ_{g0} is the microwave wavelength inside the waveguide, and δ is the shift in the minimum of the standing waves caused by introduction of the plasma. The plasma forms a column parallel to the electric field in the guide such that $a/b < 1/b$; the guide is shorted $\lambda_{g0}/4$ behind the plasma and the position of the standing wave minimum is read on a slotted line. The ratio $2\pi\delta/\lambda_{g0}$ is, by assumption, $<< 1$. Because the afterglow plasmas can be produced by a current pulse several electron diffusion time constants long, a fundamental electron-density mode has been established in the J_0 $(2.4r/R)$ distribution at $t = 0$ sec in the afterglow; this should be maintained, for the warm electron gas cools and diffuses at low density. The effective plasma post diameter is about two-thirds the plasma diameter when averaging over the electron distribution is considered.

Electron concentration is measured by passing the plasma through a C-band waveguide (frequency $\omega/2\pi = 4200$ MC) and determining the microwave impedance of the system; at this frequency the plasma may be represented by a low-loss dielectric constant $\epsilon = 1 - \omega_p{}^2/(\omega^2 + \nu_m{}^2)$, where $\omega_p{}^2/\omega^2 = 4\pi n_e e^2/m$, and ν_m, the collision frequency for momentum transfer of electrons with neutrals, is much less than ω. With these observations it is possible not only to choose the physical conditions of the afterglow, but also to measure σ_i, the cross section for ionization.

The microwave technique is usually employed for $m\omega^2/4\pi e^2 > n_e > 10^{10}\text{cm}^{-3}$. If it becomes necessary to measure electron density for $n_e < 10^{10}$, the technique must be modified. A longer (about $\frac{1}{2}$m) quartz discharge tube can be enclosed in a x-band waveguide ($\omega/2\pi = 9400$ MC) and the electron density obtained by measuring the attenuation and phase shift of the microwave[67]. If λ_g is the microwave wavelength in the plasma, and γ is the attenuation coefficient expressed in nepers/m, then in MKS units the electron concentration may be found from

$$\frac{e^2}{\epsilon_0 m}\frac{\int n_e(r)E^2\,dS}{\int E^2\,dS} = \left(\frac{c}{\omega}\right)^2\left[\gamma^2 + \left(\frac{2\pi}{\lambda_{g0}}\right)^2 - \left(\frac{2\pi}{\lambda_g}\right)^2\right] \tag{5.147}$$

where the electron distribution is averaged over the transverse guide dimension in the integrals involving E^2. Peak (axial) values for electron concentration are thus obtained.

An error of about 10 percent can result from an incorrect choice of α in (5.144). Lamp and detector noise may be reduced relative to the absorption signal by using a boxcar integrator on the signal output[68]. A signal-to-noise ratio limitation is observed at an absorption of about 1 percent. A minimum metastable density of about 10^8/cc can be observed in apparatus of convenient size. The reliability of the lifetime measurement may be improved by using several absorbed probing lines which terminate on the same metastable level and by limiting the total absorption to less than 50 percent of any line to insure accuracy in interpreting Figure 5.12 and in the representation of the spectral line shapes. Electronic deactivation of metastables may be reduced by minimizing the pulsed discharge current or by recourse to a low-energy, electron-beam bombardment system. The limited collimation of the probing light beam means that only an average metastable density may be determined along the absorption path, and hence the technique is somewhat sensitive to the geometry of the metastable concentration. For further elaboration of these and other practical details the reader is referred to the literature[69].

When an impurity is purposefully added to the gas, the altered decay rate of the metastables may be observed by a simpler method, should the impurity fluoresce in deactivating the metastables (e.g. neon in a helium-neon laser gas mixture). The intensity of the impurity's fluorescence light is directly proportional to the metastable concentration and may be detected in the afterglow with a spectrometer and a photocell. Spectral resolution is desirable, owing to interference from charge-recombination light in the afterglow. The latter radiation is easily recognized, for its decay follows the decrease of charge concentration rather than the metastable density.

5.3.4 Technique for Measuring Fluorescence Decay Times in Liquids and Solids

The fluorescence decay time or spontaneous-emission time of a material is one of the most important parameters to be considered in determining the laser application of that material. Fortunately, in the case of the solid or the liquid laser materials, this parameter is also one of the easiest to measure because only materials with a relatively long decay time are useful for laser purposes.[1] Normally the decay time for radiation in gases, due to allowed transitions in the optical region, is of the order of 10^{-8} sec. In solid or liquid laser materials, however, use is made of a partially allowed transition, usually within a transition element (including the rare earths). Decay times ranging from a microsecond to a few milliseconds are observed. (Measurement of decay times in this range is rather simple when modern experimental techniques are used.)

The radiative decay (or spontaneous emission) time τ_{ij} of an excited state is given by

$$\tau = \frac{1}{\sum_i A_i} \tag{5.148}$$

[1] Semiconductor laser materials are not discussed here.

where the values of A_i are the transition probabilities of the Einstein A coefficients for each of the individual transitions, and i represents the various lines in the emission spectrum arising out of transition from that excited state. It is inferred that the decay time of the various lines arising out of transition from the same level should be the same, although the individual oscillator strengths of the transition may be different.

If the excited-state decay involves only one transition to the ground state, the decay time can be related to the absorption coefficient by the following relation[70, 71]:

$$\frac{1}{\tau} = \frac{8\pi n^2}{N\lambda^2}(g_j/g_i)\int k_\nu\, d\nu \tag{5.149}$$

where N = number of absorbing species per cc,
λ = wavelength in centimeter,
n = refractive index of the medium,
g_i, g_j = degeneracies of the lower and upper states, respectively,
k_ν = absorption coefficient per centimeter,
ν = frequency in cycles per second

When the excited-state decay involves more than one transition, but one of the transitions is to the ground state so that its absolute strength can be estimated from the absorption measurement, ΣA_i can be calculated from the ratio of the emission intensities of the various lines and from the absolute strength of the known line. The accuracy of such calculations is limited by the fact that the observed emission intensity due to transition to the ground state is modified by absorption. For partially allowed transitions, the absorption coefficient is usually too low to make a significant contribution.

So far only the radiative decay time has been discussed. In practice the observed decay time τ_0 is determined by the probabilities of radiative and nonradiative decay from the excited state, so that

$$\frac{1}{\tau_0} = \frac{1}{\tau_r} + \frac{1}{\tau_q} \tag{5.150}$$

where $1/\tau_q$ is the rate of quenching per second. The observed decay time is, therefore, related to the quantum yield η by

$$\tau_0 = \eta\tau_r. \tag{5.151}$$

The technique for the measurement of fluorescence decay time in solid and liquid laser material is straight forward. A typical experimental arrangement[72] is illustrated in Figure 5.14. A small flash tube (FX-12), having a decay time of a few microseconds, is used for excitation. The flash tube is provided with a filter jacket to cut off light in the fluorescence region. The detection is accomplished by a photomultiplier and an interference filter that transmits only the fluorescence radiation of interest. A suitable lens and an aperture are used to reduce the background radiation further. The decay curve is recorded on an oscilloscope using a Polariod camera.

Cathode-follower impedance matching between the photomultiplier and the oscilloscope is very useful, especially for weaker intensities where a larger load resistance is necessary. The decay curves may be transferred onto semilog graph paper; the decay time is usually given by the slope of the curves because the curves are usually pure exponential in nature. Alternatively, the photomultiplier output can be fed through a logarithmic converter and the exponential decay displayed as a straight line on the oscilloscope. The curves can be recorded on Polaroid transparencies, which can subsequently be projected for a more accurate measurement of data points. Precautions should be taken not to saturate the detection system. Deviations from the exponential nature of the curve may often be taken as an indication of saturation. This can be remedied by inserting more neutral density attenuators in front of the photomultiplier (or by reducing the photomultiplier voltage).

For ruby and neodymium-doped materials, a Dumont 6911 or an RCA 7102 photomultiplier with an S1 surface can be used. For longer wavelength radiation, indium-arsenide or gold-doped germanium should be used.

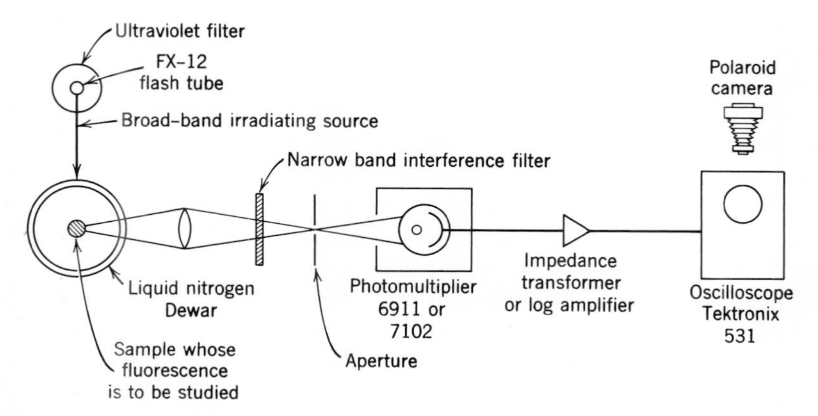

Figure 5.14. Arrangement of experimental equipment for measurement of fluorescence decay times in solids and liquids.

Details of the application of the technique described above to the study of lifetimes in neodymium-doped silicate glasses are presented in the literature[74].

A stroboscopic method has been developed that has a higher sensitivity[73]. In this method the high voltage of the photomultiplier is gated on for a short time at a desired delay time following the flash excitation. The operation is cyclic at a rate of 40 to 80 Hz, and the photomultiplier signal is integrated and amplified. The delay time can be changed and simultaneously plotted along the x-axis of a recorder by using a ganged potentiometer. The signal from the photomultiplier is plotted along the y-axis. This technique enables the relative decay curve to be recorded on chart paper. Because the photomultiplier is pulsed, a higher dynode voltage can be applied, and because this increases the signal and the signal-to-noise ratio, higher sensitivity is made possible.

5.3.5 Interferometric Phase-shift Technique for Measuring Short Flourescent Lifetimes in Semiconductors

Measurement of fluorescent-decay times by optical-pulsing techniques is limited to times of about 10^{-8} sec, although an interferometric phase-shift technique has been developed that permits lifetime measurements in the region of 10^{-8} to 10^{-11} sec[75]. A measurement on gallium arsenide was carried out to illustrate the method, and a fluorescent lifetime of $\tau = 3.0 \times 10^{-9}$ sec was obtained for this semiconductor.

The beam from a gaseous laser, when projected on the cathode of a photomultiplier, will produce a signal at the difference frequency between axial modes of the laser cavity[76]. Excitation of a fluorescent material by the same gaseous laser beam produces a fluorescence which is also amplitude modulated at just this difference frequency. (This fluorescent modulation will, however, gradually disappear as the period of the difference frequency becomes less than that of the fluorescent lifetime.)

Using a gaseous laser as a source for a two-beam interferometer, and a photomultiplier as a detector, a null in the difference signal can be obtained when one beam travels an additional distance equal to just half a wavelength of the difference frequency. Detailed consideration of the combination of the optical beams at a photocathode shows optical interference and interference of the modal difference frequency as it modulates the photoelectron current, the former occurring only when the beams are coincident. Hence, the null technique is a sensitive measure of the phase shift in the modal difference frequency produced in a fluorescent material by virtue of the fact that it emits with some characteristic time τ. To obtain a null in the interferometer, the fluorescent sample is placed in one beam and the change in the null position noted.

The experimental apparatus required for the experiment above consists of a two-beam interferometer, a gas-laser source, and a photomultiplier detector (schematically shown in Figure 5.15). The helium-neon laser is operated at 633 nm with a mirror spacing that gives stable optical modes 155 MHz apart. The receiver is tuned to accept the 155 MHz signal, which is detected by the audio beat between it and a beat-frequency oscillator. First, the audio signal is nullified using a glass-slide reflector, in place of the sample, and by moving the prism along an optical bench. The filters are such that both 633 nm beams reach the detector but with suitable attenuation. The polarizer is a convenient means of attenuating one beam to make the amplitudes of both beams identical; otherwise the null will not be complete. The glass slide is then replaced by a fluorescent sample with the appropriate filter to block the 633 nm radiation and allow the fluorescent radiation to reach the detector. The null position of the prism will now have been altered by the phase delay associated with the fluorescent lifetime.

The relation between the fluorescent lifetime and the phase shift, discussed above, may be found by considering a simple single-lifetime process. The rate equation for the fluorescent process is given by

$$\frac{dn}{dt} = \frac{-n}{\tau} + g \cos \omega t \tag{5.152}$$

where n is the density of excited states and τ is the lifetime of the states.

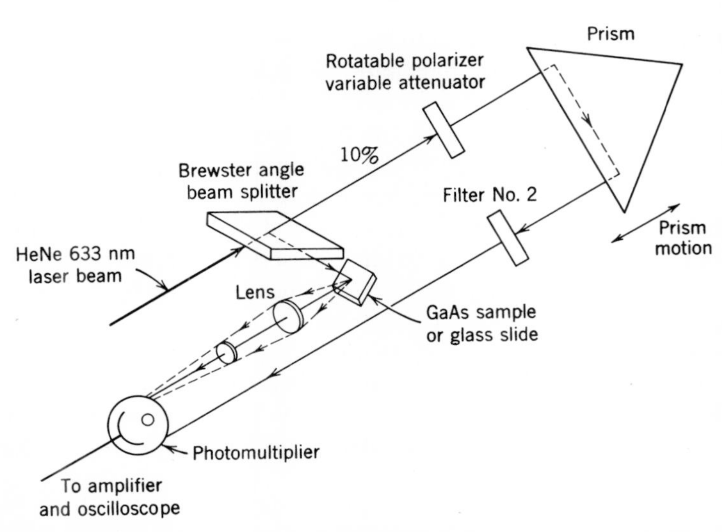

Figure 5.15. Schematic diagram of a phases-shift interferometer. (Reproduced with permission from *Applied Physics Letters*.)

The term $g\cos\omega t$ is the modulation of the population density in the fluorescent level at ω, the difference frequency of the optical modes. Because both optical modes appear in each interferometer beam and the fluorescent population is proportional to the square of the optical fields a signal will always occur at the difference frequency that will drive the fluorescence.

A solution to (5.152) is

$$n = g\tau(1 + \omega^2\tau^2)^{-1/2} \cos(\omega t - \phi) \tag{5.153}$$

where

$$\phi = \tan^{-1} \omega\tau, \tag{5.154}$$

indicating that the phase shift must lie between zero and $\pi/2$.

The phase shift produced by a change in the null position of the interferometer is given by

$$\phi = \frac{2\omega x}{c}, \tag{5.155a}$$

which for small $\omega\tau$ becomes

$$\tau = \frac{2x}{c}. \tag{5.155b}$$

It should be pointed out that in multistep-decay processes, the individual characteristic times involved cannot, in general, be separated. In fact, phase shifts greater than $\pi/2$ are possible. Some care must be taken that multiple fluorescent bands, whose lifetimes may differ, are examined separately by proper filtering. The modulation, or the difference frequency between optical modes, of the source determines the longest fluorescent time that can be measured by the method described above, although the limit can be extended by going to longer lasers (lower modulation frequencies). Also an externally modulated light beam can be used. The shortest lifetime that can be measured is determined by how accurately the interferometer null can be set, provided that the beams are not allowed to wander on the detector surface while the measurement is being made. Compensation must be made for the filter thicknesses. The technique allows the nulls to be set to about 1/1000 of the modulation wavelength, corresponding to a decay time of 10^{-11} sec.

Using the technique described above, lifetime of 3.0×10^{-9} sec have been measured at room temperature for single-crystal GaAs *n*-doped with Te. The value may be compared with an experimental value of $\tau = 6 \times 10^{-9}$

sec determined from minority-carrier storage measurements in PIN diodes[77]. Fluorescent spectra of the same sample of GaAs, excited by the 633 nm laser, show that the fluorescent radiation is contained totally in a single band 24 nm wide, peaked at 869 nm.

5.4 TECHNIQUE FOR DETERMINING THE DEGREE OF POPULATION INVERSION IN *Q*-SWITCHED LASERS

The theory of *Q*-switching in lasers has been treated in more or less detail by a number of authors[78–85]. It can be shown that the behavior of a *Q*-switched laser is described reasonably well by a set of two, coupled, nonlinear rate equations:

$$\begin{aligned} \dot{\eta} &= \xi_i \epsilon \eta \\ \dot{\epsilon} &= -(\xi_0 + \xi_e)\epsilon + \xi_i \eta \epsilon \end{aligned} \tag{5.156}$$

Here,

$$\eta(t) = \left(\frac{1}{N}\right)[n_2(t) - n_1(t)]$$

and

$$\epsilon(t) = \left(\frac{2hv_{21}}{N}\right) u(t) \tag{5.157}$$

and the following definitions apply:

$n_2(t)$ = population of the upper-laser state.
$n_1(t)$ = population of the lower-laser state.
$u(t)$ = number of photons, at the laser frequency ν_{21} in the cavity with propagation vectors in the output beam of the cavity.
N = $n_2(t) + n_1(t)$, a constant.
ξ_i = induced emission coefficient for the laser active material (sec^{-1}).
ξ_0 = intrinsic, radiation-loss coefficient for the laser, exclusive of the loss due to coupling into the laser beam (sec^{-1}).
ξ_e = radiation-loss coefficient for the laser due to coupling into the beam (sec^{-1}).

In a *Q*-switched device, ξ_e and, therefore, the lasing threshold

$$\eta_t = (\xi_0 + \xi_e)\xi_i \tag{5.158}$$

are functions of time.

The *Q*-switched laser is pumped with ξ_e^I sufficiently high that $\eta(t)$ can not exceed η_t^I. When $\eta(t)$ reaches a steady value η_0, ξ_e is rapidly switched to a lower value ξ_e^{II}, such that the new threshold η_t^{II} is well below η_0. The laser then generates a so-called giant pulse of optical radiation. The laser-rate equations have been solved to predict the peak power, duration, and soon, of this giant pulse.

In most theoretical treatments, it has been assumed that ξ_e is switched instantaneously from ξ_e^{I} to ξ_e^{II} and that the solution of (5.156) evolves from initial conditions $\eta(0) = \eta_0$ and $\epsilon(0) = \epsilon_0$, which are achieved during the pump interval prior to the instantaneous switching of ξ_e at $t = 0$.

Although numerical methods for obtaining exact solutions of the rate equations above have been considered[83, 85], it is convenient, for the present discussion, to consider approximate solutions obtainable in closed form. It can be shown that, to a good approximation,

$$\rho(t) = \frac{\gamma_r \mu(t)}{\gamma_r + \mu(t)} - (\gamma_r - \gamma_t) \ln\left[\gamma_r \frac{1 + \mu(t)}{\gamma_r + \rho_0}\right] \tag{5.159a}$$

and

$$\gamma(t) = \gamma_r - \frac{\gamma_r}{1 + \mu(t)} \tag{5.159b}$$

represent solutions of (5.156) with step-function switching of ξ_e. Here,

$$\begin{aligned} \mu(t) &= (\rho_0/\gamma_r) \exp(\gamma_r \xi_t \eta_0 t); \\ \rho(t) &= \epsilon(t)/\eta_0, \text{ from which } \rho_0 = \epsilon_0/\eta_0; \\ \gamma(t) &= 1 - [\eta(t)/\eta_0]; \\ \gamma_r &= 1 - (\eta\ /\eta_0); \end{aligned}$$

and

$$\gamma_t = 1 - (\eta_t/\eta_0).$$

η_t is the threshold value of η for $t > 0$, that is for the condition $\xi_e = \xi_e^{\mathrm{II}}$. The parameter η_r represents the residual population inversion in the laser cavity when the intensity in the giant pulse has returned to zero. η_r can be shown[84, 85] to be a solution of

$$\eta = \eta_0 + \eta_t \ln\left(\frac{\eta}{\eta_0}\right). \tag{5.160}$$

One solution is clearly $\eta = \eta_0$; The other is $\eta = \eta_r$, and (5.159*b*) indicates that

$$\lim_{t \to \infty} v(t) = \gamma_r.$$

If the threshold in a laser is held high enough to prevent oscillation with a given pump, the population excess that we could attain should be very nearly

$$\eta_0 = \frac{\xi_p - \xi_S}{\xi_p + \xi_S} \tag{5.161}$$

where ξ_p is the pump-transition rate per active ion and ξ_S is the spontaneous-emission rate per ion. Pumps are readily available for which $\xi_p = 10\xi_S$, from which $\eta_0 = 0.8$ should be attainable. If this value of η_0, and reasonable values of the rate parameters ξ_0, ξ_i, and ξ_e are used in (5.159), however, peak powers well in excess of those observed experimentally may be predicted.

A suggested explanation for the discrepancy discussed above is as follows: ξ_S in equation 5.161 represents the spontaneous-emission rate of isolated, excited, laser-active ions. In (5.161), however, this should be replaced by a parameter $\xi_S\ (\eta_0)$, which represents the decay rate of an excited ion in an assembly of ions with an inverted population. That is $\xi_S\ (\eta_0)$ can be considerably greater than ξ_S, and a laser pump with $\xi_p = 10\xi_S$ may not be able to pump the system to $\eta_0 = 0.8$. In experimental situations η_0 may be less than 0.8 for a number of reasons, but it is clear that for any *Q*-switched laser it is useful to be able to determine the value of η_0 experimentally. Experimental details of this technique are found in the literature[84].

The peak value ρ_p or $\rho\ (t)$ is related to the peak power in a giant pulse by

$$P_p = \xi_e^{\text{II}} \eta_0 \rho_p U_S \tag{5.162}$$

where $U_S = Nh\nu_{21}/2$. Now, ρ_p can be obtained from (5.159*a*) and can be displayed graphically as a function of γ_t. Similarly, if we define $\Delta T = \xi_i \eta_0, \Delta t$, where Δt is the pulse duration at half-maximum power, ΔT can be displayed as a function of γ_t with the help of (5.159*a*), and a graph of γ_r versus γ_t can be obtained from (5.160), recast in the form

$$\gamma_t = 1 + \frac{\gamma_r}{\ln(1 - \gamma_r)}. \tag{5.163}$$

Experimentally, ξ_e^{II} is at the discretion of the observer, for it can be varied over a wide range. Thus it is possible to obtain experimentally, for any rapidly *Q*-switched laser, a plot of peak-output power as a function of ξ_e^{II}, as well as a plot of pulse duration Δt as a function of

ξ_e^{II}. In all cases we must be careful to use a narrow-band-pass filter around the laser wavelength in order to ensure that only the laser output, and not radiation which has been Raman scattered in the Q-switch, is measured.

Because

$$\epsilon_p = \eta_0 \rho_p = P_p(\xi_e^{\mathrm{II}})/\xi_e^{\mathrm{II}} U_S, \tag{5.164}$$

G_p can be plotted as a function of ξ_e^{II} to obtain ξ_{e0}^{II}, the value of ξ_e^{II} for $\epsilon_p \to 0$ at large values of ξ_e^{II}.

Because $\epsilon_p = 0$ when $\eta_t^{\mathrm{II}} = \eta_0$,

$$\xi_{e0}^{\mathrm{II}} = \xi_i \eta_0 - \xi_0 . \tag{5.165}$$

The plot of Δt versus ξ_e^{II} will generally exhibit a minimum Δt_m for some value of ξ_e^{II}, as does the theoretical curve of ΔT versus γ_t. Indeed, ΔT has a minimum value of 4.6 for

$$\gamma_t = 1 - \left(\frac{\eta_t}{\eta_0}\right) \cong 0.67. \tag{5.166}$$

Thus

$$\xi_i \eta_0 = \frac{4.6}{\Delta t_m}, \tag{5.167}$$

and

$$\xi_0 = \left(\frac{4.6}{\Delta t_m}\right) - \xi_{e0}^{\mathrm{II}}. \tag{5.168}$$

Because $\xi_i \eta_0$ and ξ_0 are now known, we can derive a plot of

$$\gamma_t = 1 - \frac{\xi_e^{\mathrm{II}} + \xi_0}{\xi_i \eta_0} \tag{5.169}$$

as a function of ξ_e^{II}. This, together with the plot ρ_p versus γ_t, yields a graph of ρ_p versus ξ_e^{II} for a laser with known ξ_0 and $\xi_i \eta_0$. This graph should be compared with the experimental curve for ϵ_p versus ξ_e^{II}, and the comparison should yield a mean value of $\eta_0 = \rho_p (\xi_e^{\mathrm{II}})/\epsilon_p (\xi_e^{\mathrm{II}})$.

This technique, used with data obtained with a representative Q-switched laser, resulted in the following values for the laser parameters[84]:

$$\xi_0 \cong 7.8 \times 10^8 \text{ sec}^{-1},$$

$$\xi_i \cong 7.8 \times 10^9 \text{ sec}^{-1},$$

and $\eta_0 \cong 0.10$. The rate coefficients were in fair agreement with those expected for the device. The initial inversion, in contrast, was only about 12 percent of what was expected.

5.5 DETERMINATION OF CAVITY LOSSES

That the Fabry-Perot interferometer provides a practical means for obtaining the mode separation necessary to produce oscillation in one or a few modes, at optical frequencies, while providing the optical feedback necessary for low-gain transitions, is made evident if we consider the following example of a helium-neon laser operating at 1.15 μm. The Doppler-broadened line width of the 1.15 μm transition is approximately $\Delta\nu_D = 800$ MHz, whereas the natural line width obtained from the spontaneous emission time and equation (5.48) is approximately $\Delta\nu_N = 80$ MHz. The axial mode spacing in a laser with 1 m mirror separation ($c/2d = 150$ MHz) is in excess of the natural line width and enables five or six dominant modes across the full line width in a cavity without conducting walls. If the cavity had metal walls (5.5) would show that the mode density would approach 10^{10}.

The number of axial modes that will oscillate in an optical cavity without walls is determined for a given degree of inversion by the cavity losses. The power output that can be obtained is determined by the amount of energy that can be coupled from the cavity and still sustain stable oscillation. There are two sources of loss that must be considered in the optical cavity: diffraction losses due to the finite number of Fresnel zones formed with respect to some center of symmetry and mirror losses. The first source is imposed by physical optics and is a function of the geometry of the plasma tube (or laser rod) and the configuration of the mirrors. The second is more complex and includes, in addition to transmission, the absorption and scattering losses of the dielectrically coated surfaces and the optical quality of the mirror substrates, both with respect to surface smoothness and departure from ideal surface geometry.

Consider first the losses that can be controlled by proper optical design. The discussion outlined below assumes that the only operating mode of interest is TEM_{00q}. This mode is of great practical interest because it is capable of producing a Gaussian beam and will yield the desired optical properties in diffraction-limited optical systems. (The other higher-order modes are usually only of academic interest.)

The selection of laser mirrors appropriate for high-power, TEM_{00q}-mode operation requires an accurate knowledge of the diffraction losses of the various modes and their interrelation to the geometrical constraints of the plasma tube. This problem has been subject to intensive study, and theoretical results are now available to assist in the parameter-selection problem[87–102]. Once the optimum plasma tube geometry for desired gain and power output is known, the Fresnel number can be chosen that will yield, for a given mirror geometry, the necessary amount of discrimination between the lowest- and the second-order modes. The latter selection will determine the diffraction loss per transit. The larger the diffraction loss must be to force single-mode operation, the higher the mirror reflectivity must be to sustain oscillation. These two constraints reduce the useable output power.

5.5.1 Diffraction Losses

Space does not permit a detailed treatment here of the diffraction loss versus cavity geometry. However a few pertinent results will be useful and will be stated without proof.

The diffraction loss per transit in a stable optical cavity[2] is highest for a parallel-plane mirror geometry and may be expressed in terms of loss for the dominant mode as[93]

$$\alpha_D = 0.207\left(\frac{b\lambda}{a^2}\right)^{1.4}. \qquad (5.170a)$$

The resonator Q will be a maximum when the diffraction losses are 2.5 times the reflection (output) losses. The diffraction loss for the dominant mode in a confocal cavity may be approximated by[89]

$$\alpha_D = 10.9 \times 10^{-4.94N} \qquad (5.170b)$$

where $N = a^2/d\lambda$.

The confocal resonator Q will be maximum for a given mirror spacing when the reflection loss α_R is

$$\alpha_R = (2.30B\left(\frac{a^2}{d\lambda}\right) - 1). \qquad (5.171)$$

If $a^2/b\lambda = 0.8$, optimum Q occurs for $8\,\alpha_R \cong \alpha_D$

[2] Cavity stability is defined by the analytical conditions

$$0 \lesssim g_1 g_2 \lesssim 1$$

where $g_1 = [1 - d/b_1]$ and $g_2 = [1 - d/b_2]$ are the normalized curvature parameters, d being the mirror spacing and b_1 and b_2 being the respective mirror radii of curvature[89]. In the confocal system $2d - b_1 + b_2$. If $b_1 = 2d$ and $b_2 = \infty$, the hemispherical system is obtained.

Figure 5.16 illustrates the diffraction loss per pass for plane and confocal mirrors as a function of Fresnel number. Figure 5.17 illustrates size of the laser spot at the center of confocal and nonconfocal cavities, and Figure 5.18 relates the spot size at the mirrors to the deviation of a hemispherical cavity from the half-symmetric confocal geometry. These graphs emphasize the effect of mirror geometry and spacing on the size bore that can be used to produce laser action.

A study has been made of the effect of mirror curvature on the ratio of the average losses per transit of the two lowest-order modes in a symmetric and in a half-symmetric mirror geometry for parametric-mirror curvatures (g-values). The pertinent graphs are given in Figures 5.19, 5.20, 5.21, and 5.22. The results of this study indicate that, to obtain maximum power output from a laser, a mirror curvature should be chosen that is as close as possible to a parallel plane as is consistent with requirements

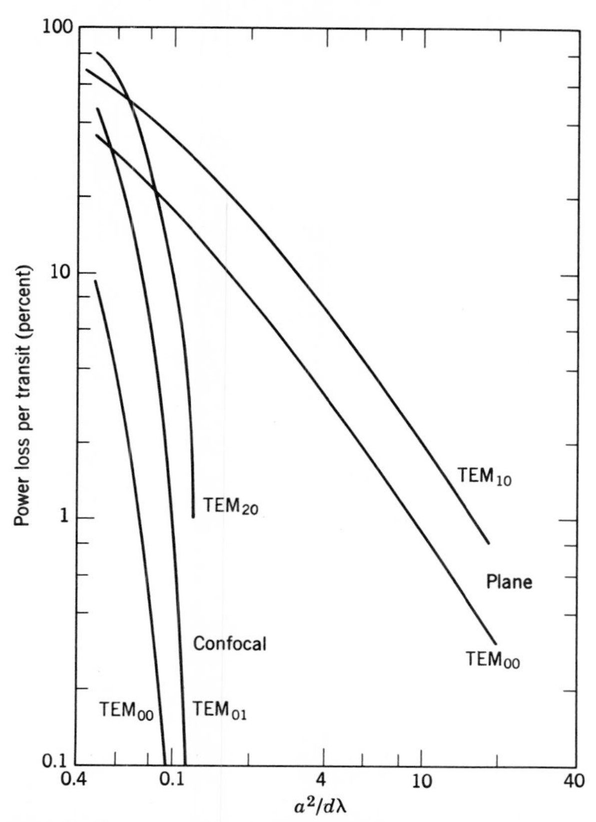

Figure 5.16. Diffraction loss per pass as a function of the Fresnel number. (Reproduced with permission from *Applied Optics*.)

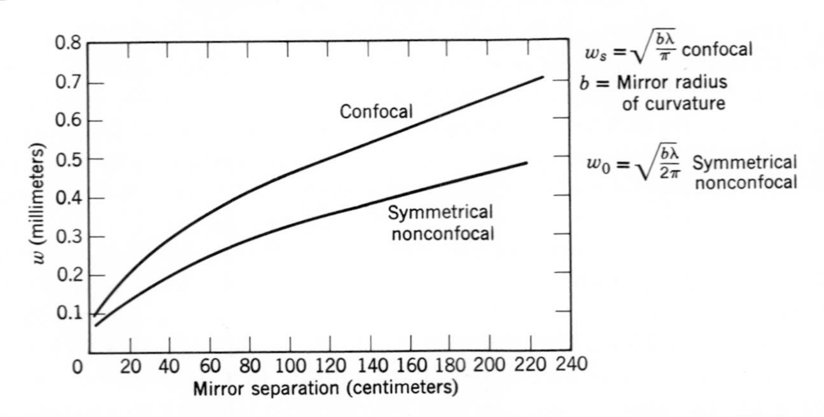

Figure 5.17. Laser spot size at the center of a spherical-mirror resonator, $\lambda = 632.8$ nm.

of good mode discrimination and mechanical stability. Practical g values range from 0.5 to 0.8.

5.5.2 Mirror Reflectivity

The second parameter of interest in cavity losses is that of mirror transparency. There are two bounds for most lasers. If the mirrors are perfectly reflecting, the laser will oscillate strongly but will produce no useful output power (neglecting scattering from windows and mirrors).

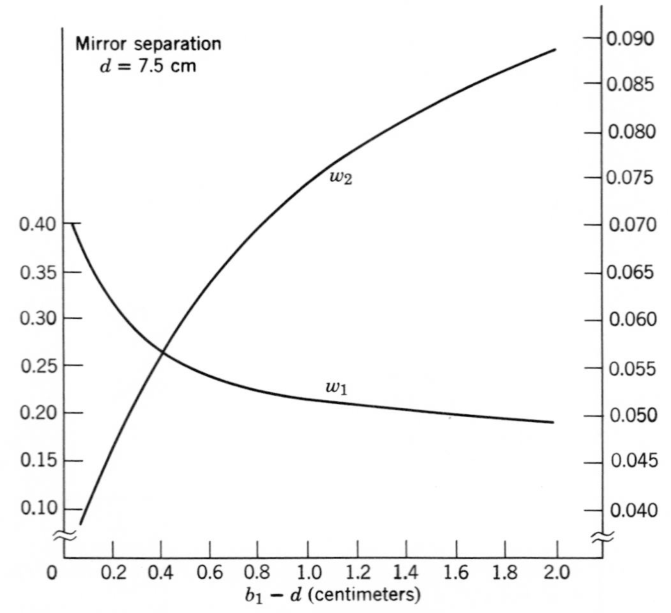

Figure 5.18. Laser-spot size in a hemispherical cavity, $\lambda = 632.8$ nm.

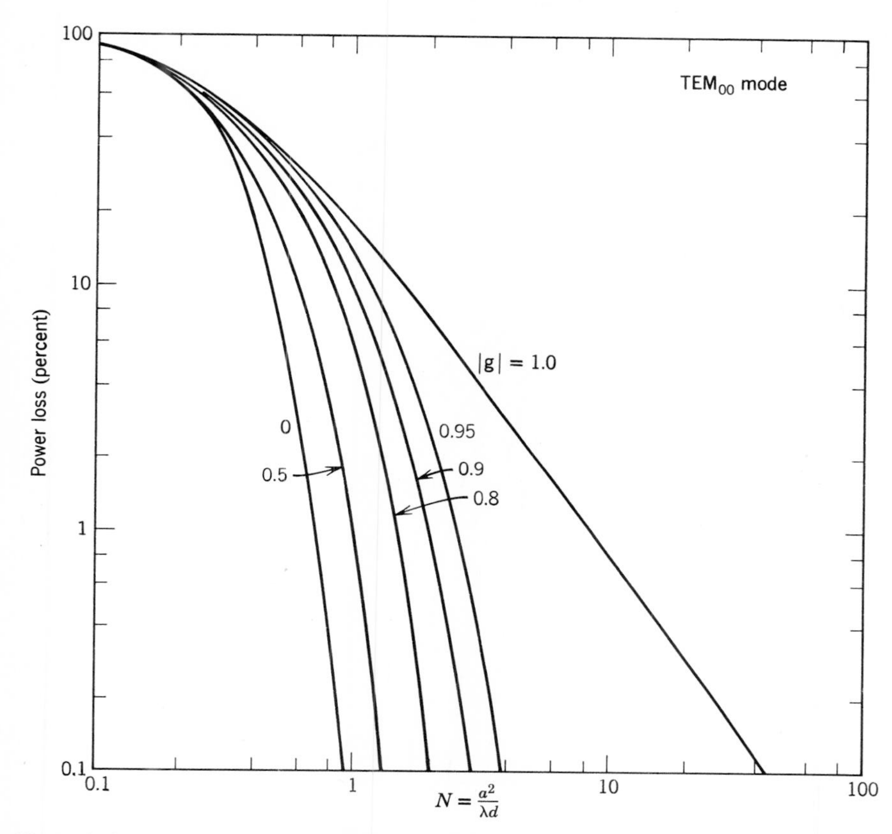

Figure 5.19. Power loss per transit of the fundamental (TEM_{00}) mode for the symmetric geometry.

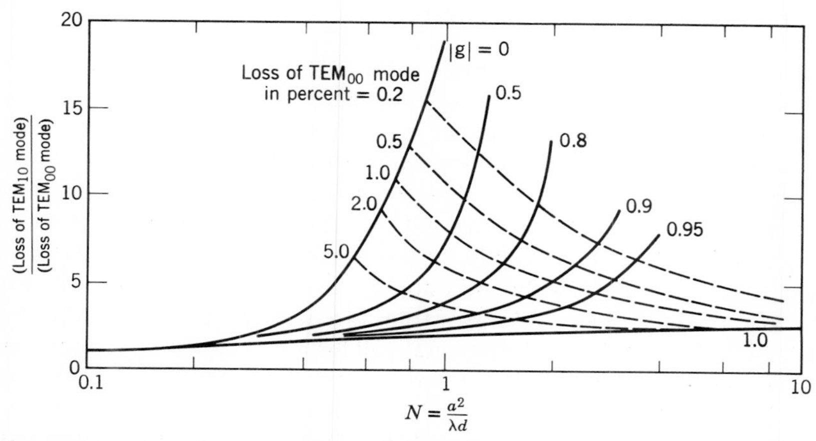

Figure 5.20. Ratio of the losses per transit of the two lowest-order modes for the symmetric geometry, the dotted curves are contours of constant loss for the TEM_{00} mode.

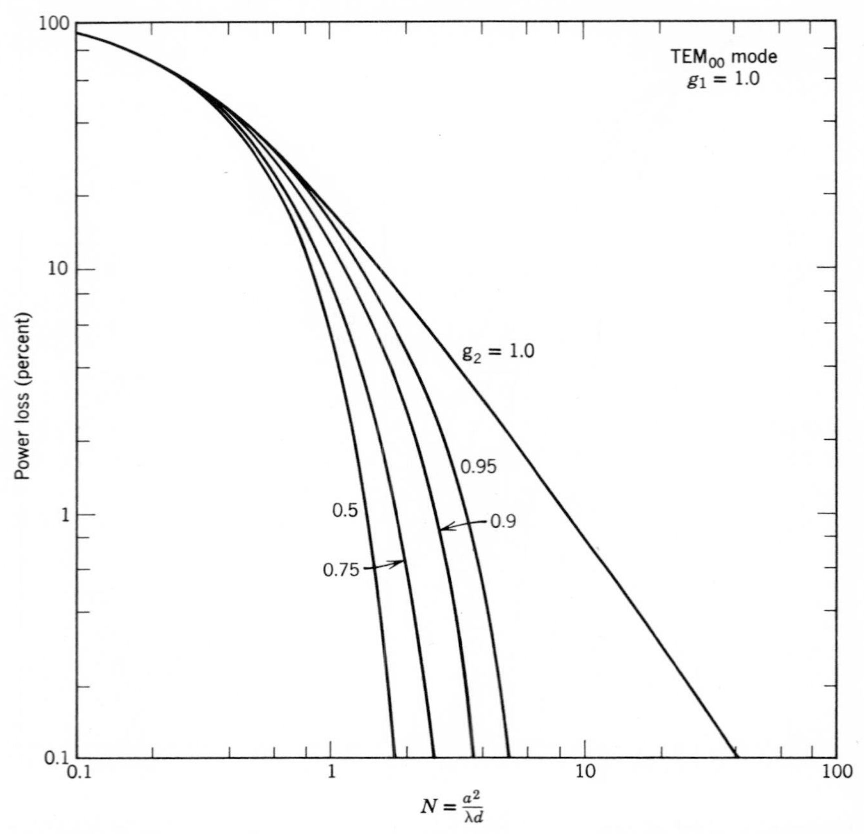

Figure 5.21. Average power loss per transit of the fundamental (TEM_{00}) mode for the half-symmetric geometry.

If the mirror reflectivity is too low, the reflection losses will exceed the net gain and the laser will not oscillate. Between these two extremes is a value of reflectance that maximizes power output.

The development of an analytical expression for that mirror transparency which will yield maximum power output has been subject to much investigation[103–108]. The selection of optimum mirror transparency has been reduced, for the much studied 633 nm transition in a helium-neon gas laser, to series of graphs and equations[107, 108]. These at least enable us to select a reduced range of mirror transparencies, that approach the ideal. Agreement between the experimentally determined performance and these calculations is better than that obtained by previous treatments[103, 104], but the discrepancy is not negligible. For the most part, mirror transparencies are determined for a given plasma tube configuration by painstaking laboratory measurements using the theoretical treatments as a guide.

The laboratory effort required to determine optimum transparency can be considerably reduced if the "variable-reflectivity" mirror technique is used. This method of measuring the mirror transparency, at which maximum power output is obtained, consists basically of measuring the power output from three reflectors simultaneously and analyzing the experimental results. One reflector serves as the output mirror. The other two serve as the surfaces of an optical flat; they are placed within the laser cavity and, first, are orientated at the Brewster angle. As the flat is rotated away from the Brewster angle, as shown in Figure 5.23, power will be coupled out of the cavity. The angle at which maximum total power is obtained can be related to the mirror transparency, as is outlined below[109, 110].

To analyze the results of the experiment described above, let I_0 represent the laser-beam intensity in the cavity as it enters the rotatable reflector R_m, T_m be the respective reflectivity and transmissivity of the exit mirror, and R by the reflectivity of the flat. From an analysis of Figure 5.23 it is easy to show that P_c, the power coupled out of the exit mirror, is

$$P_c = I_0 T_m (1 - R)^2. \tag{5.172}$$

The reflection coefficient at the surface boundary of a Fresnel reflector placed at an angle θ is

$$R(\theta) = \frac{\tan^2 [\theta - \sin^{-1}(1/n_g)\sin\theta]}{\tan^2 [\theta + \sin^{-1}(1/n_g)\sin\theta]} \tag{5.173}$$

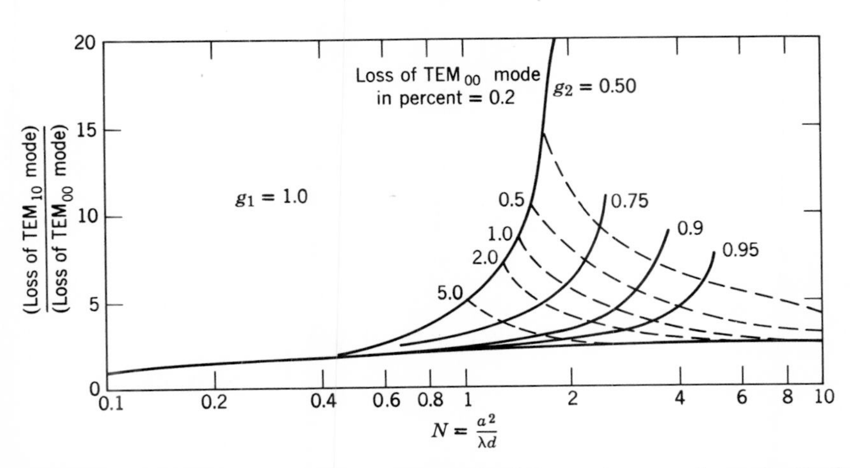

Figure 5.22. Ratio of the average losses per transit of the two lowest-order modes for the half-symmetric geometry. The dotted curves are contours of constant average loss for the TEM_{00} mode.

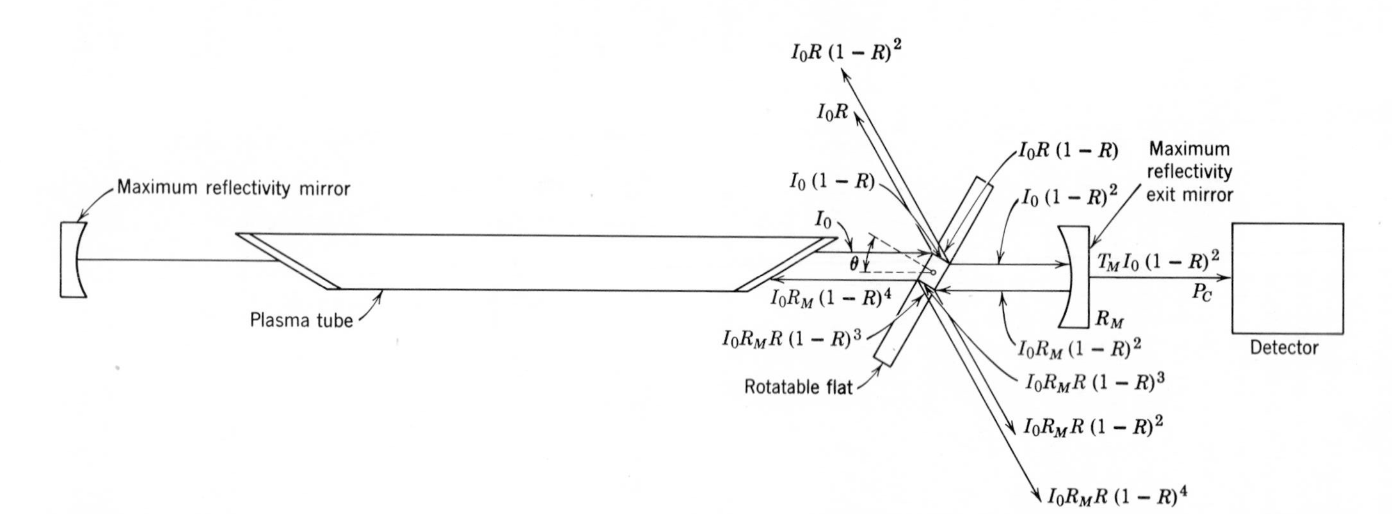

Figure 5.23. Apparatus for measuring the output power versus the Fresnel reflector rotation angle. (Reproduced with permission from the Institute of Electrical and Electronics Engineers.)

where n_g is the refractive index of the flat. The output power of a laser, having an exit mirror reflectivity equivalent to the combination of the high-reflectivity exit mirror and the rotatable reflector at some angle θ, is

$$P_s = \frac{P_c}{T_m} \frac{[1 - R_m(1 - R)^4]}{[1 - R^2]}. \quad (5.174)$$

The equivalent mirror reflectivity R_E is related to the actual exit mirror reflectivity as

$$\frac{R_E}{R_m} = [1 - R(\theta)]^4. \quad (5.175)$$

Figure 5.24 plots (5.73) for an exit mirror of approximately 99 percent reflectivity[110].

Figure 5.25 illustrates good agreement between the theory (equation 5.174) and measurement (for angles near Brewster's angle) for a high-power, pulsed, helium-neon laser[109].

The techniques outlined in Section 5.5 thus far have assumed that the output-power coupling and diffraction problem are independent. However, the cavity design can be considered from the point of view that the output power is obtained by diffraction coupling, and the diffraction loss of the cavity can be converted into useful output power[111]. The basic disadvantage of this technique, for near-field applications, is that the output beam is annular and must be converted to a spot with appropriate optics. The technique shows promise for lasers where a high total-output energy is of interest and the beam shape is of less importance. The extension of this technique to unstable resonators may prove of value in the transverse-mode control of high gain, solid-state lasers in which the active medium has less than ideal optical quality[98].

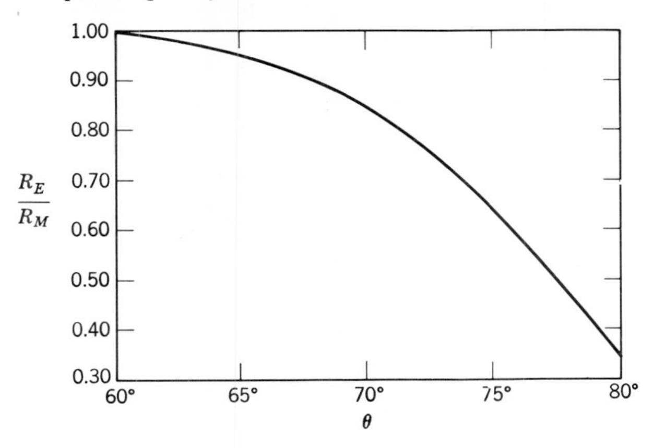

Figure 5.24. Normalized equivalent reflectivity versus angle of incidence. (Reproduced with permission from the Institute of Electrical and Electronics Engineers.)

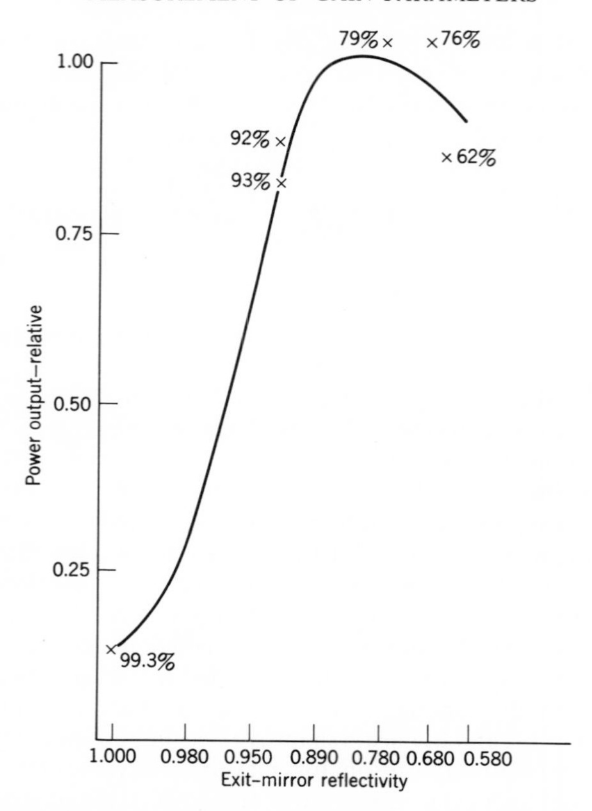

Figure 5.25. Variation in coupled-output power versus mirror reflectivity using a pulsed-infrared, helium-neon laser [109]. (Reproduced with permission from the Institute of Electrical and Electronics Engineers.)

Another technique of output coupling has recently found use in the far infrared. Sometimes referred to as hole or disc coupling, the method employs a centrally located hole or an elliptical metal disc rotated at approximately 45 degrees to remove energy from the optical cavity. Whereas hole coupling discriminates against the lowest order TEM_{00q} mode, disc coupling discriminates against modes other than TEM_{00q}. Both techniques are well suited to CW-gas lasers operating in the power range above 10-100 W where the heating of high-quality dielectric mirrors becomes intolerable.

5.5.3 Measurement of Mirror Reflectivity

The reflectivity of mirrors is usually inferred from measurements of transmissivity for ≤ 0.99. In conventional measuring systems the loss

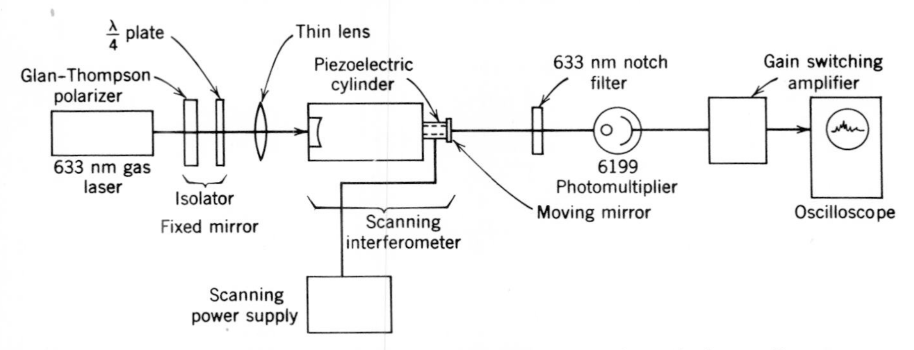

Figure 5.26. Scanning interferometer for measuring large values of mirror reflectance.

(or reflectance) is calculated by comparing the electrical output of a photodetector for two different optical conditions, first with the unknown in the system and then with the unknown removed. As the magnitude of the optical loss decreases, the technique, which essentially measures small differences between the large photodetector outputs, becomes more and more inaccurate. A number of methods[112] have been devised to decrease errors (which are principally optical) in these measuring systems, however, the comparison method still lacks the desired accuracy as $R \to 1$.

An approach relying on a frequency-spectrum analyzer that can be operated at optical frequencies has been developed to make reflectance measurements as high as 0.995 with a reproducibility of $\pm$ 0.001[113]. The measuring technique utilizes the equipment arranged as shown in Figure 5.26. When spherical mirrors are used in the interferometer, the axial ray of incident light must be aligned with the centers of the curvature of the mirrors. (A thin lens can be used to vary the incident angle so that the cavity can be aligned and minimize off-axis modes[114, 115].) The periodic transmission of the scanning interferometer, properly isolated from the gas laser, as shown, is employed on an oscilloscope via a phototube and a narrow-band (20 nm at 633 nm) filter. The photocell is followed by a gain-switching amplifier that reduces the signal output 6 dB at a 1MHz clock rate. The oscilloscope presentation thus yields, on a 20 Hz sweep basis, a display of the cavity finesse. The theory of the method follows[113].

Define the following symbols:

R_i, R_j = power reflectance of the two mirrors expressed as a ratio.
g = power loss per single pass through any material within the cavity, also expressed as a ratio.
t_c = fundamental pulse-group spacing.
t_p, t_m = half-power, output-pulse width, pulse width.
v = velocity of the moving mirror.
d_o = mirror spacing.

t = time.
λ = free-space optical wavelength.
c = free-space velocity of light.

$$\beta_0 = \frac{4\pi d}{\lambda}\frac{v}{c}.$$

$$x = g\sqrt{R_i R_j}.$$

$$\alpha = \frac{4\pi v t}{\lambda}.$$

$$g_0 = [g(1 - R_i)(1 - R_j)]^{1/2}.$$

Then

$$I_t \cong \left[\sum_{n=0}^{\infty} g_0 x^n \cos(\alpha n + \beta_0 n)^2\right]^2 + \left[\sum_{n=0}^{\infty} g_0 x^n \sin(\alpha n + \beta_0 n)^2\right]^2. \quad (5.176)$$

As $x \to 1$, the number of terms becomes large for a maximum error of ϵ.

The first-order effect of mirror velocity is to decrease the maximum response, increase the half-power, pulse width and make the response unsymmetrical about the maximum response. It can be shown that for a 20 Hz linear-mirror drive the first-order velocity effects can be neglected if the optical losses are 0.5 percent or greater.

The relation between optical losses and finesse, neglecting mirror velocity effects, becomes

$$g(R_i R_j)^{1/2} = 1 - \frac{\pi t_p}{t_c} + \frac{1}{2}\left(\frac{\pi t_p}{t_c}\right)^2 + \cdots \quad (5.177)$$

where t_c/t_p is defined as the finesse. Thus the interferometer method measures how much the combined loss $g(R_i\ R_j)^{1/2}$ deviates from unity. For $x = g(R_i\ R_j)^{1/2} \cong 0.99$, a 10 percent experimental error contributes only .001 to the value of optical loss.

The fraction of the incident light transmitted through the etalon is given by[113]

$$I_m = \frac{g(1 - R_i)(1 - R_j)}{[1 - g(R_i R_j)^{1/2}]^2}. \quad (5.178)$$

If $g = 1$ and $R_i = R_j$, $I_m = 1$ for any reflectance. If, however, $R_i \neq R_j$, I_m will be less than unity. For example if $R_i = 0.995$, $R_j = 0.97$, and $g = 1$, $I_m = 0.49$, and 51 percent of the light will be returned to the source laser. Obviously, $R_i = R_j$ is a desirable operating mode.

The time-dependent shape of the etalon transmission is given by[113]

$$I(t) = \frac{I_m}{(1 + 4t^2/t_p{}^2)}, \tag{5.179}$$

and the frequency spectrum of this pulse is

$$F(\omega) = \frac{\pi}{2} t_p t_m \exp\left(-\frac{t_p}{2} |\omega|\right). \tag{5.180}$$

The value of t_p may be chosen such that, for a 1 percent total cavity loss, $t_p = 30\ \mu$ sec, and the frequency spectrum is 1 percent below its low frequency value at 50 kHz. The required band width of the photodetector and the associated circuits is increased by a decrease in the total optical loss, for this decreases the pulse width t_p. A system capable of measuring losses of 0.1 percent must have an overall 1 percent bandwidth of 1 MHz.

The method of mirror reflectance measurement described above, though sensitive, is subject to the usual frailties of a long interferometer. The loss-measuring technique is quite sensitive to building and floor vibrations, as well as air currents. Small random vibrations in the mirror spacing both in the laser and in the interferometer cavity, will cause the output pulse to have large time-position variation on the oscilloscope and will preclude accurate measurements.

5.5.3.1 Comparison of Mirror Losses with Measured Reflectivity A confocal Fabry-Perot cavity 300 ft long has been used to measure the difference between the measured reflectance of two 99.5 percent reflectivity mirrors and the value determined from the decrement of a trapped light beam[116]. The cavity was designed for a Fresnel number of 2. Diffraction losses, both as calculated and measured, were small compared with the observed 2 percent loss per round trip. Sampling the decrement with light transmitted through one end mirror, it was possible to examine the internal-beam amplitude after several hundred reflections. Measurements indicated that the beam decrement was not due to the transmission medium (air) but that a 1 percent loss occurred on each reflection. A loss of 0.5 percent was attributed to scattering of coherent light upon reflection from the mirror surface.

5.6 REFERENCES

[1] H. G. Heard, *Nature*, **200**, 667 (1963).
[2] H. G. Heard, *Proc. IEEE*, **53**, 173 (1965).
[3] H. G. Heard, *Bull. Am. Phys. Soc.*, Paper FE4 (1965).
[4] W. Kaiser and M. J. Keck, *J. Appl. Phys.*, **33**, 762 (1962).
[5] A. Yariv and J. P. Gordon, *Proc. IEEE*, **51**, 4 (1963).
[6] A. Einstein, *Phys. Z.*, **18**, 121 (1917).
[7] C. H. Corliss and W. R. Bozman, *Experimental Transition Probabilities for Spectral Lines of Seventy Elements*, Natl. Bur. Std. Monograph No. 53 (1961).
[8] R. W. Ditchburn, *Light*, Interscience, New York, 1957, 1st ed., pp. 455, 460, and 576.
[9] A. C. G. Mitchell and M. W. Zemansky, *Resonance Radiation and Excited Atoms*, Cambridge University Press, London and New York, 1961.
[10] A. L. Schawlow and C. H. Townes, *Phys. Rev.*, **112**, 1940 (1958).
[11] D. L. Dexter, *Solid State Phys.*, **6**, 353 (1958).
[12] R. H. Silsbee, *Phys. Rev.*, **128**, 1726 (1962).
[13] D. E. McCumber and M. D. Sturge, *J. Appl. Phys.*, **34**, 1682 (1962).
[14] F. C. Hoyt, *Phys. Rev*, **36**, 860 (1930).
[15] F. Reiche, *Verh d. D. Phys. Gas.*, **15**, 3 (1913).
[16] E. I. Gordon, A. D. White, and L. D. Rigden, Proceedings Symposium on Optical Masers, Polytechnic Press, Brooklyn, N.Y., 1963, Vol. XIII, p. 309. (1963).
[17] J. W. Klüver, *J. Appl. Phys.*, **37**, 2987 (1966).
[18] W. R. Bennett, *Phys. Rev.*, **126**, 580 (1962).
[19] C. K. N. Patel, W. L. Faust, and R. A. McFarlane, *Quantum Electronics* in P. Grivet and N. Bloembergen (eds.), Paris 1963 Conference, Columbia University Press, New York, 1964, p. 508.
[20] C. K. N. Patel, W. L. Faust, and R. A. McFarlane, *Appl. Phy. Letters*, **1**, 84 (1962).
[21] H. G. Heard and J. Peterson, *Proc. IEEE*, **52**, *1049* (1964).
[22] H. A. H. Boot, D. M. Clune, and R. S. A. Thorn, *Nature*, **198**, 773 (1963).
[23] L. E. S. Mathias and J. T. Parker, *Appl. Phy. Letters*, **31**, 16 (1963).
[24] H. G. Heard and J. Peterson, *Proc. IEEE*, **52**, 1258 (1963).
[25] W. T. Watler, *et al.*, (to be published).
[26] W. Doyle, *J. Appl. Phys.*, **35**, *1348* (1964).
[27] C. B. Zarowin, M. Schiff, and G. R. White, *Proc. of Sym. Opt. Masers*, Polytechnic Press, Brooklyn, 1963, Vol. XIII, p. 425.
[28] A. Kastler, "Atoms a L'Interieur d'un Interferometre Perot-Fabry," *Appl. Optics*, Supplement on Optical Masers (1962).
[29] S. E. Frish and O. P. Bochkova, *Fiz. i Khim.*, **6**, 40 (1961); Corrections in *Fiz i Khim.* **1**, 173 (1962). (An English translation of Frish and Bochkova's article has been published by AFCRL office of Aerospace Research, Hanscom Field, Bedford, Mass. Performed under Contract No. AF 19(628)-2924.
[30] P. Mauer, *Appl. Optics*, **3**, 433 (1964).
[31] B. A. Lengyel, *Lasers*, Wiley, New York, 1962, pp. 10-12.
[32] R. L. Fock, et al., *Bull. Phys. Soc.*, II, **8**, 380 (1963).
[33] H. Kogelnik, *Bur. Std. J. Res.* (U.S.), **44**, 455 (1965).
[34] H. Kogelnik, *Bur. Std. J. Res.*, **43**, 334 (1964).
[35] D. R. Herriott and H. J. Schulte, *Appl. Optics*, **4**, 883 (1965).
[36] G. D. Boyd and J. P. Gordon, *Bur. Std. J. Res.* (U.S.), **40**, 489 (1961).
[37] T. S. Chu, *Bur. Std. J. Res.* (U.S.), **45**, 287 (1966).
[38] S. A. Collins, Jr., *Appl. Optics*, **3**, 1263 (1964).

[39] T. Li, *Appl. Optics*, **3**, 1315 (1964).
[40] G. A. Deschamps and P. E. Mast, Proceedings Symposium on Quasi Optics, Polytechnic Press, Brooklyn, N.Y., 1964.
[41] J. P. Gordon, *Bur. Std. J. Res.* (U.S.), **43**, 1826 (1964).
[42] L. M. Frantz and J. S. Nodvik, *J. Appl. Phys.*, **34**, 2346 (1963).
[43] A. E. Siegman, *J. Appl. Phys.*, **35**, 460 (1964).
[44] G. L. Clark, *High Power Laser Amplifier Chain Techniques*, RADC, Tech. Rept. No. RADC-TR-64-516 (1965).
[45] W. E. Bennett, Jr., *Phys. Rev.*, **126**, 580 (1962).
[46] E. I. Gordon, A. D. White, and J. D. Rigden, *Optical Masers*, Proceedings Symposium on Optical Masers, Polytechnic Press, Brooklyn, N. Y., 1963, Vol. XIII, p. 309.
[47] A. L. Bloom, W. E. Bell, and R. E. Rempal, *Appl. Optics*, **2**, 317 (1963).
[48] H. Bateman, *Tables of Integral Transforms*, McGraw-Hill, New York, 1954, Vol. I, p. 136.
[49] J. S. Wright and E. O. Schulz du Bois, Bell Telephone Laboratory Report No. 5, Contract DA-36-039-SC-85357, 1961.
[50] D. F. Hotz, *Appl. Optics*, **4**, 527 (1965).
[51] W. W. Rigrod, *J. Appl. Phys.*, **24**, 2602 (1963).
[51*a*] E. F. Labuda and E. I. Gordon, *Microwave Determination of Average Electron Energy and Density in He-Ne Discharges*, BTL Report
[52] P. Parzen and L. Goldstein, *Phys. Rev.*, **79**, 190 (1950).
[53] G. Bekefi and S. C. Brown, *Am. J. Phys.*, **29**, 404 (1961).
[54] S. Aisenberg, *Appl. Phys. Letters*, **2**, 187 (1963).
[55] S. Kojima and K. Takayama, *J. Phys. Soc.* (Japan), **4**, 346 (1949); E. O. Johnson and L. Matter, *Phys. Rev.*, **80**, 58 (1950).
[56] S. Aisenberg, XXIII Conference on Physical Electronics, M.I.T., 1963.
[57] J. D. Cobine, *Gaseous Conductors*, Dover, New York, 1958, esp. pp. 135–138.
[58] W. Demtroder, *Z. Phys.*, **166**, *42* (1962).
[59] T. M. Holzberlein, *Rev. Sci. Instr.*, **35**, 1041 (1964).
[60] S. Heron, R. W. P. McWhirter, and E. H. Rhoderick, *Proc. Roy. Soc.* (London), **A234**, 565 (1956).
[61] R. G. Bennett and F. W. Dalby, *J. Chem. Phys.*, **31**, 434 (1959).
[62] B. M. Glennon and W. L. Wiese, *Bibliography on Atomic Transition Probabilities*, National Bureau of Standards, Washington, 1962.
[63] H. R. Griem, *Plasma Spectroscopy*, McGraw-Hill, New York, 1964.
[64] J. R. Dixon and F. A. Grant, *Phys. Rev.*, **107**, 118 (1957).
[65] R. Ladenburg, *Rev. Mod. Phys.*, **5**, 243 (1933).
[66] N. Marcuvitz, *Waveguide Handbook*, McGraw-Hill, New York, 1951, Vol. 10, p. 266.
[67] L. Goldstein, M. A. Lapert, and R. H. Geiger, *Elec. Comm.*, 243 (1952).
[68] D. F. Holcomb and R. E. Norberg, *Phys. Rev.* **98**, 1074 (1955).
[69] A. V. Phelps, *Phys. Rev.*, **114**, 1011 (1959); A. V. Phelps and J. P. Molnar, *Phys. Rev.*, **89**, 1202 (1953).
[70] R. C. Tolman, *Phys. Rev.*, **23**, 693 (1924).
[71] G. N. Lewis and M. Kasha, *J. Am. Chem. Soc.*, **67**, 994, (1945).
[72] M. L. Baumik and C. L. Telk, *J. Opt. Soc. Am.*, **54**, 1211 (1964).
[73] M. L. Baumik, *et al.*, *Rev. Sci. Inst.* (to be published).
[74] R. F. Woodcock, *Lifetimes in* N_d^{3+} *Doped Silicate Laser Glases*, O. S. A. Spring Meeting, 1963.
[75] R. J. Carbone and P. R. Longaker, *Appl. Phy. Letters*, **4**, 32 (1964).
[76] A. T. Forrester, R. A. Gudmundsen, and P. O. Johnson, *Phys. Rev.*, **99**, 1691 (1955).

[77] J. Halpern and R. H. Rediker, *Proc. IRE*, **48** (1960).
[78] R. W. Hellwarth, *Advances in Quantum Electronics*, ed. by J. R. Singer, Columbia University Press, New York (1961), p. 334.
[79] F. J. McClung and R. W. Hellwarth, *J. Appl. Phys.*, **33**, 828 (1962).
[80] R. J. Collins and P. Kisliuk, *J. Appl. Phys.*, **33**, 2009 (1962).
[81] L. M. Frantz, S. T. L. Report No. 9801-6004-RV-000, 1962.
[82] A. A. Vuylsteke, *J. Appl. Phys.*, **34**, 1615 (1963).
[83] W. G. Wagner and B. A. Lengyel, *J. Appl. Phys.*, **34**, 2040 (1963).
[84] A. A. Vuylsteke, ASD-TDR-63-812, Part 1, 1963.
[85] I. Hodes, ASD-TDR-63-812, Part 2, 1963.
[86] W. R. Bennett Jr., *Chemical Lasers*, Appl. Optics, Supplement No. 2, 3 (1965).
[87] G. D. Boyd and J. P. Gordon, *Bur. Std. J. Res.* (U.S.), **40**, 489 (1961).
[88] G. D. Boyd and H. Kogelnik, *Bur. Std. J. Res.* (U.S.), **41**, 1347 (1962).
[89] A. G. Fox and T. Li, *Proc. IEEE*, **51**, 80 (1963).
[90] W. Streifer, *J. Opt. Soc. Am.*, **54**, 1399 (1964).
[91] D. Gloge, *Arch. Elkt. Ubertragung*, **18**, 197 (1964).
[92] J. C. Heurtley, *J. Opt. Soc. Am.*, **54**, 1400 (1964).
[93] A. G. Fox and T. Li., *Bur. Std. J. Res.* (U.S.), **40**, 453 (1961).
[94] T. Li, *Bur. Std. J. Res*, **42**, 2609 (1963).
[95] L. A. Vainshtein, *Soviet Phys.-JETP*, **44**, 1050 (1963); also *Soviet Phys-JETP*, **17**, 709 (1963).
[96] J. T. Latourette, S. F. Jacobs, and P. Rabinowitz, *Appl. Opt.*, **3**, 981 (1964).
[97] S. R. Barone and M. C. Newstein, *Appl. Opt.*, **3**, 1194 (1964).
[98] A. E. Siegman, *Proc. IEEE*, **53**, 277 (1965).
[99] J. P. Gordon and H. Kogelnik, *Bur. Std. J. Res.*, **43**, 2873 (1964).
[100] D. C. Sinclair, *Appl. Opt.*, **3**, 1067 (1964).
[101] P. O. Clark, *Proc. IEEE*, **53**, 36 (1965).
[102] R. F. Soohoo, *Proc. IEEE*, **51**, 71 (1963).
[103] W. W. Rigrod, *J. Appl. Phys.*, **34**, 2602 (1963).
[104] A. D. White, E. I. Gordon, and J. D. Rigden, *Appl. Phys. Letters*, **2**, 91 (1963).
[105] E. Spiller, *Z. Phys.*, **182**, 487 (1965).
[106] P. W. Smith, *J. Quant. Elec.*, **QE1**, 343 (1965).
[107] P. W. Smith, *J. Quant. Elec.*, **QE2**, 62 (1966).
[108] P. W. Smith, *J. Quant. Elec.*, **QE2**, 77 (1966).
[109] J. Goldsmith, E. H. Byerly, and A. A. Vuylsteke, *Proceedings 8th Convention on Military Electronics*, 1964, p. 64.
[110] E. H. Byerly, G. T. McNice, and J. Goldsmith, *Proceedings 8th Convention on Military Electronics*, 1964, p. 55.
[111] J. T. LaTourette, S. F. Jacobs, and P. Rabinowitz, *Appl. Opt.*, **3**, 981 (1964).
[112] H. E. Bennett and W. F. Koehler, *J. Opt. Soc. Am.*, **50**, 1 (1960).
[113] A. J. Rack and M. R. Biazzo, *Bur. Std. J. Res.* (U.S.), **43**, 1563 (1964).
[114] D. R. Herriott, *Appl. Opt.*, **2**, 865 (1963).
[115] D. R. Herriott, H. Kogelnik, and R. Kompfner, *Appl. Opt.*, **3**, 523 (1964).
[116] O. E. DeLange, *Proc. IEEE*, **51**, 1361 (1963).

5.7. PRINCIPAL SYMBOLS, NOTATIONS, AND ABBREVIATIONS

Symbol	*Meaning*
A	Amplitude of the electromagnetic field, a numeric, see text
A^*	Complex conjugate of A
$A_n^\pm$	Electric-field amplitude in the n axial direction
A_{ji}	Probability of stimulated emission from the state $j \to i$
a	Plasma-tube radius
B	Numerics, see text
B_{ij}	Transition probability for induced absorption
B_{ji}	Transition probability for induced emission
b_i	Radius of curvature of mirror i
c	Velocity of light in free space
d	Cavity length
d_i, d_2	Axial distance along laser beam
$E(\nu)$	Lamp emissivity
e	Charge on the electron
F_σ	Free spectral range
f	Oscillator strength, focal length of lens, see text
f_0	Mode-matching length
G_0	Initial gain of a laser-pulse amplifier prior to receipt of a pulse
G_p	Total-energy gain from a laser-pulse amplifier in a single pulse
g_i	Statistical weight for the state i
$g_L(\nu)$	Normalized Lorentzian-line-shape factor
$g_G(\nu)$	Normalized Gaussian-line-shape factor
g_1, g_2	Normalized curvature parameters, degeneracy of states, see text
h	Planck's constant
$I(\nu)$	Intensity
I_{NR}, $\bar{I}_{NR}$	Output intensity of a nonresonant cavity, average value thereof
I_R, $\bar{I}_R$	Output intensity of a resonant cavity, average value thereof, see text
J_0	Bessel function of zero order
J_e	Probe-electron current density
J_p	Electron current density in the plasma
k	Boltzmann's constant
k_0	Peak value of the absorption coefficient of a gas at the center of the absorption line
k_ν	Integrated attenuation coefficient of a gas

$\ln x$	Natural logarithm of x
M	Mass of atom in amu
N	Fresnel number, volume density of absorbing ions, see text
n	Integer
$\bar{n}$	Average quanta per mode
n_e	Electron density
n_i	dn_i/dt, time rate of change of population density in state i
n_0	Electron density
P_c	Power coupled out of a cavity
P_s	Total power output of a cavity including losses on a rotatable reflectivity
P_{IN}, P_{OUT}	Instantaneous input and output pulse power in a laser-pulse amplifier
$p(v)$	Mode density per unit frequency interval
Q	Ratio of energy stored to energy dissipated per cycle
R_i	Power reflection coefficient for mirror i
R_E	Equivalent power reflectivity of a mirror
R_M	Exit mirror power reflectivity
r_i	Amplitude reflection coefficient of mirror i
r_0	Classical electron-orbit radius
$S_i(v)$	Pumping rate to state i
T	Temperature
T_e	Electron temperature
T_0, T_2	Power-transmission coefficient
t_c	Pulse-group spacing
t_i	Amplitude transmission coefficient of mirror i
t_m	Total pulse width
t_p	Duration of the output pulse at half power
U, U_i	Energy, energy of the state i, see text
$\bar{U}$	Average energy per cavity mode
v	Velocity of light in a medium of index n
V_c	Volume of microwave cavity
V_0	Probe potential
V_p	Volume of plasma charge
W_i'	Induced transition rate
W_0	Total energy stored in inverted states in a laser-pulse amplifier
W_{IN}, W_{OUT}	Integrated input- and output-pulse energies in a laser-pulse amplifier
w_1, w_2	Beam radius at points 1 and 2
α	Normalized net incremental loss in a laser cavity due to all causes

α_D	Diffraction loss coefficient
α_{nl}, α_c	Values of attenuation, see text
α_R	Reflection loss coefficient
β	Dummy variable
Γ_1, Γ_2	Ratio of observables, see text
γ	Incremental gain
γ_c	Peak value of incremental gain at line center
Δv	Incremental change in the variable v
Δv_D	Doppler broadened line width, full width at half-value
Δv_N	Natural line width
δ	Dummy variable shift in the location in the minimum of the standing-wave ratio, see text
$\delta\ (v'-v)$	Dirac delta function operator
ϵ_0	Permitivity of free space
ζ	Absorption coefficient for 1.06 μm light in a solid
ϑ	Dummy variable
η	Pulse-energy-conversion efficiency, quantum efficiency ratio, see text
θ	Cavity-emission angle
$\Lambda\ (P_1\ T)$	Lumped-loss factor, see text
λ	Wavelength
λ_g	Guide wavelength with the laser tube in a waveguide
λ_{go}	Guide wavelength in vacuum
μ_{ij}, μ_{ji}	Dipole moment matrix element
v	Frequency
v_m	Collision frequency
v_c	Frequency at line center
ξ	Average single-pass gain coefficient for a cavity of length d
$\rho(v)$	Radiation density per unit frequency interval
σ_c	Cross section for electron impact
σ_i	Cross section for ionization
σ_{ij}	Cross section for absorption
σ_{ji}	Cross section for emission
σ_p	Photoelectric cross section
τ_{ji}	Spontaneous, fluorescent, radiative lifetime for the transition $j \rightarrow i$
τ_c	Photon lifetime in an optical cavity
τ_0	Observed decay time
τ_q	Quenching time
τ_r	Radiative lifetime in a solid
v	Dummy variable, see text

ϕ	Angle with respect to the laser axis of a line of constant phase
ψ_0, ψ'_0	Standing-wave ratios of the microwave cavity
Ω	Dummy variable, see text
ω	Angular frequency $2\pi\nu$
ω_{po}	Plasma frequency
ω_m	Collision frequency for the momentum transfer of electrons with neutrals

6

The Measurement of Wavelength

6.0 INTRODUCTION

No detector is available that is fast enough to respond at optical frequencies. Even the best photomultipliers are response limited in the range of 3×10^9 Hz, whereas the frequency range of interest is $\approx 10^{14}$ Hz. Thus all absolute frequency information is obtained by measuring the wavelength λ and converting to frequency ν, or wave number $\bar{\nu} = 1/\lambda = \nu/c =$ number of wavelengths per centimeter:

$$\lambda_{\text{vac}}\, \nu_{\text{vac}} = c. \tag{6.1}$$

The conversion from the wavelength in air to a vacuum wave number has been tabulated[1].

The fundamental problem in measuring wavelength is the same as that of any length measurement, namely, accurate pointing. To determine a wavelength, a reference is produced (usually a nonlinear one) using a stable, reproducible source as a standard. The reference scale, thus calibrated, is used to measure the unknown radiation. The accuracy of this procedure is ultimately limited by the precision with which we can point to the center of the reference-scale markings and to the trace of the unknown radiation. Precision is achieved by minimizing the width of these tracings, this width being a convolution integral that includes the source function, the instrument function, and the detector function.

In contrast with the x-ray or far-infrared regions, wavelength determination is usually not limited in the optical region of the spectrum by the resolution of the detector. It is possible to design an optical system with sufficient dispersion to use the full resolution of the optics. Ordinary spectroscopic plates and photomultiplier tubes do not introduce appreciable line broadening.

The initial portions of this chapter treat some of the common causes of and ways to minimize errors of line broadening that can be attributed to light sources. Instrumental widths also are treated, including some associated optical-system parameters. (These parameters become significant in the choice of instruments when it becomes necessary to trade resolving power for intensity.) The general procedure of wavelength measurements is treated as well as a discussion of standards of measurements, reduction of data, and precautions to be observed in the use of various instruments. Absolute determination of wavelength and resolution of spectroscopic structure will also be discussed.

6.1 LINE BROADENING AND LIGHT SOURCES

6.1.1 Causes of Line Broadening

Depending upon pressure, several phenomena interact to provide a lower limit to the band width that may be attained with conventional spectroscopic light sources. The natural line width of a source is the theoretical line width of the radiation reaction on a classically emitting dipole[2]. The quantum mechanical description associates the natural line width with the transition probability to lower-lying states[3]. In principle the limiting bandwidth may be expressed in terms of the uncertainty principle

$$\Delta U \Delta t = \text{const.} \tag{6.2}$$

For atomic lifetimes of 10^{-7} sec, $\Delta \nu \approx 10^{7}$ Hz; as $\nu \sim 10^{14}$ Hz at optical frequencies, relatively narrow fractional linewidths become evident. The argument above must be carefully interpreted; for example transitions from a metastable level to a ground state might be expected to have an excessively narrow natural linewidth. This phenomena is seldom observed in incoherent emission spectroscopy, for it is masked by more dominant effects. Because the natural line profile is Lorentzian, account must be taken of the large wings of the resonance profile in absorption spectroscopy.

In most gas lasers the operating pressure is a few milliTorr, so that the major contribution to line broadening is the Doppler shift of the emitted radiation. A cursory description of a gaseous-discharge source, possessing a velocity spectrum given by the Maxwell-Boltzmann distribution, leads to an intensity distribution of the form[4].

$$I(\nu) = \text{const.} \exp\left[-\frac{\beta^2 c}{\nu^2} (\nu - \nu_c)^2 \right]. \tag{6.3}$$

Here $\beta = M/2RT$ where M is the molecular weight, R is the gas constant, and T is the absolute temperature. The full width at the half-amplitude of this profile is given by

$$\Delta\nu_D = \left(1.67\,\frac{\bar{\nu}}{c}\right)\frac{2RT}{M}. \tag{6.4}$$

This suggests that we must work at low temperatures to decrease the Doppler width and use massive elements. The extremely monochromatic Γ radiation obtained in the Mossbauer effect[5] is a special example of the use of massive emitters; the emitting nucleus has been coupled to a host matrix, so that the mass of the emitter is effectively that of the entire host crystal. Similar effects have been obtained in gases.

The phenomenon of self-absorption[6], which occurs when radiation is absorbed and re-emitted many times before it escapes from the radiating gas, can cause marked line broadening. Because the transition probability is highest for the particular radiation at frequency $(U_2 - U_1)/h$, these quanta are preferentially absorbed and partially trapped in the light source. Self absorption leads to a depression of the intensity at the center of the line profile and a consequent broadening of the intensity half-width. For larger oscillator strength, the radiation-trapping efficiency is sufficient to cause absorption of the entire center of the line. This self-reversal of the line yields an apparent spectroscopic doublet.

Hyperfine structure and isotope shift effects[7] can often lead to the appearance of a broadened spectral line. These effects are due respectively to the magnetic and electric interactions of a nucleus with its surrounding electron fields. In the case of the magnetic-dipole interaction, the degeneracy of single-energy levels is split into a number of levels, the total number of which depend upon the net angular momentum of the system. If this splitting is less than, or on the order of, the Doppler width, the consequent structure is not resolved, and radiation emitted by the system of levels appears as a broadened symmetrical line. Similarly, electrostatic interaction depends on the nuclear-charge radius. Because this parameter differs for each isotope of the same element, the emitted radiation will be a combination of the radiation of each isotope. The emitted radiation will be slightly shifted in frequency and will produce a broadened, unresolved line. These effects are typically of the order of 0.1 cm^{-1}. Isotopic broadening effects may be eliminated if one can choose a single isotope having a zero nuclear spin. Other causes of line broadening include interatomic collisions and the Stark effect[8].

Collision broadening effects play a role whenever the mean time between collision becomes shorter than, or on the order of, the atomic lifetime. Another result of interatomic collisions is the production, during the collision time, of large, inhomogeneous electric fields (due to interplay between the two electron clouds). This gives rise to a splitting of the energy levels (Stark effect), which displays itself as an asymmetrical broadening and a shift of the spectral line. This effect will be more pronounced for nonpenetrating orbits (with a high angular momentum), inasmuch as penetrating orbits are partially shielded by their own electron clouds. Both the effects described above are only prominent in high-pressure light sources, such as an arc (or spark) in the air. An exhaustive treatment of the natural causes of line broadening is given by Breene[9].

There exists the possibility of line broadening caused by improper instrumentation. It is possible, for example, to operate a spectrograph or spectrometer with a wide slit (and thereby to introduce appreciable errors in the measured wings of a line) in an attempt to speed the recording of data.

Consider the case of a prism spectrograph used to record the line spectrum of a high-pressure arc lamp which also generates a continuum[10]. Noninfinitesimal slit widths cause a disproportionate increase in the areas under the wings of a line compared to the continuum when the slit widths are of the same order of magnitude as the widths of the lines[11–18].

There are two sources of error in a two-slit system. A wide-entrance slit transmits a broad spectrum; a broad exit slit intercepts the broad spectrum and integrates it, causing further errors in the signal transmitted to the detector.

An analysis of the problem above is aided if we construct a model of the entrace slit composed of a large number of parallel, infinitesimal slits, each of which produces its own highly monochromatic spectrum at the focussing element. The spectrum of the wide slit is then visualized as an aggregate of the overlapping spectra of each individual slit. In a system with one-to-one magnification (by the collimating and focussing elements), the spectra produced by the imaginary slits are displaced with respect to one another by the same distances as the relative displacements of imaginary slits.

Consider the focal plane irradiance produced by two of the infinitesimal slit widths located at the extremes of the entrance-slit jaws. If the slit width is ϕ and if the plate factor is S (and is constant over the spectral region of the imaginary slits), the peaks of the spectral lines will be separated by ϕS. Between these two positions there is a continuous distribution of identical spectra produced by all the other infinitesimal slit widths. Thus the spectral-energy distribution at the focal plane is the sum of the infinite number of completely resolved spectra.

The spectral irradiance at the focal plane H_2 may be represented by an integral of the true spectral irradiance H_1 over the wavelength range ϕS

$$H_2 = \int_{\lambda - \phi S/2}^{\lambda + \phi S/2} H_1 \, d\lambda. \tag{6.5}$$

Let the mathematical model of a pressure-broadened emission line centered at $\lambda = \lambda_0$, having a half-intensity width of $2\sqrt{\vartheta_2}$ and a peak value of ϑ_1, be

$$H_1 = \frac{\vartheta_1 \vartheta_2}{\vartheta_2 - (\lambda - \lambda_0)^2}. \tag{6.6}$$

The second integral of (6.5) can be used to compare, over the spectral range $\lambda_a < \lambda < \lambda_b$, the effect of the entrance-slit width on the growth of areas under lines to the areas under adjacent continua for various values of ϕS

$$\psi_1 = \int_{\lambda_a}^{\lambda_b} H_2 \, d\lambda. \tag{6.7}$$

The effects of a nonzero-exit-slit width can be represented by the integral

$$H_3 = \int_{\lambda - (\phi S/2)}^{\lambda + (\phi S/2)} H_2 \, d\lambda \tag{6.8}$$

over the exit-slit-width. Equation (6.8) is formally the same as 6.7 except for the limits of integration.

Finally the combined effect of the finite entrance- and exit-slit widths on the growth of areas under lines to areas under adjacent continua in the wavelength range $\lambda_a \leqq \lambda \leqq \lambda_b$ for various ϕS can be obtained from the integral.

$$\psi_2 = \int_{\lambda_a}^{\lambda_b} H_3 \, d\lambda. \tag{6.9}$$

Figure 6.1 illustrates the relation between $H_3\lambda$ and the wavelength for parametric values of ϕS. These curves are normalized to the same value for $\phi S = 0$, which corresponds to the undistorted line shape. If the line area were not distorted by wide slits, all the curves in Figure 6.1 would grow at the same rate with an increase in ϕS; all curves would coincide. Thus the slit width that is chosen must be based upon tests of performance that demonstrate unequivocally that the instrument is not contributing to the line width or that it is operating at its maximum resolution and the line width is instrument limited.

A convenient experimental way to determine the ultimate resolution of the spectrometer is to measure the apparent line width of a gas laser operating in the wavelength range of interest. Because the spectrometer cannot resolve the line width of the gas laser, the observed line width will be instrument limited for even the smallest entrance- and exit-slit widths.

All of the considerations above apply to thermal light sources, where the dominant mode of decay is spontaneous emission. For laser light, induced emission is the dominant radiation process, and the theoretical bandwidth

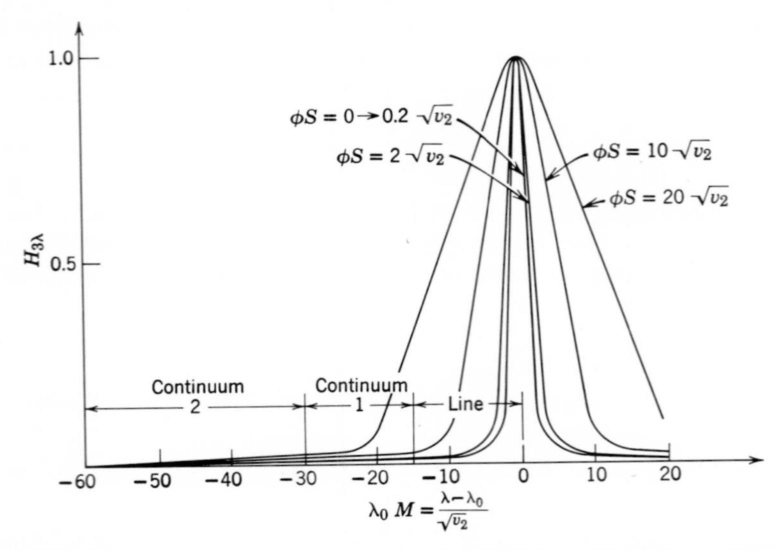

Figure 6.1. The measurement of the irradiance intercepted by the exit slit versus the wavelength for various spectral slit widths. (Reproduced with permission from *Applied Optics*.)

of a gas laser is governed only by the Q of the resonant cavity. Theoretical bandwidths based on simple theory predict band widths as narrow as a few cycles; they are not achieved in practice because mirror micromotion and laser medium nonlinearities interfere. To render the radiation highly monochromatic, it is necessary to operate the laser in a single transverse and a single longitudinal mode.

6.1.2 Monochromatic Light Sources

Many types of gaseous light sources have been constructed to produce radiation with a small band-width, a reproducible central frequency, and symmetric line shape. All of these have the common feature that they operate at sufficiently low pressures ($<$ 1 Torr) with the result that collision broadening is not a limitation.

Among the more conventional gaseous light souces are the Geissler tube [19] and the electrodeless discharge lamp[20]. The former consists of a low-pressure discharge maintained in a capillary, via electrodes. It is usually powered by a high-voltage, low-current transformer. It can typically attain coherence lengths of 30 mm for moderately heavy elements.

A much more convenient light source than the Geissler tube is the electrodeless discharge lamp. Here a small quantity of the element to be examined is sealed in an evacuated piece of quartz tubing, along with approximately 1 Torr pressure of some noble gas. The discharge is maintained by 2.5 GHz microwave radiation from a directed antenna or matched cavity[21]. Electrodeless discharge tubes have operated with as little as one millimicrogram of material. This means that we can usually obtain a sufficient quantity of a single isotope from the Oak Ridge National Laboratories to construct such a light source and thereby avoid line-width effects due to isotope broadening. An air-cooled electrodeless discharge lamp containing Hg^{198} can produce a coherence length in excess of 200 mm.

Another standard spectroscopic light source is the hollow cathode[22], cooled either by circulating water or a liquid nitrogen bath. This light source is especially free of Stark broadening effects because the plasma is essentially free of electric fields in the discharge. Although the effective ion temperature is not as low as that of the coolant, Doppler widths of .025 cm^{-1} can be attained in practice with liquid nitrogen cooling. Using these light sources at liquid helium temperatures, bandwidths of less than .010 cm^{-1} have been attained[23].

The most monochromatic of the conventional thermal light sources is the optical atomic beam[24]. Here a sharp reduction in Doppler width is attained by eliminating one component of velocity. The emitted radiation is viewed transverse to the direction of motion of a collimated beam of atoms. The beam of atoms can be excited by electron impact or by optical pumping of the atoms to an excited state. The atomic beam can also be used for absorption spectroscopy[25]. Using the resonance line of the element calcium, an emission atomic beam with a coherence length in excess of 2 m has been constructed[26]. The most precise wavelength measurements to date have been made utilizing an atomic beam of Hg^{198}[27] where a relative accuracy of 5×10^{10} has been achieved.

Methods of producing spectral lines, narrowed by some form of filtering, have come into recent prominence. One is the Zeeman filter[28] in which an absorption cell is placed in a magnetic field. The filter cell is also placed in a magnetic field. The field strength is adjusted so that the absorption dips of the σ radiation are separated, with only a narrow window of transmission between them. By properly adjusting the temperature of the cell and the field, band widths as small as .005 cm^{-1} have been achieved.

Proposals to use a Fabry-Perot interferometer as a filter are exciting. By placing an aperture at the central spot of the Fabry-Perot ring pattern, the band width can be limited to that fraction of the free spectral range intercepted by the aperture. The performance of the Fabry-Perot filter is limited by the finesse of the interferometer plates and by the inherent loss of intensity that results from increasing the resolving power. The limitation on intensity can be diminished by using spherical plates in the Fabry-Perot-interferometer[29]. Spherical Fabry-Perot instruments have already achieved coherence lengths of 10 m[30, 31], and prospects appear good for extending these results to 300 m[32].

In summary, if a high spectral-radiance source is desired, when reproducibility of the central frequency is not important, a laser is the best source available. The light source that yields the highest possible wavelength precision is still the atomic beam light. Measurements of moderate accuracy can be made with a cooled hollow cathode or an electrodeless discharge filled with isotopically pure gases.

6.2 OPTICAL-WAVELENGTH-MEASURING APPARATUS

6.2.1 Laser-Spectroscopy Requirements

In classical spectroscopy Doppler effects set a fundamental upper limit to the resolution needs of the spectroscopist. The hollow-cathode technique [35, 36] is used to investigate the hyperfine structure of spectral lines to minimize Stark broadening; pressures are kept low enough so that pressure-broadening effects are negligible. A practical limit to the reduction in Doppler broadening is reached by operating the hollow-cathode discharge in liquid nitrogen or liquid helium. The need to dissipate some power causes the measured line width in the 0.5 μm range to be 10^{-2} cm^{-1} or more even for the heaviest elements.

In all spectroscopic applications the maximum resolution of an instrument is limited by the reciprocal of the largest path difference between interfering beams. Using the largest modern gratings (25 cm) in a Littrow mounting and illuminating at a large angle of incidence, the longest optical path difference that can be made is about 40 cm. The smallest value for the resolution limit (instrumental width) or wave-numbers difference between two monochromatic radiation sources, whose wave numbers differ by the instrumental width at half-maximum transmission, is $\Delta\sigma = 2 \times 10^{-2}$ cm^{-1}. Thus grating instruments are now approaching the performance required for high-resolution spectroscopy. The Fabry-Perot interferometer, in contrast, is still unsurpassed for its resolution and luminosity in many regions of the spectrum. Its free spectral range ($\sim$ 0.1 nm) necessitates its use with cross-dispersing instruments.

The utility of grating-type instruments in laser spectroscopy is illustrated by the application of the 40 ft, MIT, high-resolution spectrograph to record the modes in the pulsed output of a ruby laser[34]. This instrument is equipped with a 25 cm grating and has an instrumental width of 2×10^{-2} cm^{-1} for 694.3 nm radiation. Photographs of the laser radiation when the laser was operated near threshold showed a group of 10 lines with a constant separation of 8×10^{-2} cm. Each line was 2×10^{-2} cm wide corresponding to, within measuring accuracy, the resolving limit of the instrument. The observed structure was consistent with the expected axial-mode spacing in a Fabry-Perot cavity having a length equal to that of the ruby.

The inhomogeneously broadened line width of a gas laser, which is barely resolved using the best conventional grating type of spectroscopic instruments, represents the maximum line width of the output spectrum of a gas laser. Instruments of higher resolution are required for analyses of gas-laser cavity modes. Grating spectrographs and monochromators are, however, of value in a detailed analysis of the wavelength characteristics of solid-state and semiconductor lasers.

Because the resolution of instruments of the grating type is inadequate for gas-laser spectroscopy, it follows that instruments of the prism type, with even poorer resolution, are similarly restricted to the broad-band lasers. Grating instruments do have significant value being applied as wave meters in surveys of the output wavelength of gas lasers and for spectroscopic identification of transitions, particularly in the far infrared where photographic techniques fail and electronic detectors must be used.

Because grating monochromators are excellent tools for solid-state and semiconductor, laser-wavelength measurement, they are treated briefly here. Emphasis is, however, given to high-resolution interferometer and heterodyne spectroscopy techniques.

6.2.2 Instrument Parameters

The heart of any instrument used to measure wavelength is the spectral redistributing element, the remainder of the instrument being designed to maximize the performance of the element. Instruments may be classified by the type element employed for wavelength distribution. Refractive instruments exploit some frequency dispersion property of a material, such as dispersion of the index of refraction or rotation of the plane of polarization. Because the dispersion depends upon the presence of some natural resonance in the material, instrument performance will depend upon the material used in the refractive element. The wavelength range of the instrument will thus be limited. Phase-sensitive instruments, including both diffractive and multiple-beam instruments, redistribute the light by virtue of a periodic ruling on a grating, or a periodic interference of multiple beams. The accuracy of these instruments depends upon the dimensional stability of the instrument's optical path length. (Grating spectrographs, Michelson, Fizeau, Fabry-Perot, and Connes interferometers belong to this class.)

A figure of merit that is frequently used to evaluate the performance of spectroscopic instruments is its resolution-luminosity product. This measure, which combines the light-gathering power and wavelength-selection capability of the instrument, is discussed in Chapter 8.

The theoretical resolving power of an instrument is defined in terms of the Rayleigh criterion[33]. This measure describes the minimum wavelength separation $\delta\lambda$ for which two monochromatic radiations, of equal intensity and centered at virtually the same wavelength λ, can be distinguished. Rayleigh resolution is accomplished when the first minimum in the diffraction pattern of one line coincides with the peak of intensity of an adjacent second time. The sensor output will display a dip of at least 19 percent at a setting halfway between the two Rayleigh-resolved lines. Grating spectrometers are often evaluated on the basis of the order necessary to resolve the mercury doublet at 313.155 and 313.183 nm.

The criterion of Rayleigh resolution is equally applicable to refractive, diffractive, and interferometric instruments, although the specific formulation of instrument-dependent factors may differ. Chapter 8 treats the theoretical resolving power of a number of instruments. It suffices here to point out that prism instruments are generally restricted to

$$(R_0)_{\text{prism}} \leq 10^5, \tag{6.10}$$

grating instruments to

$$(R_0)_{\text{grating}} \leq 10^6, \tag{6.11}$$

and passive interferometers to

$$(R_0)_{\text{interferometer}} \leq 10^7. \tag{6.12}$$

A second important parameter of any optical device is its angular dispersion

$$D_\lambda = \frac{d\theta}{d\lambda}. \tag{6.13}$$

In photographic instruments the "plate factor"

$$S = \frac{1}{f}\frac{d\lambda}{d\theta}, \tag{6.14}$$

which includes the angular dispersion, is more frequently used. This measure yields the scale factor for a photographic plate, (usually in units of Å/mm.) The plate factor can be varied by choosing the focal length f of the collimating lens or mirror.

Ordinarily, it might seem logical to adjust the plate factor so that the resolving limit of the photographic emulsion, N lines/mm, matches the limits set by the resolving power of the instrument:

$$S\left(\frac{\text{Å}}{\text{mm}}\right) = \frac{N(\text{lines/mm})\lambda(\text{Å})}{R}. \tag{6.15}$$

For example, for a large diffraction grating with a resolving power of 10^6, and an emulsion with a resolving limit of 100 lines/mm, one would think that it is only necessary to adjust the plate factor at 5000 Å to 0.5 Å/mm, to take full advantage of the resolving power of the grating. In practice, however, it is found that for large gratings the plate factor must be about twice that predicted by (6.15) in order to make full use of the instrument.

A practical consideration in the application of phase-sensitive instruments is the free spectral range F_σ, or range of wavelengths over which we can obtain spectra free of overlapping higher orders. This presents a problem in diffraction grating instruments in the far infrared. The source of the difficulty can be visualized by considering the grating equation

$$d(\sin\theta - \sin\theta') = m\lambda, \tag{6.16}$$

where d is the distance between grating grooves, θ is the angle of incidence, θ' is the angle of diffraction, and m is the integral order of diffraction. Then

$$\sin\theta' = \sin\theta - \frac{m\lambda}{d}. \tag{6.17}$$

If $m\lambda_1 = (m+1)\,\lambda_2$, overlap will occur. In the visible, for example, the second-order violet lines overlap with the extreme end of the red spectrum. This effect becomes particularly annoying in the infrared, wherein visual response cannot aid in distinguishing wavelength. Higher orders can be eliminated by prefiltering the light that is transmitted to the instrument using a source of stigmatic cross dispersion such as a prism. In grating instruments employing photographic recording, it is convenient if the F_σ can be adjusted to be limited by the camera plate.

The free spectral range of a Fabry-Perot interferometer is defined as

$$F_\sigma = \lambda^2/2nL \qquad \text{or in wavenumbers} \qquad F_\sigma = \frac{1}{2nL}, \tag{6.18}$$

where L is the mirror spacing and n is the index of the medium inside the etalon.

In interferometers the resolving power and free spectral range are inversely related. For example the Fabry-Perot interferometer has a resolution of

$$(R_0)_{FP} = \left(\frac{2nL}{\lambda}\right)\frac{\pi}{|\ln R|}, \tag{6.19}$$

where R is the power reflectance of the interferometer surfaces. Thus

$$F_\sigma \sim \frac{1}{(R_0)_{FP}}, \tag{6.20}$$

and cross-dispersion requirements become more severe.

The third parameter that must be considered in evaluating instrument performance is light-gathering capability or luminosity. The luminosity ∇, of a spectrometer is defined as the output flux per unit source of luminance. The energy incident upon a receiver in a time τ from a spectrometer of luminosity ∇ illuminated by a source of luminance B is

$$U = \tau B \nabla. \tag{6.21}$$

The two key aspects from which a spectrometer must be chosen, resolution R_0 and luminosity ∇, are conveniently expressed as a resolution luminosity product.

We show in Chapter 8 that

$$P_{(\text{Fabry-Perot})} > P_{(\text{grating})} > P_{(\text{prism})}. \tag{6.22}$$

Furthermore

$$R_{0(\text{Fabry-Perot})} > R_{0(\text{grating})} > R_{0(\text{prism})}. \tag{6.23}$$

It would be easy to reach the false conclusion that the Fabry-Perot interferometer is the best instrument for laser spectroscopy, although in high-resolution, gas-laser spectroscopy it is the only instrument with adequate resolution. The prism spectrometer is highly useful for rapid spectral surveys. A one-meter Czerny-Turner plane grating monochromator that employs a grating glazed at 6 μm is extremely useful for precise wavelength measurements. It is especially valuable where resolution of rotational lines is of interest in infrared molecular lasers.

6.2.3 Parameters for Typical Spectrographs, Monochromators, and Fabry-Perot Interferometers

A large number of wavelength-measuring instruments have been developed and refined over the years. Space does not permit treatment of any but the most general principles upon which these operate. For more details the interested reader is referred to survey articles in the literature[37, 38].

The three general types of wavelength-measuring instruments of interest in laser spectroscopy include the prism spectrograph, that is a prism spectroscope provided for photographic recording; the grating monochromator, having adjustable entrance and exit slits and provided with a variety of detectors and gratings to cover the ultraviolet through far infrared; and the Fabry-Perot interferometer, preferably with a piezoelectric scan capability with mirrors and detectors for use in the near-ultraviolet through

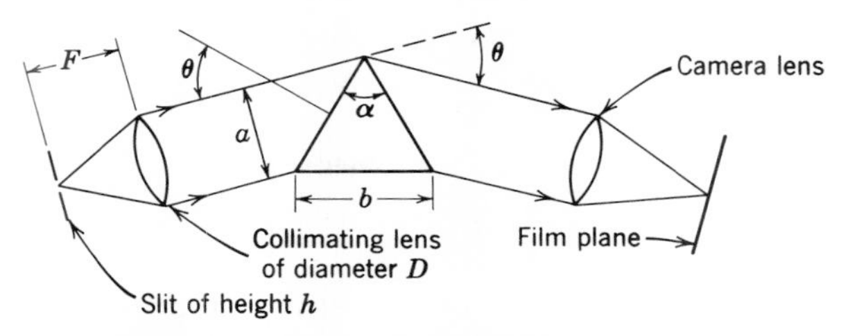

Figure 6.2. Prism-spectrograph parameters.

far infrared. Salient properties of these instruments will be discussed below.

6.2.3.1 Prism Spectrographs Prism spectrographs have a low resolving power and high photographic speed. Further characteristics of a spectrograph of interest are the angular dispersion, the resolution, and the photographic speed.

Consider an isosceles prism of base b and central angle α which at minimum deviation[33] bends the entrance beam axis through an angle θ. The beam is obtained from a slit of width B and height h and is focussed upon the prism face with a collimating lens of focal length F as shown in Figure 6.2. If the prism material has an index n and the beam is passed through the prism at minimum deviation to obtain the least image distortion, it can be shown that[38]

$$\frac{dn}{d\theta} = \frac{\sqrt{1 - n^2 \sin^2(\alpha/2)}}{2 \sin(\alpha/2)}. \tag{6.24}$$

The Hartmann formula relates the refractive index to the wavelength

$$n = n_0 + \frac{\kappa}{(\lambda - \lambda_0)}, \tag{6.25}$$

where n_0, κ, and λ_0 may be regarded as constants over a small range near λ. Differentiating (6.25) the dispersion of the prism becomes

$$\frac{dn}{d\lambda} = \frac{\kappa}{(\lambda - \lambda_0)^2}. \tag{6.26}$$

The angular dispersion D_λ is defined as

$$D_\lambda = \frac{d\theta}{d\lambda} = \frac{d\theta}{dn}\frac{dn}{d\lambda}. \tag{6.27}$$

Combining (6.24) and (6.26), the angular dispersion of a prism spectrograph becomes

$$D_\lambda = \frac{2\kappa \sin(\alpha/2)}{(\lambda - \lambda_0)^2\sqrt{1 - n^2 \sin^2(\alpha/2)}}. \tag{6.28}$$

In this formulation D_λ does not contain terms that depend upon the prism dimensions. The angular dispersion does depend upon the central prism angle α and the prism material through κ and n. The negative sign indicates an increase in deviation for a decrease in wavelength.

An alternate and perhaps more familiar derivation of D_λ is based upon the use of Cauchy's formula for the dispersion in optical glasses

$$n = A + \frac{B}{\lambda^2} + \cdots. \tag{6.29}$$

The angular dispersion is then expressed as

$$D_\lambda = -\frac{2Bd}{a\lambda^3}, \tag{6.30}$$

where a is the entrance beam width of the prism and b is the base.

The dispersion of a prism spectrograph is defined in terms convenient for measurement of spectra on a photographic plate. It is therefore usually described as reciprocal linear dispersion and is quoted in millimeters per Angstrom (mm/Å) of the photographic plate rather than angular dispersion in radians per Angstrom (rad/Å).

If the photographic plate is normal to the incident beam, it is easy to show that the reciprocal linear dispersion $d\lambda/dx$, where dx is the displacement along the photographic plate for a wavelength increment $d\lambda$, results in a plate factor S of

$$S = \frac{d\lambda}{dx} = \frac{1/F}{d0/d\lambda}. \tag{6.31}$$

It can be shown[38] that the theoretical resolution of a prism is

$$R = b\frac{dn}{d\lambda}, \tag{6.32}$$

where the whole prism face is illuminated by the slit and collimating lens.

The practical resolution of a prism is a function of slit width and prism size. When the full aperture a of the prism is used and the prism is used at minimum deviation the resolution may be written as

$$R = b\,\frac{dn}{d\lambda} = \frac{2Bb}{\lambda^3}, \tag{6.33}$$

where $dn/d\lambda$ is derived from the Cauchy's formula (6.29).

The resolution achieved in practice depends upon such factors as mode of illumination, adjustment of optics, slit width, film contrast, and relative intensities of the adjacent spectral lines being resolved[39, 40]. We comment only on slit width.

The optimum slit width to be used in laser spectroscopy is just double that used in examining the spectroscopy of incoherent sources. It can be shown that a slit width

$$\phi = \frac{2\lambda F}{D} \tag{6.34}$$

leads to an intense central diffraction maximum with negligible loss in resolving power. Here F and D are respectively the focal length and diameter of the collimating lens.

The photographic speed of a spectrograph can be expressed in terms of an f number defined in the conventional way,

$$f\,\text{number} = \frac{F}{D}. \tag{6.35}$$

Thus the optimum slit width becomes

$$\phi = 2\lambda\ f\,\text{number}. \tag{6.36}$$

The resolving power of a prism spectrograph can be increased by a number of prism arrangements, one of which is illustrated in Figure 6.3. These make it possible for us to use a larger effective prism-base width and to overcome the difficulty encountered in making large pieces of glass with adequate homogeneity. For values of $R > 50{,}000$ high cost usually dictates the use of grating instruments. (Numerous arrangements of multiple-prism trains are described in the literature.)

Where good dispersion and moderate illumination is required, the Littrow spectrograph is almost a universal choice. The principle of the

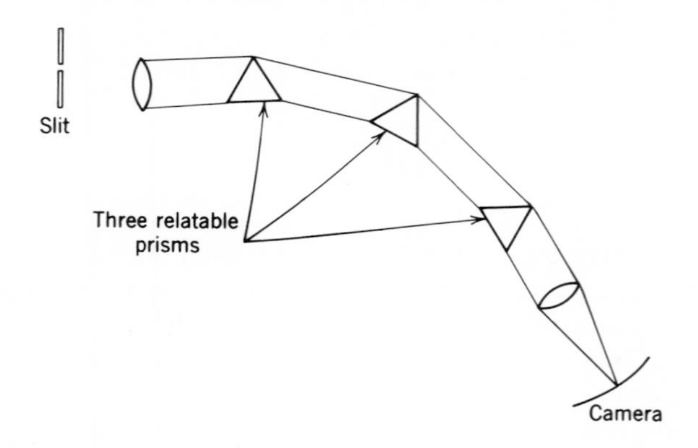

Figure 6.3. Three-prism spectrograph.

autocollimating Littrow spectrograph will be obvious from an inspection of Figure 6.4.

6.2.3.2 Grating Monochromators Grating monochromators have excellent resolution, are not limited by the range of sensitivities of photographic films, and have greater dispersions than prisms. The availability of good replica reflection gratings[41, 42, 43] enables us to use these instruments in the wavelength range from 120 nm to 40 μm. In contrast to the nonlinear prismatic spectrum, the grating monochromator produces a linear spectrum.

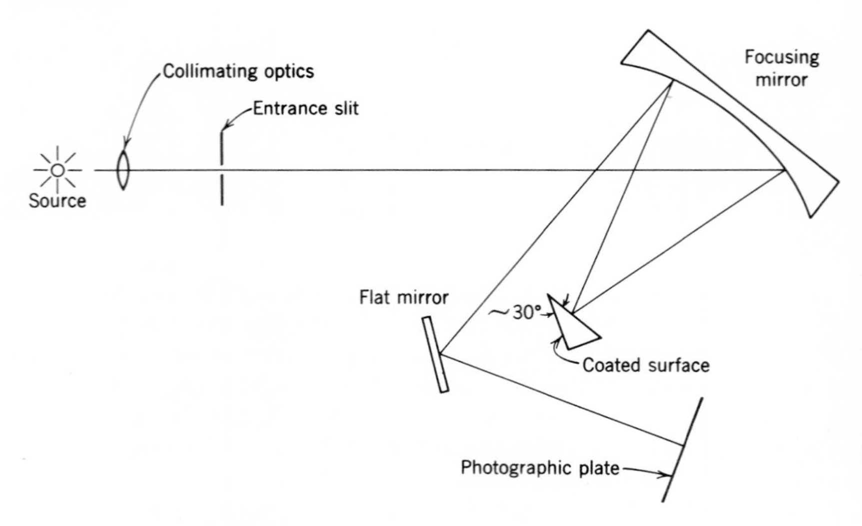

Figure 6.4. Ray diagram of a Littrow spectrograph.

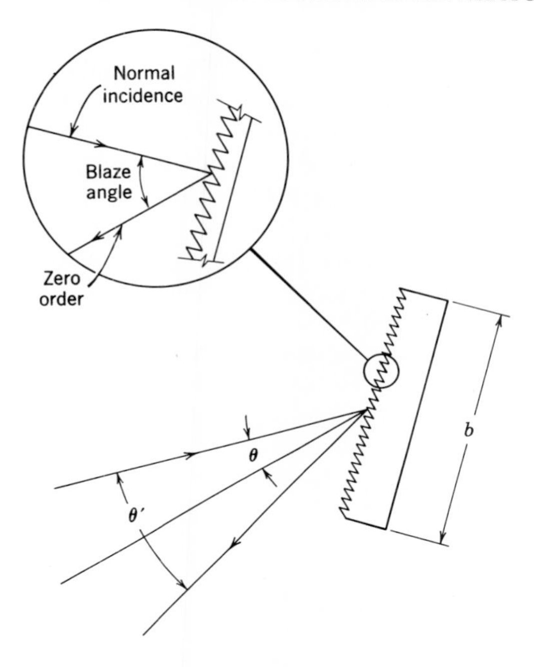

Figure 6.5. Parameters in a plane grating.

The parameters of interest in grating instruments are resolution, dispersion, free spectral range, and grating blaze and efficiency.

The theoretical resolving power of a diffraction grating may be shown to be

$$(R_0)_{\text{grating}} = mN = \frac{mb}{d} = \frac{b(\sin\theta + \sin\theta')}{\lambda} \tag{6.37}$$

where m is the order number and N is the total number of lines on the grating, b is the width of the grating, and d is the grating constant or distances between grooves on the grating. The angles are defined in Figure 6.5. From (6.37) it is clear that a high resolving power requires a wide ruled surface, and as we see below, a smaller grating constant (more lines/mm) serves only to increase the angular dispersion and tends to reduce the overlapping of orders. Note that (6.37) demonstrates that the resolving power of a grating is independent of wavelength and the grating constant.

The dispersion of a grating is just

$$D_\lambda = \frac{m}{d\cos\theta'}. \tag{6.38}$$

The angular dispersion of a grating type instrument increases with order number and is approximately linear. Knowledge of the angular-dispersion characteristics of a grating is of value in identification of spectral order.

One basic disadvantage of the grating is that it disperses the light over a large number of orders and is thus wasteful of energy. In a slit grating, for example, the intensity of the light in the *m*th-order spectra varies as $1/m^2\pi^2$ of the light in zero order. This loss in efficiency may be partially overcome in the visible or ultraviolet by increasing the number of lines per mm. The grating efficiency problem becomes more serious in the infrared where energy levels can be lower.

We can largely overcome the shortcomings of the plane grating by using a specially shaped diamond cutter that will contour or blaze the grating grooves at an angle that will concentrate the diffracted light into preferred directions with efficiencies as high as 75 percent. Gratings are blazed for a particular wavelength in a particular order and are, therefore, less efficient at other wavelengths.

Even the best of modern diffraction gratings are subject to defects, some of which are periodic and others progressive. (A perfect grating would have straight, parallel, equally spaced, identical grooves.) The periodic errors give rise to false lines or Rowland ghosts[44]. Rowland ghosts are periodic and recur at regular intervals that are calculable. They cause little problem unless very intense.

A progressive defect arising from an error of the run can occur when gratings are ruled. These errors generally cause the resolution of the grating to deteriorate. Lyman ghosts, which are periodic though not well understood, are found typically at 2/5, 3/5, 4/5, 6/5, 7/5, and so on of the wavelength of the main line; they can be distinguished by careful observation.

The limited, free-spectral range of a grating, particularly one blazed for the infrared and with a relatively large grating constant, can be overcome by using stigmatic cross-dispersion elements. These may include a prism spectroscope operated at minimum deviation, a narrow band filter, the choice of the detector, and the use of short wavelength cutoff filters, such as silicon, germanium, indium arsenide, and antimonide.

6.2.3.2.1 Grating Spectrographs Two basic grating configurations are used in grating spectrographs: concave and plane. Concave gratings are commonly ruled on mirrors (blank) having nominal radii of curvature of 1, 2, 3, and 10 m as well as 10, 21, and 31 ft. The concave grating combines the dispersing and focussing properties of the grating surface; it thereby eliminates requirements for lenses and circumvents their chromatic aberrations.

All concave-grating mountings except the Wadsworth are adaptations on the principle, originally discovered by Rowland, that the slit, the grating, and the diffracted spectrum all lie on a circle to which the grating is tangent and which has as its diameter the radius of curvature of the grating blank. Probably the most popular of the many versions of the concave-grating spectrograph is the Paschen-Runge[45] mounting illustrated in Figure 6.6. Although this mounting has nonlinear dispersion, and it is not stigmatic (the image of a point at the slit is a vertical line at the camera), it has the advantage of allowing a large portion of the spectrum to be photographed with one exposure.

Only the largest concave-grating spectrographs have adequate resolution to resolve the wavelength separation of axial modes in solid-state lasers. Because application of the scanning Fabry-Perot interferometer has almost completely supplanted the concave-grating spectrograph for high-resolution laser spectroscopy, a review of the advantages and disadvantages of the many grating mounts will be omitted. The interested reader is referred to extensive treatments in the literature[46].

The two most common plane-grating mounts in modern grating monochrometers are the Ebert[47] and the Czerny-Turner[48]. The autocollimating Littrow mount is common for grating spectrographs and

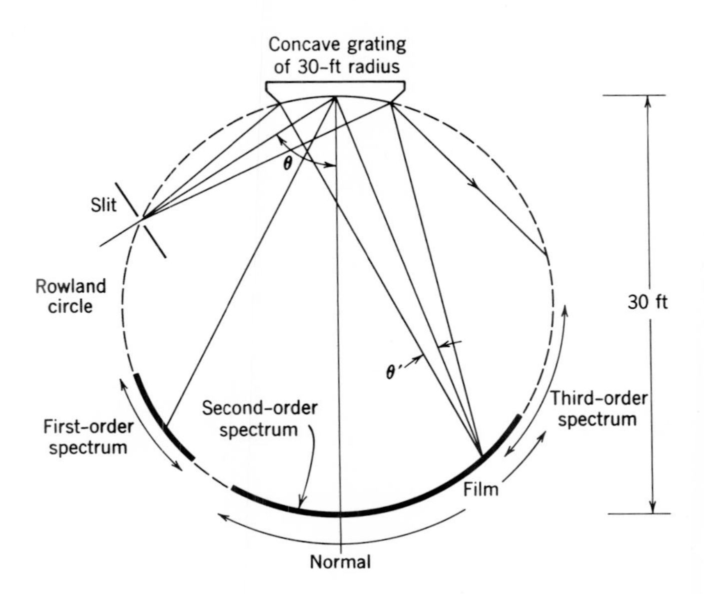

Figure 6.6. Ray diagram of a Paschen-Runge mounting.

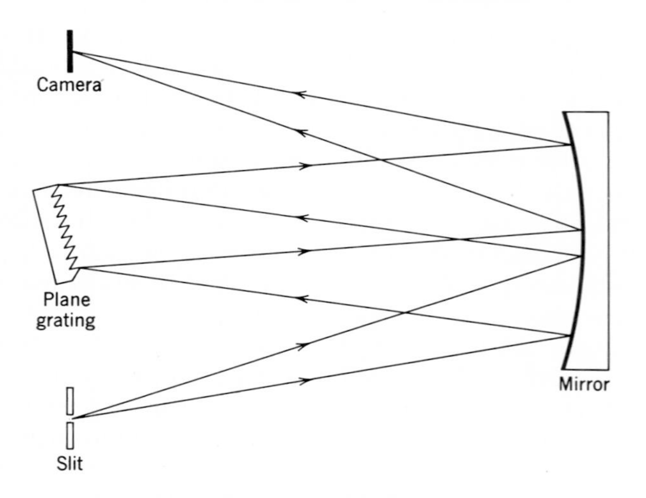

Figure 6.7. Ray diagram of the Ebert mounting.

illustrated in Figures 6.7 and 6.8. In both mountings, the light from the entrance slit is made parallel by a concave mirror, diffracted by the grating, and focussed on the camera or exit slit by another mirror (the same mirror in the case of the Ebert mounting). The virtue of the plane-grating mounting is that aberrations of the mirrors tend to cancel, and large plane gratings

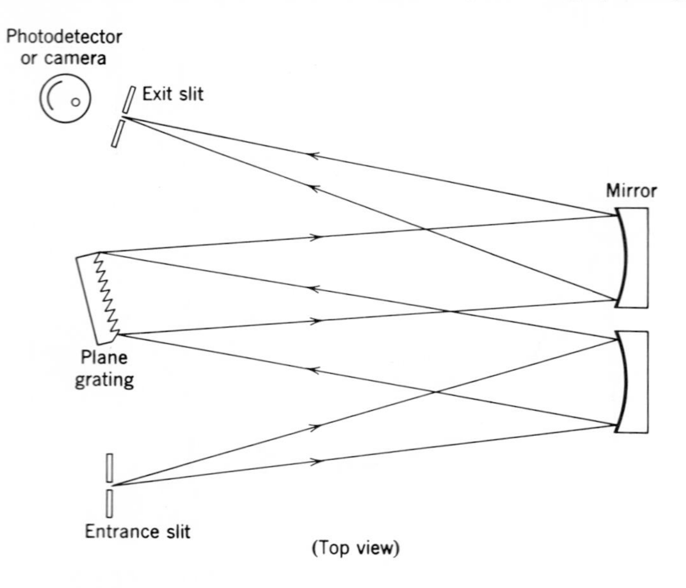

Figure 6.8. Ray diagram of the Czerny-Turner mounting.

(which can be made most accurately) may be used. The mounting is such that $\theta \cong \theta$ and large blaze angles can be used ($\theta \sim 63°$). A 10-in grating has a resolution of 8.6×10^5 at 500 nm and would require concave minors with at least a 10 m radius to take full advantage of the theoretical resolving power. Plate factors of 0.1 Å/mm are common in grating spectrographs.

The free-spectral range in a grating monochrometer can be a severe limitation because the instrument is usually operated in high orders ($10 < m < 30$) to take advantage of efficient blaze angles and to make the use of large-ruled grating areas for high resolution possible. The free-spectral range of a monochromator is defined as

$$F_\sigma = \Delta\lambda = \frac{\lambda}{m+1}. \tag{6.39}$$

For $m = 19$ at 500 nm, $F\sigma = 25$ nm, so that a stigmatic source of cross dispersion is necessary to preclude the overlap of spectra of different orders.

6.2.3.2.2 Echelons, Echelles, and Echellettes The resolving power of a grating depends upon the largest path difference between the extremes of the grating. Few gratings are made of over 10 inches in width because the error of the run due to wear of the diamond point on the scribe becomes excessive. This practical limitation can be circumvented if the grating grooves are replaced by a stack of thick plates placed in a staircase array of equal steps. Such an array, termed by Michelson[49] as an echelon, has a very high order of interference (equal to the optical-path difference between two adjacent steps). For an array of m steps of thickness d and spacing a, the optical path difference between successive rays diffracted at a small angle θ from corresponding points on two successive steps becomes

$$m\lambda = d(1 + \cos\theta) - a \sin\theta \simeq 2d - a\theta \tag{6.40}$$

because

$$\sin\theta \sim 1.$$

The order of interference of an echelon grating is very large; the free-spectral range is consequently limited. For near-normal incidence ($\theta \sim 0$) a stack of plates each 1 cm thick has an order at 500 nm of

$$m \sim \frac{2d}{\lambda} = \frac{(2)(1)}{5 \times 10^{-5}} = 4 \times 10^4.$$

If 40 plates are used, this grating has a resolution of

$$R = mN = (4 \times 10^4)(40) = 1.6 \times 10^6,$$

which rivals the Fabry-Perot. To achieve this resolution in practice, however, it is necessary to ensure that the plates are of equal thickness and are stacked together with a constant offset. The practical achievement of these requirements delayed the development of a grating with a resolution approaching the theoretical value for many years[50, 51, 52]. The angular dispersion of the reflection echelon is given by

$$D_\lambda = \frac{m}{a}. \tag{6.41}$$

Because the angular separation for $m = 1$ is just equal to the width of the diffraction pattern due to a single surface of width a, only a few orders can be observed with adequate intensity.

Echelles consist of deeply grooved sawtoothed gratings with large line spacings. Echelles are used in high order ($m \sim 500$) and correspondingly have a small free-spectral range. An auxiliary cross-dispersion element is needed to separate the various orders and can be used to develop a two-dimensional array of wavelength. This arrangement has the advantage of covering a very large spectral range on one photographic plate in a compact instrument that has a resolution as high as 8×10^5. An echelle is sufficiently stigmatic that the several different exposures on the same photograph may be separated by use of a Hartmann diagram[55].

Very coarse groove gratings, having as few as 100 lines/mm and an efficient blaze angle, have been developed to increase the luminosity of a grating instrument used in the infrared. Wood termed these gratings echellettes[53, 54].

6.2.3.3 Lummer-Gehrcke Interferometer The Lummer-Gehrcke interferometer has important applications in quantum electronics. It can be used in the ultraviolet, as a high resolution spectrograph, where the utility of the Fabry-Perot interferometer is limited by the availability of dielectric coatings of adequately low loss; it can be used as an optical delay line to reduce cavity closure times effectively in those prism-Q-switched lasers wherein electro-optic materials do not exist.

A Lummer-Gehrcke interferometer consists of a thin slab of material of thickness b, usually a quartz plate $\frac{1}{8}$ in. thick, the entrance edge of which is cut and polished at an angle such that a light beam may be injected to make multiple internal reflections in the plate. Some light is lost on each reflection and escapes at an angle θ with respect to the surface normal to the plate. The interference of these beams produces two sets of complementary fringes at a distant detector. Because the order of interference is very large, the resolution of a Lummer-Gehrcke interferometer is also very large. The order can be shown to be

$$m = \frac{2b}{\lambda}\sqrt{n^2 - \sin^2\theta}. \tag{6.42}$$

The free-spectral range is

$$F_\sigma = \lambda^2/2b\sqrt{n^2 - \sin^2\theta} \tag{6.43}$$

for small changes in n. Because the reflections within the Lummer-Gehrcke plate occur in a wavelength-dispersive medium, $n = n(\lambda)$. At a given angular position the change $\Delta\lambda$ in irradiating wavelength is, to first order,

$$\Delta m = \left(\frac{4b^2 n\, \partial n/\partial\lambda - m^2\lambda}{m\lambda^2}\right)\Delta\lambda.$$

For the source at near-grazing incidence, $\sin\theta \sim 1$, and the free spectral range becomes

$$F_\sigma = \frac{\lambda^2}{2b}\left[\frac{\sqrt{n^2 - 1}}{|\,n^2 - n\lambda(\partial n/\partial n) - 1\,|}\right].$$

For quartz the quantity in the square brackets varies from 0.6 to 0.8 in the 200 to 600 nm wavelength range.

6.2.3.4 Fourier-transform Spectroscopy The advent of computers has enabled the development of Fourier-transform spectroscopy, a technique that is expected to become important in the spectroscopy of infrared sources from which detectivity is low. The method consists essentially of decomposing a source of light in a two-beam interferometer, such as a Michelson interferometer. The two beams are recombined on a detector in the conventional way to yield an output signal proportional to the total light intensity. If this signal is recorded while one of the mirrors is moved accurately along its axis, the resulting interferogram can be reconstructed by a computer to yield the cosine Fourier transform of the source spectrum[56]. The interferometer acts as a coder that differentiates the elements of the spectrum by modulating them at different frequencies[57, 58]. The technique, though sophisticated and quite complex, is of value in that it enables us to examine an extended portion of the spectrum with the resolution-luminosity characteristics of a Fabry-Perot interferometer. This technique has been used to examine the spectrum of a helium-neon laser[59, 60, 61].

6.2.3.5 Fabry-Perot Interferometer A Fabry-Perot interferometer, illustrated in Figure 6.9, consists principally of two extremely plane and parallel surfaces coated to give a high reflectance R, a low transmittance T, and a low loss A, where $A = 1 - R - T$. The plates of this etalon are separated a distance L by a medium of index n. A circular diaphragm is usually provided to isolate the central portion of the ring system in the vicinity of the etalon and take maximum advantage of plate quality. An incident light beam is effectively decomposed into a large number of light rays that are reflected and transmitted parallel to one another.

Let m be the order of interference and $2nL \cos \phi$ be the optical retardation in an etalon of spacing L when illuminated by a beam at an angle ϕ with respect to the surface normal of the etalon. The condition for interference, when the etalon is filled with a medium of index n, is [62, 63, 64, 65]

$$m\lambda = 2nL \cos \phi. \tag{6.44}$$

A lens of focal length f is used to collect the light from a large number of beams. A series of concentric rings is observed in the focal plane of the lens.

Because constructive interference can only occur for light in an extremely narrow range of frequencies, the pattern observed in the focal plane of the lens, for a carefully aligned system, will be a series of concentric rings. If broad-band light is incident on the etalon, the interference pattern will consist of a large number of overlapping circles.

Because the order of interference is large, and the free-spectral range of the instrument is small, cross dispersion is necessary if a single spectral line is to be observed. Alternatively, the Fabry-Perot can precede the slits of a grating spectrograph so that each of a multiplicity of lines will display a section of the ring system. The net resolution of a combined Fabry-Perot

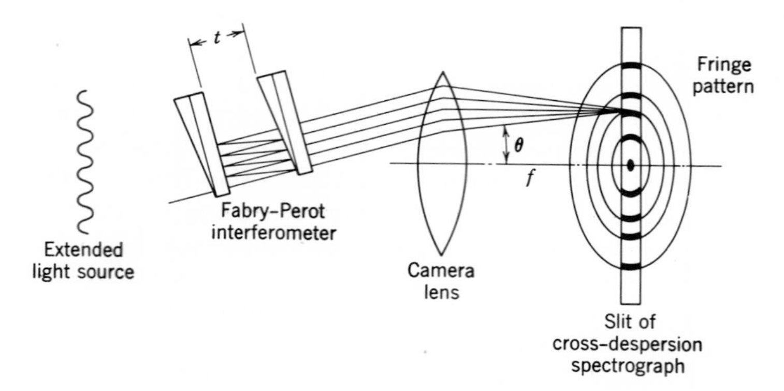

Figure 6.9. Ray diagram of a Fabry-Perot interferometer.

etalon and spectrograph is sufficient to study the average hyperfine structure of a number of spectral lines. Fortunately in laser spectroscopy, the number of lines of interest is usually one or at most a few so that cross dispersion can be accomplished with a narrow-band filter.

The free-spectral range of an interferometer may be derived from (6.44) considering $2\,nL\cos\phi$ a constant,

$$F_\sigma = \frac{\lambda^2}{2nL\cos\phi}. \tag{6.45}$$

The angular-dispersion of the etalon is

$$D_\lambda = \frac{d\phi}{d\lambda} = \frac{\cot\phi}{\lambda}, \tag{6.46}$$

considering m, n, and L as constants. Equation (6.47) shows that the angular dispersion varies with distance from the center of the fringe pattern and is infinite on axis. Advantage is often taken of this dispersion by focussing the interference pattern onto a circular diaphragm in front of a detector so that only those wavelengths for which $\cos\phi = 1$ are observed. The spectral range of the interferometer is then scanned by varying the pressure inside the etalon and using the small pressure-dependent index of refraction to scan the optical retardation of the etalon. Because the refractive index varies, the wavelengths that constructively interfere for $\cos\phi = 1$ also vary. If m, L, and $\cos\phi$ are constants, (6.44) yields

$$d\lambda = \frac{2L\cos\phi}{m}\,dn = \lambda\,dn. \tag{6.47}$$

For a small pressure change, n is a linear function of pressure:

$$n = 1 + \zeta p \tag{6.48}$$

where $\zeta = 3.57 \times 10^{-1}$/Torr for air. Thus the equation for the etalon-pressures can become

$$d\lambda = \zeta\lambda\,dp \tag{6.49}$$

The aperture that limits the etalon output also limits its resolving power. If the aperture has a radius r, the diaphragm will admit wavelengths that constructively interfere for

$$1 - r^2/2f^2 \le \cos\phi \le 1. \tag{6.50}$$

The resolution is therefore limited to

$$d\lambda_d = \frac{\lambda r^2}{2f^2}, \tag{6.51}$$

and the resolving power, as limited by the diaphragm, becomes

$$R_d = \frac{2f^2}{r^2}. \tag{6.52}$$

The resolving power, as limited by reflecting finesse, is

$$R_R = \frac{2nL}{\lambda} \mathscr{F} \cos \phi \tag{6.53}$$

where $\mathscr{F}$ is the finesse limited by the reflectivity of the etalon-plate coatings and is defined as

$$\mathscr{F} = \frac{\pi\sqrt{R}}{1 - R}. \tag{6.54}$$

A more fundamental limit on the resolution is usually imposed by our inability to fabricate etalon plates of adequate flatness. It may be shown that if x is the variation from flatness over the interferometer plate, the resolution is limited to

$$d\lambda_x = \frac{\lambda x}{nL \cos \phi} = 2N_D. \tag{6.55}$$

The corresponding resolving power is

$$R_x = \frac{nL \cos \phi}{x} = \lambda/2N_D. \tag{6.56}$$

Typical values of the finesses are

$$10 < N_D < 100,$$

$$10 < \mathscr{F} < 100.$$

Denoting N the effective number of interfering beams or effective finesse, the resolution becomes

$$R_0 = mN \tag{6.57}$$

where

$$10 < N < 50.$$

If given an effective finesse N, we increase the etalon spacing, we increase the resolving power. However because resolving power and luminosity are reciprocally related, little is achieved by increasing L beyond the value dictated by the line width to be investigated and the free-spectral range of the etalon. Fringe visibility decreases as soon as the line width becomes an appreciable fraction of the free-spectral range.

As an example of the interplay of parameters consider the requirements of an interferometer used to examine the spectrum of axial modes in a helium-neon laser operating at 632.8 nm. The Doppler broad-band line width of neon at an assumed atomic temperature of 500° K is approximately

$$\Delta\nu_D = 1700 \text{ MHz}.$$

If the laser has a 150 cm mirror spacing, axial modes will be separated by

$$\Delta\nu_a = c/2nd = 100 \text{ MHz}.$$

This corresponds to an axial mode spacing of 8.3×10^{-3} cm^{-1} and for an overall line width of 0.14 cm^{-1}.

If the free-spectral range is made equal to linewidth,

$$F_\sigma = \frac{1}{2L} = 0.14 \text{ cm}^{-1},$$

an etalon length of about 4 cm would suffice.

If the instrument width were made equal to the mode spacing, the required finesse would be[67]

$$N = \frac{F_\sigma}{\Delta k} = \frac{1.4 \times 10^{-1}}{8.3 \times 10^{-3}} \sim 17.$$

If we assume the plates are flat to 1/40 of a wave at 546.1 nm,

$$N_D = \left(\frac{40}{2}\right)\left(\frac{633}{546}\right) = 23,$$

and

$$\frac{N}{N_D} = 17/23 \sim 0.7.$$

Using the data in Figure 6.10 the required reflectivity finesse is $\mathscr{F} \sim 32$, and the required reflectivity, from Figure 6.11, is 91 percent. Assuming the scattering and absorption loss sum to be 0.5 percent, $A/T = 0.5/91 \sim 0$, for $\mathscr{F}/N_p = 32/23 \sim 1.4$. From Figure 6.12 we see that more than 62 percent

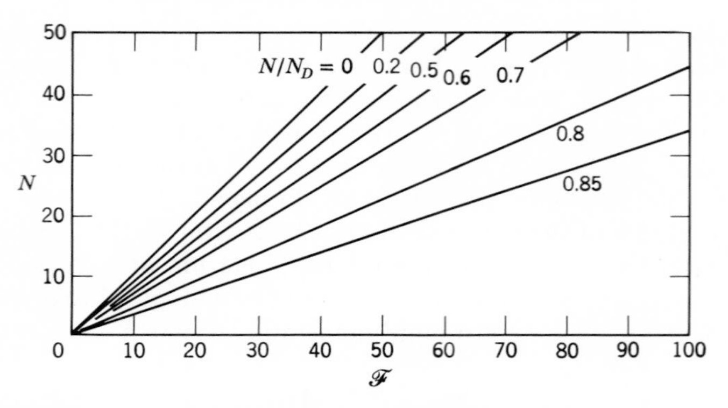

Figure 6.10. Measurement of Finesse versus reflecting finesse, as a function of N/N_D. (Reproduced with permission from *Applied Optics*.)

of the light will be transmitted. The resolving power of the spectrometer will be

$$R_k = \frac{(40)(1)(4)\cos\phi}{6.33 \times 10^{-5}} \cong 2.5 \times 10^6.$$

If a Fabry-Perot spectrograph is used, the interference fringes are recorded on a photographic plate, and if a scanning Fabry-Perot spectrometer is used, the information is obtained electrically with a photomultiplier tube. The photomultiplier is placed behind a small aperture

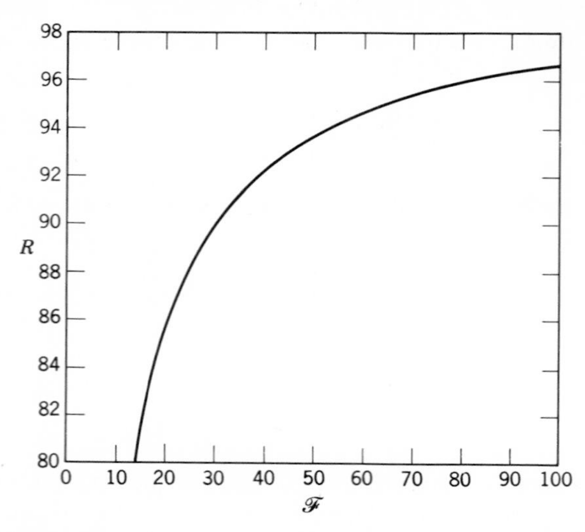

Figure 6.11. Measurement of reflection coefficient versus reflecting finesse. (Reproduced with permission from *Applied Optics*.)

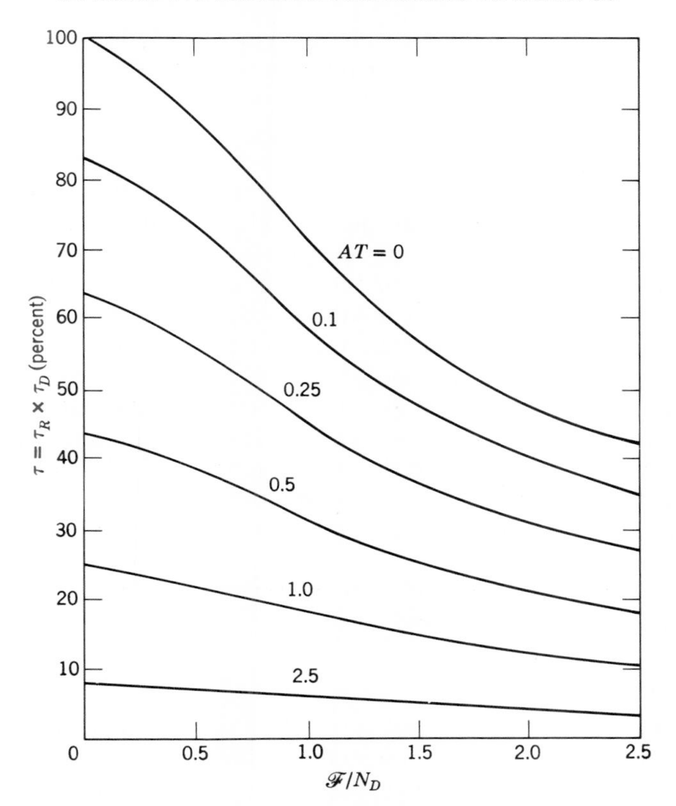

Figure 6.12. Transmission of a Fabry-Perot interferometer. (Reproduced with permission from *Applied Optics*.)

centered in the focal plane, upon which the ring pattern is produced [68]. The central spot will vary in intensity as the optical path between the reflecting layers of the etalon is changed. A scanning Fabry-Perot spectrometer is particularly suitable for low-intensity sources because the photomultiplier has a higher quantum efficiency than the photographic plate. A Fabry-Perot spectrometer of constant mirror spacing may be used to measure the time variation in the output wavelength of a laser source.

A spherical mirror or Connes configuration of the Fabry-Perot interferometer has been used extensively in laser spectroscopy to study the character of the axial modes[59]. The Connes interferometer is an afocal system of unit magnification in which the flat plates of a Fizeau or Fabry-Perot interferometer are replaced by a set of confocal mirrors. As we discuss in Chapter 8, this interferometer has a higher resolution-luminosity product and is easier to align. Also, because the diameter of an axial mode in a Connes interferometer is smaller than that in a Fabry-Perot interferometer and a smaller portion of the plates is used, the surface quality of the reflectors is less critical. The distribution of energy over the smaller diameter also reduces diffraction losses.

Because the resonances of a Connes interferometer are very different when illuminated off axis, it is necessary that the etalon be carefully aligned and that only the axial mode be illuminated. To accomplish this it is necessary to adjust the energy distribution and radius of curvature of the illuminating beam to match the cavity-mirror radius of curvature. (Techniques of mode matching are discussed in Chapter 5.) It is essential that the etalon be decoupled from the laser the spectroscopy of which is of interest. This may be accomplished with an attenuator or a polarizer and quarter-wave plate placed between the laser and the etalon.

The wavelength spectrum of axial modes may be displayed on an oscilloscope if the output of the interferometer is observed, through an optical arrangement that illuminates a photomultiplier with the central disc of the etalon. The length may be scanned repeatedly with a piezoelectric drive.

The technique may be extended to the infrared by using dielectric-coated germanium mirrors and infrared detectors.

6.2.4 Summary

The needs of laser spectroscopy cannot be supplied by a single instrument. Pulsed, solid-state lasers are best studied with spectrographs. Beyond the range of film, photoelectric detectors and scanning techniques must be employed. A 1-m Czerny-Turner monochrometer, equipped with an intermediate-ruled grating and blazed in the region of 3-8μm, and appropriate filters for cross dispersion will satisfy most requirements for the spectral range 200 nm to 40 μm. The electrically scanned Connes interferometer is an excellent tool for investigation of the detailed structure of the axial mode of CW lasers. The fixed etalon with a 1 cm spacing, when used with pulsed, solid-state lasers, gives more than adequate resolution. Table 6.1 lists the pertinent characteristics of the basic instruments. Details of many specific configurations of wavelength-measuring devices can be obtained from the literature and instrument manufacturers.

6.3 WAVELENGTH STANDARDS

For an absolute measurement of wavelength one must always have some standard ruler against which the measurement is made. This ruler may either be a natural one (i.e., a measured spectrum) or an artificial one (white light fringes, a scale, etc.). Frequently a metal-ion light source is used

TABLE 6.1 ANGULAR DISPERSION, RESOLVING POWER, AND FREE-SPECTRAL RANGE OF PRINCIPAL TYPES OF WAVELENGTH-MEASURING INSTRUMENTS

Instrument	Angular Dispersion	Resolving Power	Free-spectral Range
Prism	$D_\lambda = -\dfrac{2Bb}{a\lambda^3}$	$R_o = bdn/d\lambda$	No overlap
Grating	$D_\lambda = \dfrac{m}{d\cos\theta'}$	$R_o = mN = mb/d$	$F_\sigma = \lambda/(m+1)$
Fabry-Perot interferometer	$D_\lambda = \dfrac{\cot\phi}{\lambda}$	$R_r = 2\dfrac{nL}{\lambda}\mathscr{F}\cos\phi$ $R_x = \dfrac{nL\cos\phi}{x}$	$F_\sigma = \dfrac{\lambda^2}{2nL\cos\phi}$

to produce the calibration spectrum. Certain precautions should always be observed in the use of such a standard for calibration. Radiation from the calibrating source should be transmitted through the same optical system as the unknown radiation. In order that pressure and temperature conditions in the spectrograph will be identical for both measurements, the standard should be exposed almost simultaneously with the exposure of the unknown wavelength. A beam splitter can be used to cause the irradiance of the standard and the unknown light source to illuminate the slit simultaneously. The exposure times can be divided appropriately to preserve identical exposure conditions.

All wavelength measurements are ultimately referred to the primary standard of wavelength, the orange line of krypton, which has a vacuum length of $\lambda_{vac} = 6057.802105$ Å, and a wavelength in air (under standard conditions of temperature and pressure) of $\lambda_{air} = 6056.125253$ Å [69]. The definition of the primary standard refers to the radiation from an atom unperturbed by external influences. The recommended light source for producing this radiation is a hot-filament, krypton-discharge tube immersed in nitrogen at its triple point. The krypton-standard lamp has an accuracy limited to about one part in 10^8. Although this is sufficiently accurate for most purposes, the krypton standard may ultimately be replaced by an atomic beam standard. Atomic beams of Hg^{198} have produced radiations of greater inherent accuracy than the officially recommended primary source.

In addition to the primary standard, there exists a set of Class A secondary standards, which have been compared to the primary standard in several laboratories. Class A secondary standards, which may be used as a substitute for the primary standard in high precision measurements, include a set of Kr^{86} lines, Hg^{198} lines, and Cd^{14} lines. Table 6.2 lists the 12 wavelength standards of length.

A set of Class B secondary standards is used for more routine spectral measurements (with precisions of the order of .01 Å). Class B standards, which have been interferometrically measured in at least two, and preferably three, laboratories, include a large number of iron lines, noble gas lines, and thorium lines. The high-density lines in the thorium spectrum make it a very suitable Class B standard.

In the infrared ($1 - 3\ \mu$) a number of rare-gas lines have been measured interferometrically[70]. For ultraviolet work (2000 – 3000 Å) the Cu^{II} spectrum possesses suitable standard lines which have been interferometrically measured[71].

Use has also been made of Edson-Butler fringes as wavelength standards[72]. If a beam of white light is passed through a Fabry-Perot interferometer of spacing L, a band of bright fringes with wave number separation $< 1/2L >$ is recovered when this light is passed through a spectroscope. A standard line can be used to calibrate the scale formed by these fringes.

6.4 MEASUREMENT AND REDUCTION OF DATA FROM SPECTROGRAPHS

The measurement and reduction of data from grating spectrographs, prism spectrographs, and interferometers are considered below. Nonlinear dispersion dominates in prism spectrographs, necessitating a set of closely spaced standard lines over the photographic plate to obtain a

TABLE 6.2 WAVELENGTH FOR SECONDARY STANDARDS OF LENGTH FOR Kr^{86}, Hg^{198}, AND Cd^{114}[a]

Isotopic Source	Wavelength	Accuracy
Kr^{86}	$6458.0720 \times 10^{-10}$m	2×10^{-8}
	$6422.8006 \times 10^{-10}$m	
	$5651.1286 \times 10^{-10}$m	
	$4503.6162 \times 10^{-10}$m	
Hg^{198}	$5792.2683 \times 10^{-10}$m	5×10^{-8}
	$5771.1983 \times 10^{-10}$m	
	$5462.2705 \times 10^{-10}$m	
	$4359.5624 \times 10^{-10}$m	
Cd^{114}	$6440.2480 \times 10^{-10}$m	7×10^{-8}
	$5087.2379 \times 10^{-10}$m	
	$4801.2521 \times 10^{-10}$m	
	$4679.4581 \times 10^{-10}$m	

[a] Reproduced with permission from *J. Opt. Soc. Am.*

plot of the dispersion as a function of distance along the plate. To determine the wavelength of a specific spectral line, the distance between two standard lines and the unknown standard is measured; the nonlinear corrections are added to give the resulting wavelength of the unknown line. With more standard lines available, a better correction curve can be constructed. Because of the inherent limitations in the photographic technique and the effects of temperature and pressure on the wavelength used to measure the accuracy of a spectrograph, each photograph is unique, as is its calibration.

Corrections can be made to plates obtained from a prism spectrograph using a desk calculator. A convenient method is to use three standard lines in the region of interest and to fit the data into the Hartmann dispersion formula[73]:

$$\lambda = \lambda_0 + \frac{C}{(x - x_0)} \tag{6.58}$$

where λ_0, C, and x_0 are constants to be determined and x is the distance along the photographic plate to the line of interest. If a digital computer is available, it is more convenient to use a least-squares fit of all the measured standards to a fourth- or sixth-order polynomial approximation.

Several different methods of data reduction are used to reduce the data from grating spectrographs. At Argonne Laboratories a Paschen mounting has been constructed so nearly circular in configuration that the wavelength can be obtained to a satisfactory accuracy (.001 cm^{-1}) by using the grating formula $\lambda = d\,(\sin\theta + \theta')$. Few grating circles are this accurate, so that it is necessary to calibrate the instrument with wavelength standards. The dispersion being nonlinear, it is necessary to construct a correction curve (usually by a least-squares fit to a polynomial approximation). Caution should be exercised in selecting standards in that it is not usually possible (especially in old gratings) to compare lines in different orders (error of run). Because newer plane gratings are good enough to allow such comparison, much greater latitude can be used in the selection of standards. Essentially the same considerations apply to echelles.

The problem of wavelength measurement with a Fabry-Perot interferometer is unique, for a direct comparison is made of two wavelengths, the unknown and the standard. For this reason, only a single standard line is needed, and its wavelength can be anywhere in the spectrum relative to the unknown although it must be within the range of the film or optical detector used for intercomparison. The etalon is first calibrated. The optical length of the etalon nL is measured in terms of the wavelength of the standard line $m\lambda_0 = 2nL\cos\theta$. The etalon of known length is then used to measure the unknown wavelength.

There are several precautions to be observed in making the measurements described above. Because the index of refraction enters into the expression above, it is necessary that it remain constant for both the standard and the unknown exposures. This can be accomplished in part if the interferometer is enclosed in a pressure tight, temperature-controlled box. Even if the pressure and temperature conditions are identical, however (as they will be, for example, if one exposes both radiations simultaneously), the index of refraction of air exhibits dispersion[75], and this must be taken into account in the data reduction. Even when the precaution is taken to evacuate the etalon it is still necessary to correct for the dispersion of the etalon-surface coatings. For silver or aluminium surfaces this is a small effect and systematic methods exist for removing the error introduced. The corrections for dielectric films are larger. The reader is referred to the literature for extensive treatment of this correction[46].

The reduction of data from a Fabry-Perot interferometer photograph is tedious. If many patterns are to be analyzed, it is worth programming the results for a digital calculator. The wavelength is determined by measuring the ring diameters in the circular fringe pattern. These data are used to calculate ϵ, the factional-order number at the angle $\theta = 0$ (i.e., at the central position of the fringe pattern). The value of ϵ can usually be obtained to one part in 10^2. The integral portion of the order number must be obtained from a lower-precision measurement of the unknown wavelength (this can be done either on a grating or using a Fabry-Perot with a shorter etalon spacing). The integral order number p is added to the measured factional-order number to obtain $\xi = p + \epsilon$, the correct order number, to one part in 10^7 (p 10^5 and is known with zero error). Then, using the length of the etalon spacer I, determined by the standard wavelength (also known to better than $1{:}10^7$), the wavelength may be computed from the relation

$$\lambda = \frac{2nL}{p + \epsilon} = \frac{2nL}{\xi}. \qquad (6.59)$$

The accuracy of the measurement described above is limited ultimately by the accuracy with which nL can be measured. The latter measurement is limited by the accuracy of the wavelength standard. A very detailed description of the analysis of interferograms is given by Meissner[62, 63].

6.5 REFERENCES

[1] H. S. Coleman, W. R. Bozman, and W. F. Meggers, *Table of Wavenumber*, National Bureau of Standards Monograph 3, 1960.

[2] W. Heitler, *The Quantum Theory of Radiation*, Oxford University Press, New York, 3rd ed., p. 320.

[3] E. Fermi, *Rev. Mod. Phys.*, **4**, 87 (1932).

[4] A. Mitchell and M. Zemansky, *Resonance Radiation and Excited Atoms*, Cambridge University Press, New York, 1934, Ch. 3.

[5] D. Compton and A. Schoen, *The Mossbauer Effect*, Wiley, New York, 1961.

[6] R. D. Cowan and G. H. Dicke, *Rev. Mod. Phys.*, **20**, 418 (1948).

[7] H. Kopfermann, *Nuclear Moments*, Academic, New York, 1958, Ch. 1.

[8] H. E. White, *Introduction to Atomic Spectra*, McGraw-Hill, New York, 1934, Ch. 21.

[9] R. G. Breene, Jr., *The Shift and Shape of Spectral Lines*, Pergamon, New York, 1961.

[10] M. L. Dalton, Jr., *Appl. Opt.*, **2**, 1195 (1963).

[11] C. M. Sparros, *Astrophys. J.*, **44**, 76 (1916).

[12] W. E. Forsythe, *The Measurement of Radiant Energy*, McGraw-Hill, New York (1932).

[13] T. R. Hogness, F. P. Zscheile, and A. E. Sidwell, Jr., *J. Phys. Chem.*, **41**, 379 (1937).

[14] M. P. Lord, *Proc. Phys. Soc.* (London), **58**, 477 (1946).

[15] J. Strong, *Phys. Rev.*, **37**, 1661 (1931).

[16] J. Strong, *J. Opt. Soc. Am.*, **39**, 320 (1949).

[17] H. J. Babrov, *J. Opt. Soc. Am.*, **51**, 171 (1961).

[18] H. J. Babrov, *J. Opt. Soc. Am.*, **52**, 831 (1962).

[19] G. R. Harrison, R. C. Lord, and J. R. Loofbourow, Practical Spectroscopy Prentice-Hall, Englewood Cliffs, N.J., 1948, p. 189.

[20] Corliss, W. R. Bozman, and F. O. Westfall, *J. Opt. Soc. Am.*, **43**, 398 (1953).

[21] A. T. Forrester, R. A. Gudmundsen, and P. O. Johnson, *J. Opt. Soc. Am.*, **46**, 339 (1956).

[22] S. Tolansky, *High Resolution Spectroscopy*, Methuen, London, 1947.

[23] H. Chantrel, *Ann. Phys.*, **13**, 965 (1959).

[24] K. W. Meissner, *Rev. Mod. Phys.*, **14**, 68 (1942).

[25] D. A. Jackson and H. Kuhn, *Nature*, **134**, 25 (1934); *Proc. Roy. Soc.*, **A 154**, 679 (1936).

[26] K. W. Meissner and V. Kaufman, *J. Opt. Soc. Am.*, **49**, 942 (1959).

[27] R. L. Barger and K. G. Kessler, *J. Opt. Soc. Am.*, **51**, 827 (1961).

[28] K. G. Kessler and W. G. Schweitzer, Jr., *J. Opt. Soc. Am.*, **49**, 199 (1959).

[29] P. Connes, *J. Phys. Radium*, **19**, 262 (1958).

[30] C. Gardner and K. F. Nefflen, *J. Opt. Soc. Am.*, **50**, 184 (1960).

[31] Annual Report, 1959–60, N.S.L. University Grounds, *Chippendale*, NSW, p. 11

[32] H. Duong, *Decomposition des Raise Spectrales par Modulation en Haute Frequence*, Ph.D. Thesis, University of Paris, 1961.

[33] F. A. Jenkins and H. E. White, *Fundamentals of Optics*, McGraw-Hill, New York, 1957, 3rd ed., p. 300.

[34] M. Ciftan, A. Krutchkoff, and S. Koozekanani, *Proc. IRE*, **50**, 84 (1962).

[35] F. Paschen, *Ann. Physik*, **50**, 901 (1916).

[36] R. A. Sawyer, *Phys. Rev.*, **36**, 44 (1930).

[37] W. Ulrich, *Indust. Res.*, **28** (1963).

[38] S. Walker and H. Straw, *Spectroscopy*, MacMillan, New York, 1962, Vol. 2, pp. 22–44.

[39] P. H. Van Cittert, *Z. Phys.*, **65**, 547 (1930),**69**, 248 (1931).

[40] H. Kayser, *Handbuch der Spectroscopie*, Hirzel Leipzig, 1900, Vol. I., Ch. 5.

[41] R. Wallace, *Astrophys. J.*, **22**, 123 (1905).

[42] T. Thorp, British Patent 11460 (1899).

[43] T. Merton, *Proc. Roy. Soc.* (London), **A 201**, 187 (1950).

[44] G. W. Stroke, "Ruling, Testing and Use of Optical Gratings for High Resolution Spectroscopy," *Progress in Optics*, Interscience, New York, 1963, Vol. II.

[45] F. Paschen and C. R. Runge, *Abhand d. K. Akad. d. Wiss, Z.*, Anhang, Berlin, 1902, Vol. I.

[46] R. A. Sawyer, *Experimental Spectroscopy*, Prentice-Hall, Englewood Cliffs, N.J., 2nd ed., Ch. 6, 7.
[47] W. G. Fastle, *J. Opt. Soc. Am.*, **42**, 641, 647 (1952).
[48] M. Czerny and A. F. Turner, *Z. Phys.*, **61**, 792 (1930).
[49] A. Mickelson, *Astrophys. J.*, **8**, 37 (1898); *Proc. Am. Acad. Arts Sci.*, **35**, 109 (1899).
[50] W. Williams, British Patent 321534 (1926).
[51] W. Williams, *Proc. Opt. Conv.*, **2**, 982 (1926).
[52] W. Williams, *Proc. Phys. Soc.* (London), **45**, 699 (1933).
[53] R. Wood, *Phil. Mag.*, **20**, 770 (1910).
[54] R. Wood and A. Trowbridge, *Phil. Mag.*, **23**, 310 (1912).
[55] G. R. Harrison, S. P. Davis, and H. J. Robertson, *J. Opt. Soc. Am.*, **43**, 853 (1953).
[56] P. Connes, *Proceedings of the International School of Physics, Enrico Fermi Course 31, directed by C. H. Townes*, Academic, New York, 1964, p. 198 ff.
[57] P. Fellgett, *J. Phys.*, **19**, 187 (1958).
[58] P. Jacquinot, XVII Congress du GAMS, 1954, p. 25.
[59] D. R. Herriott, *Appl. Optics*, **2**, 865 (1963).
[60] D. R. Herriott, *J. Opt. Soc. Am.*, **52**, 31 (1962).
[61] A. Javan, W. R. Bennett, Jr., and D. R. Herriott, *Phys. Rev. Letters*, **6**, 106 (1961).
[62] K. W. Meissner, *J. Opt. Soc. Am.*, **31**, 405 (1941).
[63] K. W. Meissner, *J. Opt. Soc. Am.*, **32**, 185 (1942).
[64] P. Jacquinot, *Reports on Progress in Physics*, **23**, 267 (1960).
[65] A. C. Candler, *Modern Interferometers*, Hilger-Watts, London, 1951.
[66] R. Chabbal, *J. Rech. Centre Natl. Rech. Sci.*, **5**, 138–86 (1953).
[67] S. P. Davis, *Appl. Opt.*, **2**, 727 (1963).
[68] G. V. Deverall, K. W. Meissner, and G. J. Zissis, *J. Opt. Soc. Am.*, **43**, 673 (1953).
[69] Report of Commission des Étalons de Longueur d-Onde et des Tables de Spectres, J.C.O. (1962).
[70] C. J. Humphreys and E. Paul, Jr., *NOLC Reports*, U.S. Naval Ord. Lab., Corona, California, 1955–1960, Vols. 321, 326, 341, 352, 366, 376, 384, 390, 420, 429, 443, 454, 464, 503.
[71] J. Reader, K. W. Meissner, and K. L. Andrew, *J. Opt. Soc. Am.*, **50**, 221 (1960).
[72] F. S. Tonkins, and M. Fred, *J. Phys. Radium*, **19**, 409 (1958).
[73] R. A. Sawyer, *Experimental Spectroscopy*, Prentice-Hall, Englewood Cliffs, N.J., Ch. 9, Sec. 6.
[74] R. W. Stanley and K. L. Andrew, *J. Opt. Soc. Am.*, **54**, 625 (1964).
[75] B. Edlen, *J. Opt. Soc. Am.*, **43**, 339 (1953).

6.6 PRINCIPAL SYMBOLS, NOTATIONS, AND ABBREVIATIONS

Symbol	*Meaning*
A	Absorption loss in dielectric films, constant in Cauchy's equation, see text
B	Constant in the Cauchy dispersion equation
a	Width of entrance beam to prism spectrograph
B	Luminance of a spectral source
b	Prism base width
C	Constant

c	Free-space velocity of light
D	Diameter of a collimating lens
D_λ	Angular dispersion or dispersion
d	Distance between groves in a grating
F_σ	Free-spectral range
f number	Lens speed
f	Focal length of a lens
$\mathscr{F}$	Reflecting finesse
H, H_i	Irradiance, specific value thereof, see text
L	Spacer length in a Fabry-Perot interferometer
M	Mass in amu.
m	Order of interference
N	Grating constant in lines/mm
N_D	Limiting finesse determined by deviation from plate flatness
nm	Wavelength in nanometers (10^{-9}m)
P	Resolution luminosity product of an instrument
p	Integral order number of a Fabry-Perot ring pattern
R	Universal gas constant
$(R_0)_{\text{Interferometer}}$	Resolving power of interferometer
$(R_0)_{FP}$	Resolving power of a Fabry-Perot interferometer
$(R_0)_{\text{prism}}$	Resolving power of a prism
$(R_0)_{\text{grating}}$	Resolving power of a grating
R_d	Resolving power of an etalon as determined by an aperature
R_R	Resolving power of an etalon as determined by reflect-ivity of mirrors
R_x	Resolving power of an etalon as limited by plate flatness
r	Radius of an aperture
S	Plate factor
T	Absolute temperature
Δt	Increment in time
x	Variation from plate flatness
U, U_i	Energy, specific value thereof, see text
ΔU	Increment in energy
∇	Luminosity of an instrument
ϵ	Fractional order number of a Fabry-Perot ring pattern
h	Constant
λ, λ_i	Wavelength specific value thereof, see text
λ_{air}	Wavelength in air at STP
μm	Wavelength in micrometers (10^{-6} m)
θ, θ'	Angle of incidence or illumination, angle of diffraction

λ_{vac}	Free-space wavelength
$\Delta\nu$	Increment in frequency
$\Delta\nu_a$	Axial-mode spacing
$\Delta\nu_D$	Doppler line width
ν_{vac}	Free-space frequency of a light wave
ξ	Total-order number of a Fabry-Perot ring pattern
Δ_σ	Instrumental width
τ	Time
ϕ	Slit width

7

Measurement of Band Width and Temporal Coherence

7.0 INTRODUCTION

One of the more-subtle characteristics of a laser is its spectral band width. Measurement of the spectral character of laser radiation is complicated by the fact that the emitted energy, unless special precautions are taken, is composed of a number of simultaneous, discrete, spectral components. Ideally, these separate components correspond to the eigenmodes of the combined resonator and amplifying medium that comprise the laser. In a gas laser these spectral components are largely dependent upon the eigenmodes of the resonator; they vary in time in a reasonably slow way (due to the mechanical instability of the laser cavity). In a solid-state laser, wherein the specific gain is very high and the Fresnel number is large, and, wherein the optical characteristics of the medium vary during the output pulse in an essentially uncontrolled manner, both time resolution and considerable spectral range are required to supplement the spectral resolution if a complete history of the complex spectral output of the laser is to be recorded. In solid-state lasers the axial and angular-mode spacings may be so small that discrete spectral components may be separated as little as one hundred megacycles.

Chapter 7 concerns itself with some of the techniques used to determine the line width of the individual spectral components emitted by a quasi-monochromatic ($\Delta\omega/\omega << 1$) laser. The number of techniques available is limited by the extremely narrow spectral lines produced by lasers. In many cases the techniques that can be used are peculiar to lasers. They may use such specific laser characteristics as high spectral radiance or narrow angular beam width. In those cases wherein the spectral emission contains a large number of components, conventional spectroscopic techniques may be used both to determine the over-all band width of the laser radiation and to isolate a single component of the line width to be studied in detail.

In an ideal monochromatic wave field the amplitude of the electric field at any point is constant, whereas the phase varies linearly with time. A real laser does not produce monochromatic light because the output amplitude and phase undergo irregular fluctuations, the rapidity of which depends upon the effective spectral width $\Delta\nu$ of the laser. The laser only behaves like a monochromatic source during a time interval Δt, which is small compared with the effective reciprocal-spectral width $\Delta\nu^{-1}$.

Theoretically the band width of a spectral line, derived from a continuously operating (CW) laser, has a lower limit set by the spontaneous emission (noise) into the mode in which the laser is radiating. The limiting band width may be written as[1]

$$\Delta\nu_L \geq \frac{2\pi h\nu(\delta\nu_c)^2}{P[1 - \exp(h\nu/kT_e)]}, \tag{7.1}$$

where P is the power in the spectral line, $\delta\nu_c$ is the passive-cavity band width (assumed to be less than the flourescent line width), and T_e is the effective negative temperature defined by the population density in the upper and lower levels of the laser transition. This definition of limiting line width is the corrected form of the Schawlow-Townes[2] formula including the Shimoda[3] factor and the factor $\frac{1}{2}$ [4]. Equation 7.1 predicts theoretical laser bandwidths of the order of 10^{-3} Hz or less. In practice, at best, bandwidths of a few cycles are observed, for other factors dominate and cause a considerable increase in spectral width. In gas lasers it appears that a practical limit to an achievable line width is set by thermal fluctuations in the material used to support the laser mirrors and by inhomogeneities in the index of refraction of the amplifying medium. In a solid state, Q-switched laser, in contrast, it is the short duration of the laser pulse (through the uncertainty relation) that determines the line width. Regardless of the nature of the limiting mechanism the laser spectral width is always very narrow compared with emission from spontaneous or thermal light sources. It is this unusually narrow band width property of lasers that requires the development of special measuring techniques.

Two limitations that result from the nature of light source must be kept in mind. First, light emitted from any source is never strictly monochromatic. The radiating atoms of a thermal source emit at random for limited time durations as a consequence of the uncertainty principle. Consider one such wave train of duration $\Delta\tau$ and frequency ν_0; its power spectrum is the function

$$\left[\frac{\sin[\pi(\nu - \nu_0)\Delta\tau]}{\pi(\nu - \nu_0)\Delta\tau}\right]^2.$$

It may be shown that the total contribution from all the radiating atoms has the same power spectrum as an individual radiator, although the phase relation between the Fourier components of the frequency spectrum is random. The spectral width is related to the emission time through the uncertainty principle. The spontaneous emission or relaxation time of a typical atomic level is of the order of 10^{-8} sec; the corresponding spectral width is of the order of 10^8 Hz. In lasers excited atoms are stimulated to emit in phase so that effective wave trains of 10^{-6} sec are found in ruby lasers and 10^{-3} sec in gas lasers; the corresponding spectral widths are 10^6 and 10^3 Hz respectively.

The duration $\Delta\tau$ of the wave train and the effective spectral width $\Delta\nu$ are related for Gaussian line shape by the expression

$$4\pi\Delta\tau\Delta\nu \sim 1.$$

The notion of coherence time is associated with the fact that for intervals less than $\Delta\tau$ the source is monochromatic, and that, at a point in space, there exists a linear dependence or correlation between the amplitudes and phases of the wave train at two different times. For intervals exceeding $\Delta\tau$ two different wave trains exist, and there is no correlation.

The second limiting property of light sources is their finite extent. Natural-light sources are composed of a multitude of radiators, each one emitting a monochromatic wave of differing and random relative phase during its coherence time. Over a sufficiently small space-time interval there may be a high degree of correlation between the spatial-amplitude distribution resulting from the combined output of the radiators to form a wave front. If, however, there is no correlation of the amplitudes at successive instants of time, there is no spatial coherence. In lasers the spatially distributed radiators are forced to emit in phase and a region of spatial coherence exists. In the limiting case of a purely monochromatic source the phase relations between the radiators are constants, and the region of spatial coherence is infinite in extent.

7.0.1 Relationships between Bandwidth, Coherence Time and Coherence Length

We define an electromagnetic wave as coherent if its autocorrelation function is periodic, has the same period as the radiation, and has a constant peak amplitude. An electromagnetic wave for which the autocorrelation function is not periodic and for which the peak amplitude of the autocorrelation function decreases with time is antithetically defined as incoherent. We define a quasicoherent wave as one that has a periodic autocorrelation function for which the peak amplitude is not the same for sampling periods of arbitrary duration. These definitions are in accord with the Wiener—Khinchine theorem, which states that the autocorrelation function of a function is the Fourier transform of the power spectrum of that function[5]. Laser output radiation is thus coherent only in a very imprecise interpretation of the definitions above. The partial coherence of laser radiation is understandable from an application of the uncertainty principle. The quasiperiodic radiation of a laser is produced by processes which are described by statistical parameters only and is unlikely to have precisely the same period.

The degree of time coherence, that is the degree of linear dependence or correlation[6] between the values of the radiation field at a vector point r_1 in space at two different times t and $t + \tau$, is expressed quantitatively in terms of the autocorrelation function Γ_{11}

$$\Gamma_{11}(r_1 t) = \lim_{\tau\to\infty} \frac{1}{2T} \int_{-T}^{+T} E_1(r_1\, t + \tau) E_1^*(r_1 t)\, dt. \tag{7.2}$$

Similarly the degree of spatial coherence or cross correlation between the values of the radiation field at an instant of time at two points r_1 and r_2 in space is expressed in terms of the cross-correlation coefficient Γ_{12}:

$$\Gamma_{12}(r_1, r_2) = \lim_{\tau\to\infty} \frac{1}{2T} \int_{-T}^{+T} E_1(r_1, t) E_2^*(r_2, t)\, dt. \tag{7.3}$$

The area of spatial coherence is defined by that portion of space (or distance in one dimension) within which the cross correlation coefficient is larger than some prescribed value. Except for a constant factor, at $r_1 = r_2, \Gamma_{12} = \Gamma_{11}, (r_1,0)$ becomes the intensity at the respective points

$$I(r_1) = \Gamma_{11}(r_1, 0); \qquad I(r_2) = \Gamma_{22}(r_2, 0). \tag{7.4}$$

A highly monochromatic field may exist with little spatial coherence because spatial coherence at a point can be specified independently of the spectral-power density. The spectral-power density at a point r determines its time coherence as the spectral-power density in the Fourier transform of the autocorrelation function.

Although an incoherent source may exist, the fields radiated from the various incoherent areas will be partially coherent because there exists some correlation between field values from the incoherent areas at two points, sufficiently close in space. The time-space-coherent properties of a source may be expressed precisely as the mutual coherence function

$$\Gamma_{12}(r_1, r_2, t) = \lim_{T\to\infty} \int_{-T}^{T} E_1(r_1, t + \tau) E_2^*(r_2, t)\, dt, \tag{7.5}$$

which expresses the degree of linear dependence between two fields at two points separated by a constant time interval τ as time goes on.

It is convenient to normalize Γ_{12} with a quantity termed the degree of coherence γ_{12}

$$\gamma_{12}(r_1, r_2, t) = \frac{\Gamma_{12}}{[\Gamma_{11}(0)]^{1/2}\,\Gamma_{22}(0)]^{1/2}} \tag{7.6}$$

$$= \frac{\Gamma_{12}(r_1, r_2, t)}{I(r_1)I(r_2)}$$

where

$$0 \leq |\gamma_{12}| \leq |\tag{7.7}$$

and where $I(r_1)$, $I(r_2)$, and $\gamma_{12}(t)$ are measurables averaged over the time of observation.

7.0.1.1 Operational Definitions of Coherence We operationally define time and space coherence in terms of the visibility and persistence of fringe patterns obtained in suitable interferometer experiments, with quasimonochromatic light.

Consider an extended quasimonochromatic source behind an opaque screen that contains two narrow slits spaced by a distance a. The intensity at a point in the interference pattern due to slit $S_1(\xi_1)$ and $S_2(\xi_2)$ is given by

$$I(r) = I_1(r) + I_2\sqrt{I_1(r)}\sqrt{I_2(r)}\,\mathcal{R}\gamma_{12}(\tau) \tag{7.8}$$

where $\mathcal{R}$ is the real part of the degree of coherence $\gamma_{12}(\xi_1, \xi_2, \tau)$ the modulus of which is

$$I(r) = I_1(r) + I_2(r) + 2\sqrt{I_1(r)}\sqrt{I_2(r)} \times |\gamma_{12}(\xi_1, \xi_2, \tau)|\cos[\alpha_{12}(\tau) - 2\Pi\nu_0\tau)], \tag{7.9}$$

the argument of which is

$$\arg\gamma_{12} = \alpha_{12}(\xi_1, \xi_2, \tau) - 2\Pi\nu_0\tau, \tag{7.10}$$

and γ_{12} is the phase difference between the radiation fields at $S_1(\xi_1)$ and $S_2(\xi_2)$.

Depending upon whether $|\gamma_{12}|$ is one of two extremes, 0 or 1, the field is either completely incoherent or completely coherent for all values of ξ and τ. If $|\gamma_{12}| = 0$, $I_1(r) = I_2(r) = I_0$ and $I(r) = 2I_0$. The fields at a projection screen a distance ℓ from the slits add linearly, no fringes are formed, and interference effects are not observed. If, however, $|\gamma_{12}| = 1, \gamma_{12}(\xi_1, \xi_2 \tau) = \exp 2\pi j \nu_0 t$, the wave field is monochromatic at frequency ν_0, and

$$I(r) = 2I_0 [1 + \cos(\alpha_{12} - 2\pi\nu_0\tau)], \tag{7.11}$$

fringes are observed to vary periodically from $4I_0$ to zero.

If the field is partially coherent, $0 < |\gamma_{12}| < 1$ fringes of lower contrast will be observed:

$$I(r) = 2I_0 [1 + |\gamma_{12}| \cos(\alpha_{12} - 2\pi\nu_0\tau)],$$

$$0 < I(r) < 4I_0.$$

Michelson introduced the concept of fringe visibility V defined as

$$V = \frac{I_{max} - I_{min}}{I_{max} + I_{min}}, \tag{7.13}$$

where I_{max}, I_{min} refer to the adjacent maxima and minima in the interference pattern. It can be shown that to a good approximation

$$I_{max} = I_1(r) + I_2(r) + 2\sqrt{I_1(r)}\sqrt{I_2(r)}|\gamma_{12}(\tau)|, \tag{7.14}$$

$$I_{min} = I_1(r) + I_2(r) - 2\sqrt{I_1(r)}\sqrt{I_2(r)}|\gamma_{12}(\tau)|. \tag{7.15}$$

Hence the fringe visibility at the projection screens may be expressed in terms of the intensities of the two beams and their degree of coherence

$$V = \frac{I_{max} - I_{min}}{I_{max} + I_{min}} = \frac{2\sqrt{I_1(r)}\sqrt{I_2(r)}}{I_1(r) + I_2(r)} |\gamma_{12}(\tau)| \tag{7.16}$$

Thus γ_{12} may be deduced from measurements of $I_1(r)$, $I_2(r)$, I_{max}, and I_{min}. The argument of γ_{12} is obtained from

$$\text{agr}(\gamma_{12}) = \alpha_{12}(\xi_1, \xi_2, \tau) - 2\pi\nu_0\tau \tag{7.17}$$

by noting that the fringes produced by quasimonochromatic light are shifted laterally an amount

$$\Delta = \frac{\lambda}{2\pi} \frac{l}{a} \alpha_{12}(\tau) \tag{7.18}$$

relative to the fringes that would be formed if monochromatic cophasal light illuminates the slits. If $I_1(r) = I_2(r)$, then $V = |\gamma_{12}|$. Therefore a measurement of fringe visibility is a direct measurement of the degree of coherence.

The major disadvantage of using the technique described above for lasers is that the measure of $|\gamma_{12}|$ is limited to a length less than or equal to the diameter of the laser source[7]. Because $|\gamma_{12}| \sim 1$ for lasers when $a \sim 2$ cm, the quantity really being measured is the coherence by the double slit.[8] Measurements on a ruby laser 1/2 cm in diameter, using 0.00075 cm slits, spaced 0.00541 cm apart on the surface of a ruby rod, gave a multiple-image diffraction pattern on a screen at 32.4 cm. The number, location, and intensity of the maxima agreed well with the predicted interference pattern, even though the ruby was operated multimode.[9] Although this experiment gave confirmation of the existence of spatial coherence of the laser output over the face of the ruby, the measure of $|\gamma_{12}|$ was limited by the small value of slit separation. A simple modification of the measuring apparatus has been proposed that should remove this restriction.[10]

7.0.1.2 Measurement of Coherence Time In principle the coherence time of a laser can be measured by determining the visibility function in a Michelson interferometer, with τ representing the adjustable time delay between the two components of an amplitude-split beam[11]. It can be shown that the visibility as a function of the path difference or time delay is related to the Fourier transform of the spectral intensity[12]. Thus a monochromatic source of band width $\bar{\nu}_0 \pm \Delta\nu/2$ wherein $\Delta\nu \rightarrow 0$ will yield a fringe pattern of the form

$$I(r) \sim 1 + \cos(2\pi\nu_0\tau) \tag{7.19}$$

and a visibility function

$$V(\tau) = 1. \tag{7.20}$$

If the source is quasimonochromatic and has a rectangular intensity spectrum of width $\Delta\nu$, the fringe pattern will contain the $\sin x/x$ modulation characteristic:

$$I(\tau) \sim 1 + \left(\frac{\sin(\pi\Delta\nu\tau)}{\pi\Delta\nu\tau}\right)\cos(2\pi\bar{\nu}\tau), \tag{7.21}$$

and the visibility function will have zeros, the first of which is a measure of the coherence time, namely,

$$V(\tau) = \frac{\sin(\pi\Delta\nu\tau)}{\pi\Delta\nu\tau}. \tag{7.22}$$

Real spectral-line sources are neither of infinitesimal spectral width nor have a constant intensity spectrum; consequently the analyses above only serve as guide lines. The visibility function of a single-line, incoherent spectral source is found to decrease with time delay very nearly as a Gaussian function so that a distinct measure zero of $V(\tau)$ does not exist. The spectral-line shape may, in principle, be inferred from the $1/e$ point on the visibility function assuming a Gaussian spectral-line profile. This method of line-shape determination (and therefore coherence time measurement) is obviously inaccurate if the visibility function changes slowly with path difference (which turns out to be true for gas lasers where fringe visibility does not change appreciably over several hundred meter path differences). Thus, while we can, in principle, use a Michelson interferometer to determine the coherence time of lasers, application of the classical technique to the gas laser is precluded on practical grounds by the extremely narrow band character of laser radiation. As will be shown below, the technique can be used on pulsed, solid-state and Q-switched lasers.

A meaningful technique for measuring the coherence time of CW-gas laser has yet to be discovered. Experiments are in progress to determine whether photon-counting techniques can be successfully employed to measure the coherence properties of light fields. The experimental method is cumbersome and requires the use of counting techniques capable of recording single electrons in nanosecond intervals[16].

7.0.1.3 Measurement of Coherence Area and Length The discussion of section 7.0.1.1 suggests a way of measuring the spatial coherence and coherence length between two points $\Gamma_{12}(\xi_1, \xi_2, 0)$ in terms of the visibility function. If the visibility function of the interference pattern formed at the screen, described in section 7.0.1.1, is examined as a function of the slit separation for the path difference $c\tau = 0$, the interval over which the visibility function exceeds a preassigned value is the coherence length.

If, as in section 7.0.1.2, the source is truly monochromatic, the observed fringe pattern will be periodic in τ and the visibility function will be a constant. Similarly a quasimonochromatic source of rectangular spectral intensity will yield a fringe pattern that contains a $\sin x/x$ modulation characteristic. The visibility function will have an envelope of similar character.

Application of the principle presented above to the measurement of the spatial coherence of CW-gas lasers is once again precluded by practical considerations. Even in the case of pulsed, solid-state lasers, the source is of finite size, so that the available slit separation is limited. Furthermore, with gas lasers, it is expected that the spatial coherence is so great that the variation in the visibility function is indistinct. A meaningful technique for quantitative determination of coherence length has yet to be devised for CW gas laser sources.

7.0.2 Effect of Degree of Coherence on Beam Width

The degree of coherence $0 < \gamma < 1$ is important in that it relates to the quantity of the energy of a beam that can be focused into a region, the dimensions of which are of the order of the wavelength[14, 15]. The degree of coherence of radiation emanating from two points may be related to the diffraction pattern of the source, regarded as an aperture of specified phase and amplitude distribution through the van Cittert—Zernike theorem. This theorem equates the degree of coherence to the absolute value of the normalized Fourier transform of the intensity function of the source. A value of $\gamma > 0.88$, which corresponds to the first zero of the Bessel function that describes the degree of coherence as a function of the source, is sometimes used to determine the lower limit of spatial coherence[13]. Using this concept of quasicoherence it can be shown that the diameter d of a circular area illuminated almost coherently by a quasimonochromatic, uniform, circular, noncoherent source of angular radius θ

$$\theta = \frac{D}{2R} \tag{7.23}$$

is

$$d = \frac{0.16}{\theta} = \frac{0.32R\lambda}{D}, \tag{7.24}$$

where D is the diameter of the source at a distance R. This result is useful in determining the specifications for a source needed for studies of interference and diffraction phenomena.

To obtain the lower limit of spatial coherence $\gamma = 0.88$ over a 1-cm diameter at 500 nm, the source must be so far removed that its angular radius θ is

$$\theta = \frac{0.16\lambda}{d} = (1.6 \times 10^{-1})(5 \times 10^{-5}) = 8 \times 10^{-6} \text{ rad.}$$

Thus incoherent sources must be located at astronomical distances to yield reasonable degrees of coherence.

If a radiator is Lambertian, so that the radiated flux density varies as the inverse square of the distance from the source, the relative flux available will be proportional to $(0.16\,\lambda)^2$, or 6.4×10^{-9}, of the emitted flux density for the case considered. The relative advantages of a coherent source for focussing energy and obtaining measurements that depend upon interference effects are indeed great.

7.0.3 Effect of Amplitude and Phase Instabilities on the Shape of the Power Spectrum of a Laser

It is clear from the form of (7.2) that any periodic function has a periodic autocorrelation function. The period of the function and the autocorrelation function are the same. It is intuitively obvious that the peak amplitude of the autocorrelation function may or may not vary as the sampling time is changed over wide ranges. The dependence of the peak amplitude of the autocorrelation function on the sampling time reflects the coherence properties of the function. Clearly the degree of coherence is related to the duration of sampling time necessary to reduce the peak amplitude of the autocorrelation function by a defined factor. Most quasi-coherent functions are considered as coherent provided the coherence time is long enough to satisfy the requirements of the intended application.

The problem of interpreting the coherence properties of a wave can be expressed by an evaluation of the indeterminancy of the wave characteristics, phase (or frequency), and wave form. Both characteristics are a function of the sampling interval duration. Phase indeterminancy is a function of fluctuations in periodicity, whereas, wave form indeterminancy is related to how well the amplitude of the function can be reproduced. Both instabilities affect the autocorrelation function in exactly the same way. Any instability reduces the amplitude of the autocorrelation function for sampling intervals of the same order of magnitude as the average period of the instabilities.

Consider for example an ideal case wherein the laser emits a constant-amplitude, monochromatic, CW output. The power spectrum of this ideal source would be a constant-amplitude signal concentrated at a single frequency. Consequently the peak amplitude of the autocorrelation function would be independent of the sampling time, and the laser would be capable of forming interference fringes of maximum visibility at any distance.

Were we to resolve the power spectrum of a single-frequency laser source with a resolution adequate to measure frequency and amplitude fluctuations, we would be able to associate an autocorrelation function with the power spectrum. If the emitted laser output varied randomly in amplitude by small amounts, the power spectrum would show weak side bands distributed symmetrically on either side of the laser frequency. Provided that the frequency, as determined by the number of zero crossings counted per unit of time, remained independent of time, the phase stability of the wave would be unaffected and the quasicoherent signal would have an autocorrelation function that fluctuates both in amplitude and period.

Should the frequency of the laser exhibit fluctuations (in addition to the above-mentioned amplitude fluctuations) the power spectrum would reveal a broad spectrum consistent with the frequency fluctuations, and symmetrically disposed about the most probably frequency would be the sidebands associated with the amplitude fluctuations.

Were we to determine the autocorrelation function of a random source of radiation, for example, an ordinary thermal light source, we would note that the autocorrelation function would not be periodic. Furthermore, the envelope of the autocorrelation function would be found to decrease rapidly to zero as the sampling interval increased. This type of signal is incoherent by definition.

Laser radiation is generated by stimulated emission which results in partially timed emission of photons. The timing and number of quanta emitted at any instant are described by statistical parameters, such as the average number of photons emitted or the average emission rate. The resultant power spectrum of the laser radiation is thus more or less narrow and its autocorrelation function shows a behavior similar to that of a sinewave generator the output of which exhibits amplitude and phase instabilities.

The coherence time of a laser signal is found to be orders of magnitude shorter than that of a radio-frequency generator having the same degree of coherence. Thus it is not the absolute period of coherence but the coherence time measured in units of the period of the radiation signal that is the significant parameter. For example an extremely stable klystron oscillator can generate a 100 GHz signal with a period of coherence of 100 sec corresponding to 10^{13} periods. A laser signal with a corresponding degree of coherence would have a coherence period of the order of 10^{-3} sec.

7.0.3.1 Interferometric Measurements of Coherence Time The relationship between the shape of a spectral line and the Michelson visibility function will be used to determine the average coherence time of a solid-state laser. As previously noted a two-beam interferometer that samples the whole beam is capable of measuring the coherence time. Consider the Twyman-Green modification of the Michelson interferometer[12], shown in Figure 7.1. Light from the laser is collimated (if necessary) and amplitude divided at the beam splitter. After the two separate beams traverse the interferometer, they are recombined and yield a set of interference fringes. These fringes are localized in the vicinity of a plane P, which is located between the images of the two interferometer mirrors formed by lens L_1. The spacing of these fringes is determined by the relative alignment of mirrors M_1 and M_2. The optical paths for the two interfering beams differ by $l = l_2 - l_1$ (assuming a refractive index of unity), and the corresponding time delay is $\tau = l/c$, where c is the velocity of light. The visibility function of the fringe pattern is defined by (7.13), in which it is assumed that the intensities of the two beams are equal. Equation 7.13 can be used to determine the value of l and therefore τ by measurements of $I_{\max}$ and $I_{\min}$. It may be shown that, within

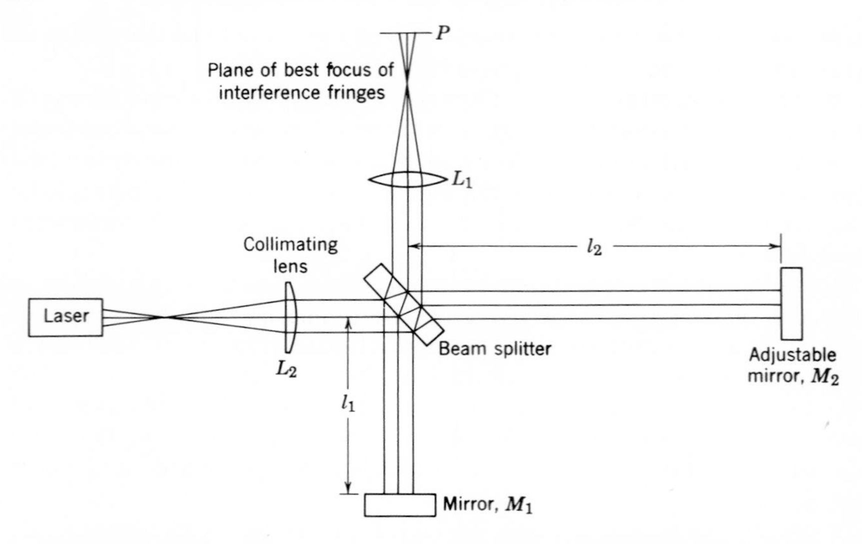

Figure 7.1. Twyman-Green interferometer for the measurement of coherence time.

certain limitations (such as the assumption of a symmetrical line shape), the spectral profile of the line may be determined from the visibility function by a straightforward Fourier inversion[23]. Figure 7.2 illustrates the relationship between the spectral profile and visibility function for specific line shapes.

In many practical applications using lasers it may be sufficient to determine the coherence time of the radiation rather than calculate the specific form of the spectral profile. As Born and Wolf show, the coherence time Δt is equal to the normalized rms width of the square of the visibility function[24]:

$$(\Delta t)^2 = \frac{\int_0^\infty \tau^2 V^2(\tau)\, d\tau}{\int_0^\infty V^2(\tau)\, d\tau}. \tag{7.25}$$

For rough calculations it is often sufficient to find the time delay for which the visibility function becomes small compared to unity and then to assume some functional form (such as a Gaussian) for the visibility function, so that the coherence time may be calculated from equation (7.25).

The major difference between the determination of the visibility function for a laser and for a conventional thermal source is the relative magnitude of the range of time delays required. Even in the case of a thermal source with a very narrow band width (such as an isotope lamp), coherence time rarely exceeds a few nanoseconds—corresponding to coherence lengths of a few feet. For a laser, in contrast, coherence times are generally somewhere between a few hundred nanoseconds (in the case of pulsed, solid-state lasers) and several milliseconds (in the case of gas lasers that are moderately well frequency stabilized). In the latter case path differences of several hundred miles or more would be needed before the fringe visibility would begin to decrease appreciably. Needless to say, this consideration renders the technique highly impractical for gas lasers. In the case of pulsed, solid-state lasers, path differences of the order of 10 m or more are adequate to measure the visibility function (an exception being the Q-switched laser,

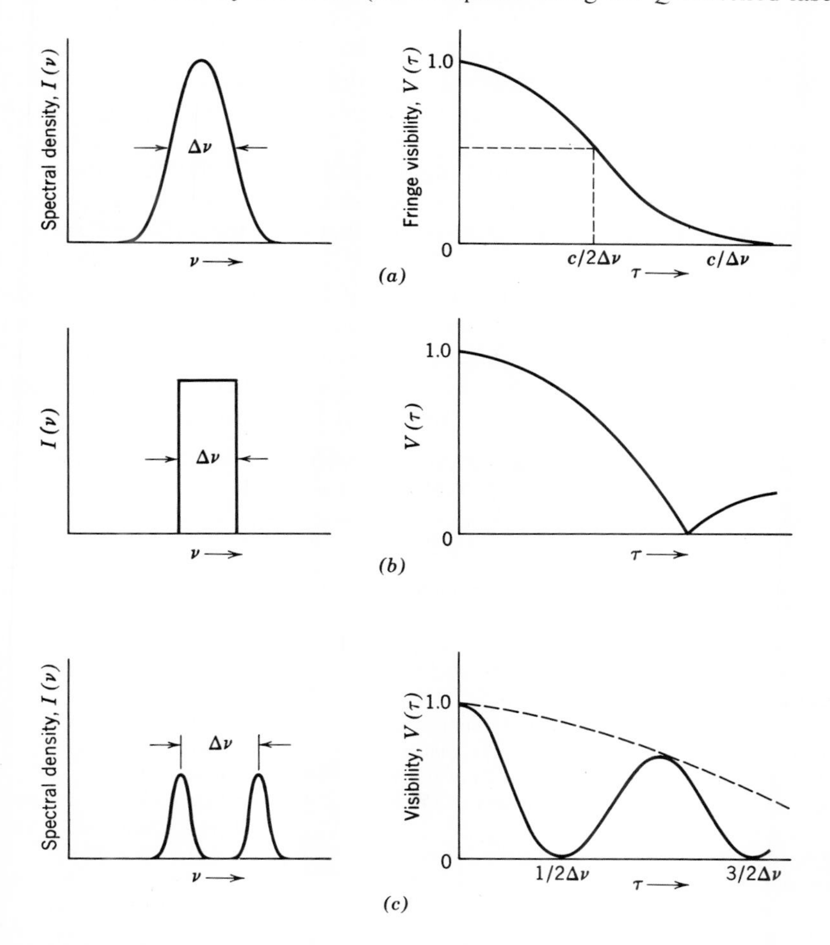

Figure 7.2. Relationship between spectral density and fringe visibility. (*a*) Gaussian line; (*b*) step-function line; (*c*) line consisting of close doublet.

which may have a coherence length of only a few feet). The coherence time of a pulsed-ruby laser has been measured using this technique[25]. It was found that a path difference of approximately 30 m was sufficient to reduce the fringe visibility to a value on the order of 1 percent. Fortunately, the small angular divergence, that is a property of a laser beam, minimizes some of the problems that arise with long path differences. Nevertheless, it is doubtful that this technique, for indirectly measuring the spectral line widths of lasers, will prove practical except in very special cases. As path differences are increased to permit the measurement of longer and longer coherence times, the problems of atmospheric turbulence and scattering become more and more troublesome, quite apart from the obvious inconveniences of excessive path lengths. Moreover, as the coherence time increases, the band width decreases, requiring even greater path differences. The heterodyning techniques become increasingly appropriate as the spectral range narrows.

When a spectral profile is multiple peaked, the visibility function is not a monotonically decreasing function of τ, but rather it goes through periodic minima, due to destructive interference between the different spectral components of the source. This effect is particularly noticeable in gas lasers, where successive axial modes are equally spaced. Caution should be exercised in interpreting coherence times measured with gas laser sources. It is generally advisable to eliminate unwanted spectral components by filtering, prior to attempting line-width measurement. The equipment listed in Table 7.1 will be found convenient for measuring the coherence

TABLE 7.1 APPARATUS FOR MEASURING THE COHERENCE PROPERTIES OF A Q-SWITCHED RUBY LASER

Item	Function	Source
Spectral light	Cadmium incoherent-wavelength standard	Ealing Corporation Hilger-Watts (FL-133)
Photomultiplier	Beam-intensity monitor	RCA (P2)
Precision dc regulated power supply	Dynode supply	NJE S-325
Lenses, apertures	Light collimation	Ealing Corporation
Notch filter	Monochromatizer for 643.8 nm and 694.3 nm	Thin Film Products, Inc., 1.0 nm filters
Oscilloscope	Intensity monitor	Tektronix Model 531 with type C plug-in unit
Twyman-Green interferometer	Coherence-time measurement	Tropel Model L-35
Pulsed-ruby laser with Lummer-Gehrcke Q-switch	Q-switched visible laser	TRG Model 104 Laser and Model 109 Lummer-Gehrcke attachment
Polaroid camera	Photograph fringes	Polaroid

time of a Q-switched ruby laser. A Q-switched (short-pulse) laser is chosen because the reciprocal line width is amenable to fring-visibility measurements with conventional laboratory equipment. If laboratory space permits, the coherence time of the same laser may be measured under ordinary (10^{-3} sec) long-pulse conditions.

If the Cd source is used, we arrange the equipment as illustrated in Figure 7.3. The Twyman-Green interferometer is used in this measurement. The collimated-light output of the cadmium lamp is filtered with the 643.8 nm narrow-band filter to remove extraneous radiation background. The standard lamp will prove invaluable to align the system under steady-state conditions. It will also serve to test the system prior to operation of the Q-switched laser. The interferometer is aligned to produce fringes with the reference and movable mirrors approximately equidistant. The telescope and Fresnel plate will facilitate alignment.

With the interferometer aligned, isolate a fringe in front of the photomultiplier with the mask and measure the light intensity at the fringe maximum and the adjacent minimum. Record the results and repeat the measurements when the movable mirror is moved in increments of approximately 5 cm. Using (7.13) we calculate the value of the visibility function

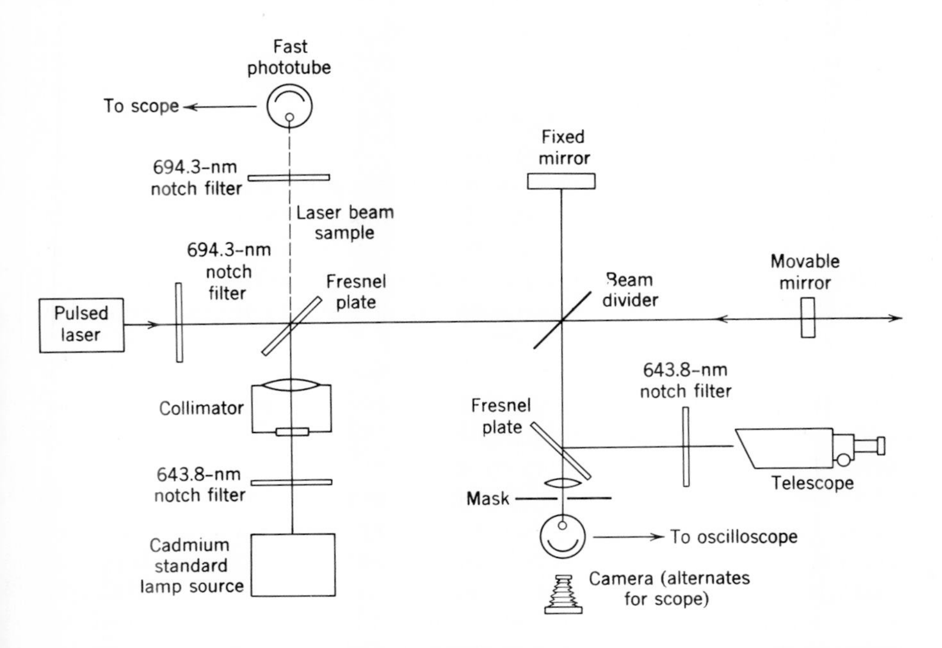

Figure 7.3. Apparatus arrangement for the measurement of coherence time with a two-beam interferometer.

and plot V as a function of the path difference between the two mirrors. The spectral line shape for this quasimonochromatic source can be computed from the data if the assumption is made that the spectral line is Gaussian in shape.

It can be shown that the visibility function reduces to half-value when the path difference is $\pi/\Delta k$, where Δk is the full width of the spectral line at half-value. The apparatus and technique may be tested against the known width of the cadmium line ($\sim 13 \times 10^{-4}$ nm). Fringes may be observed at 643.8 nm for path differences of the order of 30 cm corresponding to about 500,000 wavelengths. The coherence time of the cadmium source follows from

$$\Delta\tau \sim \frac{\lambda^2}{4\pi c\, \Delta\lambda},$$

$$\Delta\tau_{\mathrm{Cd}} \sim 10^{-9}\ \mathrm{sec}.$$

When the apparatus is aligned and tested, the cadmium source may be replaced by the laser. The photomultiplier is replaced by a camera. The movable mirror position is varied to achieve a path difference corresponding to fringe "washout." This approximately corresponds to a path difference of $2\pi/\Delta k$; the location of this point is imprecise. It should, however, correlate with the measured output pulsewidth $\Delta\tau$ of the laser because $\Delta\ell$, the total path difference, is approximately $c\Delta\tau$.

Because the Q-switched laser output pulse is composed of many discrete frequencies, the visibility pattern will be somewhat blurred. It will also be found that the apparent fringe pattern washout location will change from pulse to pulse and during a given pulse. The best results that will be obtained in these measurements is an averaged coherence time for the Q-switched laser.

An appreciation of the improved coherence time of the laser will be obtained if the fringe washout pattern photographed under Q-switched conditions is compared with a photograph for the same mirror location under long-pulse laser operating conditions.

Although it must be admitted that the analysis of coherence time by measurement of the visibility function is both indirect and somewhat uncertain, it nonetheless confirms other measurements of coherence time and yields an appreciation of, especially for solid-state lasers, the constraints that must be considered when applications are envisioned that demand the utilization of interference phenomena.

7.0.4 Average Coherence of Two Laser Sources

Stabilized CW-gas lasers have a degree of coherence such that interference phenomena can be produced over distances of several hundred meters. Furthermore, with two identical lasers operating in similar environments, it is possible to observe interference fringes between two different lasers. These fringes may be observed to move past a reference point with a frequency just equal to the difference frequency between the two lasers. This beat frequency may be thought of in terms of the correlation function between the two laser sources. An average coherence time may be expressed for the two lasers defined as twice the time interval during which the phases of the beat note remains sensibly constant. In other words the average coherence time of two independent lasers may be identified with the coherence time of the beat note[39, 40].

Some conceptual difficulties arise when we define the time coherence in terms of the correlation function of the two independent sources whose frequencies are different but whose phases have a definite relationship. It has been proposed that this difficulty can be resolved if the degree of stationarity of the coherence function is specified during the observation time T[14, 15]. Thus a mutual coherence function may be constructed for a finite T:

$$\Gamma_{T_{12}} = \frac{1}{2T} \int_{t-T}^{t+T} E_1(t+\tau) E_2(t)\, dt. \tag{7.26}$$

If $E_1\ (t) = \exp 2\pi j\nu_1 t$ and $E_2 = \exp 2\pi j\nu_2 t$, it follows that for $\Delta\nu = (\nu_1 - \nu_2)$

$$\Gamma_{T_{12}} = \exp 2\pi j(\nu_1\tau + \Delta\nu t) \frac{\sin(2\pi\Delta\nu T)}{2\pi\Delta\nu T}, \tag{7.27}$$

and interference phenomena can be observed between the two dependent lasers for observation times of

$$T = \frac{1}{4\Delta\nu}.$$

7.1 INTERFEROMETRIC MEASUREMENTS OF LINE WIDTH

In many instances it is either impractical or impossible to use more conventional spectroscopic techniques to measure the line width of a laser. Using Fabry-Perot etalons of sufficiently high spectral resolution, it has been found possible to make line width measurements of the broad-spectral laser emitters, neodymium-doped glass, ruby, and gallium arsenide. A practical dividing line between line widths, which one might reasonably hope to measure directly (using interferometric techniques), and line widths so narrow as to defy direct measurement, would be on the order of a megahertz. For line widths less than a megahertz, heterodyning techniques are used. Other secondary measures of line width that employ the statistical characteristics of laser noise are now frequently used to measure down to the kilohertz range.

7.1.1 Resolution-limiting Characteristics of the Fabry-Perot Interferometer

By far the most useful interferometric device for high-resolution spectroscopy is the Fabry-Perot interferometer, or "etalon." The Fabry-Perot interferometer is capable of high spectral resolution; it is simple to use and is relatively inexpensive. A Fabry-Perot etalon consists of a pair of partially transparent, highly reflecting, plane and parallel (slightly prismatic) glass or quartz plates. The separation between the coated surfaces of the plates is fixed by a rigid spacer element having a small temperature coefficient of expansion. In its most common spectroscopic application, the Fabry-Perot etalon is used as a filter of narrow angular aperture. When irradiated by a source with a narrow spectral range, the etalon will transmit only at certain angles; this gives rise to a series of concentric rings in the focal plane. The irradiance in the focal plane of the lens is given by[17]

$$I = \frac{I_0}{(1 + A/T)^2}\left[\frac{1}{1 + 4R\sin^2[2\pi nL\cos\theta/\lambda](1-R)^2}\right] \tag{7.28}$$

where I_o is the irradiance in the absence of the etalon, A, T, and R are respectively the absorption coefficient and geometric mean-power transmissivity and reflectivity of the plates spaced at a distance L, nL is the optical path between the plates, and θ is a measure of the angle of the incoming beam relative to the normal to the plates. The focal length of the lens is f, and r is the distance measured on the film from the center of the ring pattern ($r = f\theta$). For a monochromatic source of wavelength λ, bright fringes occur for angles θ satisfying the condition

$$m\lambda = 2nL\cos\theta \qquad (m = 1, 2, \ldots). \tag{7.29}$$

A Fabry-Perot etalon has a free spectral range F_σ, a wavelength range in which no overlap occurs in adjacent spectral orders given by

$$F_\sigma = \frac{\lambda^2}{2nL}, \tag{7.30}$$

or, in terms of wave numbers and plate spacing, $F_\sigma = 1/2nL$.

If the spectral range employed is small, unwanted spectral components must be filtered out prior to measuring widths of single components. Filtering may be accomplished with notch filters or by using an auxiliary Fabry-

Perot etalon of lower spectral range (closer spacing of plates). The minimum theoretical band width that may be resolved, that is the instrumental width (assuming perfectly parallel and plane plates of infinite extent), is given[17] by

$$\Delta\nu_{\min} = \frac{c(1-R)}{2\pi nLR^{1/2}} \approx \frac{c(1-R)}{2\pi nL} \qquad \text{for } R \approx 1, \tag{7.31}$$

or, in terms of wavelength,

$$\Delta\lambda_{\min} = \frac{\lambda^2(1-R)}{2\pi nL\sqrt{R}} \approx \frac{\lambda^2(1-R)}{2\pi nL} \qquad (\text{for } R \approx 1), \tag{7.32}$$

or, in terms of wave numbers,

$$\Delta k = \frac{(1-R)F_\sigma}{\pi\sqrt{R}}. \tag{7.33}$$

In practice, however, the minimum resolvable band width is usually limited by poor plates, misalignment, or the finite size of the aperture.

It may be shown that an rms deviation in plate surface flatness of an amount ΔL will give rise to an apparent spectral width for a monchromatic line, which is given by

$$\Delta\lambda = \frac{\lambda\Delta L}{L} \quad \text{or} \quad \Delta\nu = \frac{\nu\Delta L}{L} \tag{7.34}$$

where L is the plate spacing. Thus the spectral resolution $R_o = v/\Delta v_{\min}$ of a Fabry-Perot etalon is reduced by a factor of 2 when the deviation in surface flatness is

$$\Delta L > \frac{\lambda(1-R)}{2\pi n}. \tag{7.35}$$

For example, if $\Delta L = \lambda/100$ and $n = 1$, the theoretical resolving power is reduced by a factor of 2 when the reflectivity is approximately 94 percent. This implies that if one is to take advantage of the high reflectivities available from multilayer dielectric mirrors, one must obtain excellent plates[18], align them properly, and maintain them free of distortion.

In practice, we often attempt to avoid regions of poor plate quality by using a small aperture in the vicinity of the etalon. This will, however, prove useful only up to the point where diffraction begins to limit the spectral resolution.

It may be easily shown that diffraction at an aperture of diameter D limits the minimum resolvable band width of a Fabry-Perot etalon to a value given by

$$\Delta\lambda_{\text{DIFF}} \approx \frac{\lambda^3}{2D^2} \qquad \text{(on axis; } \theta \leq \lambda/D\text{)}. \tag{7.36}$$

or, off axis

$$\Delta\lambda_{\text{DIFF}} \approx \frac{\theta\lambda^2}{D} \qquad \text{(off axis; } \theta > \lambda/D\text{)}. \tag{7.37}$$

Obviously the theoretical resolving power is reduced by a factor of 2 when the diameter of the limiting aperture has a value

$$D = \left(\frac{4\pi\lambda nL}{1 - R}\right)^{1/2} \qquad \text{(on axis, } R \approx 1\text{)}. \tag{7.38}$$

Thus for an etalon with $nL = 10$ cm, $D = 5$ mm, and working at a wavelength of 5000 Å, the spectral resolution is predominantly diffraction limited for reflectivities greater than about 97.6 percent.

As a result of the brief consideration of the effects of plate imperfections and diffraction given above, it should be apparent that it is not possible to obtain unlimited spectral resolution from a Fabry-Perot etalon merely by increasing the reflectivity of the plates. Furthermore, we must exercise considerable caution in interpreting data obtained from a Fabry-Perot etalon; in particular, we almost never obtain the theoretical band width of the interferometer, and use of (7.31) is seldom justifiable even though we can easily obtain modern dielectric coatings with reflectivities in excess of 99 percent.

7.1.1.1 Technique for Measuring Pulsed, Solid-state Laser Line Width Three line widths are of interest in the study of solid-state lasers. The fluorescence line width, which depends upon the active ion in the laser; the cavity-mode width, which depends upon the cavity Q; the laser line width, which depends upon the number of excited modes. Usually the fluorescence line width is the broader of the lines, as shown in Figure 7.4 (Dy:CaF_2 is a notable exception). This discussion considers the common cases where the fluorescence line is the broadest, for the technique also is applicable to the narrow-line case.

Cavity modes are spaced at frequencies that have an integral number of waves in the total optical length of the laser. In the visible spectrum the band width of the fluorescence is usually broader than the cavity-mode width, so that there are usually several cavity modes oscillating simultaneously. For this case it is convenient to define the laser line width (imprecisely)

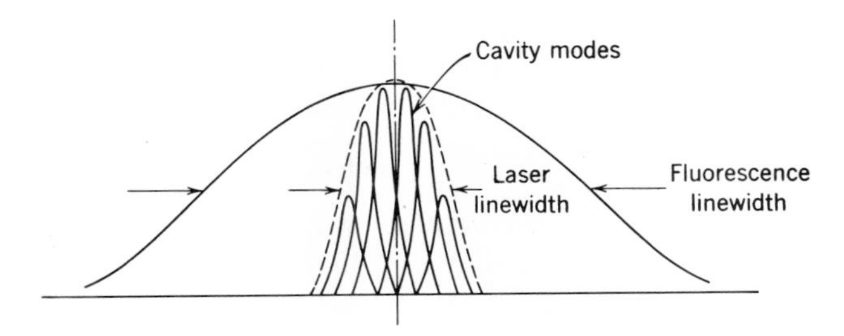

Figure 7.4. Relationship of linewidths associated with most pulsed, solid-state lasers.

as the width of the envelope of the oscillating modes. The laser line width as defined is gain sensitive and therefore depends upon the active material and the degree to which the laser is pumped above threshold. The line width also depends upon the laser medium. For example, the output from a Nd: glass laser may be as wide as 2 nm: in contrast or ruby single-mode output can be as narrow as 0.0001 nm.

The narrow line width and wavelength of a laser define the type of apparatus that may be used to examine the line width. Basically, these measurements require an element to resolve the wavelength and a suitable means of detecting the radiation.

Techniques for measuring the line width of a pulsed laser differ from those used for ordinary spectral measurement as the lines are exceedingly narrow and the wavelength often changes with time during the short observation period.

When the laser wavelength is in the visible region of the spectrum, conventional photographic techniques may be employed. The output from a single pulse or a succession of pulses can expose a strip of film in the focal plane of a grating spectrometer. The degree of detail that can be observed will depend upon the resolution of the instrument available. The maximum resolution of a grating instrument is given by (8.59):

$$(R_0)_{\text{grating}} = \frac{\lambda}{\Delta\lambda} = Nm \tag{7.39}$$

where λ is wavelength, N the order, and m the total number of lines on the grating.

Individual modes of the output from a ruby laser were observed photographically with the 40 ft, MIT high-resolution grating spectrometer[26]. The spectral band width of this instrument was 0.02 cm^{-1} at 694.3 nm. Light from as many as 10 longitudinal cavity modes exposed the photographic film.

The output from a U:CaF_2 laser operating at 2.51μm has been observed with a grating spectrometer and a PbS photocell,[27]. Observations were made by scanning the detector across the exit slits with a precision micrometer and by measuring the peak output from the detector with an oscilloscope. One laser pulse provided a single data point, and a curve was generated from the points taken as the detector was stepped across the exit slits. The dispersion of the spectrometer was 0.33 nm/mm and the spectral bandwidth was 0.05 cm^{-1}. Individual axial modes of a crystal 2.46 cm long were resolved. The slow response of the PbS detector integrated the relaxation spikes, thereby limiting the time resolution. An infrared detector with microsecond response could have been used to obtain time-resolved spectra.

A common method of measuring the spectral properties of a solid-state laser is to use a Fabry-Perot etalon as shown in Figure 7.5. When this equipment is to be used with a laser, the entering beam of light is assumed to be more or less collimated, with a diameter D. Lens L_1 serves to image the circular fringes in the plane of the film P (L_1 need not be a well-corrected lens, providing that the system is focussed using light of approximately the same wavelength as that from the laser). Lens L_2 is a field lens the optical quality of which is of no consequence and whose function it is to distribute light into the desired area in plane P. The focal length of L_2 determines the number of circular fringes that will be illuminated according to

$$N = \frac{D^2 n L}{4 f_2{}^2 \lambda} \tag{7.40}$$

where N is the number of fringes (order).

After the film is exposed, the spectral band width is determined by measuring the width of the fringes located near the center of the fringe pattern. A circular fringe of radius r and thickness Δr corresponds to a

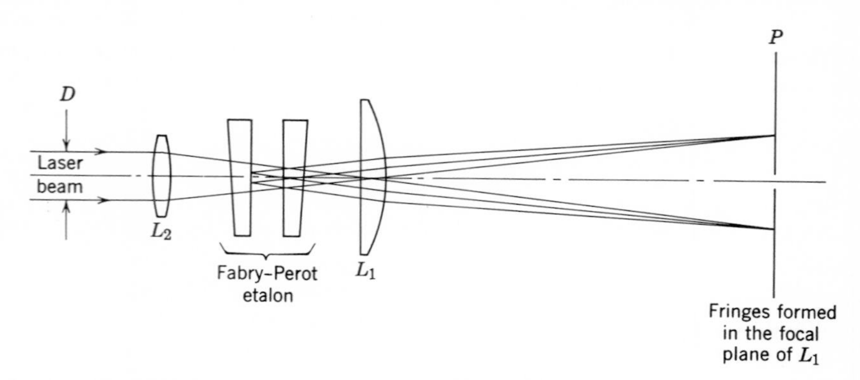

Figure 7.5. Optical arrangement for the conventional use of the Fabry-Perot etalon.

spectral band width given by

$$\Delta\nu = \nu \frac{r\,\Delta r}{f_1^{\,2}} \quad \text{or} \quad \Delta\lambda = \lambda \frac{r\,\Delta r}{f_1^{\,2}}, \tag{7.41}$$

assuming infinite spectral resolution. Of course, the measured band width will have meaning only when it is in excess of the instrumental band width as determined by plate reflectivity, plate separation, plate quality, and diffraction.

Because the laser beam is dangerous to look at and pulses from a typical dielectric laser last such a short time, it is useful to have a CW-monochromatic light source to align the etalon. A simple spectral lamp is suitable. The plates are first adjusted to center the rings on the plate. Then the plates are further brought into parallelism by moving the eye vertically and horizontally across the field and adjusting the mirrors until the fringe pattern remains stationary.

A camera is necessary to record the fringe pattern during the laser pulse. An actual photograph of the fringe field will show the average spectral character of the emission during a pulse. If a lens, the focal length of which is known accurately, is used to photograph the fringe pattern, the mode spacing or line width can be calculated from the equation[28–30].

$$\frac{\Delta\lambda}{\lambda} = \frac{d_1^{\,2} - d_2^{\,2}}{8f^2} \tag{7.42}$$

where d_1 and d_2 are the diameter of fringe rings of the same order and f is the focal length of the lens. The diameters are measured directly on the film for circles of equal intensity on either side of a bright ring or group of rings.

7.1.1.2 Technique for Measuring CW-laser Line Width In practice the method described above is most useful with pulsed lasers the short duration of which precludes the use of a scanning technique. With a continuous laser, more accurate results may be obtained by the use of a "scanning" Fabry-Perot interferometer, in which the transmitted intensity at the center of the circular fringe pattern is monitored by a photocell (see Figure 7.6) as the optical path between the plates is varied. This method has a number of advantages. The effect of diffraction is a minimum at the center of the fringe pattern, the dispersion is greatest at the center of the pattern, and because only the center of the pattern need be illuminated, this method makes more efficient use of the available optical power.

The lens used as L_2 is a very weak field lens, which serves to insure that the aperture at P is uniformly illuminated. The aperture at P should have a diameter approximately equal to the diameter of the Airy disk associated

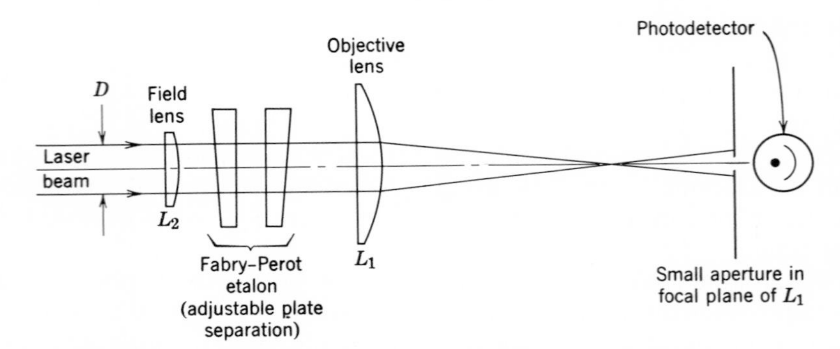

Figure 7.6. Optical arrangement for scanning the Fabry-Perot etalon.

with the focal length f_1 and the aperture D at the wavelength of interest. A larger aperture will reduce the attainable spectral resolution, provided the reflectivity of the plates is sufficiently high. If the reflectivity and plate quality are high enough to insure that the spectral resolution is diffraction limited, the spectral resolution will be given by

$$(R_0)_{\text{Fabry-Perot}} = \frac{\nu}{\Delta\nu} = \frac{\lambda}{\Delta\lambda} = \frac{2D^2}{\lambda^2}. \tag{7.43}$$

Thus, at a wavelength of 1 μm and with a beam diameter of 1 cm, the maximum spectral resolution will be approximately 2×10^8, corresponding to a minimum resolvable band width of 1.5 MHz.

The simplest method for adjusting the optical path between the plates is to vary the pressure of the gas in the etalon, and thereby change its refractive index. The wavelength interval scanned by a change in index of Δn is given by

$$\Delta\lambda = \lambda \frac{\Delta n}{n}. \tag{7.44}$$

Thus a pressure change of 1 cm of mercury, using an air-spaced etalon, will provide a spectral scan of approximately 0.02 Å or 2400 MHz (at 5000 Å). Very slow spectral scans may be obtained by partially evacuating the etalon and readmitting air slowly through a capillary tube. Faster and more conveniently controlled scanning can be obtained with somewhat reduced stability if a combined etalon is fabricated with a piezoelectric cylinder. The etalon length may then be varied with an applied voltage. A more comprehensive analysis of the electronically scanned Fabry-Perot interferometer is given in Chapter 8.

7.1.2 The Tilted-plate Interferometer

By taking advantage of the fact that laser radiation is very nearly monochromatic and easily collimated, a Fabry-Perot interferometer may be used in a modified form, which relaxes the tolerances on plate quality. Because there is only a single wavelength and single angle of incidence, the light is either transmitted or not depending upon whether (7.29) is satisfied. The tilted-plate interferometer is an adaptation of the Fizeau-Tolansky interferometer that is normally used to study plate quality[19]. Its resolving power is the same as that of a conventional Fabry-Perot interferometer, providing that the incident beam is sufficiently well collimated and that the instrument is not used at very large plate separations or with mirrors of very high reflectivities[19].

Consider a Fabry-Perot etalon in which the plates are spaced a distance L_0 and inclined to one another at a very small angle α. The spacing then becomes $L = L_0 + \alpha x$ where x is measured along the plate. Viewed as a filter for a normally-incident, collimated light beam, the etalon (to a first approximation) transmits light of wavelength λ for those zones of the etalon which satisfy the condition.

$$m\lambda = 2nL_x \qquad (m \text{ an integer}), \tag{7.45}$$

where the subscript "x" defines a zone of the etalon (see Figure 7.7). Thus, fringes of high contrast, corresponding to those zones of the etalon satisfying (7.45), will occur in the vicinity of the surface of the etalon mirror. The optimum adjustment of the interferometer requires that there be two such fringes across the face of the etalon; if there were just one fringe, it would be difficult to determine the tilt and angle α, and if there were more than two fringes, the dispersion would be less than the optimum. This means that the angle of tilt α should be on the order of λ/nD, where D is the diameter of the etalon plates.

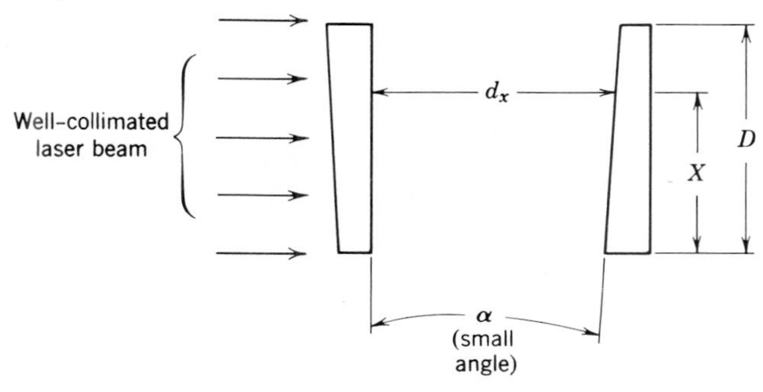

Figure 7.7. Tilted-plate interferometer.

The band width of the input source may be determined by measuring the thickness of the fringes and comparing this number to the separation of adjacent fringes of the same wavelength. If the fringe thickness is Δx, and the separation between fringes is x_0, the corresponding band width is given by

$$\Delta\lambda = \lambda \frac{\Delta x}{x_0}, \tag{7.46}$$

provided, of course, that the instrumental bandwidth has not been reached.

A tilted-plate interferometer is capable of no greater resolution than that obtainable with a conventional Fabry-Perot interferometer, but that the tolerance on plate quality is greatly reduced, and the problem of alignment is essentially eliminated. Using a pair of 98 percent reflecting mirrors having about 1–10 wave flatness, it is easily possible to resolve 10 MHz with a 10 cm plate separation and a well-collimated source.

7.1.3 The Connes or Spherical Fabry-Perot Interferometer

The properties of a Fabry-Perot etalon that employs spherical reflecting surfaces separated by a distance equal to their radius of curvature are of great interest in the measurement of laser band width[20]. The instrument offers a basic advantage for high-resolution work in that a spherical Fabry-Perot etalon has dispersion and resolution comparable to that of a plane Fabry-Perot etalon with twice the mirror separation of the spherical etalon [20]. The resultant compactness is of considerable value in maintaining stable alignment of the mirrors. In addition the spherical Fabry-Perot etalon may be shown to be less sensitive to both alignment and plate quality and suffers a reduced diffraction loss (in these respects, the spherical etalon versus the plane-parallel etalon may be compared to the confocal versus the plane-parallel laser resonator)[21].

A Connes interferometer has recently been constructed and described as a scanning Fabry-Perot interferometer with spherical mirrors[22]. This instrument is especially well-suited to measurements of laser spectra, due to the requirement that the input radiation be matched to the interferometer with regard to beam size and wave-front curvature. With a mirror separation of 50 cm and reflectivities of 99 percent, spectral resolutions of 300 million at a 1μm wavelength have been obtained. This corresponds to a minimum resolvable bandwidth of 1 MHz.

7.1.4 Techniques for Determining the Time-resolved Line Structure of a Pulsed, Solid-state Laser

A streak camera technique may be used to record the variation of the fringes with time during a laser pulse. The camera functions to sweep the image of the interference pattern across the film. An electronic streak-camera technique has been used with a 40 cm spherical etalon having a resolution of 2 to 3 MHz[31, 32] to examine the line structure of a pulsed-ruby laser. A 1P25 image tube with a linear time sweep employing a time-varying magnetic field was used to move the image across the photographic film. It was observed that the frequency of the laser shifted with temperature during a pulse.

A drum-camera technique has been used successfully with a Fabry-Perot etalon to yield the time-resolved spectral output of a pulsed-ruby laser [33]. The equipment used is illustrated in Figure 7.8. The principle upon which this technique is based is that all of the spectral information yielded by a Fabry-Perot etalon may be obtained from a radial slit, one end of which is at the center of the normal circular fringe pattern. A photographic film located just behind the slit and moved at high velocity in a direction perpendicular to the length of the slit will record a time-resolved spectral profile of the laser. The time resolution is determined by the slit width and the film velocity, whereas the spectral resolution is determined by the etalon parameters.

Typical parameters are as follows: A 0.03 mm slit width and a 10,000 cm/sec drum speed will yield a resolving time of the order of 500 ns. A Fabry-Perot etalon with a 1 cm spacer will give a free-spectral range of 0.024 nm,

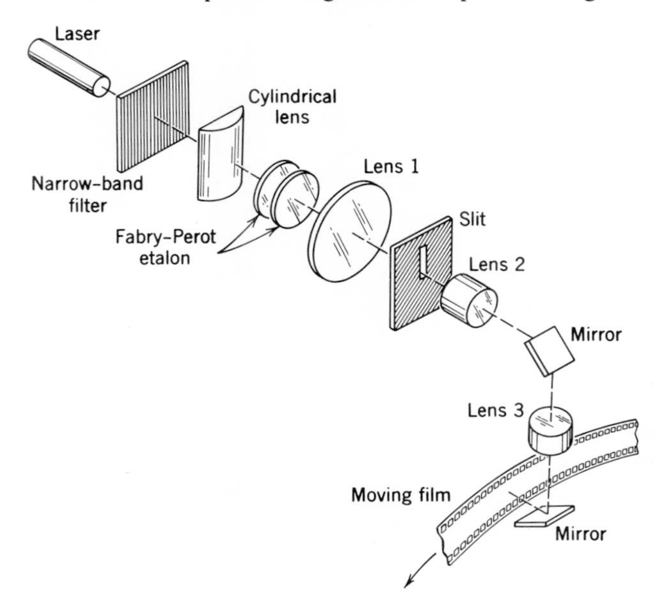

Figure 7.8. Time-resolved, drum-camera, Fabry-Perot spectrograph.

adequate for as many as 8 to 10 axial modes with a 5 cm ruby. The theoretical axial mode separation

$$\Delta\lambda = \frac{\lambda^2}{2nL} = \frac{c}{2nL} \tag{7.47}$$

is approximately 0.0027 nm.

The equipment discussed above will yield several important pieces of information regarding the time-dependent spectral performance of a solid-state pulsed laser. The exposed film essentially contains a frequency-versus-time plot of the spectral output of the laser. Two kinds of changes can be detected. If, due to temperature gradients that develop during a pulse, the center of the fluorescent line increases in wavelength but the optical length of the laser remains fixed, the axial modes of the laser will each be constant in time (while they oscillate), but the envelope of the fluorescent line will cause longer wavelength sets of axial modes to oscillate as time progresses. If, in contrast, the fluorescent-line wavelength does not change during a pulse but the optical length of the cavity increases, the wavelength of each set of axial modes, instead of being constant, will exhibit a time-dependent decrease.

Observations with ruby, pumped near threshold, indicate that the axial-mode frequencies and the fluorescent wavelength are nearly independent of time. If, however, the excess-to-threshold pumping ratio becomes large, the fluorescent wavelength increases from 693.5 nm at 100° K (with a corresponding line width of 0.015 nm) to 694.3 nm at 300° K (with a 0.3 to 0.4 nm fluorescent line width). The number of axial modes that can oscillate increases with time and the wavelength of the axial modes increase as the laser expands. This apparatus also clearly resolves the transient nature of resolved sets of axial modes, the spectral-integral nature of which yields a quasicontinuous output. That is to say different axial modes appear to oscillate essentially independently. Considerable mode hopping occurs during an output pulse.

Instead of using a mechanical streak camera to study the time-resolved spectra of a laser we can obtain much of the same information by applying electronic techniques that use fiber optics[34]. The light transmitted by the slit at the drum-camera, previously described, is directed to a linear array of fiber optic bundles, each of which illuminates a separate photomultiplier. Signals from each photomultiplier are displayed on separate oscilloscope sweeps and are simultaneously photographed. (The apparatus required is illustrated in Figure 7.9) the electronic and optical apparatus required is conventional and will not be discussed.

Using an etalon spacing adequate to cover a free spectral range of about 0.025 nm we can study the time development of the axial-modes of a

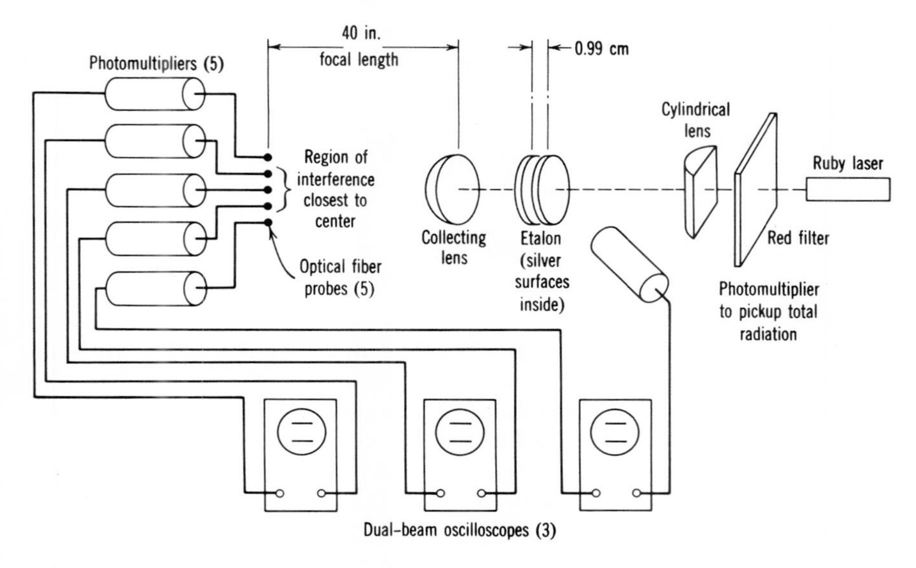

Figure 7.9. Schematic diagram of the experiment.

pulsed, solid-state laser with three to five sets of axial modes. Because the time resolution is limited only by the electronics, the existence of simultaneous oscillations on several modes during one relaxation spike can be detected. Using statistical techniques we can develop a histogram that assigns probabilities to the time-dependent appearance of various axial modes as a function of laser parameters. By using this technique we can establish, for example, that the axial-mode structure in a laser changes from spike to spike and within a relaxation spike[34].

7.1.5 Technique for Observing the Transverse and Axial Modes of a CW gas Laser

A measuring technique, which is equally applicable to observing mode conversion in optical-transmission systems, can be used to reveal the amplitude and frequency characteristics of a CW-gas laser. The method is based upon using a scanning interferometer to analyze the laser output while the latter is oscillating simultaneously on many transverse modes[35].

Consider the frequency spectrum of a laser. The laser cavity will transmit power at frequencies determined by the relation

$$\nu = \frac{c}{2L}\left[q + \frac{1}{\pi}(1 + m + n)\cos^{-1}\sqrt{g_1 g_2}\right] \tag{7.48}$$

where ν is the resonant frequency of the laser, c is the velocity of light, L is the laser-cavity length, $g_{1,2} = 1 - L/R_{1,2}$ where $R_{1,2}$ is the radius of curvature of mirror 1,2, q is the longitudinal mode number (i.e. the number of half-wavelengths in L), and m and n are the transverse-mode numbers, which give the Cartesian-coordinate, transverse-mode numbers. It can be shown[35] that if ℓ is the spacing of the scanning interferometer mirrors

$$\frac{\delta l}{\Delta \ell} = \frac{(c/2\pi L)\cos^{-1}\sqrt{(g_1 g_2)_L} - (c/2\pi \ell)\cos^{-1}\sqrt{(g_1 g_2)_I}}{c/2L} = \frac{(\delta \nu)_L - (\delta \nu)_I}{c/2L} \qquad (7.49)$$

where the subscripts I and L refer respectively to the corresponding laser-cavity and scanning-interferometer parameters and $\Delta \ell$ is the change in interferometer length required to resonate on a new mode when the longitudinal mode number q of the laser cavity is changed by one. Similarly a unit change in the transverse-mode order of the laser cavity $(m + n)$ requires an interferometer change of $\delta \ell$. The change in laser-cavity frequency for a unit change in the transverse-mode order is $\delta \nu$.

In general $L > > \ell$, so that the scanning interferometer is nearly parallel plane in configuration. The values of $g_{1,2}$ of the interferometer are chosen with a consideration of the $g_{1,2}$ of the laser cavity and the desired frequency separation of transverse modes $(\delta \ell / \Delta \ell)$ to be resolved by the interferometer. A value of $\delta \ell / \Delta \ell \sim 1/10$ has been found convenient.

An oscilloscope display of the transmission characteristics of the gas-laser beam is obtained in a conventional manner using a photomultiplier. It is necessary to isolate the interferometer from the laser to preclude mode interaction, and this is conveniently accomplished with a Glan-Thompson prism and a $\lambda/4$ plate. In general a lens or lenses are inserted in the laser beam between the cavity on the interferometer to match the beam to the interferometer. This precaution precludes the excitation of undesired interferometer modes. In the mode-matched case, each mode order in the interferometer matches the laser[36].

Much qualitative information about the operation of a laser may be obtained from a simultaneous study of the scanning-interferometer-output and the laser-output spot. Evidence of mode competition can clearly be resolved. The interferometer should be scanned over several cavity periods $mc/2L$ $(m = 1,2,3 \ldots)$ at a rate of 10 or more/sec for convenient observation. Noting the change in the display as the laser mirrors are tuned to permit oscillation on various modes, we may see that the interferometer displays transverse modes for a given axial-mode order to the right of the fundamental display. However, because the laser power appears in lower-order axial modes for higher-order transverse modes, the envelope of the spectral-line pattern shifts in the opposite direction.

7.2 MEASUREMENT OF FLUORESCENT LINE WIDTH

The foregoing discussion has been devoted to measurement of laser line width and cavity line width. Generally these properties are more difficult to measure than the fluorescence line width because the crystal need not be operating as a laser for the latter measurement to be made. When a crystal operates as a laser, adequate light is available for detection by simple means, but the dispersive element used to examine line width must have an exceptionally high resolution. Below the oscillation threshold the fluorescence from a laser crystal has a broader band width, but the intensity is very low. Fluorescence line-width measurement procedures depend upon whether the laser is of the three-level for four-level type.

The three-level laser transition terminates on the ground state, so that resonance absorption can occur without excitation. Therefore the laser transition may be examined either in absorption or emission. The four-level laser line terminates on a level above the ground state, which is unpopulated unless the crystal is excited. Therefore there is no absorption without excitation. A four-level laser can only be examined in emission.

Apparatus used to measure fluorescence is shown in Figure 7.10. A continuously operating source with appropriate spectral components to excite the crystal is focused on the crystal at about 90° to the optic axis of the spectrometer. The spectrometer is scanned through the spectrum and the detector signal displayed on a chart recorder.

When a three-level crystal is examined in absorption, it is necessary to place the sample in the dispersed beam and focus the light source on the entrance slits of the spectrometer. If the crystal is placed in the undispersed beam, there may be enough radiation at the excitation wavelength to excite fluorescence in the absorption line.

It is possible to use the laser as a monochromatic light source to measure the line width in a sample of the same material. This method will only

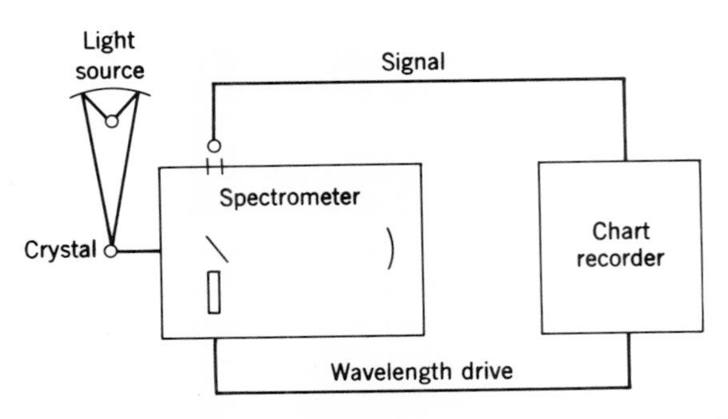

Figure 7.10. Typical apparatus and arrangement for measuring fluorescence in laser crystals.

suffice when the laser can be tuned through a sufficiently broad range to cover the width of the resonance line.

A thermally tuned pulsed-ruby laser has been used to examine the resonance absorption line in ruby at low temperature[37]. The laser beam was directed through a sample of ruby maintained at constant temperature on the cold finger of a liquid nitrogen Dewar. The laser temperature was changed after each pulse, and a thermal tuning curve was used to find the wavelength at which the laser was operating. Absorption was determined by measuring the laser-beam intensity before and after it passed through the sample, and a curve representing the spectrum was obtained by plotting these data as a function of wavelength. The R_1 line in ruby splits into two components, which present a composite spectrum at low temperature.

Overlapping lines occur frequently in the fluorescence spectra of solid-state lasers. This complicates our determination of the line width for the separate lines. A large number of two-line composite spectra have been prepared that represent the sum of closely spaced lines of differing strength, width, and separation. These curves are especially useful for finding the parameters for the individual lines.

7.3 DIRECT METHOD OF MEASURING DOPPLER LINE WIDTH

The Doppler line width of a CW-gas laser transition may be measured by using a single-frequency laser to measure the single-pass gain of an identical CW-gas laser from which the end mirrors are removed[38]. The single-frequency laser is frequency modulated (by varying the mirror spacing) to sweep through the Doppler-broadened gain curve of the amplifier. Because of frequency-dependent gain, the source-laser power output will vary with frequency. This effect must be normalized by adjusting the intensity of the source laser with appropriate attenuators to insure that the measured Doppler line width of the amplifier is meaningful. The output intensity of the source laser must be set at a low value to insure that the small-signal gain is being measured, particularly in the case of such high-gain transitions as the $5d\ [3/2]_1^0 \rightarrow 6p\ [3/2]_1^0$ transitions of xenon.

The apparatus arranged as shown in Figure 7.11 will be found convenient for making the Doppler broadened gain-curve measurements of the amplifier. The frequency calibration of the source oscillator is straightforward. Over small ranges, the displacement of the piezoelectric driver is a linear function of applied voltage. Further, the mirror spacing is known and the interorder spacing of the cavity can be determined from $c/2nL$. The change in piezoelectric drive voltage necessary to tune the source laser between two successive maxima in power output gives the desired frequency calibration.

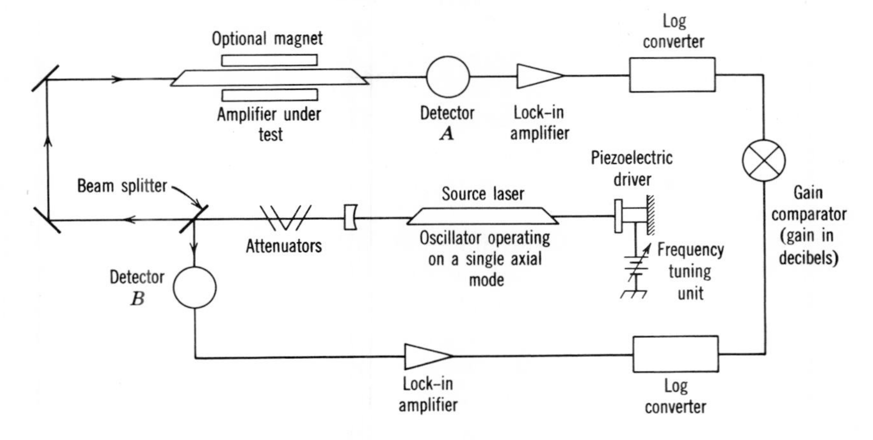

Figure 7.11. Apparatus arrangement for measuring the amplifier linewidth.

With source-laser power outputs sufficiently low ($\sim 10^{-8}$ W) to insure that at the oscillation line center the amplifier is operating in its small signal range (i.e. exponential gain), the gain measurements are made as a function of oscillator frequency for constant input power. Using the apparatus illustrated in Figure 7.11, we can conveniently measure the gain-versus-frequency curve over the tuning range for which oscillations are possible. If the gain-frequency curve is plotted in decibels, the full width of the curve, wherein the gain is reduced a factor of 2 from its line-center value, is the measured Doppler line width of the amplifier.

Typical values of Doppler-broadened gain curves at 2.026 μm in xenon are $\Delta v_D = 210 \pm 10$ MHz. Because the expression for Doppler line width contains the equivalent atomic temperature, we can solve for the latter with a measured value for Δv_D as

$$\Delta v_D = \frac{2v}{c}\left(\frac{2kT}{m}\ln 2\right)^{1/2}.$$

Using the data above and $m = 131.3$ as the average atomic mass, we obtain

$$T \sim 515^\circ K.$$

7.4 TECHNIQUE FOR MEASURING GAIN NARROWING IN A LASER AMPLIFIER

In gas lasers the line width is almost always inhomogeneously broadened. The Doppler-broadened line is much wider than the natural or radiative line width. The collision rate is much less than the spontaneous-decay rate of the excited state. The Doppler line width is a good approximation for the line width as long as the unsaturated gain is small. Although the Doppler width can be used as the line width for incremental gain $1/I(dI/dz)$, it cannot be used for net gain (I_{out}/I_{in}) at large values of unsaturated gain. The line becomes a fraction of the Doppler width because of gain variation across the line.

One technique for measuring line narrowing uses the apparatus similar to that discribed in Figure 7.11 except that the amplifier rather than the oscillator is tuned. To tune the amplifier, we can apply a solenoidal magnetic field to the plasma tube. Variations in axial magnetic field change the frequency of the σ_R-and σ_L-amplifying transitions relative to the line-center frequency ν_0 of the oscillator. Thus the Zeeman effect permits us to scan the band width of the amplifier across the relatively narrow output line of the source laser.

Consider a requirement for the determination of the gain-narrowed line width at 3.39 μm in a helium-neon laser amplifier. Because a single-frequency laser is required, a mirror spacing of 25 cm is chosen (corresponding to an interorder axial mode spacing of

$$c/2nL = 3 \times 10^{10}/(2)(1)(25) = 600 \text{ MHz}.$$

Because the dimension stated above results in an interorder spacing of less than the Doppler line width ($\Delta\nu_D \sim 340$ MHz at 3.39 μm), the average power output of the source laser will drift slowly with time as the cavity spacing changes due to thermal drift. The source-laser cavity length can be trimmed with the piezoelectric tuner to maintain a chosen power output[41]. Because the gain of the source laser is large, its line width will be narrower than 340 MHz.

The range of attenuation needed between the source-laser output and the amplifier input can be obtained by inserting 20 to 30 microscope-slide glasses in the beam. These should be tilted 3 to 5 deg with respect to the beam axis to preclude reflections into the source or amplifier. The attenuators must, of course, be calibrated, and this can easily be accomplished with the source laser and a calibrated thermopile.

A chopper wheel may be added to the beam path to permit complete source-laser intensity modulation for synchronous detection at low signal levels. The source laser must be operated under saturated-gain conditions to minimize coupling between the amplifier and source laser when the former is operated at high gain.

A typical amplifier of interest may have a 2.5 mm bore, an active discharge length of 60 cm, and a bore Fresnel number of approximately unity. The plasma tube is surrounded by an axial solenoidal electromagnetic field[44, 45] capable of generating up to 500 Oe. The coil is double wound to enable continuous axial-field variations from the maximum in one direction, continuously through zero, to the maximum in the opposite direction. The amplifier gain at line center must be measured for various levels of input signals from the source laser to facilitate the selection of an appropriately low value for the input signal. This choice is necessary to insure unsaturated amplifier operation when we measure the output line width. Some attenuators may be found necessary between the amplifier and the output detector. A typical low-input signal gain at line center for a system such as that described above is approximately 35 db.

The amplifier gain-frequency characteristic is measured by allowing current through the double-wound coil to vary slowly and continuously to sweep the amplifying transitions across the frequency of the source laser.

Care must be taken during high-gain laser measurement to insure that the measuring system geometry precludes extraneous signals. The source-laser output should be filtered with a spectral notch filter with approximately a 10 nm band width. The source laser should be far enough from the amplifier that the oscillator plasma tube bore-to-length (d/b) ratio defines the radiation cone, for spontaneous emission noise, the diameter of which is very large compared with the TEM_{00} mode diffraction-limited spot size of the source. A second spectral notch filter should, of course, be placed in front of the detector to minimize effects due to background spontaneous emission at low coherent-signal levels[42].

Because only relative signal levels are of interest at reasonable signal levels, a lens or mirror may not be needed to focus the amplified signal onto the detector. If additional sensitivity is required, a light-collecting mirror can be added. Two apertures will then be required, one at the amplifier output of area A_L and one at the detector of area A_D. These are related to the mirror focal length by[43]

$$A_D = \lambda^2 f^2 / A_L, \tag{7.50}$$

where f is the mirror focal length[43].

If, however, the detector is placed 30 to 50 cm from the output of the amplifier and on the axis of the system, and if the beam diameter of the source laser is small enough to pass through the amplifier without wall reflections, no additional spatial filtering will be needed to reduce extraneous spontaneous-emission noise. The geometry of the amplifier-tube wall limits the angles at which spontaneous-emission noise may propagate to an angular width of

$$\theta_{\substack{\text{amplifier}\\ \text{noise}}} \sim d/b,$$

where d is the amplifier-tube bore and b is its length.

To determine the line width from the measurements above, the frequency versus magnetic field characteristics must be known. This may be obtained by calibrating the change in source laser frequency that corresponds to a change in magnetic field. That is the techniques used in Section 7.3 to calibrate the source-laser frequency versus the piezoelectric transducer drive may now be used to derive the amplifier-band center versus the magnetic-field (or current) characteristic.

To ensure that the unsaturated gain is indeed small, during the measurement of a value of the line width that is in agreement with the theoretical Doppler width, it is desirable to construct a low-gain tube. Because gain and diameter are at least inversely related, in ambipolar-diffusion-limited gas lasers, the simplest solution is to construct an identical amplifier of large bore ($\sim$1 cm). For these conditions it will be found that (for an inhomogeneously broadened line) the measured line width will agree with the theoretical Doppler width providing an appropriate choice is made for the average atomic temperature[38].

A measurement of the line width versus the input intensity for a high-gain transition (such as 3.39 μm in helium-neon) will yield an apparent amplifier line width, which, for small bore tubes, is *less* than the Doppler line width and which *increases* with input intensity. The explanation for this enigmatic result is simple. At low-input levels the amplifier gain is unsaturated and large. The line width is significantly gain narrowed (as much as a factor of 3) by the gain variation across the line. As the input intensity increases and gain saturation begins (i.e. the gain decreases with intensity) and as the gain-induced line narrowing decreases, the apparent line width increases. At high-input intensities the saturated-gain amplifier band width is even broader than the Doppler line width. In the apparatus above the gain-narrowed line width is approximately 125 MHz for $G = 5000$. This is compared with the measured Doppler line width of 340 MHz and a natural or radiative line width of 25 MHz for the 3.39 μm transition. The interested reader is referred to the literature for the detailed theory of line narrowing[41].

7.5 TECHNIQUE OF MEASURING GAS-LASER SPECTRAL LINE SHAPE USING EXCESS PHOTON NOISE

It has been shown that a frequency analysis of the excess photon noise of a monochromatic light source reveals the width of the spectral source[46]. A monochromatic source radiating in a total-frequency interval or line-width $\Delta\nu$ will produce beat frequencies, due to wave-interaction effects, over the frequency spectrum from zero to $\Delta\nu$. The theory of excess photon noise has been applied to measure the linewidth of low-noise, helium-neon lasers.

The technique of using excess photon noise requires a power-spectrum measurement of the shot noise of the photodetector used. Of paramount importance is the selection of a quiet laser. Experience indicates that an rf-excited laser (or at least a hot-cathode laser with a large series resistance) is necessary if low extraneous noise is to be obtained. It is, of course, absolutely essential that the laser operate at a single frequency.

Assuming a symmetrical Doppler line shape it can be shown that the ratio of excess noise to shot noise in the photodetector is of the form

$$E(f) = (\bar{\imath}/2e\,\Delta\nu\sqrt{\pi})\exp\left[\frac{-f^2}{4\Delta\nu^2}\right] \tag{7.51}$$

where $\bar{\imath}$ is the mean photocurrent, e is the electronic charge, f is the center frequency in the band over which the noise is measured, and $\Delta\nu$ is the linewidth of the laser. As $\bar{\imath}$ is measured, $\Delta\nu$ is determined from the data by fitting the data for $E(f)$ as f goes to zero. The value obtained for $\Delta\nu$ will be different, and other relations will apply if we assume a Lorentz or a rectangular intensity distribution[47]. A typical experiment is described below.

A low-noise, single-frequency laser operating at 1.153 μm illuminates a germanium p-n junction photo detector (back biased at 30 V and provided with a 75,000Ω load resistance). The photodetector is axially placed 20 cm from the laser. Care is taken to eliminate other extraneous thermal light sources. Because the effective detector area is small (0.5 mm^2), the light source is not resolved, that is the solid angle subtended by the detector is small compared to λ^2/A_D where λ is the laser wavelength. The photodetector-output noise is amplified by a conventional dc amplifier of megahertz band width and is analyzed by an electronic synchronous noise analyzer of constant (but small $\sim$ 100 Hz) band width and continuously tunable frequency. The laser output must be constant during the measurements. The apparatus is calibrated with a conventional diode source showing full shot noise. The photodetector diode is checked for full shot-noise output with a thermal source (battery-operated flashlight). A measure is obtained of the ratio of the excess noise to the full shot-noise power. Typical results show a ratio of the excess to the shot noise of 10^8 and a frequency spectrum of 10 kHz. Because the line width depends strongly on the operating level of the laser, the data must be taken for the expected operating conditions of interest.

A discussion of more extensive noise measurements is given in Chapter 9.

7.6 TECHNIQUE OF TILTED-PLATE LINE SELECTION IN A SOLID-STATE LASER

The wavelength range over which laser action occurs in a multimode, pulsed, solid-state laser is of the order of 1 to 10 nm, depending mostly upon the ratio of the excess to threshold and the pump power. Mode selection can be induced as the result of selective filtering action if an etalon is placed either inside or outside (by replacing the output reflector) a solid-state laser cavity.

Consider the net reflectivity R at a wavelength λ at an incident angle θ to an etalon of thickness d and index of refraction n:

$$R_t = \frac{(\sqrt{R_1} - \sqrt{R_2}) + 4\sqrt{R_1 R_2} \sin^2 \phi}{(1 - \sqrt{R_1 R_2})^2 + 4\sqrt{R_1 R_2} \sin^2 \phi}, \tag{7.52}$$

where

$$\phi = \frac{2\pi n d}{\lambda} \cos \theta \tag{7.53}$$

and R_1 R_2 are the surface reflectivities. This treatment obviously assumes that both surfaces are perfectly parallel and that neither walk-off errors nor inhomogeneities in refractive index are present. If $R_1 = R_2 = R$,

$$R_t = 4R \sin^2 \phi / [(1 - R)^2 + 4R \sin^2 \phi]. \tag{7.54}$$

From (7.53) we may see that the plate is transparent at those wavelengths for which d is an even number of quarter-wavelengths thick; its reflectivity is a maximum at those wavelengths for which d is an odd number of quarter-wavelengths thick. The wavelength interval $\Delta\lambda$ between maxima (or minima) of R_t is just

$$\Delta\lambda = \lambda^2/2nd. \tag{7.55}$$

The values of R_1, R_2 in (7.52) determine the maximum reflectivity of the etalon; the values of R_1 and R_2 also effect the shape of the reflectivity curve. Higher values of R_1, R_2 correspond to broader wavelength ranges of higher reflectivity between the wavelengths that yield maximum transparency.

If the etalon is used as the end reflector of a cavity, oscillation will take place at those wavelengths for which R_t is a maximum (d is an odd number of quarter-wavelength thick). If the etalon is parallel to the end of the laser, the reflectivity of the laser rod must be considered (provided the laser-beam spread is not so large that the angle of a ray with respect to the laser axis is not well defined).

Let the reflectivity of the third surface be R_3 and the spacing be D. The total reflectivity R_{t3} at an angle θ with respect to the axis is given by[48].

$$R_{t3} = \frac{(A + R_3 B - C\sqrt{R_3})}{(AR_3 + B - C\sqrt{R_3})} \tag{7.56}$$

where

$$A = 4R \sin^2 \phi, \tag{7.57}$$

$$B = (1 - R)^2 + 4R \sin^2 \phi,$$

$$C = 4\sqrt{R} \sin \varphi[\sin(2\varphi + \phi) - R \sin(2\varphi - \phi)],$$

$$\varphi = 2\pi D \cos \theta/\lambda.$$

The remaining variables are defined as outlined above.

Two cases are of interest. If $R_3 \ll R$, for example if the laser rod ends are antireflection coated, (7.56) reduces to (7.52). If however, $R_3 = R$ and $D \gg d$, the total reflectivity will have a superimposed oscillation that results from the product of the reflectivity characteristics of the two etalons.

A second etalon may be added to the cavity between an aligned etalon and the laser rod. If the angle of the third plate with respect to the laser rod is varied from the aligned position, considerable frequency selectivity can be obtained. Tuning over a 2nm range was obtained in a Nd^{3+} glass laser with a 0.65 mm aligned plate spaced 11 cm from the end of the laser. The tilted plate was 0.11 mm thick and was placed 6 cm from the end of the laser rod[49]. As the inner plate was turned successively through 6 deg in 1-deg increments, the laser frequency was shifted over a range of nearly 2 nm.

Although the frequency selectivity of a high-reflectivity, tilted etalon is better than that of an aligned etalon, more severe constraints must be placed upon the optical quality of the laser rod to enable its efficient use. Much of the laser energy could be rejected from the cavity during mode shifts, which are characteristic of solid-state lasers. In contrast the aligned etalon is far less selective but more efficient; it cannot remove energy from the cavity during mode shifts. If an aligned and a tilted etalon are used, a compromise is obtained in selectivity and efficiency.

In applying this technique to measuring the spectral characteristics of a solid-state laser several precautions must be observed. First, the optical homogeneity of the etalon material must be so great that a plane wave front is not distorted more than $\lambda/100$ if the spectral resolution is not to deteriorate. Equation 7.32 applies. Second, the surface flatness of the plates must meet the same specifications as the homogeneity. Third, the surfaces must be parallel to the same accuracy that stipulated in Section in (7.32), or walk-off errors will cause loss of efficiency and resolution. The etalons must be held as rigidly as mirrors in a gas laser to preclude frequency sweeping and induced mode hopping during the pulse from the flash lamp.

7.7 REFERENCES

[1] J. A. Fleck, Jr., *J. Appl. Phys.*, **37**, 188 (1966).
[2] A. L. Schawlow and C. H. Townes, *Phys. Rev.*, **99**, 1264 (1965).
[3] K. Shimoda, *Inst. Phys. Chem. Research Sci. Papers*, (Tokyo), **55**, 1 (1961).
[4] P. Grivet and A. Blaquiere, "Masers and Classical Oscillators," in *Optical Masers*, Proceedings Symposium on Optical Masers, Polytechnic Press, Brooklyn, N. Y. 1963. Vol. XIII p. 69.
[5] R. C. Jennison, Fourier Transforms, Pergamon, New York, 1961, p. 85.
[6] Hodara, "The Concept of Coherence. Its Application to Lasers", in *Quantum Electronics*, Grivet and Bloembergen (eds.) Paris 1963 Conf., Columbia Univ. Press., New York, 1964, Vol. 3, p. 121.
[7] A. Okaya, "Phase Uniformity at the Ruby Laser End Surface," in *Laser and Applications*, W. S. C. Chang (ed.), Ohio State Univ. Press. 1963.
[8] D. A. Berkley and G. J. Wolga, *Phys. Rev. Letters*, **9**, 479 (1962).
[9] D. F. Nelson and F. J. Collins, *J. Appl. Phys.*, **32**, 739 (1961).
[10] W. S. C. Chang and J. W. Greg, *Discussion of Partial Coherence as Related to Lasers*, Antenna Lab Report 1579–2, Ohio State University, in 1963, p. 318.
[11] M. Born and E. Wolf, *Principles of Optics*, Pergamon, New York, 1959, p. 318.
[12] A. A. Michelson, *Studies in Optics*, University of Chicago Press, 1927, p. 34.
[13] M. Born and E. Wolf, *op. cit.*, p. 509.
[14] H. E. Neugebauer, *J. Opt. Soc. Am.*, **52**, 470 (1962).
[15] R. N. Bracewell, *Proc. IEEE*, **50**, 214 (1962).
[16] F. A. Johnson, T. P. McLean, and E. R. Pike, "Photon-Counting Statistics," in *Physics of Quantum Electronics*, P. L. Kelley, B. Lax and P. E. Tannenwald (eds.), McGraw-Hill, New York, 1966, esp. pp. 706–714.
[17] M. Born and E. Wolf, Principles of Optics, Pergamon, New York, 1959, p. 328.
[18] R. Chabbal, *J. Research Centre Natl. Research Sec. Lab Bellevue* (Paris), **24**, 138 (1953); *Rev Opt.*, **37**, 49, 336, 501 (1958); and D. R. Herriott, *J. Opt. Soc. Am.*, **51**, 1142 (1961).
[19] S. Tolansky, *Surface Microphotography*, Interscience, New York; and H. W. Moos, *et al.*, *Appl. Optics*, **2**, 817 (1963).
[20] P. Connes, *Rev. Opt.*, **35**, 37 (1956).
[21] G. D. Boyd and J. P. Gordon, *Bur. Std. J. Res.*, **40**, 489 (1961).
[22] D. R. Herriott, *Appl. Optics*, **2**, 865 (1963).
[23] M. Born and E. Wolf, *op. cit.*, p. 302.
[24] M. Born and E. Wolf, *op. cit.*, p. 539.
[25] D. A. Berkley and G. J. Wolga, *Phys. Rev. Letters*, **9**, 479 (1962).
[26] M. Cliftan, A. Krutchkoff, and S. Koozekanani, *Proc. IEEE*, **50**, 84 (1962).
[27] R. C. Duncan, Jr., Z. J. Kiss, and J. P. Wittke, *J. Appl. Phys.*, **33**, 2568 (1962).
[28] M. Born and E. Wolf, *op. cit.*, p. 334.
[29] Summer P. Davis, *Appl. Opt.*, **2**, 727 (1963).
[30] Joseph Valasek, *Theoretical and Experimental Optics*, Wiley, New York, (1956), p. 374.
[31] G. R. Hanes and B. P. Stoicheff, *Nature*, **195**, 587 (1962).
[32] S. L. Ridgway, G. L. Clark, and C. M. Yorke, *J. Opt. Soc. Am.*, **53**, 700 (1963).

[33] M. Hercher, G. Milne, and C. O. Alley, "Time Resolved Spectroscopy of the Light Emitted by a Ruby Laser," *Inst. Opt. Rpt*, 1962.
[34] C. M. Stickley, R. C. White, Jr., and R. A. Bradbury, "Time Variation of Axial Frequencies in Ruby Lasers," in W. S. C. Chang (ed.), *Lasers and Applications*, Ohio State University Press, 1963.
[35] P. W. Smith, *Appl. Optics*, **4**, 1038 (1965).
[36] R. L. Fork, D. R. Herriott, and H. Kogelnik, *Appl. Optics*, **3**, 1471 (1964).
[37] R. L. Aagard, *J. Appl. Phys.*, **34**, 3631 (1963).
[38] C. K. N. Patel, *Phys. Rev.*, **131**, 1582 (1963).
[39] R. C. Smith and L. S. Watkins, *Proc. IEEE*, **53**, 161 (1965).
[40] L. H. Enloe and J. L. Rodda, *Proc. IEEE*, **53**, 165 (1965).
[41] D. F. Hotz, *Appl. Opt.*, **4**, 527 (1965).
[42] W. B. Bridges and G. S. Picus, *Appl. Opt.*, **3**, 1189 (1964).
[43] R. A. Paananen, *et al.*, *Appl. Phys. Letters*, **4**, 149 (1964).
[44] R. L. Fork and C. K. N. Patel, *Phys. Rev.*, **129**, 2577 (1963).
[45] W. S. C. Chang (ed.), *Lasers and Applications*, Ohio State University Press, 1963, esp. p. 64.
[46] C. Th. J. Alkemade, *Physics*, **51**, 1145 (1959).
[47] P. Grivet and N. Bloembergen (eds.), *Quantum Electronics*, Columbia University Press, New York, 1964, esp. p. 193.
[48] A. W. Crook, *J. Opt. Soc. Am.*, **38**, 954 (1948).
[49] E. Snitzer, *Fiber Optic Laser*, Final Tech. Rpt., U. S. Army Research Office, USAROD Project 3209, Contract DA-19-020-AMC-0160(X), 1965, p. 14.

7.8 PRINCIPAL SYMBOLS, NOTATIONS, AND ABBREVIATIONS

Symbol	*Meaning*
A	Absorption coefficient, constant, see text
A_D, A_L	Area of a detector aperture, area of a laser aperture, see text
a	Slit spacing
c	Velocity of light
D	Diameter of a source
d	Diameter of an area of defined spatial coherence, see text
d_i	Diameter of fringe number i
E, E_i	Ratio of excess photon noise to shot noise, electric field of component i, see text
e	Electronic charge
F_σ	Free-spectral range of an etalon
f_i	Focal length of lens number i
h	Planck's constant
L, ℓ	Spacing of etalon plates, see text
$\ln x$	Natural logarithm of x
I_o	Irradiance, intensity, see text
$I_i(r)$	Intensity of component i which is a function of r

$\bar{i}$	Mean value of photocurrent
L_i	Lense number i
b	Distance from a projection screen to a pair of slits, mirror spacing, see text
M_i	Mirror number i
m	Total number of lines on a grating
m	Mass in atomic units
N	Spectral order number 0, 1, 2, . . .
n	Index of refraction
P	Power
q, m, n	Longitudinal and transverse mode numbers
g_i	Mirror-spacing-radius parameter, see text
$(R_0)_{\text{grating}}$	Maximum resolution of a grating
R, R_t	Radial distance, surface reflectivity, see text, total reflectivity of a set of composite surfaces
R_1, R_2	Radius of curvature of mirror 1, 2, see text
$\mathcal{R}$	Real part of a complex variable
$r, \Delta r$	Radius, differential radius
$S(\xi)$	Slit-distribution function
T	Transmission coefficient, atomic temperature, transmission coefficient, see text
T_e	Effective negative temperature
V	Michelson visibility function
$x_0, \Delta x$	Fringe separation, fringe thickness for parallel-fringe system
z	Axial coordinate of the laser
α	An angle
α_{12}	Phase factor in the argument of γ_{12}
$\Gamma_{11}(r, t)$	Autocorrelation function of r and t
$\Gamma_{12}(r_1, r_2, t)$	Mutual coherence function of r_1 and r_2 in t
$\Gamma_{T_{12}}$	Mutual coherence function considered over a finite time T_{12}
λ	Wavelength
$\Delta\lambda_{\text{DIFF}}$	Diffraction-limited, resolvable-wavelength range
$\Delta\lambda_{\text{min}}$	Minimum theoretical resolvable-wavelength range
γ_{12}	Complex degree of coherence
$\Delta, \Delta(\ )$	Fringe shift induced when quasimonochromatic light is replaced by monochromatic light, symbol for an increment or small part, see text
$\delta(\ )$	Infinitesimal change in a variable
θ	Angular radius
ν	Frequency
$\bar{\nu}$	Mean value of ν
$\Delta\nu_D$	Doppler line width

$\Delta\nu_L$	Limiting band width of a spectral line
ν_o	Frequency at the center of a spectral line
$\Delta\nu_{min}$	Minimum theoretical resolvable bandwidth
$\delta\nu_c$	Passive laser-cavity band width
τ_{Cd}	Coherence time of a cadmium spectral line
$\Delta\tau$	Duration of a wave train
ϕ, φ	Phase variables used in interference equations to indicate optical-path differences in radians, see text
ω	Angular frequency

8

Measurement of Frequency Stability

8.1 DEFINITIONS OF STABILITY

We begin Chapter 8 with explicit definitions of general frequency stability, short-term frequency stability, and long-term frequency stability. These terms are defined as inverse fractional variations of the measurands in a manner similar to that used to define the circuit "quality factor" Q. Thus a large number is consistent with a high degree of stability. Relative and absolute stabilities are defined in their contextual relation to the precision and accuracy of frequency measurements. The closely related quality factor frequency resettability is also defined.

8.1.1 Frequency Stability

The stability of a generator of frequency ν, as measured during an interval τ, is defined as

$$S_\nu = \frac{1}{\tau}\int_0^\tau \nu(t)\, dt \Big/ \Big[\nu_{\max} - \nu_{\min}\Big]_0^\tau = \bar{\nu}/\Delta\nu(\tau). \tag{8.1}$$

The average frequency of the generator $\bar{\nu}$ is thus related to the frequency fluctuations $\Delta\nu(\tau)$ encountered during the period of observation, or averaging time. Any definition of stability connotes an averaging time. The latter must be explicitly stated before the notion of stability has other than trivial significance.

8.1.2 Short-term Frequency Stability

We arbitrarily define short-term frequency stability in terms of an interval less than the resolving time τ_0 of the frequency measuring detector. In accord with this definition, short-term stability is denoted by

$$s_\nu(\tau) = \frac{\nu}{\delta\nu(\tau)} \qquad \tau < \tau_0 \tag{8.2}$$

Short-term stability is thus a measure of the unresolvable band width of the generator output. If $\delta\nu(\tau)$ is taken as the full width of the frequency deviation curve at the half-intensity points (half-width), the probability that the actual frequency will occur within this bandwidth is 76 percent for a Gaussian, and 50 percent for a Lorentzian statistical distribution. In the limit of very short averaging times, the short-term stability becomes

$$s_\nu^{\,0} = \lim_{\tau \to 0} s_\nu(\tau). \tag{8.3}$$

The limiting, short-term frequency stability of a generator is set by the random-noise characteristic of atomic processes.

8.1.3 Long-term Frequency Stability

Consistent with our definition of short-term stability, we define long-term stability as that inverse, fractional-frequency variation not limited by the time resolution of the detector.

$$S_\nu(\tau) = \frac{\bar{\nu}}{\Delta\nu(\tau)} \qquad \tau > \tau_0 . \tag{8.4}$$

Because long-term frequency stability is chiefly determined by macroscopic changes in the generator, long-term stabilities are frequently stated in terms of convenient averaging periods (minutes, hours, days) and under stipulated extremes in environmental conditions (pressure, warm-up time, temperature, applied line voltage, etc.) For very long averaging times, a border value of long-term stability,

$$S_\nu^{(\infty)} \lim_{\tau \to \infty} S_\nu(\tau), \tag{8.5}$$

is often set by secular changes, or aging, of the generator.

If random fluctuation is assumed and if $\Delta\nu(\tau)$ is taken as the standard deviation, the probability that the frequency of the generator will be within $\nu \pm \Delta\nu$ is 67 percent—reasonably consistent with the half-width definition of short-term stability above.

Although it is important to distinguish between short-and long-term stabilities, it is well to realize that the distinction is defined and not fundamental. The dividing line, as defined here, is established by the method of stability measurement. The method of stability measurement chosen will frequently be based upon the desire to simulate the intended application of the generator (which, in turn, dictates what is to be considered "short" or "long").

8.1.4 Frequency Resettability

A quantity very closely related to stability is frequency resettability R_ν. It is defined as

$$R_\nu = \frac{\nu}{\Delta\nu} \tag{8.6}$$

a quality factor describing the precision with which the frequency ν can be reproduced after the generator was perturbed and readjusted. In accordance with the definitions of stability above, $\Delta\nu$ represents here the standard deviation of a series of trial settings.

8.1.5 Wavelength Stability and Resettability

The distinction between frequency and wavelength stability or resettability is operational and is made here according to whether $\Delta\nu$ or $\Delta\lambda$ is the actually measured quantity, that is whether microwave or optical techniques are used to measure stability. The definitions of wavelength stability and resettability are derived from (8.1) through (8.6) by use of the fundamental relation

$$c = \nu\lambda. \tag{8.7}$$

It is evident that

$$\frac{\lambda}{\Delta\lambda(\tau)} = \frac{-\nu}{\Delta\nu(\tau)}. \tag{8.8}$$

Thus, except for the insignificant minus sign, these quantities are analogous; wavelength stability is theoretically synonymous with frequency stability. Consequently, we adopt analogous definitions of stability and resettability. Short-term wavelength stability becomes

$$S_\lambda = \lambda/\delta\,\lambda(\tau) \qquad \tau < \tau_0 \tag{8.9}$$

wherein $\delta\lambda(\tau)$ is the half-width of the unresolved spectral line.

Long-term wavelength stability becomes

$$S_\lambda = \lambda/\Delta\lambda(\tau) \qquad \tau > \tau_0. \tag{8.10}$$

Wavelength resettability is similarly defined as

$$R_\lambda = \lambda/\Delta\lambda. \tag{8.11}$$

In both (8.10) and (8.11) the $\Delta\lambda$'s represent standard deviations in the usual statistical sense.

8.1.6 Absolute and Relative Stabilities

There is, of course, no theoretical distinction between frequency and wavelength stabilities or resettabilities. The current absence of any frequency standards for the optical spectrum, and the state of the art of frequency measurement lead to an important distinction, which may be paralleled to the distinction of "precision" and "accuracy" made in metrology[1].

Absolute stability, or stability with respect to an established standard is the only true measure of accuracy. Relative stability, which is unrelated to a standard, is often an extremely valuable, though less meaningful quantity; it is, of course, only indicative of precision. Short-term stability, or line width, is always a relative measure, for it is largely independent of the exact frequency or wavelength of the line center. The concepts of long-term stability and resettability, however, require the distinction above.

The limited time-resolution capability of today's best photoelectric detectors ($\approx 10^{-10}$ sec, corresponding to detectable frequencies using techniques of comparing microwave-heterodyne-frequencies in the 10 GHz range, i.e., $\approx$ 1 Å in the visible), coupled with the current lack of laser transitions sufficiently close to primary or secondary wavelength standards, precludes the direct comparison of frequencies. As long as these limitations prevail, measurements of laser-frequency-stability will remain relative. Measurements of the relative frequency stability of two lasers can be made by using photoelectric mixing techniques, and the measurements enable us to determine the relative stability of one laser with respect to the other; they do not exclude the possibility of frequency drifts common to both lasers. Once a laser transition is discovered, or a wavelength standard is established, such that a heterodyne comparison can be made, absolute frequency measurements will become routine.

The accuracy of a measurement cannot exceed that of the standard upon which it is based, even though its precision may be considerably higher. Optical-wavelength standards are presently only accurate to one part in 10^8 at best. Wavelength stability may, therefore, represent accuracy or absolute stability up to that value only. Thus, even wavelength-stability measurements beyond one part in 10^8 become relative.

New standards are chosen on the basis that they must equal the old ones in accuracy and surpass them in precision. Once it has been demonstrated that the absolute wavelength stability of a laser matches that of wavelength standards, its relative stability (which has already been shown to exceed the inherent precision of these standards) may become a highly significant figure of merit.

8.2 THE FREQUENCY STABILITY OF LASERS AND ITS MAGNITUDE

8.2.1 Short-term Frequency Stability

The theoretical short-term stability limit, defined by (8.3), is set by spontaneous emission of photons of random phase and frequency into the oscillating laser mode. For the special case in which the center of the cavity resonance coincides with the center of the (broader) atomic line, we obtain[2]

$$s_\nu^0 = P\,\frac{[(n_2/g_2 - n_1/g_1)/n_2/g_2]}{4\pi h(\Delta\nu_c)^2} \tag{8.12}$$

where P is the power output, h is Planck's constant, n_1/g_1 and n_2/g_2 are the respective statistically weighted atomic-number densities in the lower (1) and upper (2) states, and $\Delta\nu_c$ is the frequency resolution of the (passive) Fabry-Perot laser cavity. The latter is defined as

$$\Delta\nu_c = c/2nL\ \mathscr{F} \tag{8.13}$$

wherein n is the refractive index, L the mirror spacing, $\mathscr{F}$ is the cavity finesse defined in the usual way.

$$\mathscr{F} = \pi(R)^{1/2}/(1 - R). \tag{8.14}$$

It is instructive to examine the order of magnitude of equations 8-13 and 8-12 for typical gas-laser parameters. With modern ultrahigh-reflectivity coatings and precision Fabry-Perot plates, we can obtain finesse values of 100 or more. Thus passive-cavity frequency resolutions for a 1-m mirror spacing are of the order of

$$\nu_c = \frac{c}{2nL\mathscr{F}} = \frac{(3 \times 10^{10})}{(2)(1)(100)(100)} = 1.5\ \text{MHz}. \tag{8.15}$$

Experimental evidence indicates that $(n_2/g_2 - n_1/g_1)/(n_2/g_2)$ lies between 0.1 and 0.01. Choosing the upper bound, 0.1, as a conservative value for a laser generating 1 mW of output power at a wavelength of 1.15 μm, the theoretical short-term stability becomes

$$s_\nu^\circ = P\,\frac{[(n_2/g_2 - n_1/g_1)/n_2/g_2]}{4\pi h(\Delta\nu_c)^2} = \frac{(10^{-3})(10^{-1})}{4\pi(6.61 \times 10^{-34})(1.5 \times 10^6)^2}$$
$$\cong 5.4 \times 10^{15}.$$

The corresponding line width and frequency resolution obtained from (8.8) are

$$\delta\lambda = \frac{\lambda}{s^\circ} = \frac{(1.15 \times 10^3)}{(5.4 \times 10^{15})} \cong 2.2 \times 10^{-13} \text{ nm},$$

$$\delta v = \frac{v}{s^\circ} = \frac{c}{\lambda s^\circ} = \frac{(3 \times 10^{10})}{(1.15 \times 10^{-6})(5.4 \times 10^{15})} \cong 5 \text{ Hz}.$$

The relative frequency stabilities of two gas lasers have approached the theoretical limit within about 3 orders of magnitude.

Over somewhat longer averaging times than those involved above, thermal oscillation of the mirror supports (spacers) for the gas laser cause instability. For spacers of uniform cross section of material volume V, and absolute temperature T, the corresponding mechanical limit of stability is estimated to be[3].

$$s = \left(\frac{YV}{2kT}\right)^{1/2} \tag{8.16}$$

where Y is the Young modulus of the spacer material and k the Boltzmann constant. For a carefully constructed laser operating at $\lambda = 1.15\ \mu$m, the mechanical stability limit is calculated to be $s = 10^{14}$, or $\delta\lambda = 10^{-11}$ nm, $\delta v = 3$ Hz, for averaging times of the order of 1 msec[3]. As expected the cavity will set the ultimate frequency stability limit of a free-running laser.

Although (8.12) also describes the theoretical short-term stability limit of solid-state and semiconductor lasers, these lasers, in practice, exhibit line widths a great many orders of magnitude more than the calculated values. Line widths of 0.01 nm and 0.1 nm are observed respectively with ruby[4] and GaAs[5] lasers (corresponding to short-term stabilities of 10^5 and 10^4). These observe widths are instrument-limited and are mainly caused by the heating of the device during optical or conduction pumping. The thermal dependence of wavelength has been measured as[5.6]

$$\begin{aligned} d\lambda/dT &= 0.0065 \text{ nm per degree C for the ruby } R_1 \text{ line} \\ &\approx 0.1 \text{ nm per degree C for GaAs.} \end{aligned}$$

8.2.2 Long-term Stability

The long-term stability of a gas laser is set by the requirement that it operate within that portion of the Doppler-broadened line for which the gain exceeds the losses[7]. A lower limit to the stability becomes

$$S_\nu = \frac{\nu}{\Delta \nu_D} = \frac{c}{2}\left(\frac{m}{2kT}\ln 2\right)^{1/2}. \tag{8.17}$$

where m is the atomic (molecular) mass in amu, k is the Boltzann constant, and T the absolute temperature. For a helium-neon laser at an "atomic temperature" of 500° K the long-term stability will at least be

$$S_\nu \geq \frac{(3 \times 10^{10})}{(2)}\left[\frac{(20.18)(1.65 \times 10^{-24})}{(2)(1.37 \times 10^{-16})(500)(.693)}\right]^{1/2} \approx 3 \times 10^5$$

Actually a typical long gas laser operates at several discrete frequencies corresponding to the axial-mode spacing

$$\Delta\nu = c/2nL \tag{8.18}$$

where nL is the optical-path length between mirrors. If the cavity structure could be made so stable that the frequency would not shift by an amount greater than the separation of one of the axial modes, the long-term frequency stability would be increased to

$$S_\nu = \nu/\Delta\nu = (2nL\nu)c = (2)(1)(9100)(2.6 \times 10^{14})/3 \times 10^{10} \cong 1.7 \times 10^8.$$

The limits above are pessimistic because the effective band width is greatly reduced by the logarithmic gain of the medium.

Frequency drifts in gas lasers are primarily caused by mechanical or thermal instabilities in the optical-cavity length nL. Typical gross sources of instability include thermal expansion of the cavity length, refractive index variations in the air path of external mirrors, and vibrations. These effects have been shown to limit the long-term stability of gas lasers to roughly 10^7 to 10^8 under standard laboratory conditions[8]. Proportionally higher stabilities can be achieved under more-suitable ambient conditions.

With perhaps one exception, solid-state lasers have been found to exhibit poor frequency stability. A clad ruby laser operated at liquid-nitrogen temperatures and pumped to within 1 percent of threshold[11] has yielded a single line whose short-term stability, by virtue of the high gain in ruby, was better than 2×10^7. No account was taken, however, of the time variation of the output frequency due to thermal tuning.

In beat-frequency experiments, the frequency of one helium-neon laser with respect to another was found to be about $S_\nu = 5 \times 10^8$ under laboratory conditions[3] and $S_\nu = 5 \times 10^{10}$ under the conditions of low acoustic background[9]. Averaging times $\tau \approx 2$ min were considered in both cases.

By interferometric comparison with a standard Hg^{198} lamp, the absolute wavelength stability of a helium-neon laser was determined to be $S_\lambda > 10^7$, approximately, for $\tau = 2$ min and standard laboratory conditions[8].

The normal fluorescent line, or gain curve, of ruby is about 0.3 nm wide[10]. If this width is again taken as indicative of the upper limit of long-term stability, $S_\nu > 2 \times 10^3$, for the R_1 radiation of ruby at 693 nm.

Most of the long-term instabilities of the solid-state laser are caused, as stated above, by the variation of wavelength with temperature. With the value given for the ruby R_1 line and a 1°C temperature variation $\Delta\lambda = 0.007$ nm and $S_\nu = 10^5$.

The long-term stability of semiconductor lasers is very small compared to gas lasers. The width of the gain curve is about 1 nm at 839 nm for GaAs so that, roughly, $S_\nu > 7 \times 10^2$.

Shifts between modes are a common source of instability in a GaAs laser[10]. Equation 8-18 applies for the spacing of modes; a typical example is $nL = 2$ mm, $\nu = 3 \times 10^{14}$ Hz, so that $\Delta\nu_M = 1.5 10^{11}$ H_Z, and $S_\nu = 2 \times 10^3$. Temperature variations of the order of 1°C produce a wavelength shift of the oscillating mode of about 0.1 nm and, thus, limit stability to $S_\nu = 8 \times 10^3$ at $\lambda = 839$ nm.

8.2.3 Resettability

The lower limits of the long-term stability of a gas laser are also the lower limits of its resettability. That is the laser must oscillate within the Doppler curve $R > 2 \times 10^5$ and can easily be adjusted to operate in a single mode $R \geq 2 \times 10^6$.

Better resettabilities may be accomplished by various methods. Resettabilities $R > 10^9$ have been reported[3, 12]. Centering the laser mode at the center of the "Lamb dip" may be expected to yield comparable values of R for a gas laser.

The long-term stability and resettability of solid-state and semiconductor lasers are respectively of the order of 10^5 and 10^2.

8.3 MEASUREMENT OF SHORT-TERM STABILITY

8.3.1 Optical Methods

Conventional spectroscopic techniques are totally inadequate for optical measurements of the short-term stability of gas lasers. They may, however, be applied with reasonable success to determine the short term stability of solid-state and diode lasers as the latter quantum electronic devices possess inherently lower stability.

The most important consideration in choosing a spectrometric instrument for the measurement of short-term stability, that is line width, is resolving power. To measure the energy spread $\delta E = h\,\delta\nu$ in a wave train of frequency spread $\delta\nu$, the spectroscope must evaluate the signal for at least a period of time δt such that Heisenberg's principle, $\delta E\ \delta t \geqq h$, is satisfied ($h$ = Planck's constant). If ℓ is the largest optical path difference in the spectroscope, we have $\delta t = \ell/c$ and

$$\ell \geqq c/\delta\nu = 2\pi/\delta k, \tag{8.19}$$

where k is the wave number defined as

$$k = 2\pi\nu/c = 2\pi/\lambda \tag{8.20}$$

The wave number difference δk cannot be resolved if (8.19) is not fulfilled. The resolving power of the spectroscope is defined as

$$R = \mathrm{k}/\delta k. \tag{8.21}$$

Using (8.19), (8.20) and (8.21), it is evident that the resolving power may be expressed as

$$R = \ell/\lambda. \tag{8.22}$$

It is thus equal to the longest optical path difference measured in wavelengths.

8.3.1.1 Prism and Grating Spectrometers Most of the lines of solid-state and semiconductor lasers do not require greater resolutions than those available with prism and grating spectrometers. Resolutions as great as 10^6 can now be obtained with diffraction gratings ruled under interferometric control[13,15], so that line widths down to 0.001 nm at $\lambda = 1\ \mu$m, can be determined. Although the function of a spectrometer is too well known to be discussed here, the techniques of line-width measurement are less frequently discussed and will be given a more detailed treatment.

The spectrum line $f(x)$ formed in a spectrograph is a distorted image of the unperturbed signal $f_0(x)$ the shape of which is to be measured. It may be expressed as a multiple-convolution integral.

$$f(x) = \int_{-\infty}^{+\infty} dy\, dz\, f_0(z) f_D(z - y) f_T(y - x), \tag{8.23}$$

in which $f_0(x)$ is only one of the factors. Here, x, y, and z are generalized coordinates representing wavelength or wavenumber; $f_D(x)$ is the diffraction pattern formed by the spectrograph and $f_T(x)$ the diffusion or turbidity profile characteristic of the photographic plate (or an equivalent function if another method of recording is used). The problem in linewidth determination is thus to unfold the convolution $f(x)$ to obtain $f_0(x)$.

The undistorted line may, for instance, be a Gaussian or Lorentizian profile:

$$f_0(x) = G(x) = e^{-x^2/a^2}, \tag{8.24}$$

or

$$f_0(x) = L(x) = (b^2 + x^2)^{-1} \tag{8.25}$$

where a and b are related to the half-width δx of the line by $\delta x_G = 2a$ $(\ln 2)^{\frac{1}{2}}$ and $\delta x_L = 2b$, respectively. Both distributions may be regarded as special cases of the Voigt function,

$$f_0(x) = V_0(x) = \int_{-\infty}^{+\infty} \frac{\exp(-y^2/a^2)dy}{[b^2 - (x-y)^2]}, \tag{8.26}$$

which is obtained by the convolution of a Gaussian with a Lorentzian. Being bi-parametric, $V_0(x)$ can also be used in many instances to approximate unknown profiles $f_0(x)$. The two parameters a and b are then chosen so that the 0.5- and 0.1-intensity widths of the Voigtian equal those of the profile to be approximated; this is usually possible if the ratio of the two widths falls within the limits $1.8226 \geq (\delta x)_{0.1}/(\delta x)_{0.5} \geq 3.0000$, set by the Gaussian and Lorentzian distributions that generate the Voigt function.

The diffraction pattern formed in a spectrograph of rectangular aperture is

$$f_D(x) = (\sin^2 x)/x^2 \tag{8.27}$$

for monochromatic plane-wave illumination and infinitely narrow slits. For finite slits the pattern depends on the slit width as well as on the mode of illumination (coherent or incoherent); it is a more complicated function [15, 16]. The turbidity profile obtained by recording a mathematical line on a photographic plate is

$$f_T(x)\exp\frac{-|x|}{\alpha} \tag{8.28}$$

where the diffusion coefficient α is characteristic of the emulsion. The convolution of the diffraction and turbidity profiles leads to functions that cannot generally be expressed analytically, but because the ratio $(\delta x)_{0.1}$ $(\delta x)_{0.5}$ is 1.6661 for $f_D(x)$ and 3.3219 for $f_T(x)$ (below and above the range indicated above so that their average falls within that range), it can usually also be approximated by a Voigt function:

$$\int_{-\infty}^{+\infty} dy f_D(y) f_T(y-x) = V_1(x). \tag{8.29}$$

The observed profile then is a convolution of two Voigt functions.

$$f(x) = \int_{-\infty}^{+\infty} dy [V_0(y-x) - V(x)], \tag{8.30}$$

which is again a Voigtian distribution the parameters a and b of which are related to those of $V_0(x)$ and $V_1(x)$ by

$$a^2 = a_0^2 = a_1^2 \qquad b = b_0 + b_1. \tag{8.31}$$

If $V_1(x)$ can be established by independent measurement, it is thus possible to unfold the observed profile (x) so that the desired line shape $f_0(x) = V_0(x)$ is obtained.

The procedures involved are dealt with in detail in many publications, which also contain the necessary numerical tables of Voigt functions. The interested reader is referred to a recent extensive account of the subject [17].

8.3.1.2 Two-beam Interferometers The Michelson interferometer has been proposed as a method by which one might hope to obtain data on the line width of gas lasers. Probably the longest Michelson interferometer ever developed for examination of laser line widths (200 m optical-path difference and illuminated with a helium-neon gas laser) was built at the National Bureau of Standards in 1962 [18, 19]. We see from equation 8.19 that such a device is capable of measuring line widths down to $\delta k = 3 \times 10^{-4}$ cm^{-1} or, at $\lambda = 1$ nm, $\delta\lambda = 5 \times 10^{-6}$ nm. Because the fringe visibility in this instrument remained virtually constant over the full path, still longer interferometers are feasible. The resolving power of this instrument was 500 million; these data may be taken as indicative of the practical limit of performance of two-beam interferometers. The 200 m interferometer was not able to resolve the gas-laser line width. It is instructive to explore further the characteristics of the Michelson interferometer for short-term stability measurements of solid-state and diode lasers.

To illustrate the manner in which a Michelson interferometer may be applied to measure line widths, consider the interference of two quasi-monochromatic beams of equal intensity and an optical path difference ℓ. Let $f(k)$ be the spectral intensity of the fringe pattern, due to the spectral components within $(k, k + dk)$,

$$i(k)\,dk = 2\,f(k)1 + \cos(k\ell)\,dk. \tag{8.32}$$

The total intensity is

$$I = \int_0^\infty i(k)\,dk, \tag{8.33}$$

from which we obtain [20]

$$I = P + C\cos(k_0\ell) - S\sin(k_0\ell) \tag{8.34}$$

where k_0 is the mean wave number, P the total intensity of the two beams and C and S the Fourier cosine and sine transforms of $f(k)$. The visibility, $V(\ell)$ of the fringes is

$$V(\ell) = (I_{\max} - I_{min})/(I_{\max} + I_{min}) = (C^2 + S^2)^{1/2}/P. \tag{8.35}$$

If $f(k)$ is symmetrical about k_0, $S = 0$, and

$$V(l) = \frac{|C|}{P}, \tag{8.36}$$

so that (except for the constant factor $\pm P$) the visibility of the fringes $V(\ell)$ is equal to the Fourier cosine transform C of the signal $f(k)$.

Assume, for example, $f(k)$ to be a Gaussian or Lorentzian distribution:

$$f(k) = G(k) = \exp\left[-4\ln 2\,(k - k_0)^2\right], \tag{8.37}$$

or

$$f(k) = L(k) = \left[(\delta k/2)^2 + (k - k_0)^2\right]^{-1} \tag{8.38}$$

where in each case δk is the full half-width of the line. The corresponding visibility curves, normalized so that $v(0) = 1$, are

$$V_G(\ell) = \exp[-\delta k\ell/4(\ln 2)^{1/2}]^2 \tag{8.39}$$

and

$$V_L(l) = \exp\left[\frac{-\delta kl}{2}\right] \tag{8.40}$$

In either case $V(\ell)$ has essentially become zero when $\ell = 2\pi/\delta k$, which, of course, is again Heisenberg's principle.

Line width can, in principle, be measured by gradually increasing the mirror separation in the interferometer until the fringes fade out. It can be shown, however, that fringe fading occurs asymptotically; this method lacks precision.

A more accurate determination, not only of δk but of the entire function $f(k)$ as well, is obtained by measuring visibility at various path differences and by subsequent Fourier inversion. This technique was devised by Michelson[21] to measure the visibility curve of the red line of cadmium (corresponding to a Gaussian distribution of 0.0013 nm half-width).

Neither of the methods mentioned is capable of measuring the extremely narrow lines of gas lasers. For the 1.15μm line of the helium-neon laser, the fringe visibility was found to remain virtually constant over the entire 200m path length in the Bureau of Standards Michelson interferometer [18]. The only information on line width obtained was that it was well below the 5×10^{-6} nm limit resolution of the interferometer.

8.3.1.3 Multibeam Interferometers The technique of measuring line widths by determining the greatest path over which interference effects are obtainable can also be carried out with multiple-beam interferometers. Fabry and Buisson [22] have used this method to measure the widths of helium, neon, and krypton lines. In view of the folded path in the Fabry-Perot this method leads to more practical interferometer lengths even when very narrow lines are measured. In practice it is extremely difficult to keep interferometer plates sufficiently parallel if one of them is moved over an appreciable distance. The Fabry-Perot interferometer is, therefore, best used in the customary manner as an etalon with constant, or nearly constant, plate separation.

When longer averaging times are permissible, the classical method of photographing the fringes can be employed. The fringes can also be recorded on a strip chart by slightly varying the optical mirror spacing and recording the output of a photodetector viewing the center of the fringe pattern.

Pressure variations inside the etalon [23], as well as piezoelectric [24] or magnetostrictive tuning, may be used as scanning methods.

For shorter averaging times, vibrations of the mirrors and display of the photodetector output on an oscilloscope screen are used [24-26]. With piezoelectric ceramics employed as spacers, averaging times as little as 10^{-7} sec. can be accomplished[27].

For monochromatic illumination, the intensity distribution within the Fabry-Perot fringes is given by the Airy formula:

$$I = I_{max}/(1 + F \sin 2\Lambda/2) \tag{8.41}$$

where I_{max} is the intensity at the fringe center,

$$F = 4R/(1 - R)^2 \approx (2/\ln R)^2 \tag{8.42}$$

is the "coefficient de finesse," R, the geometric mean of the reflectivities of the two plates,

$$\Lambda = (2/k)nd \cos \varphi \tag{8.43}$$

is the phase difference of successive beams, with $k = 2\pi/\lambda$ = wave number, nd = optical length of etalon, and φ = angle of energy propagation inside the etalon[28]. The resolving power R_0 is

$$R_0 = knd \cos \varphi/|\ln R| \tag{8.44}$$

For a typical, high-resolution Fabry-Perot interferometer, designed for laser spectroscopy assume $nd = 100$ cm, $R = 0.95$, $\cos \varphi \approx 1$, at $\lambda = 1\ \mu$m, or $k = 6 \times 10^4\ \text{cm}^{-1}$. Then $R_0 \approx 10^8$. Hence, line widths of the order of $\delta\lambda = 10^{-5}$ nm are measurable with Fabry-Perot interferometers, which is again not quite sufficient for gas-laser spectroscopy but adequate for other types of lasers.

The laser principle itself may someday provide a further improvement of the Fabry-Perot interferometer so that it may become applicable to the measurement of the extremely narrow lines of a gas laser. By filling the interferometer with an active medium which amplifies the signal to a level just below the threshold of laser action, the resolving power of the interferometer may increase[27].

$$R_0 = knd \cos \varphi \Bigg/ \left| \ln R + \frac{\alpha d}{\cos \varphi} \right| \tag{8.45}$$

and may thus be increased by increasing the amplification coefficient α of the active medium (ln R is negative, α positive). There are some experimental difficulties yet to be overcome before the active Fabry-Perot interferometer becomes a practical device[28].

As are the spectrum lines formed in a spectograph, the fringes observed with a Fabry-Perot etalon are distorted images of the spectral distribution $f_0(k) = G(k)$. As in (8.37), the intensity distribution within the fringes may be expressed as the following Fourier series [30, 31]:

$$I = K\left\{1 + 2\sum_{n=1}^{\infty} R^n \exp[(-n\pi\delta k/2 \ln 2)^2 \cos n\varphi]\right\} \tag{8.46}$$

where the factor K depends on the desired normalization[29]. Numerical tables are available from which we may derive the true line width from the observed fringe width [30, 32, 33, 34].

Perhaps the simplest way to obtain $f_0(k)$ is through the use of Voigt functions, as is discussed in Section 8.3.1.1. The ratio of the 0.1 and 0.5 widths of the Airy distribution (8.41) is approximately 3.0 (equal to that of a Lorentzian line). The Airy function can, therefore, be well approximated by a Lorentzian distribution, so that equation 8.46 becomes a Voigt function. The calculations outlined under section 8.3.1.1 are therefore directly applicable to unfolding the time shape of Fabry-Perot fringes[33].

8.3.2 Photoelectric-mixing Methods

The beating of optical waves, leading to a wave of the average frequency modulated at the difference frequency, was demonstrated by A. Righi[35] over 80 years ago. Typical of its applications in spectroscopy include obtaining beats between spectral lines by photoelectric mixing of the Zeeman components [36, 37] of the 545 nm line of Hg^{202}. Low signal-to-noise ratios severely handicap the application of this method so long as conventional spectral sources are used; the high intensities of laser beams eliminate this handicap. Photoelectric mixing has thus become a powerful tool of laser spectroscopy.

Although there exist more rigorous treatments now[38], the original theory [36, 37] provides an adequate description of the beat-frequency technique. The treatment given herein follows the early work.

Let $f(\omega)$ be the intensity per unit of angular frequency for the spectral signal, so that the corresponding electric-field amplitude is

$$E(\omega) \approx [f(\omega)]^{1/2} \exp j[\omega t + \varphi(\omega)] \tag{8.47}$$

where the phase angle $\varphi(\omega)$ is here taken as random. The combined intensity of two frequencies, ω and $\omega + \beta$, is

$$I(\omega, \beta) = E(\omega)E^*(\omega + \beta) \approx [f(\omega)f(\omega + \beta)]^{1/2} \exp j(\beta t + \gamma), \qquad (8.48)$$

where $\gamma = \varphi(\omega + \beta) - \varphi(\omega)$ is the difference of the random phases.

If the mean-square value of $I(\omega, \beta)$ is measured, the actually observed quantity is a time average pertaining to frequency intervals $(\omega, \omega + \Delta\omega)$ and $(\beta, \beta + \Delta\beta)$, or

$$d\langle I^2(\omega, \beta)\rangle \approx \Delta\beta \, \Delta\omega f(\omega)f(\omega + \beta). \qquad (8.49)$$

The total signal is due to the contributions of all pairs of difference frequency β within $f(\omega)$, or

$$\langle I^2(\beta)\rangle \approx \Delta\beta \int_{-\infty}^{+\infty} df(\omega)f(\omega + \beta), \qquad (8.50)$$

proportional to the autocorrelation function of $f(\omega)$.

The assumption above of random phases, undoubtedly correct for spontaneously emitted light, is somewhat questionable for laser light[38]. Yet it must still serve as an approximation as long as an adequate wave representation of the laser beam does not exist.

Equation 8.50 holds true as long as the area A of the photocathode used does not exceed the size of the diffraction pattern, or area of coherence, on the cathode,

$$A \leq \lambda^2/\Omega, \qquad (8.51)$$

where λ is the wavelength and Ω the solid angular spread of the beam. Because of its high spatial coherence, the laser source can be made to satisfy this condition, whereas for conventional sources (8.50) must be multiplied by the small factor $\lambda^2/A\,\Omega$, from which, the already mentioned difficulty of exceedingly low signal-to-noise ratios arises.

We now examine three different ways to measure the widths and profiles of spectral lines by photoelectric mixing:

8.3.2.1 Homodyne Detection This method consists of beating the line $f(\omega)$ with itself, as was assumed for the derivation of (8.50), which is therefore directly applicable. For Gaussian and Lorentzian lines, as in(8.37) and (8.38), but with k replaced by ω, the observed autocorrelation functions are

$$\langle I_G{}^2(\beta)\rangle = \Delta\beta \exp[-2\ln 2\beta^2/(\delta\omega)^2] \qquad (8.52)$$

and

$$\langle I_L^2(\beta)\rangle = \Delta\beta[1 + \beta^2/(\delta\omega)^2], \tag{8.53}$$

where, in both cases, $\delta\omega$ is the full half-intensity width of $f(\omega)$, and where both functions were normalized so that $\langle I^2(0)\rangle = \Delta\beta$. The half-widths of 8.52 and 8.53 are $(\delta\beta)_G = \sqrt{2}(\delta\omega)_G$, and $(\delta\beta)_L$, respectively, so that $\langle I^2(\beta)\rangle$ is seen to be a somewhat broader distribution than $f(\omega)$.

It is also clear that the measurement of $\langle I^2(\beta)\rangle$ permits the determination of $f(\omega)$. If the procedure outlined in Section 8.3.1.1 can be applied so that $\langle I^2(\beta)\rangle$ is expressed as a Voigt function with the parameters a and b, $f(\omega)$ is again a Voigt function with parameters $a' = a/\sqrt{2}$, $b' = b/2$.

8.3.2.2 Heterodyne Detection The homodyne technique, as described above, requires the detection of beat-frequency components from $\beta = 0$ up to a few times $\delta\omega$. In view of the difficulties associated with a frequency range including dc, as well as very low ac components, the heterodyne technique appears more practical when extremely narrow lines, such as those of gas lasers, are to be detected.

In the heterodyne detection scheme we beat two spectral signals, $g_1(\omega)$ and $g_2(\omega)$. When two laser output signals are mixed, they are usually assumed to be sufficiently similar to warrant the assumption

$$g_1(\omega) = g(\omega_1 - \omega), \qquad g_2(\omega) = g(\omega_2 - \omega) = g(\omega_1 - \omega - \beta_0)$$

where ω_1 and ω_2 are the two center frequencies and $\beta_0 = \omega_1 - \omega_2$ their difference. Equation 8.50 becomes directly applicable if we let

$$f(\omega) = g(\omega_1 - \omega) + g(\omega_2 - \omega), \tag{8.54}$$

so that, with $\beta = \beta_0 + b$,

$$f(\omega)f(\omega + \beta) = g(\omega_1 - \omega)g(\omega_1 - \omega - b), \tag{8.55}$$

when either signal is zero where the other is not. Thus

$$\langle I^2(\beta)\rangle = \langle I^2(\beta_0 + b)\rangle = \Delta\beta \int_{-\infty}^{+\infty} d\omega g(\omega_1 - \omega)g(\omega_1 - \omega - b), \tag{8.56}$$

proportional to the autocorrelation function of either signal.

If the two incoming signals g_1 and g_2 are not equal, $I^2(B)$ represents their convolution integral, from which it is only possible to derive an average of g_1 and g_2 because neither signal is distinguished from the other. Aside from this restriction, the homodyne and heterodyne techniques yield the same result, so that the analysis of (8.56) is the same as that already given in Section 8.3.2.1.

However, the photoelectric signal is now located about the difference frequency β_0 of the two incoming signals and is, therefore, more readily detectable with standard frequency analyzers. When beats between axial modes of gas lasers are used in the experiment, β_0 will usually be of the order of 10^9 Hz. Analyzers with a resolving power $\beta/\delta\beta$ in excess of 10^9 are then required to measure the linewidths, $\delta\omega < 1$ Hz, expected of gas lasers. Because such equipment is not available, instrument-limited widths, rather than line widths are usually measured in these cases. Javan's technique[3] of mixing the outputs of two independent lasers, where smaller values of β_0 are obtained by tuning each laser to the center of the gain curve, comes close to the measurement of true line widths. Even then, the requirements of resolution remain rather severe, for the resettability of the two lasers will hardly exceed the value $R = 10^9$, so that $\beta_0 = R/\omega \approx 10^6$, and $\beta/\delta\beta > 10^6$.

8.3.2.3 Quasihomodyne Detection An ingenious method to overcome the shortcomings of the homodyne and heterodyne techniques has been proposed, [39] and subsequently realized [40], in which an optical beam splitter is used to divide the signal $g(\omega)$ into two coherent beams that are recombined after one of the two beams has been reflected from a moving mirror. Because reflection from the moving mirror introduces a Doppler-frequency shift in one beam, the recombination of the two beams on a photocathode is equivalent to heterodyning two signals of difference frequency

$$\beta_0 = 2\omega v/c \tag{8.57}$$

where v is the velocity of the mirror and c the speed of light. However, $g(\omega)$ is truly beating with itself; the photoelectric signal is thus equivalent to that of a homodyne experiment and reflects exactly the desired autocorrelation function of $g(\omega)$.

The frequency location of the beat spectrum may be adjusted by adjusting the mirror velocity. For $v = 0.15$ cm/sec and $\omega = 10^{15}$, the Doppler shift (equation 8.57) is $\beta_0 = 10^4$ Hz, so that the signal occurs in the audiofrequency range. Moderate values of resolving power—$\beta/\delta\beta > 10^4$ for $\delta\beta < 1$ Hz—are now sufficient to resolve even extremely narrow lines, and the difficulties of the heterodyne technique are elegantly overcome.

The mirror motion must, of course, be sufficiently constant to provide an undistorted spectral profile of the reflected beam. From (8.57), $v/\delta v = \beta/\delta\beta$, or $\delta v < 10^{-4} v = 0.15 \mu$m/sec in the numerical example above. Interferometric control, similar to that now employed in the manufacture of high-precision gratings, may be expected to provide the required constancy of motion.

8.4 MEASUREMENT OF LONG-TERM STABILITY

8.4.1 Optical Methods

To determine the long-term wavelength stability of a laser, we must measure the time dependence of the wavelength it emits. Hence, a spectrometer must be used to reveal motions of the spectrum line produced by the changing wavelength of the laser.

Although it is conceivable that some spectrometers could be used to measure wavelength and wavelength changes (by angle and distance measurements), it is simpler to measure relative changes by comparing the laser wavelength with that of a standard spectrum line. Systematic errors and instabilities of the spectrometer are then of little significance for any but the most precise measurements because they affect both wavelengths and, therefore, tend to cancel one another.

The spectrometer used must have a limit of resolution $\delta\lambda$ about equal to the amount of wavelength fluctuation $\Delta\lambda$ to be measured. Hence its resolving power $R = \lambda/\Delta\lambda$ must be of the order of magnitude of the stability $S_\lambda = \lambda/\Delta\lambda$ to be measured.

The methods employed to measure long-term stability are those of time-resolved spectroscopy. The data obtained consist of a continuous recording of wavelength versus time, or a series of individual measurements at regular time intervals. To provide the desired degree of time resolution, the luminosity U of the spectrometer, defined as its output flux per unit of source luminance, must be high enough to provide a detectable amount of energy during the averaging period τ. This energy is $E = \tau BU$, where the luminance of the source B, is in most cases critically determined by the standard source with which the laser is compared.

The two aspects from which the spectrometer must be chosen, then, are resolution R and luminosity U. They are conveniently expressed together in the resolution-luminosity product $P = RU$, a figure of merit that will be used in the following to compare the various types of optical spectrometers suitable for stability measurements.

8.4.1.1 Prism and Grating Spectrometers The theoretical resolving power of prisms and grating is

$$(R_0)_{\text{prism}} = \ell\, dn/d\lambda \tag{8.58}$$

and

$$(R_0)_{\text{grating}} = mN, \tag{8.59}$$

where ℓ is the base length, $dn/d\lambda$ is the material dispersion of the prism, and m and N the order of interference and number of rulings of the grating, respectively. Both prisms and gratings can be used throughout the ultraviolet, visible, and infrared spectrum. Simplicity and low cost are the advantages of a prism spectrometer, but its resolution rarely exceeds 10^5. The resolution of modern gratings, in contrast, may attain values of 10^6 and more[14].

The practical resolution R of a prism or grating spectrometer is a rather complex function of slit width and prism or grating size[41]. For larger slit widths, it may be written as

$$R = \lambda D_\lambda f/b \tag{8.60}$$

where D_λ is the angular dispersion of the prism or grating, f the focal length of the camera lens or mirror, and b the width of the entrance or exit slit (assumed to be equal).

The luminosity of the spectrometer is[42]

$$U = TbhWH/f^2 \tag{8.61}$$

where T is its transmission coefficient at the wavelength concerned, h the slit height, and WH the beam cross section at the prism or grating.

The resolution-luminosity product is thus obtained as

$$P = RU = \frac{\lambda TWHD_\lambda h}{f} \tag{8.62}$$

for either prism or grating spectrometers.

The angular dispersion of a prism is $D_\lambda = (\ell/W)(dn/d\lambda)$, where W is the beam width[41]. Hence,

$$P_{\text{prism}} = \lambda T \frac{h}{f} A \frac{dn}{d\lambda} \tag{8.63}$$

where $A = \ell H$ is the area of the prism base. The angular dispersion of a grating is $D_\lambda = m/d \cos\varphi$ where d is the grating constant and φ the angle of diffraction[43]. The beam cross section is $WH = A \cos\varphi$, A being the ruled area of the grating. Thus,

$$P_{\text{grating}} = \lambda T \frac{h}{f} A \frac{m}{d}. \tag{8.64}$$

For a reflection grating in a Littrow mount at $\varphi = 30°$, the grating equation yields $m\lambda/d = 2 \sin 30° = 1$, so that

$$P_{\text{grating}} = TA\frac{h}{f} \tag{8.65}$$

in this particular case.

To compare the two types of spectrometers, equal slit heights h and focal lengths f, as well as equal areas A, may be assumed without loss of significance. For gratings with efficient blaze, we may also take T as equal in both cases, so that

$$P_{\text{prism}}/P_{\text{grating}} = (d/m)(dn/d\lambda). \tag{8.66}$$

Observing that this quotient is of the order of 0.1 in most practical cases[44], we conclude that the same resolving power can be obtained with much greater luminosity by means of gratings rather than prisms[45].

Either spectrometer may be employed to measure the wavelength stability of lasers in different ways:

1. An exposure of the spectral range including the laser and the standard wavelength is made on a moving photographic plate. If the motion of the plate is linear with time and in a direction perpendicular to the spectrum, a streak photograph is produced that shows the wavelength differences of the laser and the standard source as a linear function of time. The smallest averaging time that can be obtained by this method is equal to the exposure time for a stationary spectrum.
2. The prism or grating is slowly rotated while the output flux through the exit slit is recorded on a strip chart. Each time a scan containing the laser line and the standard line is completed, the sense of rotation is reversed, so that a series of plots is obtained from which the wavelength fluctuations of the laser may be derived as a function of time. The averaging time is equal to the time required for each individual scan.
3. To obtain shorter averaging times, the prism or grating may be made to scan back and forth through the spectral range by oscillatory angular motion while the output flux is displayed on an oscilloscope screen. If the horizontal sweep of the oscilloscope is synchronized with the oscillation of the prism or grating, an undistorted display of the spectrum is produced on the screen from which the relative wavelength of the laser may be determined at short time intervals.

8.4.1.2 Fabry-Perot and Michelson Spectrometers The length of the Fabry-Perot etalon used for measuring the wavelength stability of a laser is limited by the limited-coherence length of the standard source with which the laser is compared. But a 10 cm etalon with dielectric mirror coatings of 95 percent reflectivity still has a theoretical resolving power of 25 million (at $\lambda = 500$ nm, see Equation 8.44)—many times better than what is obtainable with even the best grating.

The etalon may be used as a time-resolving spectrograph by imaging its fringe pattern onto a slit, which, in turn, is imaged onto a photographic film via a rotating mirror. Hence a streak photograph is obtained from which the diameter of the etalon rings, and thus the laser wavelength, can be derived as a function of time. This method has been used[46] to study the relative wavelength stability of ruby lasers with a spectral resolution of better than 10^6; the time resolution obtained was 1 μsec due to the high luminance of the laser (no reference source was used).

The etalon may also be used as a photoelectric spectrometer by placing a pin-hole diaphragm at the center of the ring pattern, so that only a narrow wavelength range $\delta\lambda$ is transmitted. Any change of the optical length of the etalon, such as that we describe in Section 8.3.1.3, will then change the wavelength passing through the diaphragm, and a photomultiplier tube detecting the output flux is thus made to scan through the intensity distribution within the rings. For stability measurements at longer averaging times, slow linear changes of the spacer length may be used to produce repeated recordings of the laser and reference wavelengths on a strip chart. For shorter averaging times, the relative wavelength of the laser is obtained as a function of time by vibrations of the spacer and synchronous display of the photomultiplier signal on an oscilloscope screen. Both these methods have been used[8] to measure the absolute wavelength stability of gas lasers by direct comparison with a standard Hg^{198} lamp.

The theoretical resolving power of a Michelson interferometer is from (8.22)

$$R_0 = \ell/\lambda = (2L/\lambda)\cos\varphi, \tag{8.67}$$

where ℓ is the optical-path difference, L the mirror separation, and φ the angle of observation. For $L = 10$ cm, $\cos\phi \sim 1$, and $\lambda = 500$ nm, the value obtained is $R_0 = 4 \times 10^5$—comparable to the resolution of prisms and most gratings but of course inferior to that of an etalon of equal length.

The Michelson interferometer may be used in the same manner as the etalon. It may also be used with a moving mirror to produce an output signal modulated at frequencies proportional to the optical frequencies with which it is illuminated [47,48]. The latter method should be especially useful for stability measurements in the infrared.

For either Fabry-Perot or Michelson interferometers, the condition for a bright center fringe is $2nL = m\lambda$, where n is the refractive index inside the interferometer, L the mirror separation, m the order of interference, and λ the wavelength. If a circular diaphragm of radius r is placed at the center of the pattern to pass a wavelength range $\delta\lambda$ the condition for constructive interference at the diaphragm edge is $2nL \cos\varphi = m(\lambda + \delta\lambda)$, where $\varphi \approx r/f$ is the angular half-width of the diaphragm and f the focal length of the telescope lens. Thus $m\delta\lambda = 2nL(1 - \cos\varphi) \approx L\varphi^2$, $\delta\lambda \approx \varphi^2 nL/m = \varphi^2\lambda/2$, and

$$R = \lambda/\delta\lambda = 2/\varphi^2, \tag{8.68}$$

when the angular dispersion of the interferometer and the diaphragm size are the only factors determining resolution.

For both interferometers, the luminosity is

$$U = TAa/f^2$$

where A and a are the areas of the interferometer plates and the diaphragm, respectively and where T is the average transmission coefficient of the instrument. With $a = \pi f^2/\varphi^2$ and (8.68),

$$U = 2\pi TA/R. \tag{8.69}$$

Hence the resolution-luminosity product is [44, 48]

$$P = RU = 2\pi TA \tag{8.70}$$

for the Fabry-Perot as well as the Michelson interferometer. Comparison with a grating spectrometer (a Littrow mount at 30°, as in (8.65) then leads to

$$P/P_{\text{grating}} = 2\pi f/h, \tag{8.71}$$

when the reflection coefficient of the grating is equal to the average transmission of the interferometer and when the ruled area of the grating and the interferometer plates are of equal size. The performance of the grating is thus seen to be limited by the small angular height h/f of its slits. Typical values of h/f are of the order of 1/100, so that (8.71) shows the luminosity of Michelson or Fabry-Perot interferometers to be several hundred times higher than that of a grating spectrometer of equal resolving power.

Of the two interferometers, the etalon is usually considered a better choice because of its greater simplicity and higher resolution. The performance of the etalon is chiefly determined by the attainable surface reflection coefficients and (for dielectric coatings) the flatness of its plates. Its resolution is seen from (8.42) and (8.44) to be proportional to $\sqrt{R}/(1 - R)$, where R is the reflection coefficient of either plate. The peak intensity of the fringes is proportional to $T_2/(1 - R)^2$, where $T = 1 - A - R$ is the transmission and A the absorption coefficient of the plates[49]. Hence the resolution-luminosity product of the etalon may be expressed in terms of the reflection and absorption coefficients of the plates as

$$P \propto (1 - A - R)^2 \sqrt{R}/(1 - R)^3. \tag{8.72}$$

It is easily seen to have a maximum at

$$R_{\text{opt}} = -0.25\, A + \sqrt{1 - A + 6.25\, A^2} \tag{8.73}$$

which is the optimum reflectivity with which (perfectly flat) etalon plates should be provided.

For metallic coatings with absorption coefficients $A = 0.10$ and 0.05, respectively, the optimum reflectivities are $R_{\text{opt}} = 0.73$ and 0.86, respectively; the corresponding values of the resolution-luminosity product are $P \propto 1.25$ and 2.73. For dielectric coatings with $A = 0.01$ and 0.005, the values obtained are $R_{\text{opt}} = 0.97$ and 0.985, yielding $P \propto 14.6$ and 29.4, respectively.

Dielectric coatings, however, cannot be utilized to full advantage unless sufficiently flat etalon plates are used. Imperfections in the plates determine an upper limit of reflectivity $R_{\max}$, beyond which the resolution of the etalon remains constant, so that the only result of any further increase in R is a decrease in fringe intensity, and thus etalon performance[50]. Good etalon plates have surface imperfections of 10 or 5 nm ($\lambda/50$ or $\lambda/100$ for visible light); the corresponding values of $R_{\max}$, at $\lambda = 500$ nm, are 0.96 and 0.98, respectively. Hence, the true maximal values of P that may be obtained with dielectric coatings on 10 nm plates are seen from (8.72) to be $P \propto 13.3$ and 18.8, respectively, for $R = R_{\max} = 0.96$ and $A = 0.01$ or 0.005. Despite this restriction, the advantage of dielectrically coated etalon plates is still evident.

8.4.1.3 The SISAM Spectrometer The acronym letters stand for Spectrométre Interférentiel a Sélection par l'Amplitude de Modulation, a type of spectrometer invented by P. Connes to combine the high resolution of a grating with the high luminosity of a Michelson interferometer.

The SISAM spectrometer is a Michelson interferometer in which two mirrors are replaced by two identical reflection gratings (or Littrow prisms) arranged at identical angles φ with the axis, so that incident rays for which $m\lambda/d = 2 \sin \varphi$ (m = order number, d = grating constant) return in themselves, whereas all other wavelengths are dispersed at different angles. The SISAM spectrometer thus constitutes a Michelson interferometer that is adjusted for the wavelengths that satisfy the relation above but is misaligned for all others. It produces a fringe pattern containing these wavelengths only; the spectral width of its fringes corresponds to the theoretical resolving power of the gratings, but the fringe intensity is that of the Michelson interferometer.

If the two gratings are then rotated in a synchronous fashion, the wavelengths selected are continuously changed, so that a photoelectric detector receiving the output flux receives different wavelengths at different times. The temporal dispersion thus obtained is $dt/d\lambda = D_\lambda/\omega$, where $D_\lambda = d\varphi/d\lambda$ is the angular dispersion of the gratings and $\omega = d\varphi/dt$ their angular velocity.

The SISAM spectrometer advantages are most pronounced in the infrared, a region of considerable interest for laser spectroscopy [48, 51].

8.4.1.4 Connes Spectrometers The classical concept of a spectroscope or spectrograph is based upon the spatial (angular) dispersion of light rays. To use the same optical system as a photoelectric spectrometer, we must isolate a narrow portion of the spectrum with a fine slit or a pin-hole diaphragm, whereupon a temporal dispersion may be produced by rotating the prism or grating or changing the interferometer plate separation. It is much more efficient to design a spectrometer a priori to yield temporal rather than spatial dispersion, so that the waste of energy is avoided that is inevitably associated with the use of slits and pin-holes in conventional systems. The SISAM spectrometer may be regarded as a first example of such an instrument.

The angular dispersion of the classical Michelson and Fabry-Perot interferometers $d\varphi/d\lambda = \cot \varphi/\lambda$ is entirely due to the dependence of the optical-path difference of the interfering beams on the angle of incidence φ. By means of afocal systems these path differences may be made independent of φ, so that the interferometer becomes a wavelength filter without angular dispersion [52], transmitting the same wavelength at different angles. Its luminosity is thus greatly increased, for a wide-angle detector may now be used to receive this wavelength without loss of spectral resolution. The wavelength transmitted is solely determined by the optical-mirror separation in the interferometer; the desired temporal dispersion is easily produced by varying this separation.

In a Michelson interferometer, afocal systems, consisting of a positive and a negative lens each, are placed in the two arms of the interferometer to produce a path difference independent of incidence. Motion of one of the two afocal systems is then used to obtain the desired temporal dispersion [52].

In a Fabry-Perot system, the two-plane etalon plates are replaced by concave mirrors, which is the confocal Connes interferometer [52,53]. The two concave mirrors are arranged so that the center of curvature of each is located at the vertex of the other; their outer (convex) curvatures are chosen so that each interferometer plate is a meniscus lens of zero power. The Connes interferometer thus constitutes an afocal system of unit magnification: an incident plane wave is transmitted as a plane wave again.

The optical-path difference of successive interfering beams is $\ell = 4nL + \Delta L$, where $n\ell = r$ is the optical length of the interferometer, r the radius of curvature of the concave mirrors. The small term $\Delta\ell$ is due to the spherical aberrations of the confocal system; it is negligibly small for small mirror diameters D. The Connes interferometer thus transmits any wavelength λ for which $m\lambda = 4nL + \Delta\ell$, where m is again the order number. Hence it produces a ring system of fringes of equal spherical aberration in which the center fringe is very broad because of the negligible effect of spherical aberrations at small angles.

The theoretical resolving power is

$$R_0 = (4\pi/\lambda)\, nL/|\ln R| \tag{8.74}$$

(R = mirror reflectivity), and the free spectral range is

$$F_\sigma = \lambda^2/4nL. \tag{8.75}$$

It will be noted that these values are the same as those of a plane-Fabry-Perot etalon of twice the mirror separation.

The Connes interferometer is used as a spectrometer in a similar manner as the etalon. Pressure variations, piezoelectric or magnetostrictive changes of the spacer length, and so on, are employed to vary the wavelength transmitted; the temporal dispersion of the center fringe is

$$dt/d\lambda = (nL/\lambda)/d(nL)/dt.$$

However, in contrast to the etalon, no pin-hole diaphragm is needed to obtain a monochromatic output flux. To allow only a narrow wavelength range $\delta\lambda$ to be transmitted at a time, the pattern is restricted to the center fringe by limiting the mirror diameter, so that spherical aberrations are insignificant and no outer fringes are formed.

For mirrors of diameter D, the maximum path variation due to spherical aberration is $\Delta\ell_{\max} = -D^4/16(nL)^3$ [53]. Hence the two extreme conditions for constructive interference are $m\lambda = 4nL$ and $m(\lambda + \delta\lambda) = 4nL - D^4/16(nL)^3$. The wavelength spread of the transmitted light is, therefore, equal to $\delta\lambda < -D^4/16m(nL)^3 = -\lambda D^4/64(nL)^4$, so that

$$R = 64(nL)^4/D^4 \tag{8.76}$$

is the resolution obtained with the Connes spectrometer.

The luminosity of the Connes spectrometer is

$$U = TA\,\Omega = \pi^2 TD^4/16(nL)^2, \tag{8.77}$$

where T is the average transmission coefficient, $A = \pi D^2/4$ the mirror area, and $\Omega = \pi D^2/4(nL)^2$ the solid angular spread of light inside the interferometer.

Hence the resolution-luminosity product is

$$P = \pi^2 T\,(nL)^2, \tag{8.78}$$

so that, with (8.70),

$$P/P_{FP} = 8(nL)^2/D^2{}_{FP} \tag{8.79}$$

where a Fabry-Perot etalon of plate area $A = \pi\, D^2\, FP/4$ was chosen for comparison and where the transmission coefficients T of both interferometers were taken as equal. Assuming etalon plates of 5 cm diameter, we obtain $P/P_{FP} = 32$ and 0.32, respectively, for $nL = 10$ cm and 1 cm respectively.

The Connes spectrometer is thus seen to yield a considerably higher performance than the etalon when high resolution, that is a large nL, is considered. It is also more practical than the etalon because of the less critical adjustment of the confocal mirrors. For a smaller length, $nL < D_{FP}/\sqrt{8}$, the Connes spectrometer is both inferior to the etalon and impractical.

8.4.1.5 Summary The question of which optical spectrometer to use cannot be answered in general, but Table 8.1 may nevertheless be taken as a guideline [48]. The table lists the spectrometers that can be used in various spectral regions to provide a given wavenumber resolution with the best luminosity. It includes all types of spectrometers discussed here and shows that the grating and Fabry-Perot or Connes spectrometers are the most

TABLE 8.1 PERFORMANCE OF OPTICAL SPECTROMETERS IN DIFFERENT SPECTRAL REGIONS[a]

λ / $\delta(1/\lambda)$	200–400 nm	400 nm–1 μm	1–10 μm	10–100 μm
1 cm^{-1}	4×10^4 *P, G*	2×10^4 *P, G, F*	5×10^3 *G, M, S, F*	5×10^2 *G, M, S, F*
0.1 cm^{-1}	4×10^5 *G, F*	2×10^5 *G, F*	5×10^4 *G, M, S, F*	5×10^3 *G, M, S, F*
0.01 cm^{-1}	4×10^6 *F*	2×10^6 *F*	5×10^5 *G, S, F*	5×10^4 *G, M, S, F*
0.001 cm^{-1}		2×10^7 *F, C*	5×10^6 *F, C*	

[a](P = prism, G = grating, M = Michelson, S = SISAM, F = Fabry-Perot, C = Connes spectrometer. Figures indicate average resolution.)

versatile and efficient—the former for medium-resolution and the latter for high-resolution work.

8.4.2 Photoelectric Methods

The highest resolving power that optical spectroscopy can presently provide is of the order of $R \sim 10^7$ to 10^8 in the visible and near-infrared regions of the spectrum (corresponding to a smallest detectable frequency differences of $\sim 10^6$ Hz). In less favorable spectral regions, such as the far infrared, resolving powers of $R \sim 10^4$ to 10^5 (or frequency differences of 10^8 Hz) constitute the present limitations of optical methods.

Photoelectric methods, in contrast, allow the detection of frequency differences into the 10^9 Hz range. The frequency-response limit of most photomultiplier tubes, the most widely used detectors of all, lies above $\sim 10^8$ Hz. Photoelectric mixing methods hence, take over in the measurement of long-term laser stability where optical spectroscopy leaves off. Because optical-frequency standards, with which laser frequencies may be compared by photomixing, are not yet available, the presently used technique consists of mixing the output beams of two independent lasers operating at the same atomic transition. The photodetector receiving the two beams acts as a square-law device; its output current contains a component

$$i \propto E_1 E_2, \tag{8.80}$$

where E_1 and E_2 are the electric-field vectors of the two beams. If the two fields have parallel components (which can always be accomplished with the help of a plane polarizer even if the two fields were originally polarized at right angles), we may observe all possible difference frequencies up to the frequency response limit of the detector. Variations of the beat signal obtained in this manner then indicate relative frequency instabilities of the two lasers.

The basic concepts of the photomixing method we discuss in Section 8.3.2 may be used to measure the long-term stability of lasers. The procedures are simpler than those of the short-term stability (line width) measurement because only the center frequency of the beat signal is to be determined.

The beat frequencies obtained from the photomixing method usually lie in the audiofrequency range. The signal may thus be displayed with an oscilloscope or audiofrequency spectrum analyzer. Measurements of the beat frequency can be made at regular time intervals by standard techniques (such as from photographs of the oscilloscope screen). The signal may also be recorded on magnetic tape to provide a record of the temporal variations of the beat. The frequency variations may finally be converted into amplitude variations by means of an FM discrimator and then recorded as a function of time on a strip chart. These methods have been used[3,9] to measure relative frequency drifts of two gas lasers wherein long-term drifts in relative frequency are as low as 100 Hz over measuring intervals of several minutes.

Such data are, of course, only relative, indicative of the frequency variations of one laser with respect to another. The photoelectric mixing may ultimately be expected to permit the measurement of absolute stability also. It is anticipated that lasers with an absolute wavelength will become available. These may then be used as secondary standards for heterodyne comparisons with lasers to be tested.

At least in part, the accuracy of our present wavelength standards is set by the limited resolving power of the optical methods by which they were established. The high resolution of the photomixing technique will, therefore, play an important role in choosing still better standards, for which gas lasers have, are likely candidates.

8.5 A TECHNIQUE MEASURING THE ABSOLUTE WAVELENGTH STABILITY OF A LASER

The description of a method for measuring the absolute wavelength stability of a gas laser has been postponed for inclusion in this chapter because a full appreciation of the principles and techniques builds upon significant background information contained in preceeding chapters.

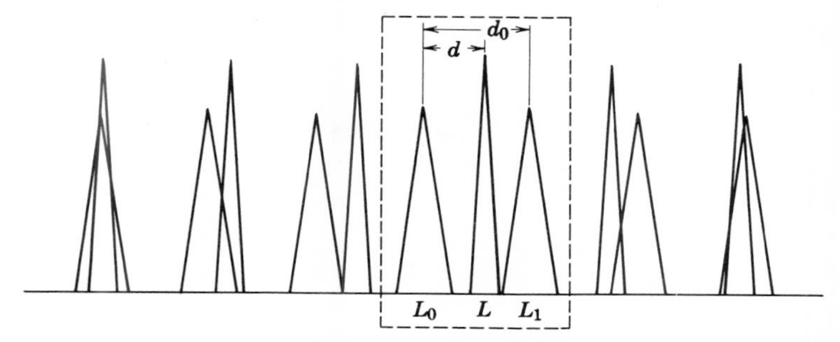

Figure 8.1. Schematic representation of an interferogram. The narrow triangles represent laser fringes, the wide triangles represent mercury fringes.

Furthermore, no present technique is capable of measuring the absolute frequency stability of a laser.

The essence of this technique is an interferometric-wavelength comparison of a helium-neon laser with a Hg^{198} standard lamp[59]. Both light sources simultaneously illuminate a 10-cm Fabry-Perot etalon, so that 546 nm mercury fringes and 633 nm laser fringes are superimposed. A photomultiplier tube, the output of which is displayed and recorded as a function of the etalon length, views the central fringe pattern through a small circular aperture. Because the ratio of wavelengths is roughly 6:7, every sixth laser fringe is nearly halfway between two mercury fringes. By periodically measuring the distance d between mercury fringes and between a reference mercury fringe and a laser fringe d, we may, through calculations presented below, determine the absolute changes of laser wavelength.

8.5.1 Physical Basis for Wavelength Comparison

Consider the characteristics of a set of fringes starting from the reference mercury fringes that are centered on either side of a laser fringe. The left-hand mercury fringe of order m_0, illustrated in Figure 8.1, corresponds to an etalon length

$$nL_o = 1/2\, m_o\, \lambda_o. \tag{8.81}$$

The right-hand fringe, of order $(m_0 + 1)$, occurs at a length

$$nL_1 = nL_o + \tfrac{1}{2}\lambda_o = \tfrac{1}{2}\,(m_o + 1)\,\lambda_o. \tag{8.82}$$

The central laser fringe in Figure 8.1, or order m_L, is at a length

$$nL = nL_o + \tfrac{1}{2}\lambda_o \epsilon = \tfrac{1}{2} m_L\, \lambda_L \tag{8.83}$$

where ϵ is

$$\epsilon = \frac{d}{d_0} \leq 1. \tag{8.84}$$

Solving for λ_L

$$\begin{aligned}\lambda_L &= \frac{2nL_0}{m_L} + \frac{\epsilon\lambda_0}{m_L} \\ &= \frac{m_0 \lambda_0}{m_L} + \frac{\epsilon\lambda_0 \lambda_L}{2nL_0 + \epsilon\lambda_0} \cong \Lambda + \frac{\epsilon\lambda_0 \,\delta\lambda_L}{2nL_0}.\end{aligned} \tag{8.85}$$

$\Lambda = m_0 \lambda_0 / m_L$ is an undetermined constant. (8.86)

For a slightly different wavelength $\lambda_L + \delta\lambda_L$ we obtain

$$\delta\lambda_L = \frac{\lambda_0 \lambda_L \delta\epsilon}{2nL_0} + \frac{\epsilon\lambda_0 \,\delta\lambda_L}{2nL_0}$$

or

$$\delta\lambda_L\left(1 - \frac{\lambda_0 \epsilon}{2nL_0}\right) = \delta\epsilon\left(\frac{\lambda_0 \lambda_L}{2nL_0}\right). \tag{8.87}$$

As $\lambda_0 \epsilon / 2nL_0 \approx 0$,

$$\delta\lambda_L \cong \delta\epsilon \frac{(\lambda_0 \lambda_L)}{2nL_0}. \tag{8.88}$$

The wavelength stability, using equations 8–10 and 8–88, becomes

$$S = \frac{\lambda_L}{\delta\lambda_L} = \frac{2nL_0}{\lambda_0 \,\delta\epsilon} = \frac{\epsilon}{\delta\epsilon}. \tag{8.89}$$

The absolute wavelength and wavelength stability of the laser can thus be obtained from periodic records of the fringe patterns. From measurements of d and d_0 we can compute ϵ and $\delta\epsilon$ and obtain a measure of S.

A plot of the function

$$\lambda_L - \Lambda = \frac{\lambda_0 \lambda_L}{2nL_0} \epsilon \tag{8.90}$$

will yield time-dependent, absolute changes in laser wavelength.

TABLE 8.2 APPARATUS FOR ABSOLUTE-WAVELENGTH MEASUREMENT OF THE 633nm, TRANSITION OF A CW GAS LASER

Item	Function	Source
Mercury light	Hg^{198} wavelength standard	Ealing Corp., Hilger-Watts FL-125
Photomultiplier	Beam-intensity monitor	RCA, 1P21 (two required)
Precision dc-regulated power supply	Dynode voltage supply and piezoelectric displacement controller	N.J.E., S-325 RM (two required)
Audio oscillator	Wavelength scan	HP, 202C
Scanning Fabry-Perot	Wavelength scanning interferometer	See text
Lenses, apertures and Fresnel plates	Light collimation and deflection	Ealing
Scanning Laser	Laser-wavelength source	*h nu systems* Model 3601
Notch filter	Monochromatizer for 546 nm mercury line	Thin Film Products, Inc., 546 nm filter
Thermocouple wire	Interferometer-temperature sampling	Van Waters and Rogers
Potentiometer	Thermocouple-voltage measurement	Leeds and Northrop, K2
Oscilloscope	Wavelength and fringe-count monitor	Tektronix Model 531 with type C plug-in unit

8.5.2 Apparatus and Arrangement Thereof

The basic instruments needed for the absolute measurement of laser wavelength are indicated in Table 8.2. A schematic arrangement of the equipment is given in Figure 8.2 to which the reader should refer in the following description.

The heart of the measuring equipment is a 10-cm Fabry-Perot etalon. The etalon consists primarily of a 7.5-cm quartz cylinder to which a 2.5 cm piezoelectric cylinder is epoxy cemented [54]. Aluminized ($R = 0.8$) flats [55] are attached to the etalon spacer. The etalon is isolated from laboratory thermal ambient by a plastic-foam blanket that is encased in an aluminum-foil reflector. The assembly is supported inside a copper box in an arrang-ment that permits optical access to the axis of the interferometer. Wires, connected to the piezoelectric driver are brought out through the case. This attention to the isolation from thermal variations will reduce the temperature change in the etalon by a factor of 20 or more for ambient temperature changes in the laboratory of the order of 1°C per five-minute interval.

Light from an Hg^{198} standard lamp is collimated and passed through an interference filter (notch) to isolate the 546 nm line. The laser output

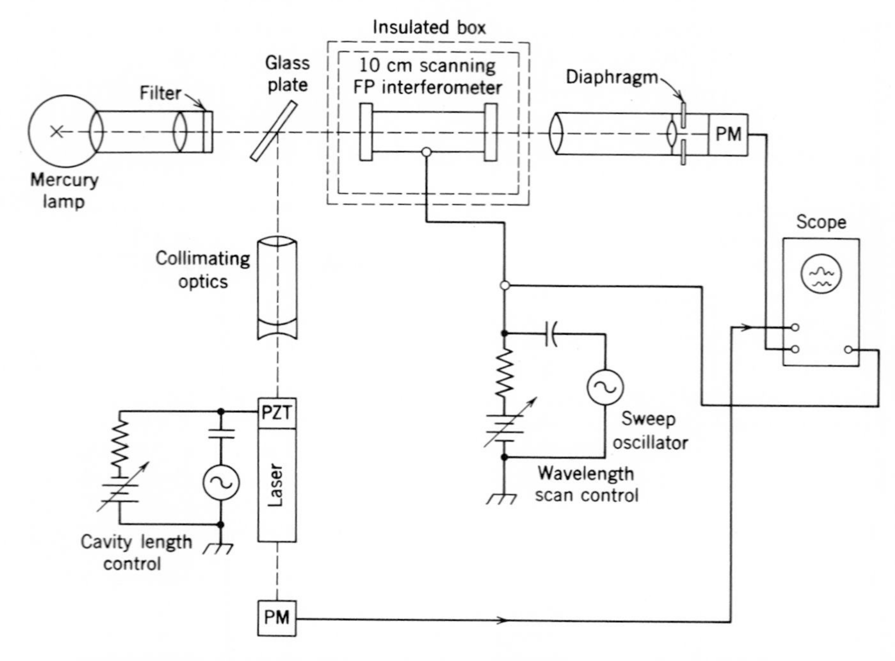

Figure 8.2. Schematic arrangement of wavelength-comparison spectrometer.

is collimated and combined with the mercury light by a Fresnel plate located near the entrance of the etalon.

The light transmitted through the etalon is focussed through an iris placed in the focal plane of an output lens, to form an exit pupil of approximately $2a = 1$ mm diamter. Light transmitted by this aperture illuminates the photomultiplier tube. The output voltage of the photomultiplier tube is displayed and recorded.

All optical components should be mounted on a heavy, vibration-free table (a granite slab mounted on inflated motor-scooter tires is adequate).

Assuming an output lens of $f = 585$ mm focal length, the resolving power of the etalon may be estimated [56] as

$$R = \frac{2f^2}{a^2} = 2\left(\frac{585}{1/2}\right)^2 \cong 2.72 \times 10^6. \tag{8.91}$$

This is in reasonable agreement with the theoretical resolving power of the etalon at 633 nm [57]:

$$R_0 = \frac{2\pi n L_0}{\lambda |\ln R|} = \frac{(2\pi)(1)(10)}{(6.33 \times 10^{-5})(.777)} \cong 1.3 \times 10^6. \tag{8.92}$$

8.5.3 Experimental Procedure

It is of paramount importance that measurements be made under steady-state conditions. The Hg^{198} light source, the laser, and all associated electronic equipment should be operated at least four hours before measurements are started. The laser must be operated on a single-transverse and a single-axial mode (single-frequency). Prior to making measurements, the laser should be peaked for maximum power output (by varying the mirror spacing with the piezoelectric drive), to operate, either at the center of the gain curve (if the plasma tube is filled with nonisotopic gas), or at the center of the Lamb dip [57] if isotopically pure gases are used. Axial tuning is facilitated by superimposing a variable-amplitude scanning voltage on the dc-bias voltage used to determine the absolute laser frequency. The output voltage of a photomultiplier that monitors the laser intensity is displayed on the oscilloscope . The horizontal oscilloscope sweep is driven by the scanning voltage.

The dc voltage of the etalon is next adjusted to pick out a mercury-fringe pair between which a laser fringe is centered. This is facilitated by observing the output-photomultiplier display of the etalon on an oscilloscope in which the horizontal sweep is driven by the scanning voltage of the etalon.

Having adjusted the equipment, a series of periodic records is obtained by photographing the etalon-output display at regular intervals. A scanning frequency of 1 to 5 Hz is adequate. Measurements of the relative mercury and laser fringe locations yield the numbers required for the calculation of stability from (8.84) and (8.89). Longer scanning periods may be obtained if the scanning voltage is derived from a motor-operated potentiometer or its equivalent. Whereas longer scanning periods are consistent with the use of strip-chart recorders and greater accuracy in the determination of d and d_0, the instability characteristics of the etalon indicate that short scanning intervals are more prudent.

8.5.4 Sources of Error in Wavelength Determination

There are three major sources of error in laser-wavelength determination. These include instabilities in the Hg^{198} source, limited precision in evaluation of the interferograms, and changes in the optical length of the etalon nL.

To insure that a Hg^{198} mercury source operates so that its absolute wavelength is significant to 5×10^{-8} over long periods, the standard lamp must be operated strictly in accord with the recommendations of LeComité Consultatif pour la Définition du Métre [60]. The interested reader is referred to this source for details.

For the short sampling intervals used, it is reasonable that a standard lamp has an instability of not greater than

$$\frac{\Delta\lambda}{\lambda_0} \leq \pm 1 \times 10^{-8}. \tag{8.93}$$

The precision with which the interferograms can be evaluated using the techniques described herein, that is the measurement of d and d_0, from recordings (wherein $d, d_0 \approx 3$ to 5 cm), will not exceed 1 percent of the free spectral range $f^2/2nL$ of the etalon. Because two measurements must be made on each recording

$$\frac{\Delta\lambda}{\lambda} = \pm \frac{\sqrt{2}\lambda}{200nL} \cong \pm 4 \times 10^{-8} \tag{8.94}$$

where $nL = 10$ cm and $\lambda = 633$ nm.

The effect of the nonlinearity of the etalon drive may be tested by measuring the variation of mercury-fringe spacing with scan voltage. It is found that this source of error is negligible.

There remains a significant source of error caused by changes in the optical-path length $\Delta\ (nL)$ of the etalon. Because these changes are slow, they are assumed to be linear during the measuring interval; only first-order effects will be considered. Changes in optical-path length of the etalon only affect the measurements to the extent that they occur during an individual sampling interval or scan. From the logarithmic derivative of (8.81), we see that

$$\frac{\Delta\lambda}{\lambda} = \frac{\Delta(nL)}{nL}.$$

Considering the time dependence of nL as linear during τ_0,

$$\frac{\Delta\lambda(\tau_0)}{\lambda} = \frac{\tau_0}{nL}\frac{d(nL)}{dt}. \tag{8.95}$$

The time-dependent drift in the composite spacer in the etalon can be determined from the first-order expansion coefficients of the etalon materials. For laboratory environments, the coefficient of thermal expansion of fused quartz is $\alpha_q \sim 4 \times 10^{-7}$ per degree C; that of the piezoelectric material is $\alpha_p = 65 \times 10^{-7}$ per degree C.

The composite temperature coefficient becomes

$$\alpha_c = \frac{L_q\alpha_q + L_p\alpha_p}{L_p + L_q} = \frac{(7.5)(4 \times 10^{-7}) + (2.5)(65 \times 10^{-7})}{(2.5 + 7.5)} \tag{8.96}$$

$$= +19.25 \times 10^{-7}/^\circ\mathrm{C}.$$

The expansion of the spacer dL becomes

$$dL = \alpha_c L\, dT \quad \text{or} \quad \alpha_c = \frac{1}{L}\frac{dL}{dT}. \tag{8.97}$$

The resonant-wavelength instability of the etalon becomes (using equations 8.95 and 8.97)

$$\frac{\Delta\lambda(\tau_0)}{\lambda} = \left(\frac{\tau_0}{nL}\right)\frac{(n\alpha_c L\, dT)}{dt} = \left(\frac{\Delta\lambda(\tau_0)}{\lambda}\right)_{L(T)} \tag{8.98}$$

$$= \tau_0 \alpha_c \frac{dT}{dt} = \pm 1.9 \times 10^{-8}/°\text{C},$$

where τ_0 the sampling time is in minutes and the rate of temperature change is

$$\frac{dT}{dt} = \pm 0.01°\ C/\text{min}.$$

Variations in the temperature T, the pressure p, and the humidity h inside the etalon will cause changes in the optical path through their respective effects upon the index of refraction n of the enclosed air volume.

Assuming again only first-order effects, for $T = 20°\text{C}$, $p = 760$ Torr, $h = 8.5$ Torr, and $\lambda = 633$ nm, the respective coefficients become[58]

$$\beta_T = \frac{1}{n}\frac{dn}{dT} = -9.3 \times 10^{-7}/°\text{C}, \tag{8.99}$$

$$\beta_p = \frac{1}{n}\frac{dn}{dp} = +3.6 \times 10^{-7}/\text{torr}, \tag{8.100}$$

$$\beta_h = \frac{1}{n}\frac{dn}{dh} = +5.7 \times 10^{-8}/\text{torr}. \tag{8.101}$$

When the respective rates of change of variables are $dT/dt = \pm 0.01°\text{C/min}$, $dp/dt = \pm 1$ Torr/hr, and $dh/dt = \pm 5$ Torr/hr, and for τ_0 in minutes, the respective effects of (8.99) through (8.101) on the etalon instability become

$$\left(\frac{\Delta\lambda(\tau_0)}{\lambda}\right)_{n(T)} = \tau_0 \beta_T \frac{dT}{dt} = \pm 9.3 \times 10^{-9}\tau_0, \tag{8.102}$$

$$\left(\frac{\Delta\lambda(\tau_0)}{\lambda}\right)_{n(p)} = \tau_0 \beta_p \frac{dT}{dt} = \pm 6 \times 10^{-9}\tau_0, \tag{8.103}$$

$$\left(\frac{\Delta\lambda(\tau_0)}{\lambda}\right)_{n(h)} = \tau_0 \beta_h \frac{dT}{dt} = \pm 4.8 \times 10^{-8}\tau_0. \tag{8.104}$$

The combined environmental effects upon the etalon under the assumed conditions become

$$\left(\frac{\Delta\lambda(\tau_0)}{\lambda}\right)_{\text{etalon}} = \pm[(19-9)^2 + 6^2 + 4.8^2]^{1/2} \times 10^{-9}\tau_0 \qquad (8.105)$$

$$\cong \pm 1.3 \times 10^{-8}\tau_0$$

For a 1-min sampling interval the etalon introduces a combined error of about one part in 10^8.

The total estimated error for the measuring apparatus, obtained from (8.93), (8.94) and (8.105) is

$$\left(\frac{\Delta\lambda}{\lambda}\right)_{\text{total}} = \pm[(1.3)^2 + (1)^2 + (4)^2]^{1/2} \times 10^{-8} \qquad (8.106)$$

$$\pm 2.6 \times 10^{-8} \qquad \text{for } \tau_0 \leq 1 \text{ min.}$$

The total estimated error could reasonably be reduced about a factor of 2 with a factor of 4 improvement in the precision with which the interferograms are evaluated (and the use of a much shorter sampling interval). The overall uncertainty is ultimately bounded by the accuracy of the present wavelength standard. The largest fluctuation in laser frequency measurable is just half the free spectral range of the etalon or

$$\Delta\lambda = \pm\frac{1}{2}\left(\frac{\lambda^2}{2nL}\right)$$

or for $nL = 10$ cm and $\lambda = 633$ nm

$$\frac{\Delta\lambda}{\lambda} \cong \pm 1.5 \times 10^{-6}. \qquad (8.107)$$

When a laser is operated in a single-transverse mode, the limit of the wavelength uncertainty is set by the separation of axial modes[61]:

$$\Delta\lambda = \pm\frac{\lambda^2}{2nL}$$

For a 10-cm laser cavity at 633 nm, this leads to an upper instability limit for the laser of

$$\Delta\lambda = \pm 3 \times 10^{-6} \qquad (8.108)$$

because only about one half the entire gain curve is usable in a short laser (8.108), the results of (8.107) are in excellent accord.

8.5.5 Sources of Wavelength Instability in the Laser

The laser may be characterized as an active interferometer with an internal light source. Application of (8.95) yields

$$\frac{\Delta\lambda_L}{\lambda_L} = \frac{\tau}{n\ell}\frac{d(n\ell)}{dt} \tag{8.109}$$

where τ is the averaging time and $n\ell$ is the time-varying, optical-path length inside the cavity. We assume that $n\ell$ varies linearly with time.

Assuming that the laser cavity is constructed in the same manner as the etalon but that the temperature excursions are $\pm 0.1°\text{C/min}$, we find from (8.97) and (8.98)

$$\frac{\Delta\lambda_L}{\lambda_L} = \tau\alpha_c\frac{dT}{dt} = \pm 1.9 \times 10^{-8}\tau. \tag{8.110}$$

The variations in the index of refraction between the laser plasma tube and the external mirrors are subject to the same effects as those we see in (8.99) through (8.101). These relations apply to the fraction q of the cavity not occupied by the plasma tube. Assuming that $q = 0.8$ and that within the air space of the laser cavity variations occur at rates $dT/dt = \pm 0.1°\text{C/min}$ $dp/dt = \pm 1$ Torr/hr and $dh/dt = \pm 5$ Torr/hr. The effects of temperature are

$$\left(\frac{\Delta\lambda_L(\tau)}{\lambda_L}\right)_{n(T)} = q\tau\beta_T\frac{dT}{dt} = \pm 7.5 \times 10^{-8}\tau, \tag{8.111}$$

$$\left(\frac{\Delta\lambda_L(\tau)}{\lambda_L}\right)_{n(p)} = q\tau\beta_p\frac{dT}{dt} = \pm 4.8 \times 10^{-9}\tau, \tag{8.112}$$

$$\left(\frac{\Delta\lambda_L(\tau)}{\lambda_L}\right)_{n(h)} = q\tau\beta_h\frac{dT}{dt} = \pm 4.6 \times 10^{-9}\tau. \tag{8.113}$$

for one-min averaging times.

The total variation in laser wavelength, neglecting expansion of the Brewster windows on the plasma tube and variation of the index of refraction of the plasma, is thus bounded to

$$\begin{aligned}\left(\frac{\Delta\lambda_L}{\lambda_L}\right)_{\text{total}} &= \pm [(1.9 - 7.5)^2 + (.5)^2 + (.5)^2]^{1/2} \times 10^{-8} \\ &\cong \pm 5 \times 10^{-8}\end{aligned} \tag{8.114}$$

or about a factor of 100 better than is predicted by the upper limit of (8.108). The major source of wavelength uncertainty is clearly the thermal environment of the laser. Under laboratory conditions this gives rise to wavelength instabilities of the order of 10^{-7} for averaging times of the order of minutes. With thermal isolation, this source of instability can easily be reduced to $\pm 0.01°C/min$ to yield a wavelength instability of 10^{-8} or better than the limit of precision of the measuring apparatus. Single-frequency lasers without any servo-control of the cavity length have attained a precision that is limited by present wavelength standards[62].

Two methods have been used to overcome cavity-length variations in gas lasers (which in turn lead to wavelength variations of perhaps 0.002 nm). One way is to adjust the cavity with a servo, so as to obtain maximum laser intensity[63]. Wavelength values for helium-neon lasers controlled by such methods have been reported[57, 64] (λ_{VAC} = 632.99141 nm). More recently developed models use $He^3 - Ne^{20}$ and use a servo control connected to the Lamb dip[65] in the center of the intensity curve. These give a wavelength λ_{VAC} = 632.99138 nm. Both methods of stabilization can cause the operating wavelength to be pulled significantly from its unstabilized value unless the cavity is set to give approximately the correct wavelength before the servo loop is locked on. It is disturbing to note that the wavelength may depend upon the fabrication techniques, for a wavelength of 632.99145 nm ($\Delta\lambda \sim 0.00007$ nm) was measured for a laser manufactured under slightly different conditions[65]. Careful wavelength measurements do, however, appear to yield values that agree within one part in 10^7 for natural as well as isotopically pure helium-neon plasmas[65].

8.6 REFERENCES

[1] A. G. McNish, *Electro-Technol., Sci. Eng. Ser.*, **71**, 113 (1963).
[2] J. P. Gordon, H. J. Zeiger, and C. H. Townes, *Phys. Rev. Letters*, **99**, 1264 (1955).
[3] T. S. Jaseja, A. Javan, and C. H. Townes, *Phys. Rev. Letters*, **10**, 165 (1963).
[4] B. A. Lengyel, *Lasers*, Wiley, New York, 1962, p. 51.
[5] N. Winogradoff, private communication.
[6] I. D. Abella and H. Z. Cummins, *J. Appl. Phys.*, **32**, 1177 (1961).
[7] A. L. Schawlow and C. H. Townes, *Phys. Rev. Letters*, **112**, 1940 (1958).
[8] K. D. Mielenz, R. B. Stephens and K. E. Gillilland, to be published.
[9] A. Javan, E. A. Ballik, and W. L. Bond, *J. Opt. Soc. Am.*, **52**, 96 (1962).
[10] O. S. Heavens, *Optical Masers*, Methuen, London, 1964, pp. 50 and 64.
[11] L. F. Mollenauer, G. F. Imbush, H. W. Moos, A. L. Schawlaw and A. D. May, *Optical Masers*, Polytech Inst. of Brooklyn, Polytechnic Press, Brooklyn, New York, Vol. XIII, p. 51 (1963).
[12] W. R. Bennett, Jr., *Phys. Rev. Letters*, **126**, 580 (1962).

[13] A. Szoke and A. Javan, *Phys. Rev. Letters*, **10**, 521 (1963).
[14] G. W. Stroke, in *Progress in Optics*, E. Wolf (ed.), J. Wiley, New York, 1963, Vol. 2, p.3.
[15] P. H. van Cittert, *Z. Physik.*, **65**, 547 (1930).
[16] K. D. Mielenz *Optik*, **13**, 437 (1956).
[17] J. Junkes and E. W. Salpeter, *Rich. Spettrosc.*, **2**, 255 (1961).
[18] T. Morokuma, *et al.*, *J. Opt Soc. Am.*, **53**, 394 (1963).
[19] *Natl. Bur. Std. Tech. News Bull.*, **47**, 80 (1963).
[20] M. Born and E. Wolf, *Principles of Optics*, Macmillan, New York, 1959, p. 319.
[21] A. A. Michelson, *Phil. Mag.*, **4**, 291 (1892).
[22] C. Fabry and H. Buisson, *Compt. Rend.*, **154**, 1224 (1912).
[23] P. Jacquinot and C. Doufour, *J. Rech. C. N. R. S.*, **2** 91 (1948).
[24] K. D. Mielenz, R. B. Stephens, and K. F. Nefflen, *J. Res. Natl. Bur. Std.* (U.S.), **68C**, 1 (1964).
[25] S. Tolansky, and D. J. Bradley, *Interferometry*, Her Majesty's Stationary Office, London, p. 375, 1960.
[26] D. R. Herriot, *Appl. Optics*, **2**, 865 (1963).
[27] J. Cooper and J. R. Greig, *J. Sci. Instr.*, **40**, 433 (1963).
[28] K. D. Mielenz, to be published.
[29] V. N. Smiley, *Proc. IEEE*, **51**, 120 (1963).
[30] K. Krebs and A. Sauer, *Ann. Physik*, **13**, 359 (1953).
[31] K. D. Mielenz, *J. Res. Natl. Bur. Std.* (U.S.), **680**, 73 (1964).
[32] P. H. van Cittert and H. C. Burger, *Z. Physik*, **44**, 58 (1927).
[33] R. Minkowski and H. Brueck, *Z. Physik*, **95**, 299 (1935).
[34] M. S. Sodha and S. S. Mitra, *Optik*, **15**, 47 (1958).
[35] A. Righi, *J. Physique*, **2**, 437 (1883).
[36] A. T. Forrester, R. A. Gudmundsen, and P. O. Johnson, *Phys. Rev.*, **99**, 1691 (1953).
[37] A. T. Forrester, *J. Opt. Soc. Am.*, **51**, 253 (1961).
[38] L. Mandel, in *Progress in Optics* E. Wolf (ed.), J. Wiley, New York, 1963, Vol. 2., p.183.
[39] H. Gamo, in *Advances in Quantum Electronics*, J. R. Singer (ed.), Columbia Univ. Press, New York, 1961, p. 252.
[40] P. J. Magill and T. Young, *Appl. Phys. Letters*, **5**, 13 (1964).
[41] K. D. Mielenz, *Optik*, **14**, 103 (1957).
[42] K. D. Mielenz, *Optik*, **15**, 10 (1958).
[43] K. D. Mielenz, *Optik*, **16**, 485 (1959).
[44] P. Jacquinot, *J. Opt. Soc. Am.*, **44**, 761 (1954).
[45] J. Strong, *J. Opt. Soc. Am.*, **39**, 320 (1949).
[46] M. Shimazu, I. Ogura, A. Hashimoto, H. Sasaki, in *Proc. Symp. Opt. Masers*, Polytech. Press, Brooklyn, N. Y., 1963, Vol. XIII, p. 405.
[47] J. Strong, *J. Opt. Soc. Am.*, **47**, 354 (1957).
[48] P. Jacquinot, Septieme Colloque Internationale de Spectroscopie, Liege, 1958.
[49] A. Steudel, *Nat. Viss.*, **44**, 249 (1957).
[50] R. Chaball, *J. Rech. C.N.R.S.*, **24**, 138 (1953).
[51] P. Connes, *J. Phys. Rad.*, **19**, 215 (1958).
[52] P. Connes, *Rev. Optique*, **35**, 37 (1956).
[53] P. Connes, *J. Phys. Radum*, **19**, 262 (1958).
[54] Clevite Corp., type PZT ceramic. A voltage change of approximately 200 V is necessary to scan between successive mercury fringes.
[55] Industrial Optics, Bloomfield, New Jersey
[56] P. Jacquinot, *J. Opt. Soc. Am.*, **44**, 761 (1954).
[57] W. E. Lamb, Jr., *Phys. Rev.*, **134**, 1429 (1964).

[58] K. E. Gillilland, *et al.*, to be published.
[59] K. D. Mielenz, *et al.*, *J. Opt. Soc. Am.*, **56**, 156 (1966).
[60] Comite Consultalif pour la Definition du Metere, 3rd Session, Gauthier-Villard, Paris, 1962, Vol. **19**.
[61] W. R. Bennett, *Jr.*, *Appl. Opt.*, **24** Supplement I (1962).
[62] K. M. Baird, *et al.*, *Appl. Opt.*, **4**, 569 (1965).
[63] W. R. C. Rowley and D. C. Wilson, *Nature*, **200**, 745 (1963).
[64] K. D. Mielenz, *et al.*, *Science*, **146**, 1672 (1964).
[65] W. R. C. Rowley and D. C. Wilson, *J. Opt. Soc. Am.*, **56**, 259 (1966).

8.7 PRINCIPAL SYMBOLS, NOTATIONS, AND ABBREVIATIONS

Symbol	*Meaning*
a, A	Area
a, b	Related half-widths of Gaussian or Lorentzian line profiles respectively.
a_0, a_1	Gaussian line widths at the 1/2 and 1/10 intensity points.
b_0, b_1	Lorentzian line widths at the 1/2 and 1/10 intensity points.
c	Velocity of light
C, S	Fourier cosine and sine transforms of $f(k)$
d	Grating constant, spacing of grating elements (cm/line)
dn	Differential change in n
D	Mirror diameter
D_{FP}	Diameter of a Fabry-Perot etalon
D_λ	Angular dispersion of a prism or grating
$E(\omega)$	Electric field of a light wave at an angular frequency ω
E^*	Complex conjugate of E
f	Focal length
$f(k)$	The total spectral-intensity distribution function
$f_0(x)$	Unperturbed spectral line shape
$f(x)$	Distorted distribution of a spectral line
$f(\omega)$	Intensity per unit of angular frequency of a spectral signal
$f_D(x)$	Diffraction pattern of spectral line as formed by a spectrograph
$f_T(x)$	Diffusion or turbidity profile characteristic of a photographic plate
F	Coefficient of finesse
$\mathscr{F}$	Cavity finesse
$G(x)$	Gaussian line profile
h	Slit height
h	Planck's constant
H	Beam height of a prism or grating

$i(k)\,dk$	Fringe pattern intensity due to spectral components in the wave number interval dk
I	Intensity
I_{max}, I_{min}	Maximum and minimum intensities of bright and adjacent dark fringes
$<I^2(x)>$	Mean-square value of $I(x)$
k	Wave number
k	Boltzmann's constant
k_0	Mean-wave number
ℓ	Largest optical path difference in an optical instrument, laser mirror spacing, see text
L, L_0	Mirror spacing, etalon spacing
L(x)	Lorentzian line profile
m	Any integer
m	Atomic mass in amu
n	Index of refraction
n_i/g_i	Statistically weighted atomic-number density for the state i
r	Radius of curvature of a mirror
P	Power output
P	Total beam intensity of both beams in a two-beam interferometer
P_i	Resolution-luminosity product of an optical instrument
q	Fraction of a laser-cavity length that is not occupied by a plasma tube
R	Reflectivity of a surface, resolution, resolving power, see text
R_0	Spectral resolving power
R_λ, R_ν	Wavelength or frequency resettability
ρ^0	Instantaneous frequency stability
s_ν, s_λ	Long term frequency, wavelength stability
S_ν^∞	Steady-state frequency stability
T	Absolute temperature (° K), temperature (° C), see text
T	Transmission coefficient
t	Time
U	Luminosity
v	Velocity
V	Material volume
$V(\ell)$	Visibility of an interference pattern as a function of the optical-path-length difference in an optical instrument
$V(x)$	Voigtian function convolved from V_1 and V_0
$V_1(x)$	Voigtian functional approximation of the convolution of the diffraction and turbidity profiles
$V_0(x)$	Voigt function

W	Beam width of a prism or grating
x, y, z	Generalized coordinates representing wavelength or wave number
Y	Young's modulus of a material
α	Amplification coefficient of a laser
β, β_L	Angular frequency, linear-expansion coefficient denoted by subscript, see text
Δ	Symbol for an increment or small part
δ	Symbol for an infinitesimal part
λ	Phase difference between two successive beams
τ	Measuring interval
τ_0	Time-resolution limit of a detector
ν	Frequency (Hz)
$\nu_{max}, \nu_{\min}$	Maximum and minimum values of ν
$\Delta\nu_C$	Frequency resolution of a passive Fabry-Perot cavity
$\Delta\nu_D$	Doppler line width
$\bar{\nu}$	Average value of frequency over the sampling interval
$\nu(t)$	Time variation of ν
λ	Wavelength (μm, nm)
λ_0, λ_L	Specific wavelengths, see text
$\varphi(\omega)$	Angular-frequency-dependent phase of a spectral signal
ω	Angular frequency
Ω	Solid angle of a beam

9

Measurement of Noise and Modulation of Laser Carriers

9.0 INTRODUCTION

We close this book with a chapter that addresses itself to the principles and techniques germane to the application of a laser carrier to the transmission of intelligence. The characterization of a laser signal for communication-type applications requires that the noise and information content be known or that it can be determined. Methods of determining the noise level in laser oscillators and amplifiers, as well as methods of measuring or detecting various types of modulation on the optical carrier, are treated.

We somewhat arbitrarily dispense with consideration of present-day, optically pumped, multimode, solid-state lasers as inappropriate for optical communication applications. Both pulsed as well as CW solid-state lasers are prone to emit radiation in the form of pulselets the character of which is determined by the currently active mode. The frequency separation of modes, governed by the time-varying crystal dimensions and refractive index, generally falls within the microwave region of the frequency spectrum. Because the output signal of a solid-state laser is multimode, the detected signal will contain cross-modulation products of great complexity. In principle the multimode properties of solid-state lasers can be removed by known mode-selection methods. When mode-suppressing restrictions are applied, however, the power output of a solid-state laser drops drastically, and the efficiency becomes so low that this embodiment of a quantum-electron device does not begin to compete in performance with CW-ionized gas lasers.

The major problems encountered in the application of high-power, CW-gas lasers to communication systems include frequency instability, availability of modulation devices of adequate band width, and noise. Techniques for intracavity frequency selection are currently being used to produce high-power, single-frequency gas lasers that operate on a single longitudinal mode within the Doppler line width. Mechanically rigid cavity designs that incorporate piezoelectric mirror position control (with associated servomechanisms) can stabilize the laser output to 10^{-9} of the base frequency. The state of the art of the laser source is thus advanced compared with that of methods of modulation and detection. Problems of pointing accuracy and distortion caused by the transmission media remain.

Assuming that the component problems will be solved with sufficient expenditure of effort and no little ingenuity, one must ultimately turn to problems of noise level in systems and in the laser itself. Because the high-power, CW-gas laser has more promising possibilities for communication-type applications the emphasis on noise in this chapter relates to this class of devices.

9.1 SOURCES OF NOISE IN COHERENT OPTICAL SYSTEMS

The gas laser is a primary source of noise in a coherent optical system. Spontaneous emission, which places a quantum-mechanical limit on the achievable signal-to-noise ratio (SNR) in the system, is by no means the only source of noise. In addition to spontaneous emission there are other laser sources of discrete and random noise. Furthermore, there are sources of noise in the system. To achieve an SNR close to the theoretical limit, sources of noise considered below must be virtually eliminated.

We classify those sources of noise as extrinsic that affect the overall SNR external to a laser. Examples include the pointing accuracy and stability of the receiver, a coherence area that is wave-front limited (caused by turbidity in the transmission medium), dark current, shot and Johnson noise in the detector and its associated amplifiers, extraneous shot-noise sources including direct or scattered sunlight, and so on. Intrinsic noise sources include discharge-current noise[1,2], spontaneous emission, potential and competing laser transitions, noise caused by the laser spreading the signal among various modes all of which have the same gain and are equally accepted by the detector, and perhaps superradiance, among others.

Space does not permit a systematic examination of the extrinsic noise sources, and these factors, which are of paramount importance in systems engineering, may be explored by the interested reader on his own[3–6]. A brief treatment of intrinsic noise sources follows the definition of some useful relations.

9.2 EXPANDED DEFINITION OF NOISE IN COHERENT SYSTEMS

We find that classical definitions of noise performance, noise figure, and effective input noise temperature, useful in the radio-frequency portion of the spectrum, must be extended to include quantum effects at laser frequencies. In the limit, when quantum effects are of importance, we cannot make simultaneous measurements of the amplitude and phase of an electromagnetic field with arbitrarily high precision; the precision is determined by the uncertainty principle[7–9]. It can be shown that the additive "white" Gaussian noise, contributed by the amplifying process that is capable of raising signals to classical power levels sufficient to permit simultaneous amplitude and phase measurements with negligible interference, ensures that the amplitude and phase of the input noise are uncertain by a minimum amount necessary to satisfy the relation[10]

$$\Delta n \Delta \varphi \geqq 1 \tag{9.1}$$

where φ is the phase of the electromagnetic field and n is the number of quanta suitably averaged over the optical band width $\Delta\nu$ of the receiver (the quantum counter) and the observation time Δt (reduced band width) of the measuring apparatus

$$n = \int_{\nu_1}^{\nu_2} \int_0^{\tau} n(t, \nu)\, dt\, d\nu. \tag{9.2}$$

The interested reader may pursue the subject in more detail[11,12].

We now characterize the noise performance of a laser amplifier through the Nyquist formula generalized to the quantum case. Consider the noise power P_e emitted by a "white" Gaussian noise source at a temperature T_e. Let the source be appropriately matched to the amplifier. Let the amplifier noise output per unit of band width prior to the connection of the noise source be P_0. Assume that P_e can be adjusted to a value sufficient to increase the amplifier output to 2 P_0. The value of P_e so determined is a measure of the noise performance of the amplifier.

Up to the highest microwave frequencies it has been customary to express the noise power density P_e in terms of an effective input-noise temperature T_e using the classical Nyquist formula:

$$T_e \approx P_e/k \tag{9.3}$$

where $k = 1.38 \times 10^{-23}\,\mathrm{J}/^{\circ}\mathrm{K}$.

At laser frequencies the quantum-mechanical expression of the Nyquist formula becomes

$$P_e = h\nu/\exp(h\nu/kT_e) - 1, \tag{9.4}$$

which yields an effective input-noise temperature

$$T_e = \frac{h\nu/k}{\ln(1 + h\nu/P_e)}. \tag{9.5}$$

Here $h = 6.61 \times 10^{-34}$ J sec, Planck's constant, ν is the laser frequency, and $h\nu/P_e$ is the number of photons per unit of time per unit of frequency.

Equation 9.5 must be further generalized to allow for different populations and degeneracies of the laser-energy levels[13].

$$T_e = \frac{h\nu}{k} \ln\left[1 + \frac{(n_2/g_2) - (n_1/g_1)}{n_2/g_2}\right] \tag{9.6}$$

where n_1 and n_2 are the populations respectively of the lower and upper levels, and g_1 and g_2 are their degeneracies. Expressed in terms of amplifier single-pass gain G, the minimum input-noise temperature of a linear amplifier (linear in the phase-preserving sense) characterized by additive "white" Gaussian noise is[12]

$$T_e = \frac{h\nu}{k\left(\ln \dfrac{2 - 1/G}{1 - 1/G}\right)^{-1}}. \tag{9.7}$$

In the limit of large gain,

$$(n_1 \to 0 \quad \text{or} \quad G \to \infty).$$

Equation 9.6 yields the limiting-noise performance of a laser

$$T_{e_{\min}} = \frac{h\nu}{k \ln 2}. \tag{9.8}$$

At 633 nm this corresponds to a minimum effective noise temperature of 32,700°K! Laser amplifiers, even in the far infrared are characteristically noisy components. Obviously any optical system is reasonably noisy by microwave standards.

Even the ideal value of noise performance (9.8) can only be approached if the spatial mode (or modes) employed in the amplifier is (are) fully excited by the source; for example when a thermal source is used, the mode(s) of interest must be excited to a level corresponding to the temperature of the source.

In that diffraction effects ultimately determine the degree of success in obtaining the minimum SNR, the manner in which a signal is injected into an amplifier deserves special attention. If SNR measurements are to be meaningful, none but the mode(s) radiated by the source should be accepted by the detector following the amplifier. In practice, the attainment of a minimum effective input-noise temperature is desired; care must be taken to see that these conditions are (at least approximately) met. The measurement techniques presented in this chapter will enable the achievement of nearly optimized operating conditions for a laser amplifier.

The pass band of a gas laser amplifier is the inhomogeneously broadened spectral line width[14]. In gases optical lines are Doppler broadened by an amount that is large compared with the natural line width[15]. The line shape is approximately Gaussian with a full width at half power

$$\Delta\nu_D = 2/\lambda_0\left(2\,\frac{kT}{m}\ln 2\right)^{1/2} \tag{9.9}$$

where m is the atomic mass in amu, T is the average atomic temperature, and λ_0 is the wavelength at the line center. Measurements, in which a tunable laser (a laser sweep-frequency generator) is used to measure the Doppler-broadened line, indicate that excited the line widths of state gas lasers corresponds to average atomic temperatures of 500°K. For example, in a He/Xe, 2.05 μm gas-laser amplifier a line width of 210 MHz/sec has been found[16], which corresponds to 515 $\pm$ 50°K.

There is a frequency-dependent gain profile that corresponds to (9.9) which effectively reduces the band width of a laser amplifier, especially when the gain is large.

The single-pass gain may be expressed as

$$G = \exp \alpha L, \tag{9.10}$$

where L is the length of the plasma and α is the gain coefficient:

$$\alpha(\nu) = \alpha(\nu_0)\exp\left[-4(\nu - \nu_0)^2\,\frac{\ln 2}{\Delta\nu^2}\right]. \tag{9.11}$$

The effective band width of a laser amplifier is inversion-density dependent and becomes much less than (9.9) in the high-gain case. This may be seen if the gain coefficient is expressed at the center of the line in terms of the transition coefficient A_{21}:

$$\alpha = \left(\frac{\ln 2}{16\pi^3}\right)^{1/2}\left[n_2 - \frac{g_2 n_1}{g_1}\right]\frac{\lambda^2 A_{21}}{\Delta\nu_D}. \tag{9.12}$$

An operational definition for the pass band of an amplifier is that frequency interval about the line center ν_0 wherein the amplifier gain is

$$G(\nu) = [G(\nu_0)]^{1/2}. \tag{9.13}$$

Considering the gain profile we can write the gain per unit of length in terms of a line shape factor $S(\nu)$ where

$$G(\nu) = \exp \alpha S(\nu) = G_0 S(\nu)$$

where

$$\int S(\nu)\, d\nu = 1 \tag{9.14}$$

normalized at line center. The line-shape factor for a Doppler-broadened line is just

$$S(\nu) = \frac{1}{\nu_D}\left(\frac{\ln 2}{\pi}\right)^{1/2} \exp\left\{-\left[\frac{\nu - \nu_0}{\nu_D}(\ln 2)^{1/2}\right]^2\right\} \tag{9.15}$$

Expanding the line-shape factor in a power series about the origin $\nu = \nu_0$, we obtain

$$S(\nu) \approx 1 - \frac{(\nu - \nu_0)^2}{\Delta\nu_D{}^2} 4 \ln 2 + \cdots, \tag{9.16}$$

so that the integral gain $\int_{\text{line}} G(\nu)\, d\nu$ becomes

$$G(\nu) = \frac{G_0\, \Delta\nu_D(\pi/\ln 2)^{1/2}}{2(\ln G_0)^{1/2}}. \tag{9.17}$$

The effective band width of the high-gain laser amplifier becomes

$$\Delta\nu_e = \frac{\Delta\nu_D(\pi/\ln 2)^{1/2}}{2(\ln G_0)^{1/2}} \tag{9.18}$$

This band width gives the net output power P_0 of an amplifier when irradiated by a source of known power density P_s which is uniform over the pass band of the amplifier

$$P_0 = G_0\, P_s\, \Delta\nu_e \qquad G_0 \gg 1. \tag{9.19}$$

The output noise power expected in a single mode is [17]

$$P_{\text{noise}} = G_0\, h\nu\, \Delta\nu_e\left[\frac{n_2/g_2}{n_2/g_2 - n_1/g_1}\right]. \tag{9.20}$$

In that they relate to the measurement of noise over the entire amplifier pass band to the spot-noise performance at the peak of the gain profile G_0, (9.18) and (9.19) may be employed to determine the spot-noise performance of the amplifier in a frequency band that is narrow compared with that of the Doppler-broadened line.

Before discussing noise-measuring techniques of interest the sources of laser noise will be treated.

9.3 LASER NOISE

The output light of a gas laser contains additional noise components that increase the noise-power output beyond the ideal level predicted by (9.7). The magnitude of these components depends upon the type of gas laser (excited, ionized), the methods of excitation (dc, rf or combined dc and rf), the excess-to-threshold ratio of excitation, and so on. There may, for example, be plasma noise caused by current fluctuations in a dc-excited plasma. There will be noise the character of which is that of excess photon noise. There may be competition between two lower-energy levels for the same inverted upper level, resulting in coherent output at more than one wavelength. Mode interference can occur, particularly in long lasers, where many axial modes oscillate simultaneously. Finally, there may be ripple "noise" (in ion lasers) wherein ripple currents in the plasma (or induced in the plasma by the magnetic field of the solenoidal current) can cause undesired gross variations in output intensity. Under certain conditions the output of a CW-gas laser may break into periodic pulsed operation as a pulse-regenerative oscillator that produces narrow pulses of light at the mode-locking frequency.

We consider first the extent and character of plasma noise.

9.3.1 Plasma Noise

Low-frequency current fluctuations occur in dc-pumped, cold-cathode glow discharges of which many helium-neon lasers are, unfortunately, fairly typical. Above a definite threshold value of current density, defined by gas pressure, discharge length, and tube diameter, variations in plasma current-density occur that couple noise into the power ouput of a dc-excited laser through macroscopic variations in gain.

Spectral measurements have been performed on the noise output of helium-neon lasers at 633 nm [1,2] in which the output signal of a laser-illuminated detector was examined (with appropriate spectrum analyzers) in the frequency interval of 14 Hz to 12 MHz. Nine lasers were examined, some of which were excited by a cold-cathode dc discharge. All lasers were operated in the lowest-order transverse mode at excitation levels compatible with single-frequency operation (i.e. no beats were found at the expected beat frequencies between axial modes).

Under cold-cathode, dc-discharge excitation, depending upon the magnitude of the discharge current, the introduction of a magnetic field, or capacitive loading of the plasma tube, the spectrum of the output noise in the zero to 300 KHz region would sometimes be as much as 40 db above background photomultiplier shot noise[1] and represented as much as 20 percent of the total laser output. For rf excitation, quiet operation with a noise output no more that 5 percent above the photomultiplier shot noise could always be obtained. The noise output of the laser in the remaining frequency interval investigated, independent of excitation, could not be distinguished from the photomultiplier output spectrum (shot noise) when the latter was illuminated with white light. The degree of noise amplitude modulation of the noise in the laser was inferred from the change in amplitude of the gain-enhanced noise spectrum above the photomultiplier shot noise[1,18].

The modulation ratio was determined as follows. Let $A(t)$ be the time-dependent electric field associated with the light output of the laser

$$A(t) = [A_0 + a(t)] \cos [\omega t + \theta(t)]. \tag{9.21}$$

The modulation ratio m^2 is defined as

$$m^2 = \frac{1}{A_0}\left[\lim_{\tau \to \infty} \frac{1}{\tau} \int_{-\tau/2}^{+\tau/2} a^2(t)\, dt\right]. \tag{9.22}$$

An approximate value for m^2 may be determined by measuring the area of that portion of the spectrum above the enhanced shot noise. With conventional measuring equipment a value for $m \approx 10^{-2}$ with a photomultiplier current of 4×10^{-6} A could be determined with an SNR of 1 for a bandwidth of 10 MHz. The noise modulation of lasers in quiet operation was found to be less than this upper limit.

The low-frequency noise in the light output of a helium-neon laser has been correlated with the dc-current noise in the discharge. A cold-cathode, helium-neon laser was operated from a dc supply with a 300,000 ohm series resistance placed between the supply and the discharge tube[2]. Noise in the light beam was detected by a silicon photodiode having a response time in the submicrosecond range. Correlation analysis performed on the photodiode and the discharge-current signals showed that the laser noise could be accounted for by the current noise in the dc discharge.

The magnitude of the cross-correlation coefficient between laser noise and discharge-current noise was 80 percent at low frequencies and decreased at higher noise frequencies. Because the current noise was found to be independent of laser action, it was inferred that current noise modulates the number of excited atoms in the laser.

Measurements of the noise amplitude distribution of a quiet laser oscillator reveal that the spectrum is not Gaussian but possesses a steady-state amplitude with a small superimposed modulation.

Plasma noise almost never occurs in rf-pumped discharges where the excitation frequency is above a few hundred kHz. This source of noise may be reduced in a dc-pumped laser by using a hot-cathode discharge. It may be further reduced in long lasers by using a modest rf field for combined dc and rf excitation.

Proper choice of the design parameters for low noise in laser tubes is paramount. If we compromise on tube gain, life, and power we will have a laser with lower noise output. If we choose a shorter laser of smaller bore, operated at lower than optimum pressure for maximum power output, we will have a laser with lower plasma-noise levels. Preliminary noise measurements indicate that high-power ion lasers, operated in the positive resistance range, have better plasma-noise characteristics than dc-pumped, helium-neon lasers[19].

9.3.1.1 Measurement of Plasma Noise The plasma-noise characteristics of a gas laser may be examined with a fast photodiode (preceded by a suitably blocked spectral-notch filter) and a spectrum analyzer. If the spectrum of the noise current is also to be examined, a precision low-noise resistor will be required. The apparatus listed below will be found convenient for a helium-neon laser at 633 nm, which employs dc pumping. The extension to the rf-pumped case is obvious.

The laser output should be directed through the filter to the photocell. Care must be taken to tilt the filter 1 to 3 degrees off axis, so that the laser beam will not be returned to the laser and induce mode coupling. The filter-photocell combination must of course be carefully shielded to preclude rf as well as stray light interference. Shielded leads must also be used between the photocell and the spectrum analyzer. It is important

TABLE 9.1 PLASMA-NOISE MEASURING APPARATUS

Function	Manufacturer	Model No. and Remarks
Photodiode	Edgerton, Germeshausen and Grier	SD – 100
Blocked spectral notch filter	Thin Film Products	632.8 + 0.2 – 0 nm Blocked for UV and IR to -40 db
Spectrum analyzer	Panoramic Instruments	Model SPA – 3A, 1 kHz – 15MHz
Current shunt	International Resistance Company	10 ohm, Type HFR

that the laser be operated significantly above threshold to ensure that plasma noise is being detected. The spectrum of the photodiode output may be measured directly from the spectrum-analyzer display. Alternatively, the noise voltage developed across a precision resistor (located in the grounded return lead of the plasma-tube power supply) may be displayed on the spectrum analyzer.

The cross-correlation factor between current and laser noise can be determined from the relation

$$C = \langle V_1^* V_2 \rangle_{\text{ave}} / \langle (V_1^2 V_2^2)^{1/2} \rangle_{\text{ave}} = |C| \, e^{j\theta}$$

Here V_1 is the rms-noise voltage developed by the photodiode in the measured frequency interval, and V_2 is the rms-noise voltage proportional to the discharge current in the same frequency interval. By measuring $< (V_1 + V_2)^2 >$ and $< (V_1 - V_2)^2 >$ we can determine the real part of C. Repeating the measurements after shifting the phase of V_2 by $\pi/4$ and combining with the previous results, we can compute $|C|$ and θ.

9.3.2 Excess-photon Noise

Photons, having unit spin, obey Bose-Einstein statistics. Furthermore, ordinary light sources give rise to highly nondegenerate light beams. Photon degeneracy, which is characteristic of laser beams, leads to fluctuations in beam intensity that are larger than predicted if we apply classical Poisson statistics to considerations of photon flow[20]. The role of the degeneracy parameter δ (the average number of photons in a light beam to be found in the same quantum state or same cell in phase space) will be evident from the discussion below.

The degeneracy of black-body radiation in an enclosure at a temperature, T may be expressed as [21]

$$\delta(\nu, T) = [\exp(h\nu/kT) - 1]^{-1} \tag{9.23}$$

where T is the source temperature (°K), h is the Planck's constant, ν is the frequency of the light the degeneracy factor of which is of interest, and k is Boltzmann's constant. As an example, consider an incandescent light source at $T = 3000$°K. In the visible range, say at $\nu = 5 \times 10^{14}$ Hz, the degeneracy factor is

$$\delta \approx 3.3 \times 10^{-4},$$

the temperature at which $\delta = 1$ in the visible spectrum is approximately 35,000°K. Because this temperature is far beyond most thermal light sources, photon may be classified as completely nondegenerate light beams.

The degeneracy of a laser may be calculated from the following considerations. Let $E_\nu \Delta\nu$ be the number of photons emitted normally per unit of area per unit of time per unit of solid angle by a source at frequency ν within a small frequency interval $\Delta\nu$ by a light source. If the source has an emitting area S, it can be shown that[22], depending upon the degree of coherence γ assigned, there will be an area, A, at a normal distance R, such that at a frequency ν.

$$A = \gamma \frac{c^2 R^2}{\nu^2 S}. \tag{9.24}$$

In making estimates of the degeneracy δ, it will be sufficient[23] to let $\gamma = 1$.

The number of photons, N, in the frequency interval $\Delta\nu$ collected by the area A at a distance R, is

$$\begin{aligned} N &= (SE_\nu \Delta\nu)(A/R^2) \\ &= (SA/R^2)E_\nu \Delta\nu \\ &= (c^2/\nu^2)E_\nu \Delta\nu. \end{aligned} \tag{9.25}$$

Associated with each frequency interval $\Delta\nu$ there is a coherence time of order $1/\Delta\nu$, for which the light remains coherent with itself. Thus the number of photons of frequency ν collected in a coherence time by a coherence area is $c^2 E_\nu/\nu^2$. Considering a polarized source this number is just the photon degeneracy

$$\delta = c^2 E_\nu / 2\nu^2$$

For unpolarized light this reduces to

$$\delta = c^2 E_\nu / 2\nu^2. \tag{9.26}$$

We now apply (9.26) to determine the degeneracy of a 1-Watt gas laser operating at 514.5 nm with an angular divergence of 10^{-3} rad, a reduced line width of 10^8 MHz, and a spot diameter of 10^{-1} cm. The radiance of this source is 1.67×10^8 W/cm²/ sr corresponding to a photon flux of 4.2×10^{18} photon/cm²/sr/sec. The light output is polarized. Therefore the degeneracy factor is

$$\delta = c^2 E_\nu / \nu^2 \sim 10^{10}.$$

Clearly a gas laser is a highly degenerate photon source. The effect of this large degeneracy is that laser photon beams behave like waves rather than particles in their fluctuation properties. The photon-bunching effect leads to wave-interaction phenomena, which may be understood by considering the radiation as composed of waves with different frequencies interfering with each other. This "beating" gives rise to wave interaction or excess photon noise in laser beams.

Photoelectric detectors emit electrons at a rate proportional to the square of the wave amplitude. Thus fluctuations of wave-interaction amplitude cause an excess noise (over and above shot noise) in the photocurrent output of the detector[23–25]. Conversely the gas laser is an excellent source for photoelectric studies of wave-interaction phenomena. Three applications include the determination of the degree of polarization[26], the detection of interference effects with incoherent light source[27], and the measurement of narrow spectral profiles[28, 29].

The excess photon or spontaneous-emission noise in a laser oscillator, which is predicted by a model that treats the gas laser as a saturated amplifier of noise[30,31], has been subject to extensive experimental investigation[32, 33]. These measurements reveal that excess noise can be detected in a low-gain gas laser operating near (above and below) the threshold of oscillation. As the laser gain and power output are raised, that is when the laser oscillator operates further above threshold, the ratio of spontaneous-emission noise to shot noise decreases rapidly and becomes less limiting than plasma noise, other sources of spontaneous-emission noise, and noise introduced by modes other than the lowest-order mode of interest.

9.3.2.1 Single-detector Technique of Measuring Excess Photon Noise In the single-detector method of excess-noise measurement the output of the laser is detected by a broad-band detector (photomultiplier) in which the input is suitably filtered to reject extraneous spontaneous-emission background. The photomultiplier output is passed through a narrow-band frequency analyzer that measures the rms noise voltage in a frequency band $\Delta\nu$. The power spectrum of the noise voltage is related to the properties of the optical source as

$$< e(\nu)\,\mathrm{rms} < \; = 2\Delta\nu\,|b(\nu)^2|[\langle I(t) \langle + \Delta P(\nu)] \qquad (9.27)$$

wherein $< I(t) >$ is the light-field intensity at the photomultiplier, $\Delta P(\nu)$ is the power spectrum of the intensity correlations of the light beam, and $b(\nu)$ is the frequency response of the detection system. When the photomultiplier is exposed to a broad-band optical source (black-body radiator) we obtain a measure of the photomultiplier shot noise, $2\Delta\nu|b(\nu)|^2 \; < \; I(t) >$, as $\Delta P(\nu) \approx 0$. The term $\Delta P(\nu)$ contains the spectral information of significance in the case of the laser oscillator.

It may be shown that the single-detector measurements on the laser yield the quantity[34]

$$\frac{\Delta P(\nu)}{\langle I(t)\rangle^2} = 2\Delta\nu\,|b(\nu)|^2\,\frac{\langle e(\nu)\rangle^2_{\text{laser}} - \langle e(\nu)\rangle^2_{\text{black body}}}{\langle e(\nu)\rangle^4_{\text{black body}}}$$

$$= 2\Delta\nu\,|b(\nu)|^2\left[\frac{\text{excess noise}}{(\text{shot noise})^2}\right]$$

$$= \frac{\Delta P(\nu)}{C(\infty)}. \tag{9.28}$$

The latter term in (9.28) is significant in the analysis of amplitude fluctuations of a laser oscillator. The term $\langle e(\nu)\rangle_{\text{black body}}$ is the rms output of the analyzer when a black body of intensity $\langle I(t)\rangle$ irradiates the photomultiplier. Theoretical expressions for a homogeneously broadened line have been derived for $\Delta P(\nu)/\langle I(t)\rangle^2$, and the results have been applied to a YAG : Nd oscillator[35].

The instrumentation necessary for the measurement of excess photon noise using the single-detector technique differs in detail from that described in section 9.3.1.1. The photodiode is replaced by a photomultiplier. It is desirable to use a second photomultiplier and electronic counter to measure the photon emission rate of the laser under test. It is convenient to include a spectral-density analyzer and a two-dimesional recorder to display the excess noise versus frequency directly on a chart. The apparatus shown in Table 9.2 will be found convenient to study a dc-excited, helium-neon laser operating at 633 nm.

In making meaningful measurements of excess photon noise it is necessary that plasma noise effects be minimized. The experimental arrangement of 9.3.1.1 should be used with the noise spectrum displayed while the laser excitation circuitry is optimized to suppress gas discharge fluctuations. The plasma should be operated at lower currents. A large value of resistance ($\approx 100{,}000\Omega$) should be placed in series with the laser exciter lead. A strong permanent magnet placed near the anode of the discharge will be found helpful in noise reduction. When relatively quiet operation of the laser is achieved the measurements can proceed. The single-frequency output of the laser should be stabilized against long-term drifts with a mirror-displacement type of servofrequency-control system.

The laser output should be passed through the polarized and thence to a beam splitter. One sample of the laser beam should feed the photomultiplier, the output of which drives the counter for measuring the photon emission

TABLE 9.2 APPARATUS FOR MEASURING EXCESS PHOTON NOISE

Function	Manufacturer	Model No. and Remarks
Blocked-notch filter	Thin Film Products	633, + 0.2, − 0 nm blocked for IR
Spectrum analyzer	Panoramic Instruments	Model SPA 3A, 1kHz–15 MHz
Spectrum density analyzer	Panoramic Instruments	Model PDA-1. 0.7–30 sec. integration interval
Strip recorder	Panoramic Instruments	RC-36
Polarizer	DBA Crystal Optics	Glan-Thompson Polarizer
Photomultiplier	h nu Systems, Inc.	Model 3227 with 1P21 Photomultiplier Tube
Phototube H. V. Supply	John Fluke	Model 405 B
Counter totalizer	Eldorado	Model 745
Wideband amplifier	Hewlett-Packard	Model 461 A. 1 kHz–150 MHz, 20, 40 db
Random-noise generator	Aerospace Research, Inc.	Model Ns-L. 100 Hz–500 MHz

rate. The second photomultiplier should drive the spectrum analysis (via a wide-band amplifier if required). The system must be calibrated to account for transmission losses through the various optical elements. The spectrum analyzer is calibrated by a random-noise source. The spectral density analyzer integration time should be set in the 1 to 5 sec range to permit convenient recording of the value of the noise output versus frequency.

Care must be exercised to preclude interference from stray light or electrical signals. Overloading of the photomultiplier tube can be precluded, when the laser is operated at high power levels, by shorting as many as half of the photomultiplier dynodes to the anode and making an appropriate reduction in the regulated-power-supply dynode voltage.

The apparatus and arrangement outlined above may be used to evaluate the spectrum of the excess photon-to-shot-noise ratio for various levels of laser excitation.

Experimentally observed spectral curves of the photomultiplier tube current (proportional to output power) show that the excess photon noise-to-shot-noise power ratio is proportional to the square of the output power below threshold and inversely proportional to the square of the output power above threshold. Deviation from the squared dependence may be observed far below threshold if more than one linearly polarized mode is allowed to contribute to the spontaneous output. The band width of the excess noise is found to vary linearly with the output power above threshold, but inversely with the power below threshold[36].

The power spectrum of the noise ratio is well approximated by a Lorentzian curve for data obtained just below threshold and above threshold. The applicability of the Van der Pol model to laser oscillations above threshold has recently been demonstrated experimentally[36, 101].

9.3.3 Other Sources of Spontaneous-emission Noise and Techniques for their Elimination

Depending upon the type of gas laser considered (excited, ionized, vibrational transfer), there are extraneous sources of spontaneous-emission noise that must be eliminated if the best SNR is to be obtained.

Consider the classical helium-neon laser. Spontaneous emission is generated by energy-level changes, whether or not the associated transitions lead to potential laser lines. It is stressed that suitable filtering be placed at the output of the laser source to ensure that noise generated into the mode of interest from this source of extraneous stimulated emission is minimized. Whereas multilayer, blocked, spectral-notch filters (~0.1 nm) are adequate for most system work, additional selectivity can be obtained through the use of a grating spectrometer (1 to 3m).

The noise power from these extraneous sources of spontaneous emission may be estimated (per mode per polarization) by considering each source of noise to be the amplified double-pass spontaneous emission attenuated by the transmittance of the output mirror at the pertinent wavelength. For the case where the gain of the transitions is well below threshold we have

$$P(\nu)\Delta\nu = h\nu\Delta\nu\left(\frac{n_2/g_2}{n_2/g_2 - n_1/g_1}\right)(G^2 - 1)(1 - R), \tag{9.29}$$

where R is the output-mirror reflectivity; the other terms are as previously defined.

Although adequate filtering will virtually remove the aforementioned sources of noise, there are two additional components of stimulated-emission noise that are present in transmitted light. These are "depolarization" noise and fluctuations in the laser-light output induced by dominance or competition of two transitions for the inverted states in the upper laser level[37]. Consider the depolarization noise. If an internal-mirror laser is used, the output of a gas laser will be essentially unpolarized. The output of spontaneous-emission noise will be almost twice that of an external-mirror laser because all polarizations are present. When an external-mirror laser is used, the laser output is partially polarized by virtue of the discrimination against light of the rejected polarization in multiple passes through the Brewster windows of the laser tube. The simplest method eliminating this noise source, when the laser output is partially polarized is to place a good polarization analyzer (Glan-Thompson prism) in front of the photodetector. If this analyzer is not used, additional noise will be present, because the Brewster windows of the laser tube do not completely eliminate the opposite polarization of the same mode; they are effective only to the extent that they keep the mode below threshold.

Dominance can exist in all forms of gas lasers. It is classically illustrated in the helium-neon laser by the competition of the high-gain 3.39 μm transition and the 633 nm transition. High-gain, helium-neon lasers that operate at 633 nm have extraordinary potential gains at 3.39 μm. (The double-pass gain of a 5-mm bore, 2-m long, helium-neon laser at 3.39 μm is approximately 90 db)[38]. Because the spontaneous-emission noise depends upon the upper-level population of the laser, fluctuations induced by dominance of the 3.39 μm transition effect the measured noise output at 633 nm near threshold. When the visible transition is operated at high power levels, this source of noise is less significant[19].

9.3.4 Mode Noise and Techniques for Its Elimination

The light output of most high-power gas lasers operated in the TEM_{00} mode will consist of a multiplicity of discrete frequencies all of which fall within the Doppler line width of a given laser transition. The spacing of these discrete frequencies is just the axial-mode separation $c/2L$. Consider a 2-m, helium-neon laser operating at 633 nm. From (9.9) we see that the Doppler-broadened line for an "atomic temperature" of 500° K is approximately 1500 MHz. The axial-mode frequency separation is 75 MHz. As many as 20 discrete frequencies could propagate in this laser if the gain would support oscillation. Practically, five to 10 of these frequencies will exist simultaneously.

Because the laser medium is highly nonlinear, beats will exist between the multiplicity of operating frequencies, (their sums as well as differences), such that a chaotic mixture of time-varying beat frequencies is produced that extends from the audio to the microwave end of the spectrum. Because the exact value of each beat frequency depends on mirror spacing, current density, and even a stray magnetic field from ac-power sources, the frequency mixture is highly unstable, often uncontrolled, and may interfere with signal processing in a communication system. Though not truly noise, these interferences must be suppressed.

Mode interference may be controlled by using a laser the cavity of which is so short ($\approx$ 10 cm or less at 633 nm) that only one axial mode will have sufficient gain to oscillate. These so-called single-frequency lasers typically have inadequate power output ($\approx$ 100 μW). Fortunately today techniques exist for producing single-frequency power outputs at substantial power levels by means of subsidiary internal cavities that discriminate against the unwanted frequencies and permit single axial-mode operation at power outputs approaching that of conventional gas lasers[39]. Phase-locking (also called supermode) techniques have also been developed which permit single-frequency operation[40].

Mode noise is a less severe problem in ion lasers, even though the Doppler line widths are three to five times that of helium-neon lasers operating in the visibile[41] because their high gain greatly facilitates the insertion of internal Fabry-Perot cavities to eliminate unwanted frequencies.

9.3.5 Excitation-dependent Interference

In contrast with helium-neon lasers, which are operated under conditions of saturated-power output, ion lasers are generally operated at power levels thermodynamically limited by power-transfer rates. The power output of a helium-neon laser power is relatively insensitive to changes in excitation current. The power output of an ion laser is a highly nonlinear function of current, being proportional to the excitation current raised to an exponent between 1.5 and 6[42]. Near threshold $P_{\text{out}} \approx I^{6}{}_{\text{dc}}$, whereas at high power levels $P_{\text{out}} \approx I^{3/2}{}_{\text{dc}}$. Thus, the light output of ion lasers is strongly affected by ripple currents in the exciter. Most high-power ion lasers employ solenoidal magnetic fields to increase laser-output power and efficiency by a factor of the order of 2. Solenoidal-field ripple obviously interacts with the laser excitation and thereby enhances low-frequency variations in the light-power output. Both sources of interference can be reduced by substantial active or passive ripple filtering.

9.3.6 Self-induced Mode Locking and Techniques for Its Detection and Elimination

In designing a communication system that uses a helium-neon, CW-gas laser it is necessary to be sure that the axial modes do not occasionally self-lock, so that the laser operates as a pulsed regenerative oscillator[43]. The mode-locked laser will generate pulses of subnanosecond width at a repetition frequency approximately equal to the axial-mode spacing $c/2L$. Conversely, to obtain a source with a high-pulse-rate and of extremely narrow pulses it may be desirable to force a CW-gas laser to operate at some multiple of the axial-mode spacing. The reader interested in the latter application is referred to already extensive literature on the subject [44, 45, 46]; our present interest is to determine whether or not incipient self-locked operation will occur in a laser designed for communication applications.

Mode locking can occur in gas lasers in which the mirror spacing is compatible with the existence of several axial modes within the Doppler line. Helium-neon lasers with mirror spacings as small as 39 cm and as large as 250 cm will self-pulse at repetition frequencies of 60 to 380 MHz. Attempts to self-lock high-power, argon-ion lasers have thus far been unsuccessful. This is consistent with the prediction that a laser operated with axial-mode separations that are either very small or very large compared with the Doppler line width will not self-lock[43]. If a CW-gas laser in a communication system becomes mode locked, the distortion resulting from the light pulses, which may have a peak power as much as 20 times the average power, will render the system useless as a light carrier.

Self-locking occurs in helium-neon lasers when the cavity Q is lowered by the insertion of losses that do not vary with time (e.g. modulation or wavelength-selection components, etc.) or when the transparency of the output mirror is increased too much. It is necessary to observe the power-output characteristics of a CW-gas laser to ensure that the light output is free of this defect. This source of interference, if found, can be eliminated by increasing the mirror reflectivity, reducing the surface reflectivity of equipment placed in the optical cavity, increasing the excitation power (if the laser is not already power-input saturated), or choosing a laser the dimensions or Doppler line-width characteristics of which are more suited to the application.

In evaluating performance characteristics of the mode-locked laser the equipment, listed in Table 9.3, will be found desirable.

To examine the output of the laser, direct the laser beam into the photodiode. Connect the back-biased output of the photodiode through a short cable to the oscilliscope input. Connect the oscillator output to the alternate sweep input of the oscilloscope. Set the oscillator frequency at the $c/2L$ frequency and establish a steady-state display on the oscilloscope. While observing the oscilloscope presentation vary the laser excitation power slowly over its full dynamic range. Look for the appearance of output pulses at the frequency of the axial-mode spacing. If no pulses appear, adjust the laser for maximum power output. Insert the optical flat in the beam. Vary the cavity losses by changing the angle of the optical flat slowly from Brewster's angle until it is normal to the beam while watching the oscilloscope display for output pulses. The amount of insertion loss required to

TABLE 9.3 EQUIPMENT FOR MEASURING SELF-LOCKING CHARACTERISTICS OF GAS LASERS

Function	Manufacturer	Model Number, Remarks
Sampling oscilloscope	Tektronix	Type 661 with 4S2 head 0.1 ns rise time, 2 mv/cm
Photodiode	Philco	L4501 Photodiode 5 GHz bandwidth
Thin optical flat	Pearson Scientific Optical	Mounted Quartz Plate, 10^{-2} cm thick $\times$ 1 cm dia. surfaces flat and parallel to 1/10 wave
Signal generator	Hewlett-Packard	Model 608D 10–420 MHz

induce self-locking at various levels of excitation is a measure of the tolerance of the laser against self-induced mode locking.

9.3.7 Spectral-interaction Interference and Techniques for Its Measurements and Elimination

A problem frequently encountered with helium-neon lasers is that a low-level audio modulation appears on the intensity detected output light. In communication systems that require constant output amplitude (particularly if low-index modulation is to be used) the presence of uncontrolled, time-varying, audio modulation is objectionable. The source of this interference may be traced to gain interference between spectral lines that oscillate simultaneously. For example the 640 nm line may be made to oscillate with the 633 nm line by varying the excitation rate of the laser. If the laser output is monitored with a photodetector, an audio amplifier, and a speaker, it will be found that the audio squeals are strongest just as the additional spectral line reaches the threshold of oscillation. Obviously, if the laser power supply is not well regulated, the squeal will be a periodic and may even switch off and on with ripple current from the exciter. This source of interference is highly microphonic because the threshold for laser action is sensitive to mirror orientation.

The astable relaxation oscillations that result from spectral-interaction interference are not restricted to the visible lines. This type of interference is also present in the near infrared (1.114, 1.118, 1.15, 1.16, and 1.77 μm). The frequency of the interference modulation is related to the relative gains and lifetimes of the various transitions involved.

Freedom from spectral-interaction interference can be obtained in at least two ways: operate the laser at an excitation level that precludes oscillation on the interacting lines, and place a dispersing or filtering element in the laser cavity to preclude operation on more than one spectral line. Although this source of interference has been most completely investigated in the helium-neon lasers, it is expected to occur in all lasers.

Having briefly reviewed the extraneous sources of noise or interference in the laser output, we consider more detailed problems of techniques that must be used to measure the ultimate noise performance of a gas laser.

9.4 GENERAL METHOD OF MEASURING NOISE PERFORMANCE OF NONRESONANT AMPLIFIERS

To obtain meaningful noise measurements, it is necessary to ensure that noise associated with modes other than the signal mode do not reach the detector. A system of apertures must be inserted between the amplifier and the detector[47,48]. As an example we consider the case in which the

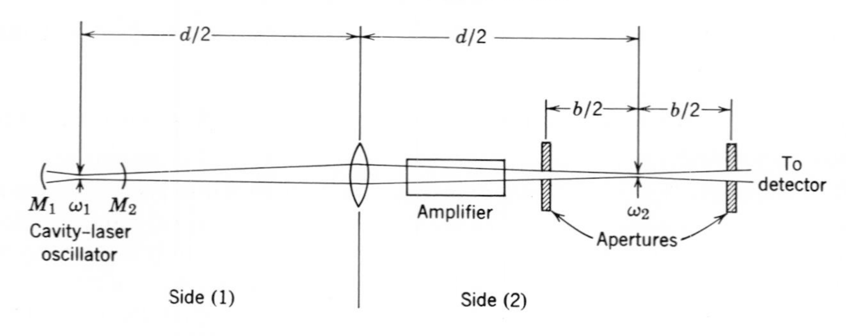

Figure 9.1. Mode structure and aperture placement.

measurements are performed on a near-confocal laser, which is used as the optical signal generator. (We discuss later an alternate case in which a source other than a cavity laser is used.)

Consider a beam with a minimum radius w, which must be transformed into a beam with minimum radius w_2 to achieve the desired mode matching. Let the signal generator have mirrors with a radius of curvature b' and spacing d. Then the minimum beam radius w_1, which occurs in the center of the resonator, can be computed from the formula[49].

$$w_1 = \left(\frac{\lambda}{2\pi}(2b'd - d^2)\right)^{1/2}. \tag{9.30}$$

Refer to the arrangement shown in Figure 9.1. The light wave emerging from the mirror M_2 passes through a lens of focal length f. Denote the distance between the waist of the resonator mode and the lens by d_1 and the radius of the waist of the light beam after passage through the lens by w_2. Then we can show that the following relations hold[46]:

$$\frac{d_1}{f} = 1 \pm w_1/w_2(1 - f_0{}^2/f^2)^{1/2}. \tag{9.31}$$

or

$$w_2 = w_1(1 - f_0{}^2/f^2)^{1/2}(d_1/f - 1), \tag{9.32}$$

$$\frac{d_2}{f} = 1 \pm w_2/w_1(1 - f_0{}^2/f^2)^{1/2}, \tag{9.33}$$

where the characteristic matching length f_0 is determined by w_1 and w_2 through the relation

$$f_0 = \pi \frac{w_1 w_2}{\lambda}. \tag{9.34}$$

Given d_1 and w_1, we may evaluate w_2 from (9.32) and then d_2 from equation 9.33. The position of the waist d_2 and its radius w_2 determine the phase-front curvatures everywhere on side (2) of the lens. The spacing b_2 of those two phase fronts, the radius of curvature of which is equal to the distance between them, is found from the formula[49] relating the waist radius w_2 to b_2.

$$w_2 = (b_2 \lambda / 2\pi)^{1/2}. \tag{9.35}$$

If we locate two apertures of radius $w_2\,(2)^{1/2}$ at the limits of these particular phase fronts, as shown in Figure 9.1, we achieve an equivalent input-noise temperature that is greater than the ideal of (9.6) by a factor of 1.64. This represents an (approximate) optimization of the noise performance.

The measurement of the noise performance now proceeds as follows. The signal generator is blanked out and a measurement of the power emitted by the amplifier is made. Then a known amount of power P_s from the previously calibrated signal generator is admitted. Denote by Y the factor by which the output power of the amplifier has increased. To make the measurement accurate, Y should not differ greatly from 2. Then

$$P_e \Delta\nu_e = P_s / Y - 1. \tag{9.36}$$

Using (9.4) or (9.5) we can compute the effective input-noise temperature.

Now consider the question of how the noise temperature T_e measured in the way described here, is realized when the laser is used to amplify a signal from a source other than a cavity laser. The measured noise temperature T_e is indicative of the noise performance of the laser amplifier under general operating conditions only if the input signal is diffraction limited. If the radius of the waist w_1 of the focused incoming signal (whose intensity distribution of a resonator field) is determined and its position is found, we may proceed with the optimization, as shown in Figure 9.1, except that the distance d_1 now refers to the distance from the lens to the waist of the incoming signal beam.

Output apertures other than those described herein can be employed. More detailed treatments, including the choice of size and location of apertures, are available[48].

9.4.1 Method of Verifying Optimal Noise Performance

The experimental technique described herein has been used to verify that lasers are capable of yielding noise levels reaching the basic limits predicted by the uncertainty principle of quantum mechanics[12].

Consider a helium-neon laser operating on the high-gain 3.39 μm transition in neon ($3s_2 - 3p_4$). The noise expected in a single mode can be calculated from (9.5) by inserting the latter into Planck's formula[17]:

$$\left(\frac{dI}{dv}\right)\Delta v = \frac{hv^3}{c^2[\exp(hv/kT) - 1]}\,\Delta v. \tag{9.37}$$

The equivalent noise-output power per mode per unit-frequency interval becomes[17]

$$E(v)\Delta v = hv\left[\frac{n_2/g_2}{n_2/g_2 - n_1/g_1}\right]. \tag{9.38}$$

The total noise output is obtained by multiplying equation 9.38 by the frequency-dependent gain $G(v)$ and integrating the result over the Doppler-broadened line. It can be shown that this yields (9.20) under the assumption of a Gaussian line shape with a half-width

$$\Delta v_D = 2v/c\left(\frac{2kT}{m}\ln 2\right)^{1/2} \tag{9.39}$$

when the line-shape factor is normalized to unity at the line center.

The factor

$$\frac{(n_2/g_2)}{(n_2/g_2) - (n_1/g_1)}$$

may be determined experimentally by measuring the spontaneous emission from the side of the laser tube. Denote by I_1 the intensity of the $(3p_4 - 1s_2)$ transition at 3.39 μm and by I_2 the intensity of the $(3s_2 - 2p_4)$ transition at 633 nm, both when the laser is oscillating (L) and when it is operating as a simple amplifier (A).

Assuming approximate population equality,

$$n_1/g_1 \approx n_2/g_2.$$

It can be shown that[17]

$$\frac{n_2/g_2}{(n_2/g_2) - (n_1/g_1)} = \frac{I_{2A}}{I_{2A} - I_{1A}(I_{2L}/I_{1L})}; \tag{9.40}$$

an experimental value of 1.56 is obtained for this ratio[17].

The amplifier gain at 3.39224 μm was measured by sending a weak signal through the mode-matched amplifier with the discharge switched on and off. A null measurement was made using calibrated attenuators to reduce the gain-multiplied signal to its original level, as sensed by a PbSe detector. The small signal gain for a 7 mm i.d. × 150 cm long amplifier was measured as $G = 1060$. Assuming an effective gas temperature of 500° K for Ne and a Doppler-broadened line (9.19) yields as 315 MHz line width; using (9.10), we obtain a total noise power per mode of 12.3×10^{-9} W. Application of (9.6) shows an effective noise temperature of 8550° K as compared with an ideal value of 6120° K.

The measured noise power, after correction for Brewster's-angle window losses and filter losses was 27.3×10^{-9} W. When it is recalled that (9.20) applies to each polarization in the amplifier output, or 24.6×10^{-9} W total, the experimental noise level is seen to approach the theoretical value. Even better agreement can be obtained if the measured shape in the gain curve is used to compute the expected noise. Experience indicates that the measured noise level will usually exceed the theoretical value if the interference filter is not sufficiently sharp to reject other closely spaced sources of noise (i.e. the $3s_3 - 3p_5$, 3.39091 μm, transition probability 2/9 and the $3s_2 - 3p_2$, 3.39123 μm, transition probability 2/9). The laser line has a relative-transition probability of 5/9. It is further necessary to control the orientation of all sources of specularly reflected light to minimize the energy scattered or specularly reflected into the mode of interest caused by noise traveling in the opposite direction to the signal.

Figure 9.2 depicts the experimental arrangement used for noise measurements. Mode matching was accomplished with a silvered concave mirror separated from the detector iris by the mirror focal length f. The area A_L of the 7 mm iris at the end of the laser tube was matched to the detector iris A_D through the relation:

$$A_D = \frac{\lambda^2 f^2}{A_L}. \tag{9.41}$$

9.4.2 Method of Measuring Amplitude Noise in Resonant (Folded) Laser Amplifiers

A resonant- or folded-gas amplifier is a conventional laser except that the level of excitation is reduced to just below the threshold of oscillation for the mode structure involved. To measure the noise performance of a resonant-gas amplifier a laser signal is injected into the desired mode and subsequently coupled out. The arrangement of apparatus is shown in Figure 9.3.

The transmission of the resonant amplifier may be expressed as [17, 30]

$$\text{Power out} = \left[\frac{(1 - R_1{}^{1/2} R_2{}^{1/2} G)^2}{G(1 - R_1)(1 - R_2)}\right]^{-1} + \left[\frac{4R_1{}^{1/2} R_2{}^{1/2}}{(1 - R_1)(1 - R_2)}\right] \sin^2 \frac{2\pi\nu L}{c} \tag{9.42}$$

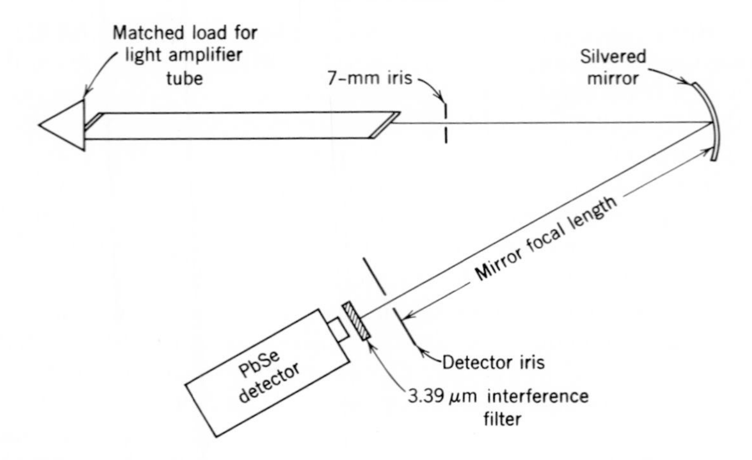

Figure 9.2. Experimental arrangement for noise measurement. (Reproduced with permission from *Applied Physics Letters*.)

wherein R_1 and R_2 are the power reflectivities of the input and output mirrors, G is the single-pass gain, and $2\pi \nu L/c$ is a phase factor that relates to the periodic-transmission properties of a Fabry-Perot cavity of length L.

The procedure of the noise-performance measurement of a resonant laser amplifier is essentially the same as that outlined in section 9.4, except that reasonable care must be taken to ensure that the amplifier does not break into oscillation. To achieve large stable gains in a resonant-laser

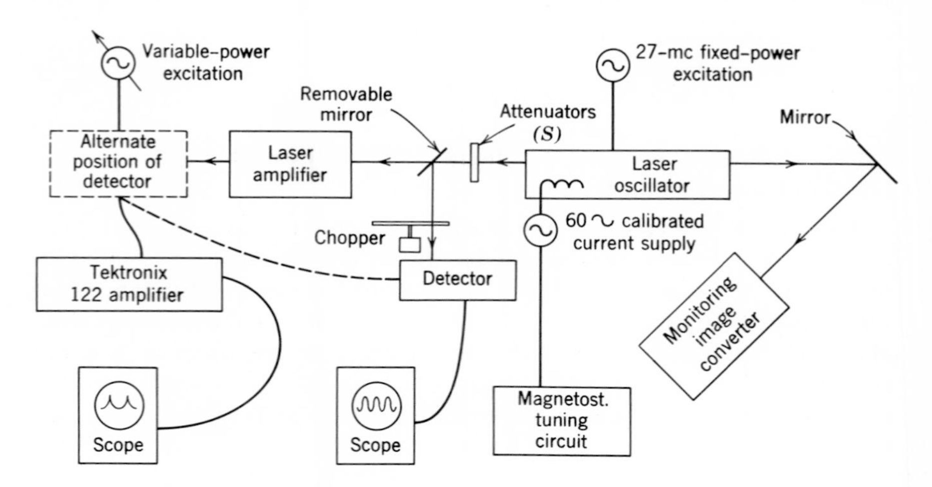

Figure 9.3. Instrumentation for the measurement of resonant-amplifier gain. (Reproduced with permission from Institute of Electrical and Electronics Engineers.)

amplifier it is necessary that the resonant cavity assembly be extremely rigid and that the laser exciter be well regulated.

The essential feature of the resonant amplifier may be seen by simplification of (9.42). Let $R_1 = R_2 = R$ and neglect the periodic phase factor. Then

$$\frac{\text{Power out}}{\text{Power in}} = \frac{GT^2}{(1 - RG)^2}. \tag{9.43}$$

The tolerance of the amplifier against oscillation is contained in the $(1 - RG)$ term. Because G is excitation dependent, so is the threshold of oscillation.

The theoretical noise performance of the resonant-gas laser amplifier may be treated from the point of view that the system is a resonant amplifier of its own spontaneous emission. The analogy of the folded amplifier is similar to that of the nonresonant amplifier with n-fold reflections wherein account is taken of all losses on reflection (including power output from either end).

9.5 LASER AMPLIFIER

If we amplify a coherent light signal with a laser amplifier, spontaneous-emission noise is added to the signal. By employing a diffraction-limited, mode-matched, spatial-filtering structure, as is described above, we can reduce the added spontaneous-emission noise to a value that approaches the theoretical minimum. The question of whether or not an improvement can be obtained in the minimum detectable signal remains. As we show below, the answer depends upon the wavelength-dependent properties of the detector. If we make a cursory analysis of system SNR based upon the use of low-gain lasers operating in the visible portion of the spectrum, wherein high-performance photoemissive detectors are available, we easily can conclude that a laser amplifier degrades the performance of most communication systems[19, 50] particularly when the laser preamplifier is compared with an optical heterodyne or homodyne system. Broader theoretical analyses, too detailed to be reproduced herein, indicate[51], however, that, depending upon the inversion level of the laser amplifier, and the wavelength-dependent quantum efficiency of the detector used, the laser preamplifier may improve the minimum detectable signal level. The results of measurements conducted at 3.508μ (one of the best atmospheric transmission windows) with a high-gain xenon laser preamplifier[52, 53] indicate that because of gain narrowing an improvement of 16 db was obtained in the minimum detectable signal. Because of difficulties encountered in making an independent measurement of the inversion level of the xenon amplifier, no direct numerical comparison could be made between theory and experiment. It was established, however, that for maximum improvement, a high inversion level was required and the amount of improvement depended upon the properties of the detector used.

The theoretical evaluation of the place of a laser amplifier in a laser communication system[51] requires answers to two questions: What is the quantum efficiency of the detector in the wavelength range of interest? If the amplifier increases the signal-to-noise ratio, how much gain is useful? In laser radar systems in which a quantum detector is followed by a threshold discriminator, the value of a laser preamplifier cannot be determined on a SNR basis alone. The statistical properties of the signal and noise must be examined. (The reader is referred to the literature[50] for more details.) It suffices here to consider a communication system from the optical region of the spectrum, wherein the high effective-noise temperature of a laser amplifier degrades the communication system performance, especially when optical heterodyning techniques are used.

Consider the SNR of a laser preamplifier followed by a photodetector for the restricted case in which a detector of low dark current (such as a photoemissive detector) is used. Without the laser amplifier the SNR, expressed in terms of the detector-output-current ratio, where the detector is shot-noise limited and has no dark current is

$$\mathrm{SNR} = [\eta n(\nu)\,\Delta\nu/2\delta\nu]^{1/2} \tag{9.44}$$

wherein η is the quantum efficiency of the photodetector, $n(\nu)$ is incident photon rate per unit of transmitted optical band width, $\Delta\nu$, and $\delta\nu$ is the reduced band width of the photodetector $(1/\tau)$.

If a laser amplifier has been placed before the photodetector, we added a spontaneous-emission noise power P_{spont}

$$P_{\text{spont}} = h\nu\Delta\nu(G-1)\left(\frac{n_2/g_2}{n_2/g_2 - n_1/g_1}\right), \tag{9.45}$$

which yields a photocurrent I_{spont}

$$I_{\text{spont}} = \eta e(G-1)\Delta\nu\left(\frac{n_2/g_2}{n_2/g_2 - n_1/g_1}\right) = \eta e\,\frac{P_{\text{spont}}}{h\nu} \tag{9.46}$$

where e is the electronic charge; the remaining symbols are as previously defined.

The signal photocurrent that existed prior to the insertion of amplifier is now multiplied by the gain G:

$$I_{sig} = \eta en(\nu)\Delta\nu G. \tag{9.47}$$

Now the shot noise of a photodetector in which a current I flows is just

$$i_n = \Gamma\sqrt{2eI\delta\nu} \approx \sqrt{2eI\delta\nu}, \tag{9.48}$$

neglecting the shot-noise enhancement factor Γ.

Although both the signal current and the spontaneous-emission current contribute to the total shot noise, their contributions do not sum directly; the spontaneous-emission noise is incoherent. When the statistical character of the excess photon noise (amplified spontaneous-emission noise) is considered in more detail, it can be shown[19] that in high-gain lasers, where in the input spontaneous-emission noise is the dominant noise source in the detection process,

$$SNR = n(\nu)\left[\eta\Delta\nu/2\delta\nu\left(\frac{n_2/g_2 - n_1/g_1}{n_2/g_2}\right)\right]^{1/2}$$
$$+ \text{ shot noise terms proportional to } n(\nu). \tag{9.49}$$

+ shot noise terms proportional to $n\ (v)$.

Thus the noise added to the signal by the amplifier does not improve the SNR, and no advantage is to be gained in using an amplifier in a system whose detector is shot-noise limited.

In an optical heterodyne system we can exchange the problem of dark current noise (by using a low-noise-laser local oscillator) for wave-front matching and critical alignment of the respective wave fronts of the received signal and the local oscillator.

When the photocell in a heterodyne system is irradiated by an incident power P_n

$$P_{in} = \frac{cE^2A}{8\pi} = h\nu n(\nu)\Delta\nu. \tag{9.50}$$

The photosurface of area A will generate a current

$$I = \eta en(\nu)\Delta\nu = \frac{\eta ecE^2A}{8\pi h\nu} = \kappa E^2 \tag{9.51}$$

where E is the rms electric field consisting of the sum of the local oscillator $E_1 \cos \omega_1 t$ and the incoming signal $E_2 \cos \omega_2 t$. Assuming a strong local oscillator, (9.51) is approximated by

$$I \approx \kappa[E_1{}^2/2 + E_1 E_2 \cos(\omega_1 - \omega_2)t] \approx \kappa E_1{}^2/2 \tag{9.52}$$

so that the photocell noise is determined almost entirely by the local oscillator-induced shot noise. Because $\omega_1 \approx \omega_2$, the heterodyne terms $(\omega_1 - \omega_2)$ and $(2\omega_1 - \omega_2)$ both contribute to the photocell noise, so that

$$i_n = [4eI\delta\nu]^{1/2} \approx E_1[2\kappa e\delta\nu]^{1/2}. \tag{9.53}$$

The signal produces a photocurrent proportional to the rms value of the heterodyned component at the difference frequency $(\omega_1 - \omega_2)$:

$$I_{\text{signal}} = \frac{\kappa E_1 E_2}{\sqrt{2}} \tag{9.54}$$

so that

$$SNR = \frac{E_2\sqrt{\kappa}}{2(e\delta\nu)^{1/2}}, \tag{9.55}$$

which simplifies, using (9.50), to

$$SNR = \left[\frac{\eta n(\nu)\Delta\nu}{2\delta\nu}\right]^{1/2}. \tag{9.56}$$

The SNR in the optical heterodyne case thus reduces to that of the photodetector (9.44) alone.

The relative practical significance of achieving the performance described above must be evaluated in terms of system requirements. Neither the laser preamplifier nor the heterodyne lasers perform as well as indicated by the simple theory. Where low-gain lasers are used in the visible range, wherein quantum efficiencies are relatively large and wherein traveling-wave photocells are available with band widths necessary to accommodate the large and relatively rapid excursions in the heterodyned difference frequency, the heterodyne system has the attractive advantage of intrinsic broad-band operation and an excellent SNR. It is assumed that problems of wave-front matching and alignment can, of course, be handled without encountering other problems from thermal drift and vibration.

Though the laser amplifier, by microwave standards, is a relatively noisy device, though it requires diffraction-limited, mode-matching, noise-reduction structure (which limits the photocell field of view), and though it has a more restricted band width, it faithfully preserves the phase, amplitude, and frequency characteristics of signals within its pass band; it has been found to provide a substantial improvement in the minimum detectable signal level in the range of the intermediate and far-infrared wavelength wherein photoemissive surfaces are practically nonexistent.

9.6 ERRORS IN INTERPRETATION OF LASER NOISE

The field of noise measurement in lasers has been developing at a rapid pace, as has been the understanding of the phenomena associated with noise in lasers. Since the advent of the laser there have been a number of treatments of noise temperature or noise figure[54, 55, 7, 56]. Most of these have treated the specific device as a quantum system and have determined limiting-noise temperature arising because of amplified spontaneous emission. These investigations have consistently derived $h\nu/k$ for the minimum device-noise temperature.

Laser amplifiers are phase-preserving linear amplifiers and may be classified principally as voltage amplifiers[12]. We avoid here any discussion of the quantum-counter type of amplifier that does not preserve the signal phase and that has a zero-limiting noise temperature[57].

It is important in surveying the early literature on lasers that the reader distinguish between an author's application of the uncertainty principle of quantum mechanics to interpreting the results of a physical measurement rather than to the signal-amplification process itself[58, 59]. Some of the early papers are also criticized for lack of rigor with respect to their treatment of statistical averaging of signals. It is similarly important that the application of the uncertainty principle not be applied directly to the detector but rather to the results of the measurement process that the detector performs[59].

It can be argued quite generally and rigorously that a noiseless linear amplifier is physically unrealizable. By characterizing the amplifier as adding white noise to a linear, noiseless amplifier, we can derive the minimum noise temperature. The latter reduces to the earlier expressions for noise temperature when the gain of the amplifier is sufficiently large[12].

In addition to problems of rigor in theoretical papers interpreting noise in the literature, there are some pitfalls in the interpretation of the early measurements of spectral line width in which the laser noise output was interpreted as excess noise[33]. More extended experiments have, in general, concluded that it is plasma noise, not wave-interaction noise, that has been responsible for some of the narrow-band phenomena reported[60].

Whether the sources be excess noise or plasma noise, an excellent means now exists for determining the spectral line width of an operating laser.

9.7 SOURCES OF MORE EXTENSIVE TREATMENTS OF LASER-NOISE THEORY

The scope of the treatment of laser noise given above has been reduced to those essentials necessary for one to gain a rudimentary understanding of laser-noise sources and to enable one to handle the noise-measurements problem. The reader is referred to the literature for more extensive theoretical work on probability-density distributions before and after linear amplification[54, 61], channel-capacity characterization[62], and the theory of spatio-temporal coherence[63, 64].

Recently progress has been reported on the fully quantum-mechanical treatment of laser noise, including nonlinear effects[98–100]. The computation of the nonlinear line width, as determined by spontaneous-emission, yields essentially the linear line width multiplied by the factor 1/2:

$$\Delta\nu = \left[\frac{\pi(\Delta\nu_c)^2 h\omega_n}{P}\right] \frac{n_2/g_2}{(n_2/g_2 - n_1/g_1)_{thd}},$$

where $\Delta\nu_c$ is the passive-cavity band width and the second fraction contains the ratio of the density of the actual upper-state population to that of the threshold population.

9.8 MEASUREMENT OF MODULATION

9.8.1 Introduction

Optical carriers can be modulated in the same variety of ways as conventional carriers, that is amplitude (intensity) modulated, phase modulated, frequency modulated, or single-side-band modulated (frequency translated). The modulation may be generated internally in a laser cavity or by an external unit in the system. The mechanism of modulation may be electro-optic retardation, acoustic interaction, or any of a number of other phenomena. With the exception of the section immediately following the measurement techniques described herein involve actual detection of the modulation by direct or heterodyning techniques; these methods are then applicable to the determination of modulation, regardless of its means of generation. A very simple dc technique is first described that affects the indirect determination of the high-frequency retardation in an external electro-optic modulator. This introductory example treats a technique which may be the most useful of all, for many optical modulators are of the external, electro-optic type.

The techniques described in this chapter are, for the most part, appropriate specifically to modulation at microwave frequencies because only at such frequencies is the information-carrying potential of an optical carrier significantly exploited. Some of the measurement techniques will be readily recognized as being directly applicable at lower modulation frequencies.

In the final section of the chapter, an important microwave noise problem, which is basically unique to optical-carrier systems, is discussed. This relates to the detection of undesired radiative microwave "leakage" signals, which often swamp the desired, optically transmitted microwave signal, especially in laboratory environments where the optical transmitter and receiver are only a short distance apart. A means of circumventing this effect is described that employs a variation on the usual "phase-sensitive detection" technique.

9.8.2 Technique for the dc Measurement of Average-modulation Index

In most current applications, light modulation is achieved through the use of an external cavity electro-optic retardation modulator. The relative retardation along the two principal axes of an electro-optic crystal may be used in several ways, depending on the input-light polarization, to generate polarization modulation, intensity modulation, or phase modulation. In most instances, a knowledge of the electro-optic retardation is sufficient to determine the degree of modulation for any given mode of operation. In this section a means of determining the amount of the radio-frequency or microwave retardation by means of dc measurements is described. The technique amounts to a determination of the performance of the modulator, rather than a measurement of modulation on a beam under actual operating conditions in a specific modulation mode.

The use of the retardation thus obtained from the technique above, that is for determining the actual degree of modulation, will be given for phase modulation, amplitude (intensity) modulation, and single-sideband modulation. Because the justification of these relationships involves a detailed discussion of the theory of operation of electro-optic light modulators, the results will be stated for brevity without derivation. The interested reader may refer to representative references on optical-modulation techniques for further inquiry.

The basic measurement system used with optical-modulation techniques is shown in Figure 9.4. A well-collimated, input-light beam is linearly polarized at an angle of 45° with respect to an arbitrary set of XY-axes,

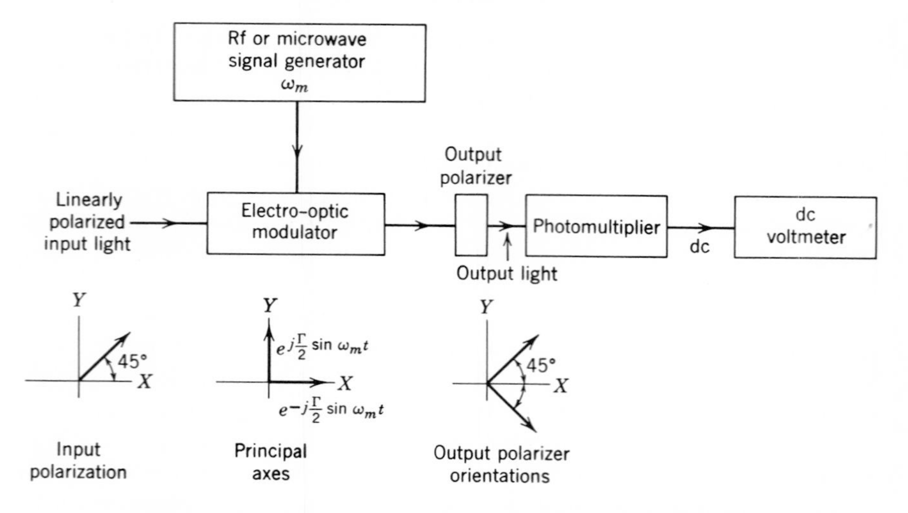

Figure 9.4. Direct-current measurement of electro-optic retardation.

and sent through an electro-optic modulator the electrically induced principal axes of which are along the XY directions.

The modulator may be of the cavity type, the capacitor type, or the travelling-wave type[65]. In principle it may utilize a cubic electro-optic material such as CuCl, ZnS, or HMT;[66, 67] or a uniaxial material such as KDP or ADP, in a longitudinal configuration (optical propagation along the c-axis). The fact that crystals of adequate optical quality do not exist today is considered transitory[71]. The modulator could consist of the more available materials KDP or ADP in a transverse configuration[68, 69]. However, because this measurement technique involves the propagation of two (perpendicular) polarization components through the crystal, even though the modulator is intended for single-polarization, phase-modulation use, we encounter the problems of natural birefringence and unstable optical bias, which limit the utility of the transverse KDP or ADP configuration for intensity modulation[70]. For the technique illustrated in Figure 9.4, the modulator is followed by a photomultiplier, which is connected to a sensitive dc voltmeter. Because the technique under discussion involves only the measurement of average intensities, the photomultiplier may be of the slow-response type.

The basic technique of measurement is as follows. With a no signal applied to the modulator, the output polarizer is rotated for a maximum indication by the photomultiplier. Call this position "a". Record the photomultiplier output voltage V_1. Because this represents "wide open" transmission of the laser beam, it will probably be necessary to use wavelength-calibrated optical attenuators in front of the photomultiplier. For instance, if an optical density 2 attenuator is used, we multiply the volt meter reading by 100, and call this value V_1. V_1 is now proportional to the total light intensity.

We next rotate the output polarizer 90°, to position "b," for a minimum indication by the photomultiplier. Ideally, the output light is now zero. Record the value of the voltmeter reading. Apply the desired modulation signal, at frequency ω_m, and measure the photomultiplier output V_2, which is proportional to the average intensity of that component of the input light that has been rotated 90° by means of the modulator. The relationship between V_1, V_2, and the retardation Γ (radians) is given by

$$\frac{V_2}{V_1} = \frac{1}{2}[1 - J_0(\Gamma)] \approx \frac{\Gamma^2}{8} \qquad (\Gamma \ll 1) \tag{9.57}$$

where J_0 is a zero-order Bessel function. Thus the ratio of the measured quantities, V_2 and V_1, may be used to determine the retardation Γ. The relationship between retardation and differential electro-optic path length L is given by

$$\Gamma = \frac{2\pi L}{\lambda} \tag{9.58}$$

where λ is the optical wavelength.

Certain practical problems have been neglected in the discussion thus far. The most important of these applies to the common situation of longitudinally-oriented KDP or ADP, and the fact that those components of the input light which are not perfectly parallel to the optic axis will encounter natural birefringence. In this case the rotation of the polarizer to position "b" will not result in the complete extinction of the output light before the modulating signal is applied. A "dark cross" pattern will be generated in the output light (which incidentally may be used to adjust the collimation and alignment of the beam and modulator crystal). Imperfect polarizers will also pass some output light under these conditions.

This residual light will usually be a negligible percentage of the total-input light. However, for low or moderate retardations, it will represent a high percentage of the total-output light with the output polarizer in position "b" and the modulator running. For this reason we should actually denote by V_2 the *increment* in output light when the modulator is turned on; this increment may be so small that it is best measured by chopping the modulator signal at a 1 kHz rate and viewing the photomultiplier output on an oscilloscope, (using a large dc-blocking capacitor). The resultant peak-to-peak square wave represents V_2. Another precaution is the obvious one of preventing extraneous light from entering the photomultipliers while making measurements.

The final problem involves the quasiperiodicity of the J_0 function in (9.57); if the retardation is large, it is necessary to increase the modulation level gradually so that the number of oscillations of J_0 (i.e. V_2) may be observed as Γ is slowly adjusted from zero to its full value.

9.8.3 Relationship between Retardation and Modulation Index

9.8.3.1 Phase Modulation In this mode of operation the modulator is used with input light that is linearly polarized along one principal electro-optic axis. Thus, if δ is the peak-phase deviation (phase-modulation index), we have

$$\Gamma = 2\delta. \tag{9.59}$$

9.8.3.2 Amplitude Modulation The relationship between the intensity modulation index and the retardation Γ is (neglecting harmonics):

$$\begin{aligned} m &= \frac{\text{ac intensity}}{\text{dc intensity}} = 2J_1(\Gamma) \qquad (9.60) \\ &\approx \Gamma \qquad\qquad (\Gamma \ll 1). \end{aligned}$$

For retardations much less than unity, we may write m directly in terms of V_1 and V_2. From (9.57) and (9.60) we have[65]

$$m = \left(8\,\frac{V_2}{V_1}\right)^{1/2}. \tag{9.61}$$

9.8.3.3 Single-side-band Modulation The relationship between the fractional intensity m, which is frequency-translated, and the retardation Γ, assuming that the modulator is a true rotating birefringent plate, is given by[66].

$$m = \sin^2\left(\frac{\Gamma}{2}\right). \tag{9.62}$$

9.8.3.4 Quasisingle-side-band Modulation The relationship between the fractional intensity m, which is frequency translated into the first upper or lower side band, and the retardation Γ in each of the two required modulators is[72]

$$m = 2\left[J_1\left(\frac{\Gamma}{\sqrt{2}}\right)\right]^2. \tag{9.63}$$

9.8.4 The Determination of Optical Alignment of Birefringent Crystals

Interference patterns obtained by passing divergent (or convergent) light through a birefringent crystal placed between two polarizers are

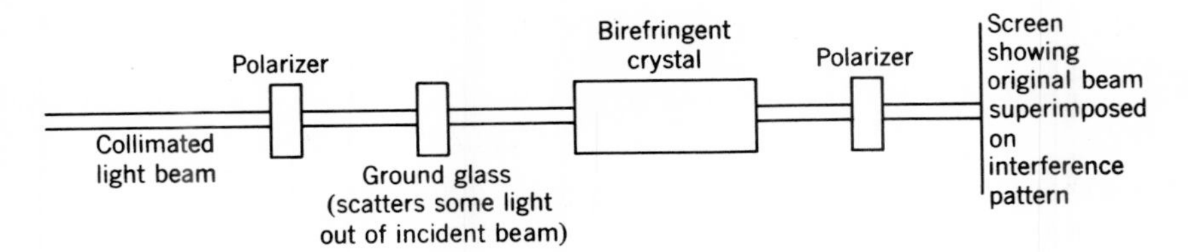

Figure 9.5. System for viewing birefringent interference.

not only interesting in their own right, but useful in aligning the crystal optic axis with respect to a collimated incident light beam (see Figure 9.5). A brief explanation of these interference patterns will make their utility apparent.

Consider a light beam incident on an optical system consisting of a polarizer, crystal, and analyzer. Let $\overline{OP}$ and $\overline{OA}$ be the direction of polarization passed by the polarizer and analyzer respectively. Let $\overline{D_1}$ and $\overline{D_2}$ be the orthogonal polarizations corresponding to the ordinary and extraordinary indices of refraction within the crystal. If ϕ is the angle between $\overline{OP}$ and $\overline{D_1}$, and χ is the angle between $\overline{OA}$ and $\overline{OP}$ (Figure 9.6), the transmitted intensity is given by[97]:

$$I = E^2\left[\cos^2 \chi - \sin 2\phi \sin 2(\phi - \chi)\sin^2 \frac{\delta}{2}\right] \tag{9.64}$$

where δ is the phase difference between the orthogonally polarized components after passing through the birefringent crystal. Note that if the crystal were absent ($\delta = 0$), the second term would be zero and therefore represents the effect of the crystal.

Two special cases are of particular interest. If the polarizer and analyzer are parallel ($\chi = 0$), the intensity becomes

$$I_{\parallel} = E^2\left(1 - \sin^2 2\phi \sin^2 \frac{\delta}{2}\right). \tag{9.65}$$

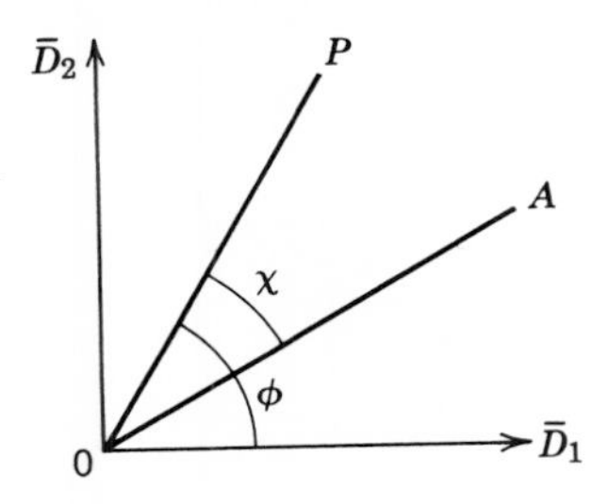

Figure 9.6. Definition of terms.

If the polarizer and analyzer are perpendicular ($\chi = \pi/2$), the intensity is

$$I_{\perp} = E^2 \sin^2 2\phi \sin^2 \frac{\delta}{2}. \tag{9.66}$$

Note that $I_{\parallel}$ and $I_{\perp}$ are complementary. It is, therefore, sufficient to consider the case for which $\chi = \pi/2$ (less pronounced interference effects are observed for $0 < \chi < \pi/2$).

The interference effects described above may be described in terms of curves of equal intensity. Curves of constant ϕ are called "isogyres," and curves of constant δ are called "isochromates." These two sets of curves are superimposed, but may be considered separately. In order to describe the interference patterns analytically, it is necessary to investigate the variation of ϕ and δ with the angle of incidence. This variation will be a function of the orientation of the crystal optic axis. For purposes of light modulation, two orientations may be used, either parallel or perpendicular to the direction of propagation. In both of these cases, the interference patterns discussed above provide an excellent means of aligning the crystal.

Consider a uniaxial crystal with plane-parallel end faces and with its optic axis perpendicular to these faces and parallel to the direction of desired optical propagation. Now examine an incident ray making an angle θ with the optic axis (Figure 9.7). The phase difference δ will be a function of θ, L, and the birefringence of the crystal. Note that all rays in the cone surrounding the Z-axis and making an angle θ with it will have the same δ upon emergence from the crystal and thus will form a circle of constant intensity, centered about the Z-axis on the viewing screen. Thus the isochromatic pattern for this case will consist of alternately light and dark concentric circles (for monochromatic light) the spacing of which decreases with increasing radii (Figure 9.8*a*). Note that the center of these circles corresponds to the direction of the optic axis. Simple geometrical considerations show that the isogyre pattern consists of a dark cross (corresponding to the directions of the polarizer and analyzer) centered at the same point as the isochromatic circles. Thus the alignment procedure, using the system illustrated in Figure 9.5, consists of adjusting the

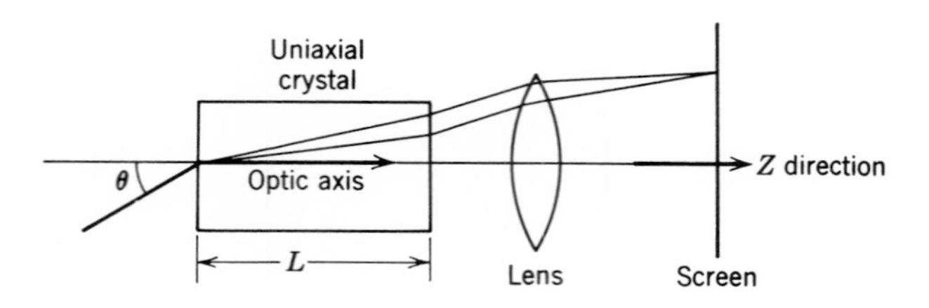

Figure 9.7. Stepup for uniaxial crystal.

Figure 9.8. Interference patterns. (*a*) the pattern obtained with the optic axis parallel to optical propagation. (*b*) The pattern obtained with the optic axis perpendicular to optical propagation. (Reproduced with permission from the Clevite Corporation.)

crystal until the transmitted beam is centered upon the interference pattern formed by the light that has been scattered out of the main beam. As a practical matter, this procedure is very easy and extremely accurate. Best results have been obtained when the end faces are parallel to 20 seconds of arc or less, this being well within the parallelism obtainable with present polishing techniques. It should also be pointed out that when the beam is aligned in this manner, the rate of change of δ with θ ($d\delta/d\theta$) is minimum. Therefore the effects of any slight divergence within the collimated beam are minimized.

When the optic axis is oriented parallel to the end faces (perpendicular to the direction of desired optical propagation), the interference pattern obtained is quite different. In this case consideration of the geometry shows that ϕ is independent of θ, the angle of incidence. There is therefore no isogyre pattern. However, ϕ should be chosen to be 45° in order that the coefficient of $\sin^2 \delta/2$ in (9.66) will be a maximum. This will give the best contrast for the isochromatic pattern. A qualitative argument to obtain the isochromatic pattern is not possible in this case. A complete treatment of the isochromatic patterns for an arbitrary orientation of the optic axis is available, and will not be repeated here[97].

For the optic axis oriented perpendicular to the desired direction of propagation, the isochromatic pattern consists of two sets of hyperbolas, as shown in Figure 9.8*b*. The center of this pattern corresponds to propagation perpendicular to the optic axis and can be used as described above for the alignment of the crystal.

The technique described above can also be used to align biaxial crystals with one optic axis parallel to the direction of optical propagation. The interference pattern of a biaxial crystal has two "centers," each corresponding to propagation along an optic axis. Again, these "centers" can be used to align one of the crystal optic axis parallel to the direction of optical propagation.

9.9 MEASUREMENT OF SINGLE-SIDE-BAND MODULATED LIGHT

Single-side-band, suppressed-carrier (SSBSC), optical modulation is advantageous for transmission of information in systems using optical heterodyne detection[66]. Using a single-frequency, high-level input, we can also obtain an efficient optical-frequency translation[73]. This has application for generating mixer input frequencies offset from, but coherent with, laser sources. In either case measurements of carrier and unwanted side-band suppression are of interest.

9.9.1 Generation of SSBSC Signals and Theoretical Light Outputs

In the first SSBSC modulator reported[72], two electro-optic crystals of KH_2PO_4 (KDP) were placed in succession along the light beam and were driven in phase quadrature with two equal signals; the crystals were oriented to fulfill the signal and optical phasing requirements for single-side-band generation. A left-circularly polarized input wave, with amplitude A results in a left-circularly polarized component in the output consisting of a carrier of frequency ω_e, and a series of even-ordered, double-side-band terms at frequencies $\omega_\ell + n\omega_m$ is the modulation frequency. The amplitudes of these left hand components are

$$L_n = \begin{cases} AJ_n(\Gamma/\sqrt{2}), & (n \text{ even or } 0) \\ 0, & (n \text{ odd}). \end{cases} \tag{9.67}$$

where Γ is the induced phase retardation in each crystal and J_n are the nth-order Bessel functions. All of these components are eliminated by passing the beam through a righthand analyzer which passes a series of odd-ordered single-side-band terms at frequencies $\omega_e + n\omega_m$ with amplitudes.

$$R_n = \begin{cases} \sqrt{2}AJ_n(\Gamma/\sqrt{2}), & (n = +1, -3, \ldots), \\ 0, & (\text{all other } n). \end{cases} \tag{9.68}$$

At small signal levels, therefore, only $\omega_\ell + \omega_m$ appears in the output with amplitude A $(\Gamma/2)$, but the higher-order harmonics must be considered at high drive levels.

A simpler SSB modulator than that described above utilizes a single cubic electro-optic crystal with the light beam travelling perpendicular to a (111) plane in which a modulating electric field of constant magnitude is made to rotate[73]. The output consists of a component at the carrier frequency circularly polarized in the same sense as the input wave with amplitude A cos $(\Gamma/2)$ and a single-side-band component oppositely polarized with amplitude A sin $(\Gamma/2)$, where A is the input light amplitude and Γ the phase retardation induced by the electric field.

If the applied rotating electric field is produced by two perpendicular field components, A and B, in the (111) crystal plane of the form

$$A = A_0 \sin \omega_m t,$$
$$B = B_0 \sin(\omega_m t + \varphi), \tag{9.69}$$

the unwanted side-band suppression will be

$$S = 10 \log_{10}\left(\frac{A_0{}^2 + B_0{}^2 - 2A_0 B_0 \sin \varphi}{A_0{}^2 + B_0{}^2 + 2A_0 B_0 \sin \varphi}\right) \text{ dB} \tag{9.70}$$

when the input light is circularly polarized. Optimum performance requires that $A_0 = B_0$ and $\varphi = 90°$.

9.9.2 Suppression of the Carrier and Its Measurement

In the suppressed-carrier mode of operation the electro-optic elements are followed by a polarization analyzer set to absorb all of the beam when no signals are applied. The degree of suppression attainable is limited by the analyzer quality and the random or inhomogeneous birefringence present in the optical elements due to strain or their naturally birefringent nature. Quantitatively, it can be defined as the ratio of the light intensity leaving the analyzer to that entering it under zero signal conditions; it is best measured in that way.

9.9.3 Measurement of Unwanted Side-band Suppression

Suppressed-carrier operation with a single modulating-frequency input should give a light output having a constant intensity. Any residual carrier or unwanted side band will appear as ac components on the intensity at frequencies ω_m and $2\omega_m$ respectively. If the ratio r of the peak ac photocurrent to the average photocurrent of a suitable detector can be measured, the suppression of either of these undesired components can be calculated using the relation

$$S = 10 \log_{10}\left(\frac{1 - \sqrt{1 - r^2}}{1 + \sqrt{1 - r^2}}\right) \text{dB}. \qquad (9.71)$$

At the higher modulation frequencies the measurement of the ratio r above is complicated by the variation of detector sensitivity with frequency, and a more satisfactory method for the measurement at higher temperatures is as follows. With the modulating signals off, the circularly polarized optical carrier leaving the electro-optic elements is passed through a plane analyzer and into a photodetector (such as a flat-faced photomultiplier). The dc photocurrent should, therefore, be independent of the analyzer orientation. Application of the signals produces an oppositely circularly polarized single side-band, which causes the total polarization state of the light entering the plane analyzer to have constant ellipticity with axes rotating at $\omega_m/2$.

An ac photocurrent at ω_m should be detected, the phase, but not amplitude, of which is dependent upon the analyzer orientation. If there is a minimum and maximum value of detected ac amplitude and the ratio of these two values is R, it results from an unwanted side band suppressed by

$$S = 10 \log_{10}\left(\frac{1 - R}{1 + R}\right)^2 \text{dB} \qquad (9.72)$$

below the wanted side band.

In the case of microwave-frequency operation of an SSB optical modulator, the side-band suppression can often be determined by isolating the side-band components using a spectrometer or scanning interferometer and measuring their intensities directly.

9.10 DIRECT DETECTION OF AMPLITUDE-MODULATED LIGHT

9.10.1 Introduction and Definitions

The key element in a system that demodulates light that has undergone amplitude modulation (AM) is the photodetector, the device that produces an electrical output in response to the optical input. Whether we are studying noise which appears as AM of a laser, evaluating the performance of AM light modulators, or designing a communications or radar system employing AM of an optical carrier, we must know the transfer function of the photodetector used. This transfer function is the crucial link between the detector output, which we observe, and the AM of the light-beam input, which we desire to know. Much of this section therefore, treats the determination of photodetector-transfer functions.

The limited space available here will be restricted to a discussion of direct detection of AM light. Much of what is said is, however, appropriate to optical heterodyning and homodyning or "coherent detection" in which the received light beam is photomixed with an optical "local oscillator." Optical heterodyne techniques are presently in a rapid state of development and should prove of great importance in future laser systems (see Section 9.12).

9.10.1.1 Modulation Index Photodetectors are square-law devices. The amplitude of the detector output occurs over a reasonable range of intensities directly proportional to the light intensity and, hence, varies as the square of the input amplitude. It is convenient to use an intensity modulation index m, which is defined for cosinusoidal modulation at a radian frequency ω_m, by

$$P_L = P_{0L}(1 + m \cos \omega_m t) \tag{9.73}$$

where P_L and P_{0L} are the instantaneous and average light powers, respectively. We note that $m \approx 2M$ where M is the familiar amplitude-modulation index. That is if the optical electric field is $E = E_0 (1 + M \cos \omega_m t)$, the intensity is proportional to

$$E^2 = E_0{}^2\left(1 + \frac{M^2}{2} + 2M \cos \omega_m t + \frac{M^2}{2} \cos 2\omega_m t\right).$$

9.10.1.2 Photodetector-transfer Function Photodetection may be conveniently divided into two parts:

1. The conversion of the incident light to "current" (e.g. photoelectric current in a vacuum tube or electron-hole pairs in a solid-state device).

2. Conversion of this current to output power at the modulation frequency.

In the first step, the current produced is

$$I = P_L \eta e/h\nu \tag{9.74}$$

where η is the quantum efficiency of the current-generating process (electrons out per photon in), e is the electronic charge, h is Plancks' constant, and ν is the optical carrier frequency. The amplitude of the ac current, which represents the modulation, is thus

$$i = mP_{0L} \eta e/h\nu. \tag{9.75}$$

The second step may involve a number of complex processes, depending upon the particular detector. It may involve, for example, the effects of current multiplication, current-carrier transit time, recombination, trapping, and so on. All these can be lumped into a single factor, an "equivalent resistance," R_{eq}[74], defined by

$$P_{\text{out}} = \tfrac{1}{2} i^2 R_{\text{eq}} = \tfrac{1}{2}(\eta e/h\nu)^2 R_{\text{eq}}(mP_{0L})^2 \tag{9.76}$$

where P_{out} is the signal power output of the photodetector at ω_m. The transfer function, which is in general a function of ω_m, is seen to be $(\eta e/h\nu)^2 R_{\text{eq}}$. R_{eq} is not, in general, the detector load resistance, although the two are identical in some cases. Consider, for example, a simple planar photodiode; at sufficiently low modulation frequencies R_{eq} is the same as the load resistor R_L, but at higher ω_m electron-transit-time effects can make R_{eq} much smaller than R_L.

9.10.2 Measurement Apparatus

9.10.2.1 Low Modulation Frequencies For AM at frequencies low enough that the photo-generated, ac current i can be measured directly, the modulation index m is given by the ratio of the measured peak ac current i to the average current

$$I_0 = P_{0\text{L}} \eta e/h\nu.$$

9.10.2.2 High Modulation Frequencies The major emphasis in AM-laser detectors has been on broad-band (and hence high-frequency) devices, chiefly because of the desire to exploit the broad modulation, band-width

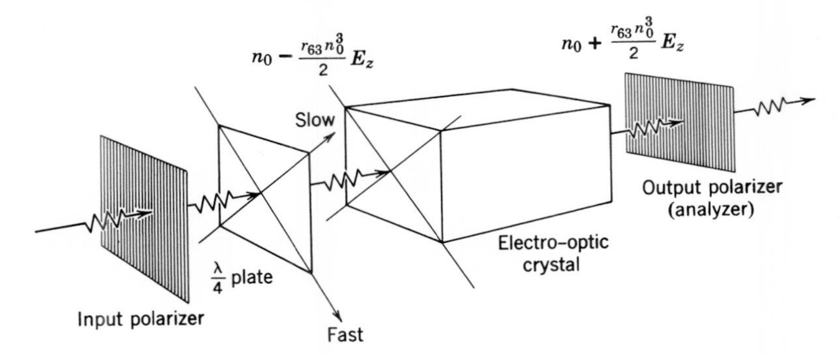

Figure 9.9. AM-light modulator employing the linear-electro-optic effect. The time-varying modulating voltage $E_z = E_{oz} \cos \omega_{mt}$ induces birefringence in the crystal, which is typically KDP or ADP. The output intensity is $P_L \sim P_{oL} (1 + m \cos \omega_{mt})$, where $m = 2 J_1(2\delta) \sim 2\delta$ for $2\delta < 0.7$. To measure m, the $\lambda/4$ plate is removed and the output polarizer rotated by 90°, whereupon the ratio of average output and input intensities with E_z applied becomes $P_{out}/P_{in} = 1/2 - 1/2 J_0(2\delta)$.

potential of optical carrier frequencies. At microwave frequencies, where most of the work has been done, direct measurement of i is not possible, and the photodetector transfer function $(\eta e/h\nu)^2 R_{eq}$ must be determined before quantitative measurement of m can be made.

The most direct approach to finding $(\eta e/h\nu)^2 R_{eq}$ is to measure the response to a light beam that contains a known amount of AM at the modulation frequency of interest[74]. Equation 9.71 tells us that

$$(\eta e/h\nu)^2 R_{eq} = 2P_{out}/(mP_{0L})^2, \tag{9.77}$$

so that three quantities must be known: P_{out}, m, and P_{0L}. A heterodyne radio receiver is usually convenient to determine the detector output P_{out}. Methods of measuring the average laser power P_{0L} are discussed above. For the popular AM modulator, shown in Figure 9.9 the modulation index m can be determined by measuring the ratio of average output and input intensities with the modulating voltage applied, the $\lambda/4$ plate removed and the output polarizer rotated by 90°[75]. (This technique is discussed in Section 9.8.2.) The major short-coming of this measurement procedure is that broad-band light modulators are not generally available, so the transfer function cannot be readily determined over broad ranges of modulation frequency.

A more versatile way of measuring R_{eq} is to use the so-called "shot-noise technique"[76, 77], which takes advantage of the fact that the mean-square, short-noise current per unit of band width is

$$< i_s^2 > = 2eI_0 \tag{9.78}$$

for many important photodetectors. Current multiplication complicates this expression slightly because of the statistics of the multiplication process. For n stages with gain M_0 per stage (9.78) should be multiplied by

$$M_0^{2n}(M_0 - M_0^{-n})/(M_0 - 1).$$

For large M_0 and n, this is just M_0^{2n}, representing simple multiplication of the initial ac-shot-noise current. This current produces a shot-noise power output

$$P_{\text{shot}} = 2eI_0 \Delta \nu R_{\text{eq}} \tag{9.79}$$

where $\Delta \nu$ is the receiver bandwidth. Thus a measurement of shot-noise power output versus frequency gives the valve of R_{eq} as a function of frequency. Obviously this method is not appropriate for detectors to which (9.78) does not apply. For example the generation-recombination noise in bulk photoconductors prevents the use of this approach.

Measurement of the quantum-efficiency factor $\eta e/h\nu$ is straightforward for those detectors for which R_{eq} may be determined by the shot-noise method; we simply measure the photocurrent generated by a light beam of the wavelength of interest and of known intensity P_{0L} and use (9.74). For detectors whose generated current cannot be determined (because of an unknown current multiplication factor, for example) the measurement of the direct-transfer function described earlier must be employed.

A typical substitution type of shot noise measuring apparatus is shown in Figure 9.10. The light chopping serves both to permit synchronous detection and to eliminate thermal noise, which is not square-wave modulated, and is therefore not detected. A point sometimes overlooked is that the substitution measurement should be made at both $f_{\text{LO}} + f_{\text{IF}}$ and $f_{\text{LO}} - f_{\text{IF}}$, where f_{LO} and f_{IF} are the local oscillator and intermediate-frequencies. This is because shot-noise at both these frequencies contributes to the IF output, and the mixer characteristic may differ at the two frequencies.

9.10.3 Examples

Our examples treat three microwave AM photodetectors which have received considerable attention[74, 76, 77, 78]: the fast PIN silicon photodiode, the CdSe bulk photoconductor, and the traveling-wave phototube (TWP).

9.10.3.1 Demodulation Technique Measurements of the photodetector transfer functions may be made by demodulating an amplitude-modulated,

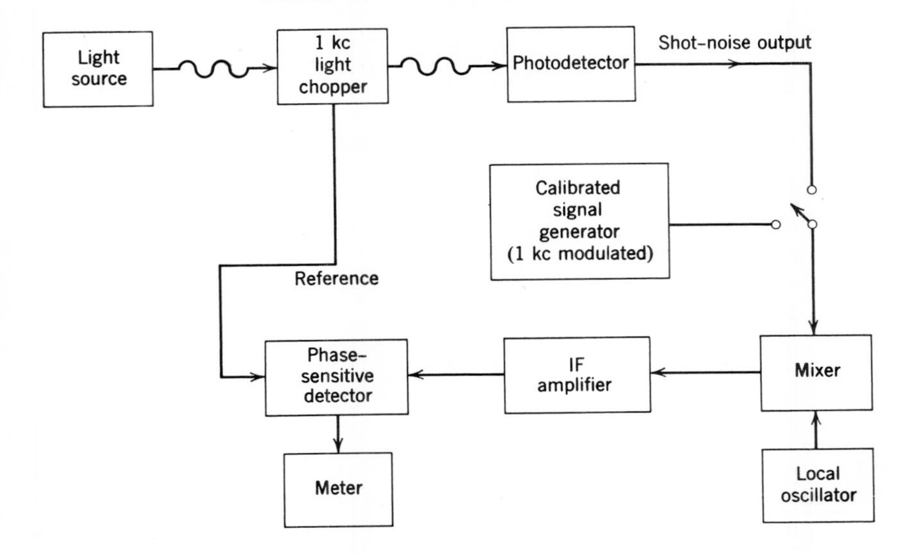

Figure 9.10. Shot-noise measurement apparatus.

633 nm, helium-neon, gas-laser beam[74,78]. In both cases, a narrow-band, 3 GHz, cavity-type, KDP modulator was used[75]. Space does not permit description of the three detectors here, but they were, briefly:

1. A PIN silicon photodiode operated in a tuned circuit[78].
2. A CdSe platelet operated in a tuned circuit[78].
3. A commercial S-1 cathode TWP, the Sylvania SYD-4302A, and an octave-band-width device that did not require tuning[79].

The data are summarized in Table 9.4; the reader is invited to check these numbers using (9.77). Newer TWP's with S-20 cathodes and longer interaction regions have $(\eta e/h\nu)^2 R_{eq}$ of about 780 (see Table 9.5).

9.10.3.2 Shot-noise Technique Shot-noise measurements of the transfer functions of PIN diodes and S-band TWPs with S-1 cathodes have also been made at 3 Gc with S-1 cathodes[78, 79]. (The reader is invited to use (9.79) to check the numbers in Table 9.4.) Once again the 633 nm, helium-neon laser wavelength is used in the calculations.

9.11 DIRECT DETECTION OF FREQUENCY-MODULATED OR PHASE-MODULATED LIGHT

This section deals with methods for the direct detection of frequency modulation (FM) or phase modulation (PM) on a laser beam. We limit the

TABLE 9.4 DEMODULATION-TECHNIQUE MEASUREMENTS OF PHOTO-DETECTOR-TRANSFER FUNCTIONS

Detector	P_{oL} (measured)	m (measured)	P_{out} (measured)	$(\eta e/h\nu)^2 R_{eq}$ [a]
PIN Diode [b]	1.5 mW	0.02	−57 dBm	4.4
CdSe Bulk Photo-Conductor [c]	1.5 mW	0.02	−104 dBm	8.9×10^{-5}
SYD–4302A TWP [d] (S–1 Cathode)	0.42 mW	0.12	−73 dBm	4×10^{-2}

[a] Calculated from (9–77).
[b] See reference [78].
[c] See reference [78].
[d] See reference [74].

discussion to techniques that do not require the use of an optical local oscillator, thereby excluding optical heterodyne and homodyne detection. These are discussed in Section 9.12. The methods described here are appropriate both for measuring undesired FM or PM occurring due to vibrations, and so on and for demodulating a laser beam that has been intentionally modulated.

The techniques and devices required for direct demodulation of FM or PM light are quite different from those used for demodulating AM light. The methods proposed to date may be placed in two general categories. The first is analogous to that used for demodulating angle-modulated signals at radio and microwave frequencies; namely, the FM signal is converted to an AM signal, which is then detected by an AM photodetector. The second method provides demodulation by converting the FM signal into a spatially

TABLE 9.5 SHOT-NOISE TECHNIQUE MEASUREMENTS OF PHOTO-DETECTOR-TRANSFER FUNCTIONS

Detector	P_{shot} (measured)	I_o (measured)	B (measured)	R_{eq} [a]	η (measured)	$(\eta e/h\nu)^2 R_{eq}$ [b]
PIN Diode [c]	−118 dBm	30μA	2.3 MHz	72 Ω	0.4	2.9
TWP with S–1 Cathode [d]	−92 dBm	1 μA	6 MHz	3.3×10^5 Ω	2×10^{-3}	0.3
TWP with S–20 Cathode				3×10^5 Ω	10^{-1}	780

[a] Calculated from (9–79).
[b] Calculated from the two preceding columns.
[c] See reference [76].
[d] See reference [77].

modulated signal (i.e. an optical beam the position of which in space depends upon the modulating signal), with subsequent detection by a photodevice which responds to a spatially varying signal.

9.11.1 Demodulation of FM Light by Conversion to AM Light

The current, most practicable method for the demodulation of FM light involves conversion of the FM to AM with subsequent detection by a conventional photodetector. Because photodetectors are readily available, the problem becomes one of performing conversion of FM to AM (i.e. discrimination) at optical frequencies[80].

At microwave-carrier frequencies, FM discriminators, based on the differential delay between two optical paths of different electrical length, have been successfully used. Two analogous schemes have been proposed for optical frequencies. The first of these utilizes a birefringent device termed the "birefringent disciminator"[81] wherein natural birefringence in a crystal produces the paths of different electrical length. In the second scheme[82], the incoming optical beam is split by a half-silvered mirror into two beams, which then travel different path lengths in air before being recombined by a second half-silvered mirror.

These two approaches are essentially just different embodiments of the same concept. We therefore focus our attention on the birefringent discriminator, keeping in mind that the same principles of operation apply to the beam-splitter discriminator.

The basic components of the birefringent discriminator are shown in Figure 9.11. They include an input polarizer, a naturally birefringent crystal,

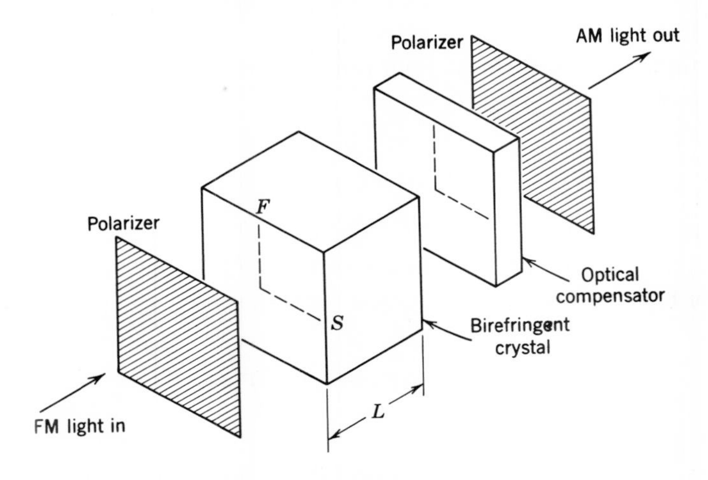

Figure 9.11. Birefringent discriminator.

and an output polarizer. An optical compensator will usually also be necessary immediately preceding or following the crystal. The crystal should have end faces that are optically flat and parallel, and its principal axes should lie in the plane of the faces.

The incoming frequency-modulated light passes through an input polarizer and is incident upon the crystal. The crystal's principal axes (denoted by the *S*- and *F*-axes of Figure 9.11) are at 45° with respect to the transmission axis of the input polarizer. The components of the light polarized along the *S*- and *F*-axes of the crystal are therefore equal in amplitude. Due to the birefringence of the crystal, the *F* and *S* components are no longer in phase when they emerge from the crystal. The output polarizer is oriented identically to the input polarizer. The amplitude of the signal transmitted by the output polarizer depends upon the relative phases of the emerging *F* and *S* components, which in turn depend upon the frequency of the light.

The frequency-transfer function of the birefringent discriminator is

$$C(\omega) = \cos(a\omega/2); \tag{9.80}$$

the quantity $a\omega$ is the difference in electrical-path length mentioned earlier and is directly proportional to the birefringence Δn and length L of the crystal:

$$a\omega = \frac{L\Delta n\omega}{c}. \tag{9.81}$$

To demodulate frequency-modulated light, the birefringent discriminator is followed by a square-law photodevice in which the photocurrent is proportional to the incident light power (and, therefore, to the square of the magnitude of the electric field):

$$I_{\text{out}}(t) = \frac{k}{2} E(t)E^*(t) \tag{9.82}$$

where k is a constant of the photodevice and $E^*(t)$ is the complex conjugate of $E(t)$. For an incoming FM optical signal given by

$$E_{\text{in}}(t) = \exp j(\omega_c t + \delta \sin \omega_m t) \tag{9.83}$$

where ω_c is the optical carrier frequency, ω_m is the modulation frequency, and δ is the modulation index ($\delta = \Delta\omega/\omega_m$), the output current from the photodevice will be

$$I_{out}(t) = \frac{k}{4}\left[1 - 2J_1\left(2\delta \sin \frac{a\omega_m}{2}\right)\cos \omega_m t + 2J_3\left(2\delta \sin \frac{a\omega_m}{2}\right)\cos 3\omega_m t \cdots\right] \tag{9.84}$$

where J_n is the nth-order Bessel function of the first kind. Equation 9.84 assumes that the optical compensator has been adjusted so that cos $a\omega_c = 0$. An equivalent statement is that the optical compensator should be adjusted so unmodulated light of frequency ω_c will be circularly polarized when it reaches the output polarizer. Note that 50 percent of the input optical intensity has been lost in the output polarizer.

From (9.84) we see that the photocurrent contains a dc term plus an infinite number of harmonics. We are interested in the cos $\omega_m t$ term because this is the component that corresponds to our input signal. The other harmonic terms represent distortion. We can minimize the distortion terms by making the crystal length L small enough so that

$$\frac{a\omega_m}{2} \ll 1, \tag{9.85}$$

and

$$a(\Delta\omega) \ll 1. \tag{9.86}$$

If the inequalities (9.85) and (9.86) are satisfied, we can replace sin $(a\omega_m/2)$ by a $\omega_m/2$ in (9.84), and each Bessel function can be approximated by the first term of its series representation. Equation 9.84 then becomes

$$I_{out}(t) \approx \frac{k}{4}(1 - a\Delta\omega \cos \omega_m t), \tag{9.87}$$

and the output amplitude is proportional to $\Delta\omega$, as desired.

The elimination of distortion terms is accompanied by a decrease in output amplitude. In order to make the terms small, $J_3, J_5, J_7 \ldots$ it is necessary to make J_1 quite small. Therefore, we should choose L no smaller than is necessary to justify the approximations used in deriving (9.87). (This is possible, of course, only if $\Delta\omega$ is known.)

The choice of a photodevice for a particular application should be carefully considered. The frequency and band width of the modulation will influence this choice to a large extent. In general, we should choose the device with the greatest sensitivity, provided it will satisfy the frequency response and band width requirements.

The birefringent discriminator is probably most useful for modulation frequencies in the microwave range, where convenient crystal lengths result. The lower the modulating frequency, the greater the length of birefringent crystal that is required. For sufficiently low frequencies, the required delay can no longer be supplied by a single crystal of reasonable length. It may then be necessary to use several crystals in series, or some form of the beam-splitter discriminator, mentioned earlier.

Let us return to (9.84) and suppose that we are concerned with maximizing the amplitude of the cos $\omega_m t$ term, rather than with minimizing distortion. Let us assume that the modulation index δ of the incoming signal is small. The choice of crystal length will be different for this situation than the previous one. To maximize the amplitude of the cos $\omega_m t$ term, L should be chosen so that

$$\sin \frac{a\omega_m}{2} = 1. \tag{9.88}$$

Substituting (9.88) into (9.84), we obtain

$$\begin{aligned} I_{\text{out}}(t) &= \frac{k}{4}\left[1 - 2J_1(2\delta)\cos \omega_m t + 2J_3(2\delta)\cos 3\omega_m t - \cdots\right] \\ &\approx \frac{k}{4}\left[1 - 2\delta \cos \omega_m t + \cdots\right] \quad (\delta \ll 1). \end{aligned} \tag{9.89}$$

According to (9.89), the output photocurrent now contains dc term, the desired cos $\omega_m t$ term, and an infinite number of odd harmonics. However, the coupling circuit that transfers energy from the photocurrent to the output will probably couple out only a few (if any) of the higher-order harmonics, especially at microwave frequencies.

Note that (9.88) limits the band width over which (9.89) is approximately correct. It may also be noted that the amplitude of the cos $\omega_m t$ term is no longer proportional to $\Delta\omega$. It is thus apparent that this choice of crystal length should be used only when the modulating signal contains a single frequency or reasonably narrow band of frequencies.

Finally, we note that it is possible to obtain optical discriminators, having performance characteristics superior to those of the birefringent discriminator, by using more complicated arrangements of polarizers and birefringent crystals. This was made possible by the development of a technique of optical-network synthesis utilizing birefringent crystals.[83].

With the technique described above it is possible to realize any desired transfer function having the form

$$C(\omega) = C_0 + C_1 e^{-ja\omega} + C_2 e^{-2ja\omega} + \cdots + C_n e^{-jna\omega}. \tag{9.90}$$

The required network will consist of n birefringent crystals and two polarizers. The quantities calculated from the synthesis procedure are the angles to which the crystals and the output polarizer are rotated.

9.11.1.1 Procedure for Measuring the Modulation Index of FM Light If the FM on the light signal consists of a single, known frequency, and if the modulation index is known to be small ($\delta < 1$), the following procedure should be used to measure δ:

1. The birefringent discriminator, shown in Figure 9.11, is followed by an appropriate photodevice. Adjust the optical compensator so that unmodulated light of frequency ω_c will be circularly polarized upon reaching the output polarizer. The length L of the birefringent crystal should be $L = \pi c/\Delta n\omega_m$. (If this results in an impractically long crystal, some type of beam-splitter discriminator with $a = \pi/\omega_m$ should be substituted for the birefringent discriminator.)

2. Pass the FM light through the birefringent discriminator and onto the photodevice. Measure the amplitude of the component of the photocurrent at ω_m and the dc photocurrent. If it is not possible to measure photocurrent at ω_m directly it should be calculated from

$$P_{\text{out}} = \tfrac{1}{2} i^2 R_{\text{eq}},$$

where P_{out} is the output power from the photodevice at ω_m, i is the amplitude of the photocurrent at ω_m, and R_{eq} is the "equivalent resistance" of the photodevice. Methods of measuring R_{eq} for various photodevices are discussed in section 9.10.

3. Divide i, the amplitude of the current at ω_m, by the amplitude of the dc current. The magnitude of this ratio is seen from (9.89) to be equal to $2J_1(2\delta)$, which enables us to determine δ.

If the FM on the light signal does not fall in the category above, that is if it contains more than one frequency, or if the modulation index is greater than unity (or completely unknown), the following steps should be used to measure δ:

1. The same procedure is used that we used in step 1, except the crystal length is changed so that $L << 2c/\Delta n\omega_m$ or $L << c/\Delta n\Delta\omega$, whichever is smaller. Because $\Delta\omega$ is unknown, the second inequality cannot be evaluated with certainty. It is, therefore, best to choose a conservative value for L using an estimated value of $\Delta\omega$. If two birefringent crystals are available, it would be desirable to perform the experiment twice, using different values of L (each of which is believed to satisfy the inequalties above), checking to see whether the same results are obtained.

2. The same procedure is used that we used in Step 2.

3. Divide i (the amplitude of the current at ω_m by the amplitude of the dc current). The magnitude of this ratio is seen from (9.87) to be equal to $a\Delta\omega$. Because a is known from (9.81), this gives Δw, which permits the calculation of δ.

9.11.2 Demodulation of PM Light by Conversion to AM Light

A variation on the system we described of the previous section will convert phase-modulated (PM) light to amplitude-modulated light[84]. In this approach a second long birefringent crystal is introduced immediately following the optical compensator illustrated in Figure 9.11. This second crystal is oriented with its principal axes at 45° with respect to those of the first crystal, and the output polarizer is omitted. If the output light is incident on a photodevice, the output current corresponding to (9.84) is given for the present case by

$$I_{\text{out}}(t) = \frac{k}{2}\left[1 - 2J_1\left(2\delta \sin\frac{a\omega_m}{2}\right)\sin\frac{b\omega_m}{2}\sin\omega_m t + \cdots + (\text{odd harmonics})\right] \tag{9.91}$$

where k, δ, and a are defined as in the preceding section, and $b\omega$ is the difference in optical-path length corresponding to the second crystal. We assume that the optical compensator is adjusted as in the preceding section, that is $\cos a\omega_c = 0$. Note that we have regained the 3 dB light loss which resulted from the use of an output polarizer in the FM case.

Now let us suppose that the modulation index δ —which is now independent of modulating frequency—is much less than unity and that we wish to achieve a maximum AM output while tolerating some harmonic distortion. We then choose a and b such that the following relationships are satisfied:

$$\sin\frac{b\omega_m}{2} = 1,$$

and

$$\sin\frac{a\omega_m}{2} = 1. \tag{9.92}$$

This results in an output photocurrent given by

$$\begin{aligned} I_{\text{out}}(t) &= \frac{k}{2}[1 - 2J_1(2\delta)\sin\omega_m + \cdots] \\ &\approx \frac{k}{2}[1 - 2\delta\sin\omega_m t] \qquad (\delta \ll 1). \end{aligned} \tag{9.93}$$

The latter expression in (9.93) represents ideal total conversion of PM to AM light, to the extent that (9.92) is satisfied over the range of frequencies present, and harmonic distortion may be neglected. The frequency characteristic introduced by the breakdown of the conditions (9.92) results in a limitation to an octave band width[84].

It is instructive to compare (9.93) with (9.89). Both relationships exhibit the dependence on δ(rather than $\delta\omega_m$), which is characteristic of phase modulation. Thus it may logically be asked, In what sense is (9.89) a representation of FM conversion, while (9.93) is asserted to represent PM conversion? The answer is to be found in the phase of the photocurrent at the modulation frequency(s); for an input PM signal given by (9.83), (9.89) represents a 90° phase distortion in all components of the modulation.

Now, let us suppose that the modulation index δ is not small. In order to hold the harmonic distortion to a reasonable level, we choose the length L of the first crystal so that

$$\frac{a\omega_m}{2} \ll 1, \qquad \delta a\omega_m \ll 1, \tag{9.94}$$

while b is chosen as in (9.92). The output photocurrent is then given by

$$I_{\text{out}}(t) \approx \frac{k}{2}\,[1 - \delta a\omega_m \sin \omega_m t]. \tag{9.95}$$

Because the sin $\omega_m t$ term is now proportional to the modulation frequency, this scheme must be used only with a narrow band of frequencies. In fact, the dependence on $\delta\omega_m$ is like that for an FM signal, but we consider this to be a PM converter rather than an FM converter for reasons of phase distortion as discussed above.

A practical problem in the use of the double-birefringent crystal scheme is the instability of effective optical bias due to thermal effects in the crystals. One solution to this problem is to maintain the temperature of the crystals constant, for example to within 10^{-2} °C in calcite. Otherwise, the polarization of the output optical signal will vary in a random fashion, and the photodevice giving rise to the output current in (9.91), (9.93) or (9.95) must be insensitive to polarization. Variations in the effective optical bias of the first crystal are assumed to be adjusted out by means of the optical compensator illustrated in Figure 9.11; an alternative means involves a quarter-wave plate and rotation of the second crystal, Senarmont compensation[84].

9.11.3 The FM Phototube

Direct detection of microwave frequency-modulated light, without conversion to AM light, has been demonstrated[85, 86] using a transverse-

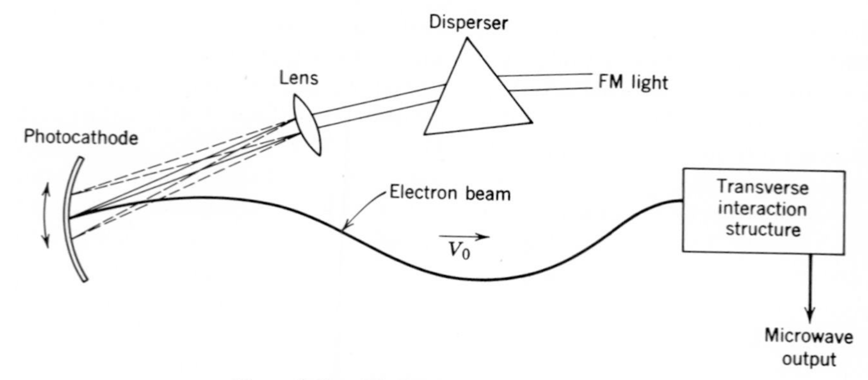

Figure 9.12. The FM phototube system.

wave phototube in conjunction with a dispersing system. Because such devices are not yet commercially available, a short description of their design and operation is appropriate.

The basic system is illustrated in Figure 9.12. A frequency-modulated light beam is incident on an optical dispersing element (prism, diffraction grating, and so on), so that, in the simplest viewpoint, the output beam is swinging back and forth with an instantaneous angle that is proportional to the instantaneous frequency. This angle-modulated beam is then incident on a photocathode, so that a transverse "synchronous" electron-beam mode is generated[87], which then couples power into a transverse-wave interaction circuit.

9.11.3.1 The Dispersing System In the original proposal for an FM phototube system[88], it was recognized that the instantaneous-frequency viewpoint must be applied with caution. A frequency-modulated signal is rigorously describable in terms of a Bessel series of steady-state side bands, and from this viewpoint, we would expect the output of the disperser in Figure 9.12 to be composed of continuous beamlets at discrete angles. This is, in fact, the case, if each monochromatic component of the input light (i.e. each side band) is sufficiently resolved by the disperser that it does not overlap (in the far field) with other monochromatic components. It is, therefore, necessary to design the disperser so that side bands overlap appreciably on the photocathode and so that their mixing produces the net effect of a spot that is swinging back and forth.

For simplicity, let us assume a single-frequency modulating signal, so that the modulated-input light may be expressed as

$$E(t) = \sqrt{2}\exp(j\delta \sin \omega_m t)\exp j\omega_c t \tag{9.96}$$

where ω_c is the optical-carrier frequency and ω_m is the modulating frequency. In a properly designed interaction structure, the output will depend only on the first moment of the net light intensity or photocurrent. It may be shown[88] that a necessary and sufficient condition for the first moment to duplicate the modulating signal is that adjacent FM side bands overlap appreciably, that is that

$$k\omega_m \ll \Delta\chi \tag{9.97}$$

where $\Delta\chi$ is the nominal width of the disperser line shape on the photocathode for any given side band and k is the dispersion constant (e.g. in cm per Hz) relating the separation of sidebands to their frequency difference. If this condition is satisfied, the first moment of the photocurrent $m(t)$ may be written

$$m(t) = Kkf'(t) = Kk\delta\omega_m \cos \omega_m t \tag{9.98}$$

where K is a constant involving the quantum efficiency of the photosurface, and so on. Note that (9.98) represents perfect reproduction of the frequency modulation by the first moment.

A more severe condition on spectral overlap is required if we are to ensure that the spot mentioned above is substantially constant in size and shape as it swings back and forth in accordance with the modulation. This may be important because of the finite aperture of the phototube or interaction circuit. The necessary condition for strict applicability of the instantaneous-frequency viewpoint is[85]

$$k\delta^{1/2}\omega_m = k\sqrt{\omega_m \omega_d} \ll \Delta\chi \tag{9.99}$$

where $\omega_d = \delta\omega_m$ is the total frequency-deviation.

The choice of a disperser thus depends on the modulation frequency and the total frequency deviation. For a diffraction grating, the dispersion $\delta\chi$ on the photocathode is given by

$$k = \frac{\partial\chi}{\partial\omega} = \frac{\ell M\lambda^2}{cd2\pi} \tag{9.100}$$

where ℓ is the lever arm, M is the order of the interference, c is the velocity of light, and d is the separation of the grating elements. The nominal spot width for a given side band is

$$\Delta\chi = \frac{\ell\lambda}{Nd} \tag{9.101}$$

where N is the number of grating elements. To satisfy (9.97), therefore, we require that

$$NM \frac{\omega_m}{\omega_c} \ll 1 \tag{9.102}$$

whereas, if aperturing is a problem requiring a spot that is invariant in size and shape, we must satisfy 9.99:

$$NM\delta^{1/2} \frac{\omega_m}{\omega_c} \ll 1. \tag{9.103}$$

The conditions (9.102) and (9.103) place definite limits on the allowable resolving power NM of the grating. These same relationships apply for other types of dispersers, such as a prism. To obtain a large linear spot movement, we desire a large M and d; however, it may be shown that the signal-to-noise output of the tube, when limited by shot noise, depends only on δ [85, 86].

It should be noted that a combination of two lenses (a telescope) may be used in place of the single lens to magnify the angular swing and spot size. That is, the use of a telescope of power P_0 will increase the effective lever arm by a factor P_0. This may be a great convenience.

9.11.3.2 The Interaction Structure Several configurations are possible for achieving a slow-wave curcuit with transverse electric fields that will interact with the transverse-electron modes. It is important to note that, although the excitation of the photocathode is such that the beam is apparently undulating back and forth, the individual electrons have no transverse velocity, and therefore they undergo rectilinear motion through the interaction structure, with uniform velocity v_0. A schematic drawing of the phototube is shown in Figure 9.13, where the large, longitudinal dc

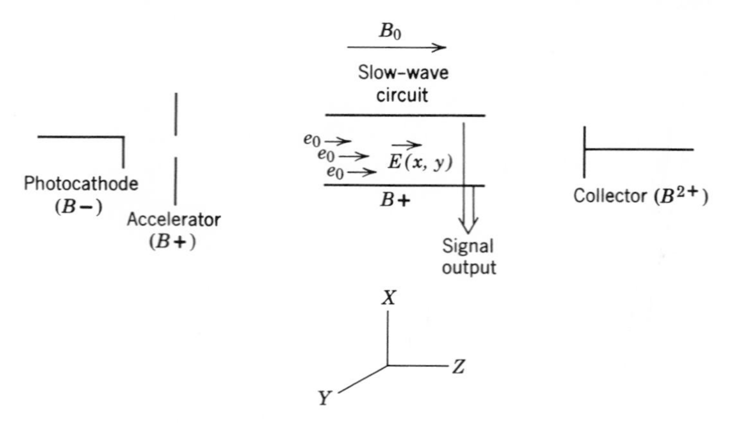

Figure 9.13. Schematic diagram of an FM phototube.

magnetic field B_0 assures this rectilinear motion. The light may be passed through a hole in the collector, and hence through the interaction region.

Some of the more important configurations include a helix-on-rod circuit[89], the use of the -1 mode in a conventional hollow helix[90], and the use of two parallel helices[91]. The helix-on-rod circuit is wide band, has a large interaction impedance, and is relatively easy to construct. It has the disadvantage, however, that there is no symmetry point at which the longitudinal field is zero, which means that AM light and shot noise are not suppressed in an optimum manner[85]. Various means for parametric amplification also exist but will not result in an impoved signal-to-noise ratio, when limited by shot noise.

Under usual conditions, space-charge interaction may be neglected in a transverse-wave phototube. In this case the power output is given by

$$P = \frac{z^2 \beta_e^4 K_t}{4} \left[\underline{R}_t \middle| \cdot \int_A \int \underline{r}_t J_1(x, y)\, dz\, dy \middle| \right]^2 \tag{9.104}$$

where z is the length of the circuit, $\beta_e = \omega_m / r_0$ is the electronic-propagation constant, $\underline{R}_t$ is the unit vector in the direction of the transverse field (which is assumed uniform), $J_1\ (x, y)$ is the peak ac current density and $\underline{r}_t = x\underline{x} + y\underline{y}$ is the position vector at point (x, y). K_t is the transverse-interaction impedance, given in terms of the peak-transverse field by

$$P = \frac{|E_x^2| + |E_y|^2}{4\beta_e^2 K_t}. \tag{9.105}$$

It is important to note that the integral in (9.104) is simply the first moment represented by (9.98), so that the ac-current density J_1 need not be explicitly determined. Thus the power output may be written

$$P = \frac{z^2 \beta_e^4 K_t}{4} (Kk)^2 \delta^2 \omega_m^2 = A^2 \delta^2 \omega_m^2 I_0^2, \tag{9.106}$$

where I_0 is the dc photocurrent, and we assume that the field and moment are colinear; A^2 is a calibration constant that must be determined. Because A^2 involves many parameters that are difficult to predict with accuracy, it is best determined experimentally, using a known value of modulation index δ at each modulating frequency of interest. A^2 may also then be considered to contain the effects of loss on the helix. Once A^2 is determined, (9.106) is used with an FM light source with an unknown index of modulation to determine δ.

A major precaution is that aperturing of the light or electron beam, due to the use of too large a spot or misalignment and due to the effective amplitude modulation, which results from the cutting-off of part of the reciprocating beam, will result in the generation of a spurious signal. Also the helix output must be impedance matched at the microwave frequency of operation, either by the design of a broad-band matching structure, or the use of a stub-tuner arrangement at each frequency of operation. Finally, the accelerating voltage must be carefully adjusted and highly regulated, so that the velocity of the circuit wave will match that of the electrons.

It may be mentioned that a conventional AM, traveling-wave phototube might possibly be used in a -1 mode for the narrow-band detection of the transverse-electron modes. In this case the signal output is at the photocathode end, and the matching structure which is designed into the tube for the 0th mode, must be supplemented by external tuners.

9.11.4 Aperture Method

The approach of dispersing an FM light signal to create an angle-modulated beam (see Section 9.11.3) suggests a routine technique for detecting FM light. If the dispersing element is chosen so that the spot moves a small amount compared to its own size and has an invariant size and shape as it moves, then we may aperture part of the spot, creating an effective intensity modulation beyond the aperture. This intensity modulation will be linearly related to the frequency modulation because of the small spot movement and may be detected by means of a conventional microwave AM light detector (Section 9.10). The AM modulation index will be inherently small, but easily detectable signals are readily obtained.

It is apparent from the preceding section that we must satisfy both of the following conditions:

$$NM\delta^{1/2}\frac{\omega_m}{\omega_c} \ll 1, \tag{9.107a}$$

$$NM\delta\frac{\omega_m}{\omega_c} \ll 1. \tag{9.107b}$$

It is suggested that a spot movement of 1 percent of the spot size is appropriate. Hence the diffraction element should be chosen so that the resolving power is approximately given by

$$NM \approx 10^{-2}\omega_c/\delta\omega_m, \tag{9.108}$$

providing that δ is not so small that this violates (9.107a).

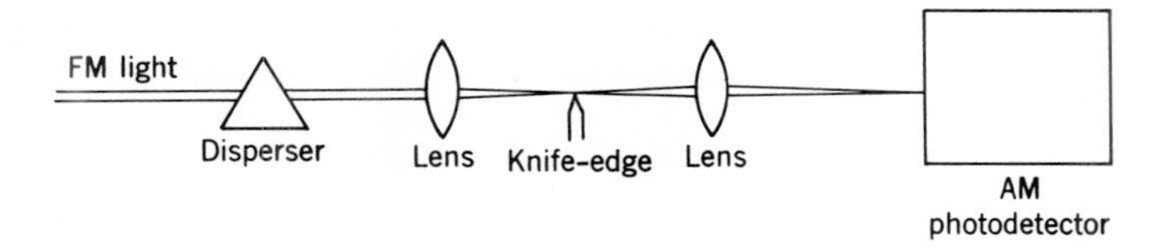

Figure 9.14. Aperture detection of FM light.

The experimental arrangement for applying the aperture method is shown in Figure 9.14. The aperture is conveniently achieved using a razor blade mounted on a micropositioner. It is inserted approximately halfway into the light beam, with the linear dimension of the razor blade perpendicular to the direction of the spot movement. This insertion is done at the focal point of a telescope, so that the aperture will case a quasigeometric shadow on the (far-field) photodevice, free of diffraction effects. This supposes that redefraction by the second lens is negligible, which is easily achieved with laser light.

If the conditions above are satisfied, the output current from the AM photodevice may be written

$$\begin{aligned} i(t) &= I_0(1 + A'\delta\omega_m \cos \omega_m t)^2 \\ &\approx I_0(1 + 2A'\delta\omega_m \cos \omega_m t) \end{aligned} \tag{9.109}$$

where $A'\delta\omega_m \ll 1$ and A' must be determined by using a known δ.

9.12 MEASUREMENT OF OPTICAL MODULATION BY HETERODYNE DETECTION

9.12.1 Introduction

Heterodyne detection of optical modulation is an interference or mixing technique made possible by the square-law characteristic of power detectors. In optical heterodyne detection the signal wave is combined with a second wave generated within the receiver, and the combined waves are caused to interfere on the surface of the photodetector. The second wave is known as the local oscillator (LO) wave because of the obvious similarity between this technique and conventional mixing at radio frequencies. If the signal and local oscillator waves are of amplitudes E_1 and E_2 and frequencies ω_1 and ω_2, respectively, the resulting field on the detector is

$$E(t) = E_1 \cos \omega_1 t + E_2 \cos(\omega_2 t + \phi) \tag{9.110}$$

where ϕ is the relative optical-phase angle between the waves at one point in time. Because the detector-output current is proportional to input-optical power, the current varies as the square of the field, and thus

$$i(t) \propto E^2(t) = (E_1 \cos \omega_1 t)^2 + [E_2 \cos(\omega_2 t + \phi)]^2 + 2E_1 E_2 \cos \omega_1 t \cos(\omega_2 t + \phi). \tag{9.111}$$

The first two squared terms are the values of $E^2(t)$ with either the signal or the local oscillator alone illuminating the detector; the third term contains the products of the mixing that occurs when both waves are present simultaneously. Because the detector output does not follow temporal variations at optical frequencies, the first terms represent a dc-integrated output. The mixing term contains equal spectral components at the sum and difference frequencies. It is apparent that the detector response at the sum frequency will be integrated because it falls in the optical band; however, the difference frequency can be much lower and may be detected as a time-varying electrical signal. The expression for this difference or intermediate frequency (IF) signal is

$$i_{\text{IF}} \propto E_1 E_2 \cos[(\omega_1 - \omega_2)t - \phi]. \tag{9.112}$$

Therefore, if I_1 and I_2 are the integrated dc outputs obtained with the signal and the local oscillator separately, the intermediate-frequency current becomes

$$i_{\text{IF}} = 2\sqrt{I_1 I_2} \cos[(\omega_1 - \omega_2)t - \phi]. \tag{9.113}$$

The rms magnitude of the IF current is $\sqrt{2I_1 I_2}$. It is clear that a weak signal, corresponding to I_1, can be amplified greatly if detection is performed at the difference frequency with a relatively powerful local oscillator, corresponding to I_2. The ultimate limit in sensitivity to be obtained in this way is reached when the shot noise in the photocurrent becomes larger than other sources of noise in the receiver. If the local oscillator is much stronger than the signal, this shot-noise current can be approximated by

$$i_n \approx \sqrt{2eI_2 \Delta v} \tag{9.114}$$

where e is the charge on the electron and Δv is the electrical bandwidth of the receiver. The power signal-to-noise ratio obtained in such an ideal case would be

$$\text{SNR} = \frac{i_{\text{IF}}^2}{i_n{}^2} \approx \frac{2I_1 I_2}{2eI_2 \Delta v} = \frac{I_1}{e\Delta v}. \tag{9.115}$$

Evidently, an ideal heterodyne receiver should provide unity signal-to-noise ratio with only one signal photoelectron per cycle of receiver bandwidth[92].

9.12.2 Modulation Detection

If a modulated signal wave is present in the receiver instead of the simple single-frequency wave of (9.110), the heterodyne-detection process can be analyzed by considering the modulated signal as a carrier and an array of side bands. Each spectral component of the signal produces an IF current, described by (9.113), with I_1 corresponding to the dc level that would be produced by each spectral component alone. If the spread of modulation side bands is less than the offset or frequency difference between the local oscillator and the carrier signal, the total effect of the heterodyne process is to reproduce the modulation spectrum of the optical signal in the photocurrent at a much lower intermediate frequency. As (9.112) indicates, the relative amplitudes and phases are preserved through this transformation. The problem of detecting modulation at the intermediate frequency is generally less difficult than detection of the modulation directly on the optical carrier because the IF signal may be observed on a spectrum analyzer or detected by any of a number of highly developed electronic techniques. Narrow-band filters, envelope detectors, and discriminators are available for receiving and measuring the modulated information. Thus the heterodyne technique is optically similar for amplitude, phase, or frequency modulation on the signal because the demodulation apparatus is electronic rather than optical.

9.12.3 Advantages and Disadvantages of Heterodyne Demodulation

Several advantages of heterodyne demodulation are pointed out above. The method theoretically provides noise-free amplification of the signal and permits demodulation at a more convenient frequency. Electronic filtering may be used to narrow the receiver band width far below that obtainable with practical optical filters. Straightforward electronic processing of the signal is possible. Another feature of the optical heterodyne is its ability to limit the angular field of view of the receiver. This follows from the fact that the relative phase between signal and LO waves is reproduced in the photodetector current. If the phase fronts of both waves are not parallel at all points on the detector, the components of the IF current from different areas of the detector will have different phases and will tend to cancel. The geometrical manner in which the LO illuminates the detector can be used to select the angle over which signal radiation is efficiently detected. This selectivity can be very sharp. Suppose the detector is 1mm in diameter and illuminated by plane local oscillator waves 1μm in length. Then the angular misalignment of a plane signal wave, which would produce a phase shift of 180° in the output signal from opposite sides of the detector, is 0.5μm per millimeter, or 0.5 milliradian. This is illustrated in Figure 9.15.

The disadvantages of optical heterodyning as a means of detecting modulated light derive from the sensitivity of the technique. The sensitivity to the optical phase makes it necessary to take great care in aligning the components in the receiver, so that the signal and local oscillator waves mix in phase across the detector. All of the lenses or mirrors used to process the signal and LO light before they are combined must be perfect to within a fraction of an optical wavelength for efficient mixing. This requirement of phase uniformity also extends to the atmosphere which may lie in the optical path of either or both beams before combination.

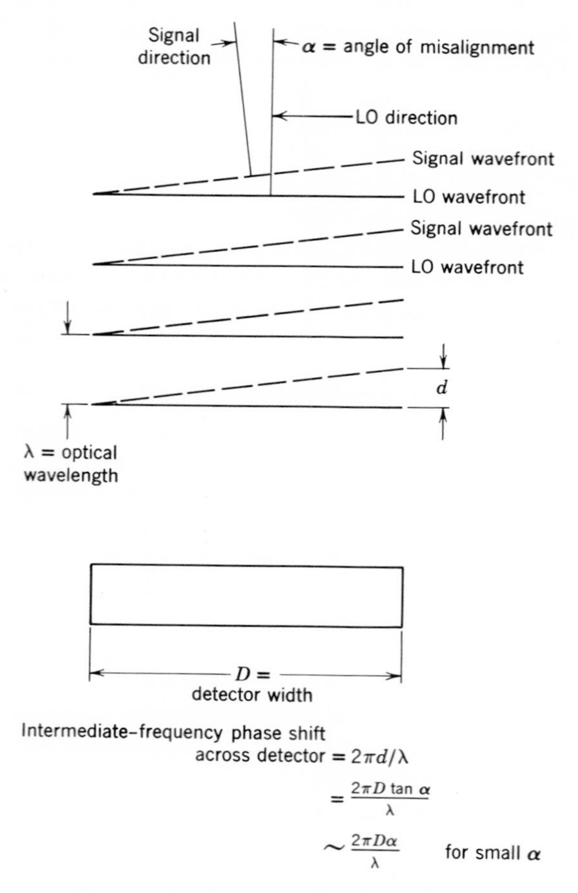

Figure 9.15. Heterodyne angular sensitivity.

A more difficult problem arises because of the large ratio that exists between the optical frequencies to be detected and the IF electrical band width. The frequency of the LO and signal sources must be stable to within a small fraction of the electrical band width of the receiver, or the beat frequency will drift out of the pass band. Sophisticated frequency-stabilization systems are necessary to control the laser drift even to within a few megacycles. Frequency translating devices[72, 93, 94] can also be used on the local oscillator to provide automatic frequency control[95].

Another difficulty is encountered in heterodyne detection if the lasers providing the signal or local oscillator waves are oscillating in several frequency modes. In that case, an IF signal will appear at the intermode frequency, and overlapping "images" of the signal spectrum may lie in the receiver pass band. It should be noted that in practice there may be excess noise above the shot noise on the output of the local oscillator laser, due to mechanical vibrations, discharge noise, and other effects. If the noise falls within the receiver pass band, the heterodyne signal-to-noise ratio will be lower than the theoretical optimum value.

9.12.4 Experimental Procedure for Heterodyne Detection

Because of the difficulties we mention in Section 9.12.3, it is not possible to outline the steps necessary for successful heterodyne detection for all cases. Equipment in general use is not sufficiently refined at present to permit heterodyning to be carried out in a straightforward manner over a wide range of modulation frequencies, frequency differences, and receiver band widths. However, there are two extreme cases in which the experimental procedure can be described.

One case is detection of a microwave-modulated signal using a local oscillator offset by a greater microwave-difference frequency. The offset may be obtained by passing the local oscillator wave through an electro-optic frequency translator or single-side-band modulator. Sufficient receiver band width is necessary to avoid problems of laser drift. Modulation can be detected in this case as follows:

1. Align the signal and local oscillator light beams so that the wave fronts are parallel. This may be accomplished by superposition of the images of the two sources in an external magnifying optical system.
2. Select a detector with sufficient electrical band width to receive the modulation side bands. The detector should be followed by a wide-band amplifier providing enough gain to allow the signal modulation to be detected by a discriminator, if the phase or frequency is modulated, or by an average or envelope detector, if amplitude modulation is present.
3. Adjust the level of the local oscillator power to increase the IF-signal strength until the level of the receiver noise begins to increase. At this

point, maximum heterodyne sensitivity is obtained. If insufficient power is available, increase the IF signal to the largest value possible.

4. Analysis of the modulation can now be made using conventional radio-frequency techniques. If the modulation can be temporarily removed from the signal, it is possible to check the heterodyne efficiency by measuring dc photocurrents and comparing the experimental results with (9.113).

The second case is one in which the signal and local oscillator waves are produced from the same laser, as in an optical radar experiment. In this case drift is not a problem, but the modulation band width and offset frequency must be much less than the intermode spacing if ideal results are to be obtained using a multimode laser. Frequency translation is necessary to avoid distortion of the modulation spectrum, which occurs if the offset is less than the largest important side-band frequencies. The procedure for heterodyne demodulation in this case is similar to the one described above, but the band width of the detector and receiver electronics need not be wider than the modulation to allow for drift.

9.13 REJECTION OF COHERENT INTERFERENCE

A common problem in the laboratory testing of optical modulation-demodulation schemes is the leakage of microwave power from the modulating system and its pickup by the detection system. The usual scheme of audio-rate "chopping" of the light or optical detector does not discriminate against such "coherent interference" because of the cross product between the chopped signal that is desired and the leakage pickup, which is generated in the crystal-mixer or detector elements of the system. The techniques of shielding that are usually used to circumvent this problem are costly, time consuming, and rarely successful enough to eliminate the effect entirely.

A combined phase-sensitive-detection (PSD) and homodyne system has been used to eliminate the problems of leakage[96]. The system is shown in Figure 9.16. Due to the use of a modulating type of mixer, which presents a linear load, cross products are not generated. The system has successfully rejected coherent interference and chopped and unchopped noise several orders of magnitude larger than the desired signal, resulting in precise quantitative measurements free of "hand effect," and other disturbances.

Figure 9.17 shows a simpler system, in which the coherent interference signal is swamped by a deliberately introduced, controllable interference signal. This "pseudo-homodyne" system has the disadvantage that impedance changes in the optical detection element, which may be due to chopping of the light beam, result in a chopped component of that part of the deliberately introduced interference signal that enters the line to the

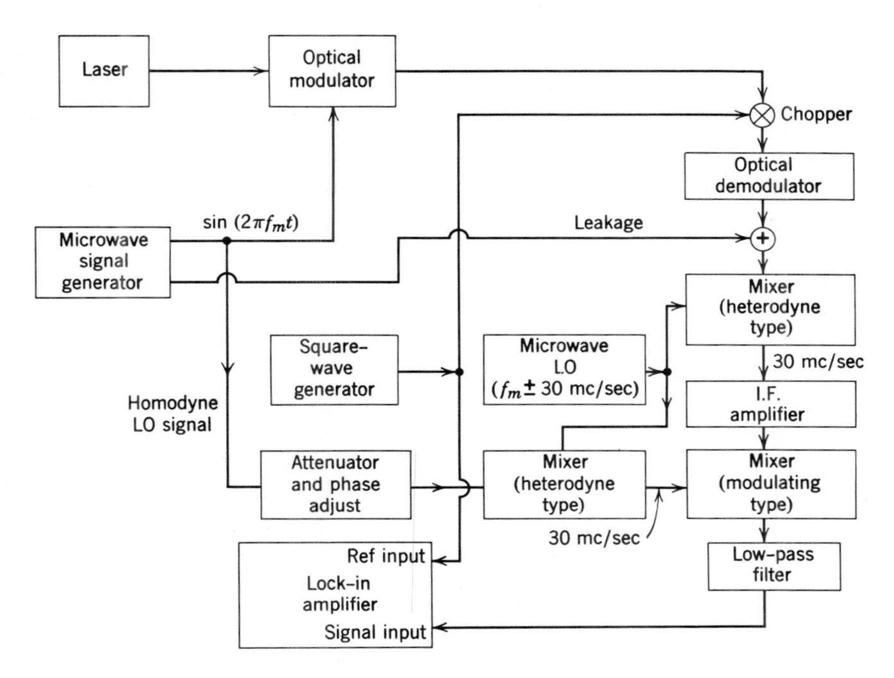

Figure 9.16. Pseudohomodyne system.

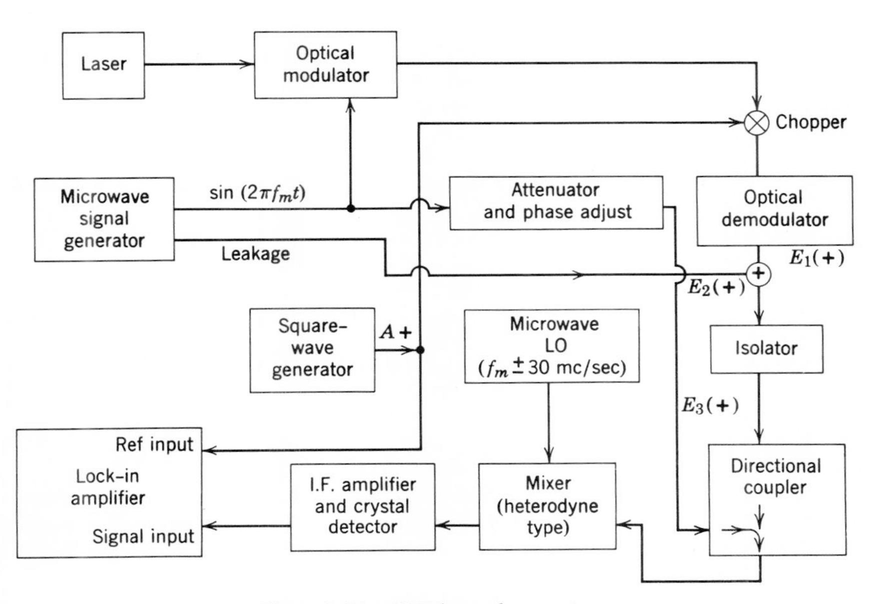

Figure 9.17. PSD homodyne system.

optical detector and is reflected; this may be reduced to a negligible level using an isolator, as shown. A further disadvantage is the possible saturation of the amplifying system. Nevertheless, this system has been used with great success, and is simpler to instrument than the scheme of Figure 9.16.

9.14 REFERENCES

[1] J. A. Bellisio, C. Freed, and H. A. Haus, *Appl. Phys. Letters*, **4**, 5 (1964).
[2] L. T. Prescott and A. van du Ziel, *Appl. Phys. Letters*, **5**, 48 (1964).
[3] D. G. C. Luck, *RCA Rev.*, **22**, 359 (1961).
[4] K. W. Otten, "*Principles of Optical Communications*," *Electro-Technology*, **69**, 111 (1962).
[5] D. S. Bayley and J. P. Campbell, *Spectral Suitability, Modulation and Detection Techniques in Communication with Wavelengths between 30 and 10,000 Angstroms*, ASTIA Document 282, 726, (1962).
[6] G. L. Knestrick, T. H. Cosden, and J. A. Curcio, *Atmospheric Attenuation Coefficients in the Visible and Infra-Red Regions*, U.S. Naval Research Lab Report NRL-5648 (1961).
[7] M. W. Muller, *Phys. Rev.*, **106**, 8 (1957).
[8] M. W. P. Strandberg, *Phys. Rev.*, **106**, 617 (1957).
[9] R. V. Pound, "Spontaneous Emission and Toise Figure of Maser Amplifiers," *Am. Phys.*, **I**, 23 (1957).
[10] H. A. Haus and J. A. Mullen, "Noise in Optical Maser Amplifiers," in *Optical Masers*, Vol. XIII *Proc. Symp.*, Polytechnic Press, Brooklyn, N.Y., 1963.
[11] W. Heitler, *The Quantum Theory of Radiation*, Oxford University Press, London, 1954.
[12] H. Heffner, *Proc. IRE*, **50**, 1604 (1962).
[13] A. Yariv and J. Gordon, *Proc. IEEE*, **51**, 4 (1963).
[14] W. R. Bennett, "*Optical Masers*", *J. Appl. Opt.* Supplement, p. 24 (1962).
[15] A. C. G. Mitchell and M. W. Zemousky, *Resonance Radiation and Excited Atoms*, Cambridge University Press, New York, 1961, Ch. 3.
[16] C. K. N. Patel, *Phys. Rev.*, **131**, 1582 (1963).
[17] R. A. Paananen, et al., *Appl. Phys. Ltrs.*, **4**, 149 (1964).
[18] C. Th. J. Alkemode, *Physica*, **25**, 1145 (1959).
[19] A. L. Bloom, *Noise in Lasers and Detectors*, Tech, Bull. **4**, Spectra-Physics, Mountain View, Calif., 1965.
[20] L. Mandel, *J. Opt. Soc. Am.*, **51**, 797 (1951).
[21] A. Einstein, Congres Solvay (1912).
[22] M. Born and E. Wolf, *Principles of Optics*, Pergamon, New York, 1959, p. 509.
[23] H. H. Brown and R. Q. Twiss, *Proc. Roy. Soc.* (London), **A242**, 300 (1957) and **A243**, 291 (1958).
[24] P. Fellgett, C. R. Jones, and R. Q. Twiss, *Nature*. **184**, 967 (1959).
[25] L. Mandel, *Proc. Phys. Soc.* (London), **25**, 1037 (1958).
[26] E. Wolf, *Proc. Phys. Soc.* (London), **76**, 424 (1960).
[27] L. Mandel, *J. Opt. Soc. Am.*, **50**, 1131 (1960).
[28] M. Parker Givens, *J. Opt. Soc. Am.*, **51**, 1032 (1961).
[29] P. T. Bolwign, C. T. Alkemode, and G. A. Boschloo, "Excess Photon Noise and Spectral Line Shape of Laser Beams," in P. Grivet and N. Bloembergen (eds.), *Quantum Electronics*, Paris 1963 Conference, Columbia University Press, New York, (1964), Vol. III, pp. 194–197.
[30] E. I. Gordon, *Bur Std. J. Res.* (U.S.), **43**, 507 (1964).

[31] A. Van der Ziel, *Proc. IEEE*, **52**, 1738 (1964).
[32] L. J. Prescott and A. Van der Ziel, *Phys. Letters*, **12**, 317 (1964).
[33] P. T. Bolwign, C. Th. J. Alkemode, and G. A. Bsochloo, "Excess Photon Noise and Spectra Line Shape," in P. Grivet and N. Bloembergen (eds.), *Quantum Electronics III*, Columbia University Press, 1964, pp. 193.
[34] D. E. McCumber, to be published.
[35] J. E. Geusic, *Solid State Maser Research*, 7th Quarterly Report, Contract DA-36-039-AMC-02333 (E), 1965.
[36] C. Freed and H. A. Haus, *Appl. Phys. Letters*, **6**, 85, (1965).
[37] A. L. Bloom, *Appl. Phys. Letters*, **2**, 101 (1963).
[38] E. K. Gordon, A. D. White, and J. D. Rigden, "Gain Saturation at 3.39 Microns in the He-Ne Maser" *Optical Masers, Proc. Symp.*, Polytechnic Institute of Brooklyn Press, New York, 1963, Vol. XIII, pp. 309–319.
[39] N. Kumage and M. Matouhara, "Design Consideration for Mode Selective Fabry-Perot Laser Resonator," *J. Quant. Elec.*, **1**, 85 (1965).
[40] G. A. Massey, M. K. Oshman, and R. Targ, *Appl. Phys. Letters*, **6**, 10, (1965).
[41] E. I. Gordon, Bell Telephone Laboratory, Private Communication.
[42] E. I. Gordon, E. F. Labuda, and W. B. Bridges, *App. Phys. Letters*, **4**, 178 (1964).
[43] M. H. Crowell, *J. Quant. Elec.*, **1**, 12 (1965).
[44] W. E. Lamb, *Phys. Rev.*, **134**, 1429a (1964).
[45] DiDomenico, *J. Appl. Phys.*, **33**, 2870 (1964).
[46] L. E. Hargrove, R. L. Fork, and M. A. Pollack, *Appl. Phys. Letters*, **5**, 4 (1964).
[47] H. Kogelnik, *Bell System Tech. J.*, **43**, 334 (1964).
[48] H. Kogelnik and A. Yariv, *Proc. IEEE*, **52**, 165 (1964).
[49] G. D. Boyd and H. Kogelnik, *Bell System Tech. J.*, **41**, 1347 (1962).
[50] H. Steinberg, *Proc. IEEE*, **52**, 28 (1964).
[51] H. Steinberg, *Proc. IEEE*, **51**, 943 (1963).
[52] W. B. Bridges and G. S. Picus, *Appl. Opt.*, **3**, 1189 (1964).
[53] P. O. Clark, *J. Quant. Elect.*, **1**, 109 (1965).
[54] K. Shimoda, H. Takehashi, and C. H. Townes, *J. Phys. Soc.* (Japan), **12**, 686 (1957).
[55] J. Weber, *Phys. Rev.*, **108**, 537 (1957).
[56] W. H. Louisell, A. Yariv, and A. E. Siegman, *Phys. Rev.*, **124**, 1646–1654 (1961).
[57] J. Weber, "Masers," *Rev. Mod. Phys.*, **3 1**, 681 (1959). See also N. Bloembergen, "Solid-State Infrared Quantum Counter," *Phys. Rev. Letters*, **2**, 84–85 (1959).
[58] R. Serber and C. H. Townes, "Amplification and Complementarity," in C. H. Townes (ed.), *Quantum Electronics*, Columbia University Press, New York, 1960, pp. 233–255.
[59] H. Friedburg, "General Amplifier Noise Limit," in C. H. Townes (ed.), *Quantum Electronics*, Columbia University Press, New York, 1960, pp. 228–232.
[60] P. T. Bolwijn, "Proceedings International Symposium on Laser-Physics and Applications," *Z.A.M.P.*, **16**, 85 (1965).
[61] W. H. Louisell, J. P. Gordon, and L. R. Walker, *Phys. Rev.*, **129**, 481 (1963).
[62] J. P. Gordon, *Proc. IRE*, **50**, 1898 (1962).
[63] E. Wolf, "Basic Concept of Optical Coherence Theory," *Optical Masers*, Proceedings Symposium on Optical Masers, Polytechnic Press, Brooklyn, N.Y., 1963, Vol. XIII, pp. 29–43.
[64] R. J. Glauber, Phys. Rev., **130**, 2529 (1963).
[65] Kenneth M. Johnson, *Microwave J.*, **51**, (1964).
[66] C. F. Buhrer, *et al.*, *Appl. Opt.*, **2**, 839 (1963).
[67] G. H. Heilmeier, *Appl. Opt.*, **3**, 1281 (1964).
[68] W. W. Rigrod and I. P. Kaminow, *Proc. IEEE*, **51**, 137 (1963).

[69] C. J. Peters, *Proc. IEEE*, **51**, 147 (1963).
[70] C. J. Peters, "Session L-24," *NEREM Record* (1964).
[71] J. E. Geusic, *et al.*, *Appl. Phy. Letters*, **4**, 141 (1964).
[72] C. F. Buhrer, *et al.*, *Proc. IEEE*, **50**, 1827 (1962).
[73] C. F. Buhrer, *et al.*, *Appl. Phy. Letters*, **1**, 46 (1962).
[74] B. J. McMurtry, *IEEE Trans. El. Dev.*, **ED 10**, 219 (1963).
[75] R. H. Blumenthal, *Proc. IRE*, **50**, 452 (1962).
[76] L. K. Anderson, *Proc. IEEE*, **51**, 846 (1963).
[77] R. Targ, D. E. Caddes, and B. J. McMurtry, *IEEE Trans. El. Dev.*, **ED 11**, 164 (1964).
[78] M. DiDomenico, Jr. and L. K. Anderson, *Proc. IEEE*, **52**, 815 (1964).
[79] A Comparison of a number of photodetectors are gicen in D. E. Cadde and B. J. McMurtry, "Evaluating Light Demodulators," *Electronics*, 54–61 (1964). (Corrected reprints are available from the authors, at Sylvania Electronic Systems, West, Box 205, Mountain View, California.)
[80] C. L. Ruthroff, "Microwave-to-Baseband Discriminator," Unpublished Memorandum, 1954; R. W. Pound, in C. G. Montgomery (ed.), *Techniques of Microwave Measurements*, McGraw-Hill, New York, 1947, Vol. XI; R. J. Mohr, *IEEE Trans. Microwave Theory Techniques*, **MTT-11**, 263 (1963).
[81] S. E. Harris, *Appl. Phy. Letters*, **2**, 47 (1963); *Demodulation of Frequency-Modulated Light*, Tech. Report No. 0576–5, Stanford Electronics Laboratories, Stanford, Calif., Chaps. III and V.
[82] I. P. Kaminow, *Appl. Opt.*, **3**, 507 (1964).
[83] S. E. Harris, E. O. Ammann, and I. C. Chang, *J. Opt. Soc. Am.*, **54**, 1267 (1964).
[84] S. E. Harris, *Proc. IEEE*, **52**, 823 (1964).
[85] J. R. Kerr, Doctoral Dissertation, Hansen Microwave Laboratories, Stanford University (to be published).
[86] J. R. Kerr, "Theory and Experiments with an FM Phototube" (to be published).
[87] A. E. Siegman, *J. Appl. Phy.*, **31**, 17 (1960).
[88] S. E. Harris and A. E. Siegman, *IRE Trans. El. Dev.*, ED-9, 322 (1962).
[89] G. S. Kino and G. W. C. Mathers, *Some Properties of a Sheath Helix with a Center Conductor or External Shield*, TR No. 65, Stanford Electronic Laboratories, Stanford, Calif., 1952.
[90] D. A. Watkins and E. A. Ash, *J. App. Phy.*, **25**, 782 (1954).
[91] R. Adler, O. M. Kromhout, and P. A. Calvier, *Proc. IRE*, **44**, 82 (1965).
[92] B. M. Oliver, *Proc. IRE*, **49**, 1960 (1961).
[93] H. Z. Cummins, N. Knable, *Proc. IEEE*, **51**, 1246 (1963).
[94] R. Targ, G. A. Massey, and S. E. Harris, *Proc. IEEE*, **52**, 1247 (1964).
[95] R. Targ, *Proc. IEEE*, **52**, 303 (1964).
[96] J. R. Kerr, *Rejection of Coherent Interference in Modulation-Demodulation Experiments*, Submitted to PGMTT, IEEE.
[97] M. Born and E. Wolf, *Principles of Optics*, Permagon, New York, 1964, 2nd ed., pp. 694–702.
[98] H. Hanken, *Phys. Rev.*, **13**, 329a (1964).
[99] H. Hanken, *Z. Phys.*, **181**, 96 (1964).
[100] J. A. Fleck, *J. Appl. Phys.*, **37**, 188 (1966).
[101] J. A. Armstrong and A. W. Smith, *Phys. Rev. Letters*, **14**, 68 (1965).

9.15 PRINCIPAL SYMBOLS, NOTATIONS, AND ABBREVIATIONS

Symbol	*Meaning*
A	Area normal to a source vector, amplitude of a general optical electric field component, see text
A_D	Area of a photodetector that is mode matched to a laser
A_L	Area of an iris used to limit the defraction a beam aperture and minimize noise
A^2	Calibration constant of FM phototube
A_{21}	Einstein transition coefficient
$A(t)$, $a(t)$	Time-dependent electric field associated with the laser output
A, B	Time-dependent amplitude of two perpendicular, optical, electric-field components in a circularly polarized wave
A_0, B_0	Peak amplitude of the perpendicular, optical, electric-field components in a circularly polarized wave
$a\omega$	Difference in electrical-path length in a birefringent crystal
b'	Radius of curvature of mirror
$b\,(v)$	Frequency response of a detection system
$C, \lvert C \rvert$	Cross-correlation coefficient between current and laser noise
$C(\omega)$	Frequency-transfer function of a birefringent discriminator
c	Velocity of light
d	Grating-line spacing
d_i	Mirror spacing
$E_0, E\,(t)$	Magnitude and amplitude of the optical electric field
E_1, E_2	Amplitude of the optical electric field
$E_v \Delta v$	Number of photons emitted /sec/sr/cm^2 by a source of frequency v in a range Δv
e	Electronic charge
f, f_0	Focal length of a lens or mirror, characteristic matching length of a laser
$f_{\rm LO}, f_{\rm IF}$	Local oscillator and intermediate frequencies
G	Amplifier single-pass gain
g_i	Degeneracy factor of state i
h	Planck's constant
I_{dc}	Direct current
I_0	Average value of the phototube current
I_i	Intensity for the case i, see text
$I_{\rm out}$	Photomultiplier-output current
$I_{\rm sig}$	Phototube current due to a laser signal

I_{spont}	Phototube current caused by illumination with a source of spontaneous emission
I_1, I_2	Phototube-output currents when illuminated by (1) the optical signal and (2) a local optical oscillator
$I_\perp, I_\parallel$	Intensity of light transmitted in two orientations of a polarizer analyzer pair
$<I^2(t)>$	Root-mean-square light intensity at a photodetector
i_n	Shot-noise current
$<i_s^2>$	Root-mean-square shot-noise current per unit of band width in a phototube
$J_i(x)$	Bessel function of order i and argument x
k	Photodevice constant (device-dependent quantity)
k	Boltzmann constant
K_t	Transverse-interaction impedance
L	Length of the birefringent crystal
L	Length of a plasma
L	Optical-path length, cavity-mirror spacing, see text
L_n	Left circularly-polarized electric field components in an optical wave
ln	Natural logarithm of a function
M	Amplitude modulation index
M	Order of interference, see text
M_0	Current gain per stage in a photomultiplier
M_1 M_2	Lens designated as M_i
m	Intensity modulation index $m = 2M$
m	Ratio of output ac to dc intensity in an amplitude-modulated laser
$m(t)$	First moment of the photocurrent in a photodetector
$N(\Delta\nu)$	Number of photons in a frequency interval $\Delta\nu$
N	Number of grating elements, number of quanta, see text
nm	Wavelength, 10^{-9} meters
$n(\nu)$	Incident photon rate per unit optical band width, see text
n_i	Population density of the state i
Δn	Birefringence of a crystal
P_L, P_{0L}	Instantaneous and average light powers
P_s	Source of standard power density
P_e	Noise power of a "white" Gaussian-noise source
P_0	Laser-amplifier-noise output
P_{noise}	Single-mode, output-noise power
P_{out}	Phototube-signal-power output
P_{out}	Total coherent-power output of a laser
P_{spont}	Spontaneous-emission-noise power

$\Delta P(\nu)$	Power spectrum of the intensity correlations of a laser beam
R	Radial distance from a source
R_i	Power reflectivity of a surface i
R	Ratio of minimum to maximum ac-photocurrent amplitudes when a phototube is illuminated by an elliptically polarized light source
R_t	Unit vector in the direction of the transverse field
R_1, R_2	Surface reflectivities (0 to 100 percent)
R_n	Components of an electric field right circularly polarized in the optical wave
R_{eq}	Phototube-equivalent resistance, see text
r	Ratio of ac photocurrent to dc photocurrent
S	Source-emitting area
S	Unwanted side band suppression (dB), see text
SNR	Signal-to-noise ratio
$S(\nu)$	Normalizing spectral line-shape factor
T	Transmissivity of a mirror
T_e	Effective input-noise temperature of an amplifier
t	Time
$\langle V_2{}^2 \rangle$	Root-mean-square noise voltage proportional to plasma-tube-current noise.
$\langle \mathrm{V}_1{}^2 \rangle$	Root-mean-square noise voltage developed by a photodiode when irradiated by a coherent light source
V_i	Voltage developed across phototube-load resistor, see text
w_i	Laser-beam waist radius, $i = 1, 2$
Y	Scaling factor
Z	Axis of laser beam
z	Length of the optical circuit
x	Gain coefficient of a gas laser
β_e	Electronic-propogation constant
Γ	Retardation of phase, induced retardation
γ	Degree of coherence, see Chapter 7
Δ	Symbol for an increment or small part
$\delta(\ )$	Infinitesimal change in a variable
δ	Degeneracy parameter, see text
δ	Peak phase deviation (phase-modulation index)
δ	Modulation index $\Delta\omega/\omega_m$, see text
η	Quantum efficiency of photodetector
θ	Angle of incidence with respect to the Z-axis of a system
κ	Constant that relates the photosurface current density to the incident optical electric-field intensity

λ_1 λ_0	Wavelength, wavelength at band center
ν_1, ν_0	Frequency, frequency at band center
$\Delta\nu$	Bandwidth
$\Delta\nu_D$	Doppler line width
$\Delta\nu_e$	Effective band width of a high-gain laser amplifier
μm	10^{-6} meter
τ	Observation interval
φ	Phase of the electromagnetic field
χ	Angle between the direction of polarization pressed by a polarizer and an analyzer pair
$\Delta\chi$	Nominal width of the disperser line shape at the photocathode
ω_i	Angular frequency, see text
ω_c	Optical-carrier angular frequency
ω_d	Total frequency deviation
ω_ℓ	Angular frequency of radiation that is left circularly polarized.
ω_m	Modulation angular frequency
$3s_2$, $3p_4$, etc.	Spectroscopic notation of the $3s$ and $3p$ states in a gas

Index